Geophysical Monograph Series

Geophysical Monograph Series

231 **Bioenergy and Land Use Change** *Zhangcai Qin, Umakant Mishra, and Astley Hastings (Eds.)*
232 **Microstructural Geochronology: Planetary Records Down to Atom Scale** *Desmond Moser, Fernando Corfu, James Darling, Steven Reddy, and Kimberly Tait (Eds.)*
233 **Global Flood Hazard: Applications in Modeling, Mapping and** *Forecasting Guy Schumann, Paul D. Bates, Giuseppe T. Aronica, and Heiko Apel (Eds.)*
234 **Pre-Earthquake Processes: A Multidisciplinary Approach to Earthquake Prediction Studies** *Dimitar Ouzounov, Sergey Pulinets, Katsumi Hattori, and Patrick Taylor (Eds.)*
235 **Electric Currents in Geospace and Beyond** *Andreas Keiling, Octav Marghitu, and Michael Wheatland (Eds.)*
236 **Quantifying Uncertainty in Subsurface Systems** *Celine Scheidt, Lewis Li, and Jef Caers (Eds.)*
237 **Petroleum Engineering** *Moshood Sanni (Ed.)*
238 **Geological Carbon Storage: Subsurface Seals and Caprock Integrity** *Stephanie Vialle, Jonathan Ajo-Franklin, and J. William Carey (Eds.)*
239 **Lithospheric Discontinuities** *Huaiyu Yuan and Barbara Romanowicz (Eds.)*
240 **Chemostratigraphy Across Major Chronological Eras** *Alcides N.Sial, Claudio Gaucher, Muthuvairavasamy Ramkumar, and Valderez Pinto Ferreira (Eds.)*
241 **Mathematical Geoenergy: Discovery, Depletion, and Renewal** *Paul Pukite, Dennis Coyne, and Daniel Challou (Eds.)*
242 **Ore Deposits: Origin, Exploration, and Exploitation** *Sophie Decree and Laurence Robb (Eds.)*
243 **Kuroshio Current: Physical, Biogeochemical and Ecosystem Dynamics** *Takeyoshi Nagai, Hiroaki Saito, Koji Suzuki, and Motomitsu Takahashi (Eds.)*
244 **Geomagnetically Induced Currents from the Sun to the Power Grid** *Jennifer L. Gannon, Andrei Swidinsky, and Zhonghua Xu (Eds.)*
245 **Shale: Subsurface Science and Engineering** *Thomas Dewers, Jason Heath, and Marcelo Sánchez (Eds.)*
246 **Submarine Landslides: Subaqueous Mass Transport Deposits From Outcrops to Seismic Profiles** *Kei Ogata, Andrea Festa, and Gian Andrea Pini (Eds.)*
247 **Iceland: Tectonics, Volcanics, and Glacial Features** *Tamie J. Jovanelly*
248 **Dayside Magnetosphere Interactions** *Qiugang Zong, Philippe Escoubet, David Sibeck, Guan Le, and Hui Zhang (Eds.)*
249 **Carbon in Earth's Interior** *Craig E. Manning, Jung-Fu Lin, and Wendy L. Mao (Eds.)*
250 **Nitrogen Overload: Environmental Degradation, Ramifications, and Economic Costs** *Brian G. Katz*
251 **Biogeochemical Cycles: Ecological Drivers and Environmental Impact** *Katerina Dontsova, Zsuzsanna Balogh-Brunstad, And Gaël Le Roux (Eds.)*
252 **Seismoelectric Exploration: Theory, Experiments, and Applications** *Niels Grobbe, André Revil, Zhenya Zhu, and Evert Slob (Eds.)*
253 **El Niño Southern Oscillation in a Changing Climate** *Michael J. McPhaden, Agus Santoso, and Wenju Cai (Eds.)*
254 **Dynamic Magma Evolution** *Francesco Vetere (Ed.)*
255 **Large Igneous Provinces: A Driver of Global Environmental and Biotic Changes** *Richard. E. Ernst, Alexander J. Dickson, and Andrey Bekker (Eds.)*
256 **Coastal Ecosystems in Transition: A Comparative Analysis of the Northern Adriatic and Chesapeake Bay** *Thomas C. Malone, Alenka Malej, and Jadran Faganeli (Eds.)*
257 **Hydrogeology, Chemical Weathering, and Soil Formation** *Allen Hunt, Markus Egli, and Boris Faybishenko (Eds.)*
258 **Solar Physics and Solar Wind** *Nour E. Raouafi and Angelos Vourlidas (Eds.)*
259 **Magnetospheres in the Solar System** *Romain Maggiolo, Nicolas André, Hiroshi Hasegawa, and Daniel T. Welling (Eds.)*
260 **Ionosphere Dynamics and Applications** *Chaosong Huang and Gang Lu (Eds.)*
261 **Upper Atmosphere Dynamics and Energetics** *Wenbin Wang and Yongliang Zhang (Eds.)*
262 **Space Weather Effects and Applications** *Anthea J. Coster, Philip J. Erickson, and Louis J. Lanzerotti (Eds.)*
263 **Mantle Convection and Surface Expressions** *Hauke Marquardt, Maxim Ballmer, Sanne Cottaar, and Jasper Konter (Eds.)*
264 **Crustal Magmatic System Evolution: Anatomy, Architecture, and Physico-Chemical Processes** *Matteo Masotta, Christoph Beier, and Silvio Mollo (Eds.)*
265 **Global Drought and Flood: Observation, Modeling, and Prediction** *Huan Wu, Dennis P. Lettenmaier, Qiuhong Tang, and Philip J. Ward (Eds.)*
266 **Magma Redox Geochemistry** *Roberto Moretti and Daniel R. Neuville (Eds.)*
267 **Wetland Carbon and Environmental Management** *Ken W. Krauss, Zhiliang Zhu, and Camille L. Stagg (Eds.)*
268 **Distributed Acoustic Sensing in Geophysics: Methods and Applications** *Yingping Li, Martin Karrenbach, and Jonathan B. Ajo-Franklin (Eds.)*
269 **Congo Basin Hydrology, Climate, and Biogeochemistry: A Foundation for the Future (English version)** *Raphael M. Tshimanga, Guy D. Moukandi N'kaya, and Douglas Alsdorf (Eds.)*
269 **Hydrologie, climat et biogéochimie du bassin du Congo: une base pour l'avenir (version française)** *Raphael M. Tshimanga, Guy D. Moukandi N'kaya, et Douglas Alsdorf (Éditeurs)*
270 **Muography: Exploring Earth's Subsurface with Elementary Particles** *László Oláh, Hiroyuki K. M. Tanaka, and Dezso˝ Varga (Eds.)*
271 **Remote Sensing of Water-Related Hazards** *Ke Zhang, Yang Hong, and Amir AghaKouchak (Eds.)*
272 **Geophysical Monitoring for Geologic Carbon Storage** *Lianjie Huang (Ed.)*
273 **Isotopic Constraints on Earth System Processes** *Kenneth W. W. Sims, Kate Maher, and Daniel P. Schrag (Eds.)*
274 **Earth Observation Applications and Global Policy Frameworks** *Argyro Kavvada, Douglas Cripe, and Lawrence Friedl (Eds.)*
275 **Threats to Springs in a Changing World: Science and Policies for Protection** *Matthew J. Currell and Brian G. Katz (Eds.)*
276 **Core-Mantle Co-Evolution: An Interdisciplinary Approach** *Takashi Nakagawa, Madhusoodhan Satish-Kumar, Taku Tsuchiya and George Helffrich (Eds.)*
277 **Compressional Tectonics: Plate Convergence to Mountain Building (Tectonic Processes: A Global View, Volume 1)** *Elizabeth J. Catlos and Ibrahim Çemen (Eds.)*
278 **Extensional Tectonics: Continental Breakup to Formation of Oceanic Basins (Tectonic Processes: A Global View, Volume 2)** *Ibrahim Çemen and Elizabeth J. Catlos (Eds.)*
279 **Strike-Slip Tectonics: Oceanic Transform Faults to Continental Plate Boundaries (Tectonic Processes: A Global View, Volume 3)** *Ibrahim Çemen and Elizabeth J. Catlos (Eds.)*
280 **Landscape Fire, Smoke, and Health: Linking Biomass Burning Emissions to Human Well-Being** *Tatiana V. Loboda, Nancy H. F. French, and Robin C. Puett (Eds.)*

Geophysical Monograph 280

Landscape Fire, Smoke, and Health

Linking Biomass Burning Emissions to Human Well-Being

Editors

Tatiana V. Loboda
Nancy H. F. French
Robin C. Puett

This Work is a co-publication of the American Geophysical Union and John Wiley and Sons, Inc.

AGU
ADVANCING EARTH AND SPACE SCIENCES

WILEY

Published under the aegis of the AGU Publications Committee

Matthew Giampoala, Vice President, Publications
Carol Frost, Chair, Publications Committee
For details about the American Geophysical Union visit us at www.agu.org.

Registered Office
John Wiley & Sons, Inc., 111 River Street, Hoboken, NJ 07030, USA

Editorial Office
111 River Street, Hoboken, NJ 07030, USA
For details of our global editorial offices, customer services, and more information about Wiley products visit us at www.wiley.com.

Wiley also publishes its books in a variety of electronic formats and by print-on-demand. Some content that appears in standard print versions of this book may not be available in other formats.

Library of Congress Cataloging-in-Publication Data
Names: Loboda, Tatiana V., editor. | French, Nancy H. F., editor. | Puett, Robin C., editor.
Title: Landscape fire, smoke, and health : linking biomass burning emissions to human well-being / editors Tatiana V. Loboda, Nancy H.F. French, Robin C. Puett.
Description: Hoboken, NJ: Wiley, 2024. | Series: Geophysical monograph series; 280
Identifiers: LCCN 2023015053 (print) | LCCN 2023015054 (ebook) | ISBN 9781119757009 (hardback) | ISBN 9781119757016 (adobe pdf) | ISBN 9781119757023 (epub)
Subjects: LCSH: Particulate matter–Environmental aspects. | Smoke–Environmental aspects. | Air quality–Health aspects. | Wildfires–Health aspects. | Wildfire forecasting. | Wildfires–Simulation methods.
Classification: LCC RA577.P37 F57 2024 (print) | LCC RA577.P37 (ebook) | DDC 613/.11–dc23/eng/20230710
LC record available at https://lccn.loc.gov/2023015053
LC ebook record available at https://lccn.loc.gov/2023015054

Cover Design: Wiley
Cover Image: Wildfire smoke east of Los Angeles, USA. © E4C/Getty Images

Set in 10/12pt Times New Roman by Straive, Pondicherry, India

CONTENTS

Part III From Concentrations to Health Outcomes

LIST OF CONTRIBUTORS

Ravan Ahmadov
Cooperative Institute for Research in Environmental Sciences
University of Colorado Boulder
Boulder, Colorado, USA
and
NOAA Global Systems Laboratory
Boulder, Colorado USA

Nicole Asencio
Météo-France
Toulouse, France

Merched Azzi
Department of Planning and Environment
Government of New South Wales
Sydney, Australia

Alexander Baklanov
World Meteorological Organization
Geneva, Switzerland

Angela Benedetti
European Centre for Medium-Range Weather Forecasts
Reading, United Kingdom

Tarik Benmarhnia
Scripps Institution of Oceanography
University of California San Diego
San Diego, California, USA

Partha Bhattacharjee
I. M. Systems Group
NWS/NCEP/EMC
College Park, Maryland, USA

Allison E. Bredder
Department of Geographical Sciences
University of Maryland,
College Park, Maryland, USA

Melissa E. Brooks
Met Office
Exeter, United Kingdom

Christopher P. Camacho
Naval Research Laboratory
Marine Meteorology Division
Monterey, California, USA

Jack Chen
Environment and Climate Change Canada
Ottawa, Ontario, Canada

Boon Ning Chew
Centre for Climate Research Singapore
Meteorological Service Singapore
Singapore

Peter R. Colarco
NASA Goddard Space Flight Center
Greenbelt, Maryland, USA

Martin Cope
CSIRO Climate Science Centre
Aspendale, Victoria, Australia

Natalie Crnosija
School of Public Health
University of Maryland
College Park, Maryland, USA

Arlindo Da Silva
NASA Goddard Space Flight Center
Greenbelt, Marlyand, USA

Didier Davignon
Environment and Climate Change Canada
Dorval, Quebec City, Canada

Savannah M. D'Evelyn
Department of Environmental and Occupational Health Sciences
University of Washington
Seattle, Washington, USA

Evan Ellicott
Department of Geographical Sciences
University of Maryland
College Park, Maryland, USA

Johannes Flemming
European Centre for Medium-Range Weather Forecasts
Reading, United Kingdom

Nancy H. F. French
Michigan Tech Research Institute
Michigan Technological University
Ann Arbor, Michigan, USA

Christopher Gan
Centre for Climate Research Singapore
Meteorological Service Singapore
Singapore

Carlos Perez Garcia-Pando
Barcelona Supercomputing Center
Barcelona, Spain and Catalan Institution for Research and Advanced Studies
Barcelona, Spain

Emily M. Gargulinski
National Institute of Aerospace
Hampton, Virginia, USA

Louis Giglio
Department of Geographical Sciences
University of Maryland
College Park, Maryland, USA

Jonathan Guth
Météo-France
Toulouse, France

Rosie Howard
Earth, Ocean and Atmospheric Sciences Department
The University of British Columbia
Vancouver, British Columbia, Canada

Andrew T. Hudak
Forestry Sciences Laboratory
Rocky Mountain Research Station
United States Forest Service
Moscow, Idaho, USA

Michael L. Humber
Department of Geographical Sciences
University of Maryland
College Park, Maryland, USA

Edward J. Hyer
Naval Research Laboratory
Marine Meteorology Division
Monterey, California, USA

John Innis
EPA Tasmania
Hobart, Tasmania, Australia

Eric James
Cooperative Institute for Research in Environmental Sciences
University of Colorado Boulder
Boulder, Colorado, USA
and
NOAA Global Systems Laboratory
Boulder, Colorado, USA

Oriol Jorba
Barcelona Supercomputing Center
Barcelona, Spain

Christopher O. Justice
Department of Geographical Sciences
University of Maryland
College Park, Maryland, USA

Robert E. Keane
Rocky Mountain Research Station
United States Forest Service
Missoula, Montana, USA

Zak Kipling
European Centre for Medium-Range Weather Forecasts
Reading, United Kingdom

Adam K. Kochanski
Department of Meteorology and Climate Science
San Jose State University
San Jose, California, USA

Patricia D. Koman
Department of Environmental Health Sciences
University of Michigan
Ann Arbor, Michigan, USA

Rostislav Kouznetsov
Atmospheric Composition Unit
Finnish Meteorological Institute
Helsinki, Finland

Narasimhan K. Larkin
Pacific Northwest Research Station
United States Forest Service
Seattle, Washington, USA

Tatiana V. Loboda
Department of Geographical Sciences
University of Maryland
College Park, Maryland, USA

E. Louise Loudermilk
Southern Research Station
United States Forest Service
Athens, Georgia, USA

Duncan Lutes
Rocky Mountain Research Station
United States Forest Service
Missoula, Montana, USA

Angus MacNeil
Forest Practices Authority
Hobart, Tasmania, Australia

Derek V. Mallia
Department of Atmospheric Sciences
University of Utah
Salt Lake City, Utah, USA

Miriam E. Marlier
Fielding School of Public Health
University of California Los Angeles
Los Angeles, California, USA

Jeffrey McQueen
NOAA National Centers for Environmental Prediction
College Park, Maryland, USA

Kerryn McTaggart
Department of Environment, Land, Water and Planning
Government of Victoria
Melbourne, Victoria, Australia

Christopher Migliaccio
Department of Biomedical and Pharmaceutical Health Sciences
University of Montana
Missoula, Montana, USA

Luke Montrose
School of Public and Population Health
Boise State University
Boise, Idaho, USA

M. Talat Odman
School of Civil and Environmental Engineering
Georgia Institute of Technology
Atlanta, Georgia, USA

Susan M. O'Neill
Pacific Northwest Research Station
United States Forest Service
Seattle, Washington, USA

Amy M. Padula
Department of Obstetrics, Gynecology and Reproductive Sciences
Program for Reproductive Health and the Environment
University of California San Francisco
San Francisco, California, USA

Radenko Pavlovic
Environment and Climate Change Canada
Dorval, Quebec City, Canada

David A. Peterson
Naval Research Laboratory
Marine Meteorology Division
Monterey, California, USA

Susan J. Prichard
School of Environmental and Forest Sciences
University of Washington
Seattle, Washington, USA

Robin C. Puett
Institute for Applied Environmental Health
School of Public Health
University of Maryland
College Park, Maryland, USA

Camille Raynes-Greenow
Sydney School of Public Health
The University of Sydney
Sydney, Australia

Jeffrey S. Reid
United States Naval Research Laboratory
Marine Meteorology Division
Monterey, California, USA

Fabienne Reisen
CSIRO Climate Science Centre
Aspendale, Victoria, Australia

Eric Rowell
Tall Timbers Research Station
Tallahassee, Florida, USA

David P. Roy
Department of Geography, Environment and Spatial Sciences, and Center for Global Change and Earth Observations
University of Michigan
East Lansing, Michigan, USA

Pablo E. Saide
Department of Atmospheric and Oceanic Sciences, and Institute of the Environment and Sustainability
University of California Los Angeles
Los Angeles, California, USA

Elizabeth A. Satterfield
Naval Research Laboratory
Marine Meteorology Division
Monterey, California, USA

Adam Schuller
School of Public and Population Health
Boise State University
Boise, Idaho, USA

Mikhail Sofiev
Atmospheric Composition Unit
Finnish Meteorological Institute
Helsinki, Finland

Amber J. Soja
National Institute of Aerospace
Hampton, Virginia, USA and NASA Langley Research Center
Hampton, Virginia, USA

Roland Stull
Earth, Ocean and Atmospheric Sciences Department
The University of British Columbia
Vancouver, British Columbia, Canada

Taichu Tanaka
Meteorological Research Institute
Japan Meteorological Agency
Tsukuba, Japan

Daniel Tong
Department of Atmospheric, Oceanic and Earth Sciences
George Mason University
Fairfax, Virginia, USA

Joseph K. Vaughan
Department of Civil and Environmental Engineering
Washington State University
Pullman, Washington, USA

Alan Wain
Australian Bureau of Meteorology
Melbourne, Victoria, Australia

Elizabeth B. Wiggins
NASA Langley Research Center
Hampton, Virginia, USA

Peng Xian
United States Naval Research Laboratory
Marine Meteorology Division
Monterey, California, USA

Maria Zubkova
Department of Geographical Sciences
University of Maryland
College Park, Maryland, USA

PREFACE

Smoke from biomass burning is known to be a significant source of air pollution, and smoke inhalation is well understood to be detrimental to human health. Developing approaches to quantify the impacts of biomass burning smoke on health and well-being began with the need to protect wildland firefighters and the general public from smoke. In the United States, the Forest Service initially, and the Environmental Protection Agency and National Weather Service in more recent years, developed the science, measurements and analysis frameworks for responding to the threat of air pollution coming from landscape fires. Similar agencies in other developed regions of the world, such as Canada's FireWork system and the Emergency Management Services in the European Copernicus program, provide important information for dealing with biomass burning pollution. More recently, advanced approaches using satellite-derived information combined with in situ measurements and advanced data assimilation and analysis techniques have allowed for improved retrospective assessments of air pollution from biomass burning as well as predictions on where and when smoke can be expected to intersect with human populations.

The focus of this volume is on describing the observational and modeling approaches that are currently used in fire science, smoke characterization, and health assessment, some of which are operationally employed, while others are at the forefront of research. The concepts, analytical approaches, and models that have been developed in these disciplinary camps provide exceptional capability to answer questions and solve problems related to the topic of biomass burning smoke exposure and health. Their application now, however, is required across the broad transdisciplinary space in order to address a complex set of concerns. With the transdisciplinary reach of scientific inquiry and modeling efforts comes the necessity of building a foundational understanding of approaches, methods, and tools to craft seamless and robust chains of data analysis. This book is written by leading experts (researchers, modelers, practitioners) from several disparate disciplines who have worked across disciplines, providing synthesis of activities from an international perspective through inclusion of lead authors and coauthors from several countries outside of North America. Our diverse international and multidisciplinary community of authors and reviewers highlights the "common language" challenge faced when connecting across established divides. One clear focus of this book is on gaining an improved vocabulary and transdisciplinary knowledge-set for considering solutions. The authors have written the text to be comprehensible to experts in the other fields of the modeling chain, with the dual aims of making a subset of this book of value to all experts currently working in this domain, as well as making all components of this book of value to newcomers to the field. Scientific discovery and technological development often outpace scientific publishing and, thus, books focused exclusively on cutting edge knowledge tend to become outdated before they come out of print. For this book, we include information that will provide the scaffolding for accelerated scientific growth.

The book is simplistically structured to cover a broad set of related topics, presented to be accessible outside of an expert's basic knowledge field, and provides foundational concepts along with new research and applications. The book is divided into three parts, which broadly address fire science (Part I: From Fires to Emissions), atmospheric chemistry and dynamics (Part II: From Emissions to Concentrations), and human health research (Part III: From Concentrations to Health Outcomes). Each part contains four chapters, which are designed to cover the foundational knowledge within the field; highlight recent advancements; describe commonly used methods; and outline existing data sets, models, or systems in general use by the respective communities. The three parts of this volume represent three distinct research communities, with members who are often focused on topics adjacent to the topics of fire, smoke, and health. Included are in-depth reviews of the state of the art within each topic aimed at understanding health outcomes of biomass burning emissions (smoke).

Since the AGU's Fall Meeting in 2019 when we first convened a session related to health outcomes of biomass burning emissions, we have evidenced the widespread attention and the rapidly growing interest in this subject among the scientists, resource managers, and public health practitioners. However, it also became clear that the community was largely involved in advancing the science within its disciplinary camps with limited communication and collaboration among scientists involved across various components of the modeling chain (starting with fire detection, through emissions modeling, atmospheric transport, and constructing epidemiological models of health outcomes). As a result, researchers along the modeling chain have struggled to identify and incorporate the most appropriate developments from adjacent disciplinary links into developing their methodologies. Moreover, the rapid proliferation of methods for each of these components has also created challenges evaluating the validity and linkages among reported results leading to opaque science. In this book, we attempt to synthesize and harmonize foundational as well as state-of-the-art information across all components of the modeling chain and to build a common language to advance transdisciplinary work. While far from comprehensive, the primary aim of this book is to build a collaborative community with a common understanding and to promote further scientific inquiry and applied approaches.

A book like this, with a diverse set of authors and broad audience, takes dedication and patience from the people involved. We would first like to acknowledge the effort that each of our chapter lead authors and coauthors put into writing excellent and valuable material that makes up the bulk of this volume. In addition to thanking these dedicated authors, we thank Jenny Lunn at AGU, Geeta Persad from the AGU Books Editorial Board, and the production team at Wiley including Keerthana Govindarajan, Noel McGlinchey, Rituparna Bose, and Lesley Fenske. They have been quite gracious in guiding us along the path to completion. Also, in the background making sure materials flowed efficiently was Julie Carter, who read each chapter draft for consistency, a task of great value and payoff for making a good product. In addition, the quality of the work is due in large part to the set of reviewers who graciously provided their time and expertise to critiquing each chapter. A full-fledged peer review takes time and effort on the part of many people. Finally, a thank you to the AGU Geohealth Section leaders, session organizers over the past three years, and section members for encouraging us to take on this project. The AGU embraces the philosophy of diversity of thought and development of transdisciplinary bridges. The Geohealth Section, in particular, has shown the way to improving health and well-being with this approach.

Tatiana V. Loboda
University of Maryland, USA
Nancy H. F. French
Michigan Technological University, USA
Robin C. Puett
University of Maryland, USA

ACRONYMS AND ABBREVIATIONS

Acronym	Description	Chapter
AATSR	Advanced Along-Track Scanning Radiometer	3
ABI	Advanced Baseline Imager	3
ACCESS	Australian Community Climate and Earth-System Simulator	9
ACCP	Aerosol and Cloud Convection and Precipitation	2
ADME	Absorption, Distribution, Metabolism, and Excretion	11
ADRD	Alzheimer's Disease Related Dementias	12
AERONET	Aerosol Robotic Network	7
AFAC	Australian National Council for Fire and Emergency Services	2
AFDRS	Australian Fire Danger Rating System	9
AFIS	Advanced Fire Information System	2
AHI	Advanced Himawari Imager	3
AhR	Aryl Hydrocarbon Receptor	11
AIRPACT	Air Indicator Report for Public Awareness and Community Tracking	8
ALS	Airborne Light detection and range Scanning	4
AMS	Autonomous Modular Sensor	3
AOD	Aerosol Optical Depth	6
AOS	Atmosphere Observing System	2
API	Application Programming Interface	6
AQ	Air Quality	6
AQFx	Air Quality Forecasting	9
AQHI	Air Quality Health Index	9
AQI	Air Quality Index	11
ARI	Acute Respiratory Illness	11
ASEAN	Association of Southeast Asian Nations	9
ASMC	ASEAN Specialized Meteorological Centre	9
ATS	American Thoracic Society	10
ATSR-2	Along-Track Scanning Radiometer 2	3
AVHRR	Advanced Very High Resolution Radiometer	3
BAER	Burned Area Emergency Response	2
BB	Biomass Burning	9
BC	Black Carbon	9
BEIS	Biogenic Emission Inventory System	9
BLANkET	Base-Line Air Network of EPA Tasmania	9

BMI	Body Mass Index	13
BOM	Bureau of Meteorology (Australia)	9
BSC	BlueSky Canada	9
BSC-S	Barcelona Supercomputing Center (Spain)	9
CAAQMS	Continuous Ambient Air Quality Monitoring Stations	6
CALIOP	Cloud-Aerosol Lidar with Orthogonal Polarization	7
CALIPSO	Cloud-Aerosol Lidar and Infrared Pathfinder Satellite Observations	2
CAMS	Copernicus Atmosphere Monitoring Service	9
CANSAC	California and Nevada Smoke and Air Committee	9
CAP	Criteria Air Pollutants	5
CASTNet	Clean Air Status and Trends Network	6
CAWFE	Coupled Atmosphere-Wildland Fire Environment	2
CCI	Climate Change Initiative	3
CDF	Cumulative Distribution Function	7
CEOS	Committee on Earth Observation Satellites	2
CFD	Computational Fluid Dynamics	4
CFFEPS	Canadian Forest Fire Emissions Prediction System	9
CH4	Methane	5
CI	Confidence Interval	13
CMAQ	Community Multiscale Air Quality	10
CNEMC	China National Environmental Monitoring Center	6
CNES	Centre National d'Etudes Spatiales (France)	2
CO	Carbon Monoxide	5
CO2	Carbon Dioxide	5
COHb	Carboxyhemoglobin	11
CONAFOR	Comisión Nacional Forestal (Mexico)	2
CONUS	Continental United States	7
COPD	Chronic Obstructive Pulmonary Disease	10
CRF	Concentration-Response Functions	12
CrIS	Cross-Track Infrared Spectrometer	7
CSA	Canadian Space Agency	2
C-SEM	CSIRO Smoke Emissions Model	9
CSIRO	Commonwealth Science and Industrial Organization (Australia)	9
CSMS	Coordinated Smoke Management System	9
CTM	Chemical Transport Model	10
CWD	Coarse Woody Debris	4
CWFIS	Canadian Wildland Fire Information System	2
DEASCO3	Deterministic and Empirical Assessment of Smoke's Contribution to Ozone Project	9
DEHM	Danish Eulerian Hemispheric Model	9

DELWP	Department of Environment, Land, Water, and Planning (Victoria, Australia)	9
DEP	Diesel Exhaust Particulate	11
DMSP	Defense Meteorological Satellite Program	3
DOI	Department of the Interior (United States)	2
DPIE	Department of Planning, Industry, and the Environment (New South Wales, Australia)	9
DRI	Desert Research Institute	9
DTM	Digital Terrain Models	5
ECCC	Environment and Climate Change Canada	9
ECMWF	European Centre for Medium-Range Weather Forecasts	9
EEA	European Environment Agency	5
EF	Emission Factor	5
EFFIS	European Forest Fire Information System	2
EJScreen	Environmental Justice Screening	2
EMEP	European Monitoring and Evaluation Programme	5
ENEA	Energia Nucleare ed Energie Alternative (Italy)	9
ENVISAT	Environmental Satellite	3
EOS	Earth Observing System	3
EOSDIS	Earth Observing System Data and Information System	2
EPA	Environmental Protection Agency	2
ERS-2	European Remote Sensing 2	3
ETM	Enhanced Thematic Mapper	3
EU	European Union	5
EURAD-IM	EURopean Air pollution Dispersion-Inverse Model	9
EVT	Existing Vegetation Type	5
EWS	Early Warning Systems	12
FASMEE	Fire and Smoke Model Evaluation Experiment	2
FBP	Fire Behavior Prediction	9
FCCS	Fuel Characteristic Classification System	5
FEER	Fire Energetics and Emissions Research	5
FEMA	Federal Emergency Management Agency (United States)	2
FEPS	Fire Emission Production Simulator	10
FFDI	Forest Fire Danger Index	9
FINN	Fire INventory from NCAR	8
FireMARS	Fire Monitoring Accounting and Reporting System	2
FIREX-AQ	Fire Influence on Regional to Global Environments and Air Quality	2
FIRMS	Fire Information for Resource Management System	2
FIS	Fire Information System	9
FLAMBE	Fire Locating and Modeling of Burning Emissions	9
FLEXPART	Flexible Particle dispersion model	8

FMI	Finnish Meteorological Institute	9
FNMOC	Fleet Numerical Meteorology and Oceanography Center	9
FOFEM	First Order Fire Effects Model	5
FPA	Forest Practices Authority (Tasmania, Australia)	9
FRE	Fire Radiative Energy	3
FRM	Federal Reference Method	10
FRP	Fire Radiative Power	3
FT	Free Troposphere	8
FWI	Fire Weather Index	9
GA	Gaussian Anamorphosis	7
GAFIS	Global Air Quality Forecasting and Information System	9
GAM	Generalized Additive Model	6
GBA	Global Burnt Area	3
GBBEP-Geo	Global Biomass Burning Emission Product from Geostationary Satellites	5
GBBEPx	blended Global Biomass Burning Emissions Product	9
GDPFS	Global Data-Processing and Forecasting System	9
GEDI	Global Ecosystem Dynamics Investigation	5
GEFS	Global Ensemble Forecast System	9
GEM	Global Environmental Model	9
GEMS	Geostationary Environmental Monitoring Spectrometer	2
GEOS	Goddard Earth Observing System	9
GEOS-Chem	Goddard Earth Observing System with Chemistry	2
GFAS	Global Fire Assimilation System	5
GFED	Global Fire Emissions Database	5
GFMC	Global Fire Monitoring Center	2
GFS	Global Forecast System	8
GFWED	Global Fire Weather Database	2
GHG	Greenhouse Gases	5
GMS	Geostationary Meteorological Satellites	2
GOES	Geostationary Operational Environmental Satellite	2
GOFC-GOLD	Global Observation of Forest Cover - Global Observation of Land Dynamics	2
GPS	Global Positioning Systems	10
GSL	NOAA Global Systems Laboratory (United States)	9
GTOS	Global Terrestrial Observing System	2
GWIS	Global Wildfire Information System	2
GWR	Geographically Weighted Regression	6
HAPs	Hazardous Air Pollutants	12
HAQAST	Health and Air Quality Applied Sciences Team	2
HMS	Hazard Mapping System	2

HONO	Nitrous Acid	8
HRRR	High Resolution Rapid Refresh	9
HRRR-Smoke	High Resolution Rapid Refresh Model with Smoke	8
HYSPLIT	Hybrid Single-Particle Lagrangian Integrated Trajectory Model	8
IASI	Infrared Atmospheric Sounding Interferometer	7
IAWF	International Association of Wildland Fire	2
ICAP-MME	International Cooperative for Aerosol Prediction Multi-Model-Ensemble	9
ICESat-2	Ice Cloud And Land Elevation Satellite	5
IDEQ	Idaho Department of Environmental Quality (United States)	9
IDW	Inverse Distance Weighting	6
IEK	Institute of Energy and Climate Research (Germany)	9
IEP-NRI	Institute of Environmental Protection – National Research Institute (Poland)	9
IMPROVE	Interagency Monitoring of Protected Visual Environment (United States)	6
INERIS	Institut National de l'Environnement Industriel et des Risques (France)	9
INPE	Instituto Nacional de Pesquisas Espaciais (Brazil)	2
IPCC	Intergovernmental Panel on Climate Change	5
IS4FIRES	Integrated Monitoring and Modeling System (IS) for wildland fires	9
ISA	Integrated Science Assessment	10
ISRO	Indian Space Research Organization	2
IWFAQRP	Interagency Wildland Fire Air Quality Response Program	9
JAXA	Japan Aerospace Exploration Agency	2
JFSP	Joint Fire Science Program (United States)	2
JMA	Japan Meteorological Agency	3
JPSS	Joint Polar Satellite System	2
KNMI	Royal Netherlands Meteorological Institute	9
LANCE	Land, Atmosphere Near-real-time Capability for EOS	3
LAR	Laboratory for Atmospheric Research	9
LCAQS	Low-Cost Air Quality Sensors	6
LPDMs	Lagrangian Particle Dispersion Models	8
LRTAP	Long-Range Transboundary Air Pollution	5
LUR	Land Use Regression	6
LWIR	Long Wavelength Infrared	3
MAIA	Multi-Angle Imager for Aerosols	2
MAIAC	Multi-Angle Implementation of Atmospheric Correction	7
MASINGAR	Model of Aerosol Species in the Global Atmosphere	9
MCE	Modified Combustion Efficiency	5
MEGAN	Model of Emissions of Gases and Aerosols from Nature	9
MERIS	Medium Resolution Imaging Spectrometer	3
MFLEI	Missoula Fire Laboratory Emission Inventory	8

MINNI	Italian National Integrated Assessment Model	9
MISR	Multiangle Imaging SpectroRadiometer	2
MODIS	Moderate Resolution Imaging Spectroradiometer	3
MONARCH	Multiscale Online Nonhydrostatic AtmospheRe CHemistry model	9
MONCAGE	Météo France Modèle de Chimie Atmospherique à Grande Echelle	9
MOPITT	Measurement of Pollution in the Troposphere	7
MOVES	EPA MOtor Vehicle Emission Simulator (United States)	9
MOZART	Model of Ozone and Related Chemical Tracers	8
MPLNet	Micropulse Lidar Network	7
MSC	Meteorological Service of Canada	9
MSI	MultiSpectral Instrument	3
MSS	Multispectral Scanner System	3
MSS-S	Meteorological Service Singapore	9
MSS-UKMO NAME	MSS-United Kingdom Meteorological Office Numerical Atmospheric-dispersion Modeling Environment	9
MWIR	Medium Wavelength InfraRed	3
NAAPS	Navy Aerosol Analysis and Prediction System (United States)	7
NAAQMN	National Ambient Air Quality Monitoring Network (United States)	6
NAAQS	National Ambient Air Quality Standards (United States)	10
NAM	North American Mesoscale Forecast System	8
NAPS	National Air Pollution Surveillance (Canada)	6
NAQFC	National Air Quality Forecast Capability (United States)	9
NASA	National Aeronautics and Space Administration (United States)	2
NCAP	National Clean Air Program (India)	6
NCAR	National Center for Atmospheric Research	8
NEI	National Emissions Inventory (United States)	5
NEPMs	National Environment Protection Measures (Australia)	6
NESDIS	National Environmental Satellite, Data, and Information Service (United States)	12
NFPA	National Fire Protection Association	2
NH_3	Ammonia	5
NICC	National Interagency Coordination Center (United States)	2
NIFC	National Interagency Fire Center (United States)	2
NIROPS	National Infrared Operations (United States)	3
NISAR	NASA-ISRO Synthetic Aperture Radar	2
NO	Nitrogen Monoxide	5
NO_2	Nitrogen Dioxide	5
NOAA	National Oceanic and Atmospheric Administration (United States)	2
NOx	Nitrogen Oxides	10
NRC	National Research Council (United States)	10
NRCan	National Resources Canada	9

NRL	Naval Research Laboratory (United States)	9
NW-AIRQUEST	Northwest International Air Quality Environmental Science and Technology	9
NWCG	National Wildfire Coordination Group (United States)	2
NWP	Numerical Weather Prediction	9
NWS	National Weather Service (United States)	2
O_2	Molecular Oxygen	8
O_3	Ozone	5
OC	Organic Carbon	9
OLCI	Ocean and Land Color Instrument	3
OLI	Operational Land Imager	3
OLS	Operational Linescan System	3
OM	Organic Matter	9
OMI	Ozone Monitoring Instrument	7
OMPS	Ozone Mapping Profiler Suite	7
OR	Odds Ratio	13
PAHs	Polycyclic Aromatic Hydrocarbons	11
PBL	Planetary Boundary Layer	8
PDF	Probability Distribution Function	7
PM	Particulate Matter	5
PM_{10}	Particulate Matter with a diameter of 10 μ m or less	7
$PM_{2.5}$	Particulate Matter with a diameter of 2.5 μ m or less	7
PMAp	Polar Multisensor Aerosol product	9
POLDER	Polarization and Directionality of the Earth's Reflectances	7
PPE	Personal Protective Equipment	10
PRH	Pseudo Relative Humidity	7
PTSD	Post-Traumatic Stress Disorder	10
pyroCb	Pyrocumulonimbus	8
pyroCu	Pyrocumulus	8
QFED	Quick Fire Emissions Data set	9
RAP	Rapid Refresh	9
RAQDPS-FW	Regional Air Quality Deterministic Prediction System with near-real-time wild Fire Emissions	9
RF	Random Forest	6
RR	Relative Risk	13
RSMC	Regional Specialized Meteorological Centers (WMO)	9
SAR	Synthetic Aperture Radar	3
SASEM	Simple Approach Smoke Estimation Model	8
SBG	Surface Biology and Geology	2
SDS-WAS	Sand and Dust Storm Warning Advisory and Assessment System	9

SERA	Smoke Emissions Repository Application	5
SEVIRI	Spinning Enhanced Visible and Infrared Imager	3
SfM	Structure-from-Motion	4
SILAM	System for Integrated modeling of Atmospheric coMposition	9
SLSTR	Sea and Land Surface Temperature Radiometer	3
SMARTFIRE v2 (SF2)	Satellite Mapping Automated Reanalysis Tool for Fire Incident Reconciliation	9
SMHI	Swedish Meteorological and Hydrological Institute	9
SMOKE	Sparse Matrix Operator Kernal Emissions	8
S-NPP	Suomi National Polar-orbiting Partnership (United States)	3
SO2	Sulfur Dioxide	5
SOA	Secondary Organic Aerosols	8
SPOT	Satellite Pour l'Observation de la Terre	3
STILT	Stochastic Time-Inverted Lagrangian Transport	8
SWIR	Short Wavelength InfraRed	3
TEMPO	Tropospheric Emissions: Monitoring of Pollution	2
3D	Three Dimensional	4
TIROS-N	Television Infrared Observation Satellite - Next generation	3
TLS	Terrestrial Light detection and range Scanning	4
TM	Thematic Mapper	3
TNO	The Netherlands Organization for applied scientific research	9
TPM	Total Particulate Matter	5
TRAP	Traffic-Related Air Pollution	11
TRMM	Tropical Rainfall Measuring Mission	3
TROPOMI	Tropospheric Monitoring Instrument	7
TRP	Transient Receptor Potential	11
TSP	Total Suspended Particles	5
UAS	Uncrewed Airborne Systems	3
UKMO	UK Met Office	9
UNECE	United Nations Economic Commission For Europe	5
UNFCCC	United Nations Framework Convention On Climate Change	5
USDA	United States Department of Agriculture	9
USFS	United States Forest Service	8
USGS	United States Geological Survey	2
VACES	Versatile Aerosol Concentration Enrichment System	11
VAQUM	Verification of Air QUality Models (ECCC)	9
VAS	Visible Infrared Spin Scan Radiometer	3
VFSP-WAC	Regional VFSP-WAS Centers	9
VFSP-WAS	Vegetation Fire and Smoke Pollution Warning Advisory and Assessment System	9
VIIRS	Visible Infrared Imaging Radiometer Suite	3

VIRS	Visible and Infrared Scanner	3
VOC	Volatile Organic Compounds	9
WACCM	Whole Atmosphere Community Climate Model	9
WFAS	Wildland Fire Assessment System	2
WFDS	Wildland-urban interface Fire Dynamics Simulator	8
WFEI	Wildland Fire Emissions Inventory	5
WFEIS	Wildland Fire Emissions Inventory System	5
WFRT	Weather Forecast Research Team	9
WHO	World Health Organization	2
WIMS	Weather Information Management System	10
WMO	World Meteorological Organization	9
WRAP	Western Regional Air Partnership (United States)	9
WRF	Weather Research and Forecasting model	2
WRF-Chem	Weather Research and Forecasting model with Chemistry	8
WRF-SFIRE	Weather Research and Forecasting model with Sfire Fire Spread Model	8

1

Bridging Geophysical and Health Sciences to Study the Impacts of Biomass Burning on Human Well-Being

Tatiana V. Loboda[1], Nancy H. F. French[2], and Robin C. Puett[3]

ABSTRACT

Biomass burning in natural and management fires is a known source of air pollution that impacts millions of people worldwide. However, quantifying this impact and establishing definitive linkages between fire smoke and adverse health effects is a highly complex problem, which requires collaborative work between researchers across numerous disciplines within geophysical and health sciences. This chapter introduces the framework for this book and how we lay out the components of modeling chain from fire through smoke transport to health outcomes. Most of the concepts and models used in this modeling chain have been developed within disciplinary camps but are applied within a broad transdisciplinary research space. Our primary goal for this monograph is to build the foundation of common understanding of the entire process for nonspecialists in the field. And to achieve that, we aim to create a shared language, which interdisciplinary, transdisciplinary, and multidisciplinary teams of investigators might use to make their research efforts more robust and accelerate the pace of new knowledge development.

1.1. INTRODUCTION

Fire has been a part of the Earth system for at least 420 million years (Glasspool et al., 2004). Ever since levels of atmospheric oxygen produced by terrestrial vegetation rose enough to sustain fire propagation, burning of vegetation (biomass), ignited mostly by lightning strikes and volcanic activity, became an integral part of many global land ecosystems. Humankind evolved alongside naturally occurring fires to eventually develop masterful and extensive techniques of fire management for its benefit. The importance of fire to the development of our species and the societal evolution can hardly be overstated (Gowlett, 2016). Although over time, industrial and, in some parts of the world, domestic use of fire shifted toward other sources of fuel (e.g., coal, oil, and gas), to-date anthropogenic use of fire (biomass burning) continues to present a critical part of life and well-being for people worldwide, a robust landscape management tool, and a potent weapon (Bowman et al., 2011). As anthropogenic use of biomass burning has expanded, humanity's tolerance of naturally occurring fires dwindled, which led to the development of policies and practices in some countries, mostly notably the United States, aimed at near-complete fire suppression about a century ago (Forest History Society, n.d.). The subsequent shift in recent decades toward more severe fires, and a broad appreciation of the value of natural fire to ecosystem health, slowly brought about a more nuanced perspective,

[1] *Department of Geographical Sciences, University of Maryland, College Park, Maryland, USA*
[2] *Michigan Tech Research Institute, Michigan Technological University, Ann Arbor, Michigan, USA*
[3] *Institute for Applied Environmental Health, School of Public Health, University of Maryland, College Park, Maryland, USA*

Landscape Fire, Smoke, and Health: Linking Biomass Burning Emissions to Human Well-Being, Geophysical Monograph 280, First Edition.
Edited by Tatiana V. Loboda, Nancy H. F. French, and Robin C. Puett.

DOI: 10.1002/9781119757030.ch01

one that acknowledges fire's benefits within the framework of an adaptive resilience approach (Schoennagel et al., 2017). The perception and acceptance of wildfire, however, have ebbed again as the large number of extreme fire events swept across North America, Southern Europe, Australia, and Northern Eurasia over the last decade. Brought on by the combined impacts of climate (Flannigan et al., 2013) and extensive land-use change (Archibald et al., 2009; Doerr & Santín, 2016), these events have raised concern among policy makers and the general public about the threat to life and economic damage caused by uncontrolled fires. These concerns have grown larger as evidence of negative health outcomes associated with fire-produced air pollutants began to emerge in recent decades.

1.2. CONNECTING THE MODELING CHAIN

Although the number of articles in the scientific literature that examine linkages between air pollution resultant from biomass burning and human health and well-being continues to grow exponentially, this branch of scientific inquiry is still in its infancy partly owing to the complexity of the processes that necessitate a close collaboration between several branches of geophysics and public health fields. The process of establishing and quantifying causal relationships between biomass burning and health outcomes is extremely challenging and requires input from a large number of different experts. Where and when did the fire happen? What was burned and how much? How many pollutants were produced and what kind? How long did they remain the air and how far did they travel? How did they change while they were transported? Who inhaled them and how much? What effect on human bodies do they have? What polices, resources, and intervention strategies are needed to support well-being and resilience to biomass burning? The scope of the inquiry brings together experts in fire ecology, satellite remote sensing, forestry and natural resource management, atmospheric chemistry, atmospheric dynamics, space-time modeling, environmental epidemiology, toxicology, and public health.

The focus of this volume is on describing the observational and modeling approaches that are currently used in fire science, smoke characterization, and health assessment related to biomass burning smoke. The concepts and models that have been developed in these disciplinary camps provide exceptional capability to answer questions and solve problems related to the topic of biomass burning smoke exposure and health. However, their application now is required across the broad transdisciplinary space in order to address a complex set of concerns. And with the transdisciplinary reach of scientific inquiry and modeling efforts comes the necessity of building a foundational understanding of approaches, methods, and tools to craft seamless and robust chains of data analysis. While this foundational understanding is not a substitute for expert knowledge, it will help ensure that the data flow along the modeling chain is not controlled by asking "what we can do" or "what we have always done" in typical siloed approaches but begins solving the more complex interdisciplinary questions of "where are the gaps" and "what and who are needed to fill these gaps". Understanding the fundamental concepts and the limitations of the modeling chain segments will allow research teams to fine-tune their methodological approach at an early stage of inquiry and make research efforts more robust.

1.3. BUILDING A SHARED LANGUAGE

Our diverse international and interdisciplinary community of authors and reviewers highlights the "common language" challenge faced when connecting across established divides. One clear purpose of this book is focused on gaining an improved vocabulary and transdisciplinary knowledge set for considering solutions, which include mitigation actions and adaptive strategies, for avoiding adverse outcomes from biomass burning smoke exposure. The word *fire* is an incredibly commonplace and yet very complicated term. In the *Glossary of Wildfire Terminology* (National Wildfire Coordinating Group, 2021), *fire* is defined as "rapid oxidation, usually with the evolution of heat and light" and while it is technically our subject, the definition is substantially broader than the focus of this book. Here we focus on fires of both natural and anthropogenic origin, intended and unplanned but only those that consume alive and dead plant matter or biomass and occur on the landscape. We thus refer to *fire* here as "biomass burning," a term that incorporates events that can be found in the literature as *wildland fire*, *wildfire*, *bushfire*, *grassland fire*, *peat fire*, *forest fire*, *crop residue fire*, *prescribed fire*, *management fire*, or *landscape fire,* among other terms. While it does capture the absolute majority of potential instances of fire on the land, there are some very important types of fire that we are not considering here. Those include structural fires, trash fires, fossil fuel burning, and any other types of burning of human-made or nonbiomass materials as well as wood-burning for heating and indoor or outdoor cooking. Although all those instances also represent sources of air pollution and richly deserve an in-depth assessment, they are not considered here.

With the widely anticipated increase in biomass-burning driven air pollution during the 21st century (Intergovernmental Panel on Climate Change, 2019), capacity building in assessing the health burden and projecting health outcomes of air pollution arising from biomass burning (natural or anthropogenic) has become not only necessary but also urgent. It takes decades of study and practice to develop sufficient expertise in subfields of each of these broader disciplines. Meanwhile, an interdisciplinary team looking at this subject needs to have basic understanding of fundamental components of the complex process, a scientific equivalent of a common language, to be able to ask the right questions and collaborate effectively or, in the now famous words of Steven Pinker, to connect "the members of a community into an information sharing network with formidable collective powers" (Pinker, 1995, pp. 2–3). Traditionally, successful interdisciplinary teams undergo a multiyear multiproject coevolution where the team members learn these fundamentals through frequent interactions and continuous exposure to ideas and expertise of their collaborators. However, as scientists are challenged with addressing complex and pressing societal issues, streamlining the knowledge base for this subject will pave the way for accelerated interdisciplinary team building.

1.4. STRUCTURE, SCOPE, AND AIMS

In this book we aim to create a foundational knowledge base across a suite of disciplines, which will enable interdisciplinary teams to interact more effectively in addressing impacts of biomass burning air pollution on human health. The book is divided into three sections, which broadly address fire science (Part I: From Fire to Emissions), atmospheric chemistry and dynamics (Part II: From Emissions to Concentrations), and human health research (Part III: From Concentrations to Health Outcomes). Each section contains four chapters, which are designed to cover the foundational knowledge within the field, highlight recent advancements, describe commonly used methods, and outline existing data sets, models, or systems in general use by the respective communities. The three sections of this volume represent three distinct research communities, with members who are often focused on topics adjacent to the topics of fire, smoke, and health. In addition to what is presented here, many other aspects of the broad topic include improvements in geophysical data collection and analysis, predictive and retrospective modeling and data assimilation, as well as socioeconomic aspects of dealing with existing and future biomass burning smoke events. As noted, this is a global problem with far-reaching implications for human health. The chapters here are meant to provide some of the knowledge base for seeding further exploration by the international interdisciplinary community.

The authors have written the text to be comprehensible to experts in the other fields of the modeling chain, with the dual aims of making a subset of this book of value to all experts currently working in this domain, as well as making all components of this book of value to newcomers to the field. Scientific discovery and technological development often outpace the scientific publishing and, thus, books focused exclusively on cutting edge knowledge tend to become outdated before they come out of print. For this book, we include information that will provide the scaffolding for accelerated scientific growth. We have made it a priority to include information that represents the state-of-the-knowledge across the globe with the chapters coauthored and reviewed by members of the international scientific community. However, because we are drawing information exclusively from English-language literature, our scope tends to focus on geographic regions where biomass burning emissions are recognized as a public health concern and covered by research teams that publish their results in English language peer-reviewed journals. Having said that, we believe that the information collected here is applicable to conditions in other geographic regions and should be of help to all national and international scientific and management communities.

1.4.1. Elements of Fire Science

The multiple aspects of fire science (fire ecology, fire behavior, combustion, and more) are studied with a wealth of observational and modeling tools and approaches that draw from extensive field campaigns examining ongoing wildfire and fire management events, surveying post-fire impacts to assess biomass consumption, laboratory and field experiments quantifying consumption and emissions across different types of biomass and under various weather conditions, linking field and laboratory results to satellite observations, and finally convolving the knowledge about properties and spatial extent of biomass burning events and meteorological conditions into emissions modeling (see Part I of this book, "From Fires to Emissions"). Fire regimes (a cumulative description of the frequency, timing, extent, and severity of biomass burning) vary drastically across the globe. Chapter 2, "Biomass Burning as an Integral Force," delves into the mutual impacts of biomass burning, landscape properties, atmosphere, and climate systems that form the basis for understanding when, where, and how biomass burning is likely to impact air quality and how extensive this issue is across various regions of the globe. Much of our current understanding of contemporary fire regimes, the extent of biomass burning events, and our ability to monitor ongoing events globally has been shaped by satellite observations. Chapter 3, "Mapping and Characterizing Fire," describes the kind of information available about biomass burning from Earth observing systems, how this information is extracted, and what gaps and uncertainties these data sets have. However, biomass burning emissions are determined not only by the characteristics of the burning process but also by the composition, structure, and volume of surface fuels. Because fuels are so diverse and complex across global ecosystems, a certain level of standardization and simplification is a must to enable development of modeling approaches. Chapter 4, "Wildland Fuel Characterization Across Space and Time," synthesizes the current approaches to fuel mapping and highlights remaining gaps and future development needs. Finally, Chapter 5, "Biomass Burning Fuel Consumption and Emissions for Air Quality," reviews approaches to quantifying how much fuel is consumed during a particular biomass burning event and defining the composition and strength of resultant emissions, generally referred to as smoke.

1.4.2. Elements of Atmospheric Sciences

Biomass burning is a well-known source of gases and particulates that impact the composition of the atmosphere, causing changes in radiative forcing relevant to climate as well as air quality when smoke is found at surface levels (Larsen et al., 2018; Liu et al., 2014). The broad array of methods to sense, model, and map smoke in the atmosphere come out of a rich history of atmospheric science disciplines that include atmospheric physics, dispersion, and chemistry. Part II of the book, "From Emissions to Concentrations," provides basic concepts of atmospheric smoke concentration sensing, mapping, modeling, and prediction and methods to use information on fire emissions to quantify exposure of people to the pollution resulting from biomass burning. Chapter 6, "Surface Monitoring of Fire Pollution," provides a broad review of in situ smoke monitoring systems, including advanced sensor networks deployed and maintained by government agencies as well as newer low-cost monitoring technologies that provide data on air pollution concentrations in areas previously void of this information. In Chapter 7, "Data Assimilation for Numerical Smoke Prediction," concepts behind the methods that use advanced analytical tools to assimilate in situ and remote sensing-based measures of air quality are introduced, providing a glimpse into the complexities of smoke concentration modeling. Chapter 8, "A Review of Modeling Approaches Used to Simulate Smoke Transport and Dispersion," adds to this by reviewing coupled atmospheric and fire models that allow for integration of fire with atmospheric processes. In Chapter 9, "Profiles of Operational and Research Forecasting of Smoke and Air Quality Around the World," the various modeling and data assimilation methods are presented in the context of currently operational forecasting systems around the world that are used to inform decisions related to biomass burning smoke encroachment to locations far from the fires themselves. In sum, Part II provides the reader with an understanding of the complexities of smoke concentration mapping and forecasting that help inform decisions and feed knowledge used for assessing biomass burning pollution exposure.

1.4.3. Elements of Health Sciences

The linkage of ambient air pollution exposure with cardiovascular and respiratory morbidity and mortality is well established in the scientific literature (Johnston et al., 2012; Rajagopalan et al., 2018; Schraufnagel et al., 2019a). Studies of additional adverse health effects (i.e., reproductive, perinatal, metabolic, neurological) have accumulated rapidly (Klepac et al., 2018; Schraufnagel et al., 2019b). However, research of physiological and psychological impacts on human well-being specifically related to biomass burning is in its infancy. Part III, "From Concentrations to Health Outcomes," describes the current state of research, highlights important gaps, and provides guidance for building an interdisciplinary body of work specific to biomass burning risk and exposure assessment, toxicology, and life-span epidemiology. This section is particularly aimed at providing an overview of biomass burning health research findings and needs to (1) promote collaborations between the fire science and smoke modeling communities with health researchers and (2) draw the attention of the ambient air pollution health researchers to the unique and complex issues specific to biomass burning. Chapter 10, "Assessing Smoke Exposure in Space and Time," provides a general introduction to biomass burning and health research with a focus on exposure and risk assessment science, providing useful paradigms to guide future work. Moving toward more etiologic questions, Chapter 11, "Wildfire Smoke Toxicology and Health," outlines foundational principles of exposure science and toxicology, and describes the complexities and questions particular to the investigation of human health effects from biomass burning. Chapters 12, "Wildfire Smoke Exposures and Adult Health Outcomes," and 13, "Health Effects of Wildfire Smoke During Pregnancy and Childhood," summarize current findings from the limited epidemiologic literature particular to biomass burning, while drawing from the larger body of evidence on the impacts of particulate matter exposures to highlight potential pathways for future work. Chapter 13 is focused on perinatal and early life impacts, and Chapter 12 addresses adulthood.

1.4.4. Geospatial Science Elements

Space-time modeling (a component of geospatial data science) provides a common structure for linking our three subjects of biomass burning, smoke concentration, and exposure of human populations to smoke. Although we do not have a separate section devoted to the geospatial data science elements, various aspects of this rapidly evolving field, from satellite monitoring of biomass burning and atmospheric composition to models of atmospheric dispersion to spatially explicit assessments of vulnerable populations, play a crucial role in enabling this research agenda. Concepts of geospatial data science are not only central to the individual fields of inquiry we are considering in this book but also are absolutely essential for linking these three heretofore disparate topics. The development of powerful tools for studying spatiotemporal subjects in the past three or so decades introduced a mechanism to effectively connect the modeling chain. Despite the general applicability of geospatial methods and modeling approaches among the three fields, the specifics of spatial and temporal scales, sources of error and uncertainties, and overall suitability or compatibility of deliverables for specific studies vary greatly and are not well understood by scientists outside the specific communities. One of our implicit goals in preparing this book was to highlight not only the tremendous opportunities afforded by the contemporary models but also to outline their limitations. While most scientists are acutely aware of how uncertain the information provided by modeled and remotely sensed data sets is within their field, by and large, they are nearly completely blind to accuracy and reliability of data sets emerging from other disciplines. With the exponential growth of geospatial data sets and methods, development of deep understanding of uncertainties across the entire modeling chain is entirely unfeasible. However, a general awareness of where the greatest errors and uncertainties are likely to emerge within this very long chain is of paramount importance in building the most robust scientific inquiry that can support meaningful policy development.

1.5. CONCLUSION

This book presents the first attempt to build a common base for a multidisciplinary community of researchers and practitioners who strive to understand the impact of biomass burning on health and well-being of people worldwide. Individual chapters introduce core concepts, principles, methods, and terminology across a suit of disciplines to bolster multidisciplinary team engagement and raise the starting point for scientific inquiry. Our ultimate goal is to flatten the learning curve and accelerate the rate of knowledge building across various disciplines to enable the experts from various domains speak a common language based upon common understanding. But this book is only the beginning. There are still many concepts, methods, and considerations that we did not cover, and more knowledge is emerging every day. With the observed and expected trends of increased biomass burning under the ongoing and anticipated climate change and the land management projections, associated air pollution will continue to gain prominence as an environmental health hazard. This is the first step toward bringing experts together, making sure that all expertise needed is at the table, and starting the conversation at the next level, asking more sophisticated questions, and building more robust methodologies to enable the strongest possible science.

REFERENCES

Archibald, S., Roy, D. P., van Wilgen, B. W., & Scholes, R. J. (2009). What limits fire? An examination of drivers of burnt area in Southern Africa. *Global Change Biology*, *15*(3), 613–630.

Bowman, D. M., Balch, J., Artaxo, P., Bond, W. J., Cochrane, M. A., D'Antonio, C. M., et al. (2011). The human dimension of fire regimes on Earth. *Journal of Biogeography*, *38*(12), 2223–2236.

Doerr, S. H., & Santín, C. (2016). Global trends in wildfire and its impacts: Perceptions versus realities in a changing world. *Philosophical Transactions of the Royal Society B: Biological Sciences*, *371*(1696), 20150345.

Flannigan, M., Cantin, A. S., De Groot, W. J., Wotton, M., Newbery, A., & Gowman, L. M. (2013). Global wildland fire season severity in the 21st century. *Forest Ecology and Management*, *294*, 54–61.

Forest History Society (n.d.). *U.S. Forest Service fire suppression*. https://foresthistory.org/research-explore/us-forest-service-history/policy-and-law/fire-u-s-forest-service/u-s-forest-service-fire-suppression/#:~:text=In%201933%2C%20the%20federal%20government,day%20following%20its%20initial%20report.

Glasspool, I. J., Edwards, D., & Axe, L. (2004). Charcoal in the Silurian as evidence for the earliest wildfire. *Geology*, *32*(5), 381–383.

Gowlett, J. A. (2016). The discovery of fire by humans: A long and convoluted process. *Philosophical Transactions of the Royal Society B: Biological Sciences*, *371*(1696), 20150164.

Intergovernmental Panel on Climate Change (2019). *Summary for policymakers*. https://www.ipcc.ch/site/assets/uploads/sites/4/2020/02/SPM_Updated-Jan20.pdf

Johnston, F. H., Henderson, S. B., Chen, Y., Randerson, J. T., Marlier, M., DeFries, R. S., et al. (2012). Estimated global mortality attributable to smoke from landscape fires. *Environmental Health Perspectives*, *120*(5), 695–701.

Klepac, P., Locatelli, I., Korošec, S., Künzli, N., & Kukec, A. (2018). Ambient air pollution and pregnancy outcomes: A comprehensive review and identification of environmental public health challenges. *Environmental Research*, *167*, 144–159.

Larsen, A. E., Reich, B. J., Ruminski, M., & Rappold, A. G. (2018). Impacts of fire smoke plumes on regional air quality, 2006–2013. *Journal of Exposure Science and Environmental Epidemiology*, *28*(4), 319–327.

Liu, Y., Goodrick, S., & Heilman, W. (2014). Wildland fire emissions, carbon, and climate: Wildfire-climate interactions. *Forest Ecology and Management*, *317*, 80–96.

National Wildfire Coordinating Group (2021). NWCG glossary of wildland fire, *PMS 205*. https://www.nwcg.gov/publications/pms205

Pinker, S. (1995). *The language instinct: The new science of language and mind* (Vol. *7529*): U.K.: Penguin.

Rajagopalan, S., Al-Kindi, S. G., & Brook, R. D. (2018). Air pollution and cardiovascular disease: JACC state-of-the-art review. *Journal of the American College of Cardiology*, *72*(17), 2054–2070.

Schoennagel, T., Balch, J. K., Brenkert-Smith, H., Dennison, P. E., Harvey, B. J., Krawchuk, M. A., et al. (2017). Adapt to more wildfire in western North American forests as climate changes. *Proceedings of the National Academy of Sciences*, *114*(18), 4582–4590.

Schraufnagel, D. E., Balmes, J. R., Cowl, C. T., De Matteis, S., Jung, S.-H., Mortimer, K., et al. (2019a). Air pollution and noncommunicable diseases: A review by the Forum of International Respiratory Societies' Environmental Committee, Part 1: The damaging effects of air pollution. *Chest*, *155*(2). https://doi.org/10.1016/j.chest.2018.10.042

Schraufnagel, D. E., Balmes, J. R., Cowl, C. T., De Matteis, S., Jung, S.-H., Mortimer, K., et al. (2019b). Air pollution and noncommunicable diseases: A review by the Forum of International Respiratory Societies' Environmental Committee, Part 2: Air pollution and organ systems. *Chest*, *155*(2), 417–426.

Part I
From Fires to Emissions

2

Biomass Burning as an Integral Force

Amber J. Soja[1,2], Emily M. Gargulinski[1], and Elizabeth B. Wiggins[2]

ABSTRACT

Biomass burning has been an integral, interactive process on our landscapes and in our atmosphere since long before humans evolved. While biomass burning can be devastating, the benefits of biomass burning on our landscapes have long been recognized in fire and land management communities. Biomass burning acts to maintain flora and fauna diversity, cycle nutrients, and control fuel accumulation, without which leads to devastating and severe fires that are impossible to suppress. At the same time, smoke is an unequivocal pollutant that results in poor air quality and serves to directly and indirectly harm human health in ways we do not fully comprehend. The aim of this chapter is to provide an overarching objective view of the benefits and potential hazards of biomass burning. One fundamental goal of this chapter is to provide a basic introduction to biomass burning, fire terminology and classifications, and a description of the fundamental interactive nature of biomass burning on our landscapes and in our atmosphere, in tandem with our larger communities and weather and climate systems. It is our hope that we provide the building blocks and threads necessary to connect the essential interactive nature of biomass burning to the detailed chapters that follow to provide further meaning to this publication.

2.1. BIOMASS BURNING AS AN INTEGRAL FORCE AND PROCESS ON OUR LANDSCAPES

In this chapter, we provide an overarching, balanced introduction to biomass burning and the potential beneficial and negative consequences of biomass burning on landscapes. The chapters that follow will provide an in-depth analysis of the biomass burning topics broached in this introduction. Biomass burning is neither good nor bad, but perhaps biomass burning should be viewed as both a management tool and as one of the fundamental operational processes that acts on Earth to provide equilibrium and stability to flora and fauna, the atmosphere, and climate systems. Figure 2.1 plays a primary role in guiding these chapters.

Natural fire was on our landscapes and evolved with vegetation long before humans learned to use the power of fire to control our environs (Pyne, 2019). Hominids have used fire as a tool, studied fire, and endeavored to control fire for the last 50,000 to 400,000 years (Bar-Yosef, 2002; Karkanas et al., 2007). Consequently, we have a moderate understanding of the complexity of fire on a wide range of spatial and temporal scales, as well as fire-weather-climate feedbacks. At the same time, we recognize that we do not fully comprehend the processes through which fire and smoke interact with our environs and our health, at micro- to synoptic-scales.

[1] *National Institute of Aerospace, Hampton, Virginia, USA*
[2] *NASA Langley Research Center, Hampton, Virginia, USA*

Landscape Fire, Smoke, and Health: Linking Biomass Burning Emissions to Human Well-Being, Geophysical Monograph 280, First Edition.
Edited by Tatiana V. Loboda, Nancy H. F. French, and Robin C. Puett.

DOI: 10.1002/9781119757030.ch02

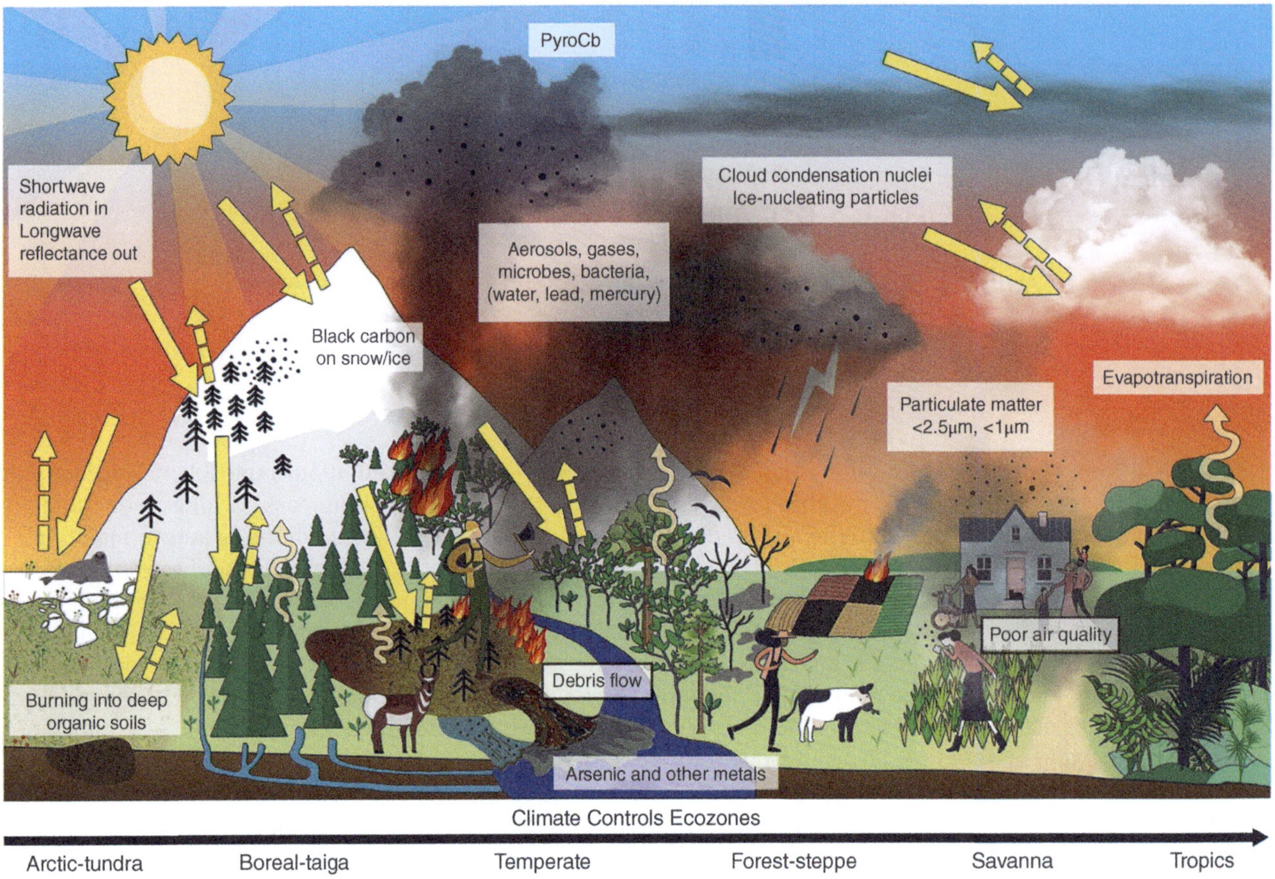

Figure 2.1 Feedbacks between biomass burning, ecosystems, humans, weather, and climate systems. Across the bottom row and figure are primary representative ecozones, from colder to warmer environments. Earth system feedbacks shown here include (1) ecosystems (e.g., altered albedo (relative energy/radiation reflected) due to burn scars, younger more reflective vegetation and vegetation structure change, especially in snow/ice-covered environs; aerosol deposition on snow/ice at altitude and Arctic-boreal environs; transition to agricultural lands; animals); (2) hydrologic systems (e.g., hydrophobicity; post-fire overland and debris flows, as opposed to filtered through the landscape; reduced evapotranspiration; inhibited near-field and changes in patterns of precipitation); (3) soils (deep burning in soil organic matter (SOM), peatlands, and permafrost can release carbon stored for millennia, often associated with methane releases; long-term severity-dependent alteration; resuspension of metals; suspension of microbes); and (4) atmospheric {transport/redeposition of metals and microbes; emissions and smoke transport that affects atmospheric albedo; smoke injection affecting weather (e.g., pyrocumulonimbus (pyroCb), pyrocumulus (pyroCu)), air quality, and human health (e.g., firefighters, downwind communities). The pronghorn represents fire-adapted-dependent species (require grasses, young shrubs, burned trees), and the seal represents the importance of ice (negative fire feedback). Poor Air Quality (AQ) can be measured in the atmosphere by the size and amount of particulate matter (PM) that is <2.5 μm or <1 μm in diameter (e.g., $PM_{2.5}$ or PM_1). Arrows show the relative energy balance: solid arrows depict shortwave energy from the sun, dashed arrows depict longwave reflected energy, and curved arrows depict evapotranspiration. The size of the arrows is equivalent to the relative reflectance or evapotranspiration, showing the differences between burned and unburned ecosystems and aerosol-burdened clouds (figure created by Katherine Froiland).

2.1.1. Fire Definitions, Terminology, and Processes

Biomass burning is a globally and nationally significant process that strongly interacts with our biosphere, atmosphere, and climate systems. Biomass burning can be broadly categorized as wildfire, wildland fire, prescription, land clearing, and agricultural. Wildfire is any undesired fire, regardless of the ignition source, where the objective is to put the fire out (National Wildfire Coordinating Group, 2022). Wildland fire is any nonstructure fire that burns in wildlands, where development is minimal to nonexistent, and this category of burning includes wildfire and prescribed fire. Prescribed fires are human-ignited fires that are set to achieve land management objectives when fire danger conditions are controllable. Objectives can include minimizing fuel loads (living and dead vegetation), maintaining the diversity and stability of flora and fauna, and controlling invasive species. Cultural, Indigenous, Native American, First Nation, and Aboriginal burning are types of prescribed fires that are increasingly respected and are also used to maintain cultural and/or spiritual connections with the landscape. Land-clearing fires are ignited by humans and used to clear lands for agricultural use, silviculture, industrial establishment, construction, and maintenance. Agricultural fires are human-ignited fires that are used to burn orchards, baled residue, and croplands for clearing and/or maintenance.

Biomass burning terminology is broad, because fire has been integral to our evolution with our landscapes, across continents and ecosystems. Even though biomass burning terminology can vary internationally, there is more similarity than difference. Several of these fire terms are worthy of definition to enable readers to fully comprehend the chapters that follow. Generally, biomass burning can be described by the way a fire burns across landscapes as crown, surface, and/or ground fires. A crown fire burns into the forest canopy. Crown fires can intermittently crown and may or may not be connected to a surface fire. Generally, crown fires burn a larger proportion of flaming emissions, relative to less-efficient smoldering emissions that can result in increased lung toxicity (Altshuler et al., 2020; Kim et al., 2018b). Surface fires burn the surface fuels on an ecosystem floor, which can include needles, leaves, cones, grasses, brush, stems, downed woody debris, and downed logs. These fires are typically less intense and less severe.

Fire intensity is a physical measurement of the rate of heat energy released per unit area, which is related to fuel conditions, largely fuel amount and structure, topography, and fire weather (Byram, 1959). Fire severity is the degree to which a site is altered and could be measured in terms of the amount of organic matter consumed, hydrophobicity of the burned surface, or the potential for ecosystem recovery (Keeley, 2008). There is a range of surface fire intensities that can serve to remove a portion of the fuels on a landscape leaving the forest overstory undamaged. However, under extreme fire danger, surface fires can burn deeply into duff (partially decomposed organic material) and, at times, to mineral soil, damaging seed beds and forest roots, resulting in 100% tree mortality. Surface fires are difficult to quantify at a large scale, because they are less intense and are often understory fires, which are challenging to remotely detect (Giglio et al., 2016; Loboda et al., 2011; Ponomarev & Shvetsov, 2015; Rogers et al., 2015; Roy et al., 2008; Soja, et al., 2020). For example, official Russian burned area estimates can differ by a mean of 35% in comparison with those estimated by Moderate Resolution Imaging Spectroradiometer (MODIS) products, which is likely due to the dominance of surface (understory) fire in Siberia (Kukavskaya et al., 2013; Soja et al., 2020).

A ground fire burns below the surface, deep into organic material and typically burns in peatlands or permafrost, potentially releasing carbon that has been stored for tens to thousands of years. These fires can overwinter, even below the snowpack, to return in the spring. These biomass burning events have recently been sensationalized as "zombie" fires, even though they are not a new type of burning (McCarty et al., 2020; Wein & MacLean, 1983; Witze, 2020), but this does bring attention to our currently altered fire regimes. In fact, there is a ground fire that has been burning in Delburne, Alberta, Canada for at least 140 yr that has flared in the last few years. Ground fires can alter landscapes for decades or longer, and they can be particularly dangerous, because they are not always evident from the surface. Burning these moist soils is a type of inefficient burning, which results in a larger concentration of hazardous pollutants that are typically released into the Planetary Boundary Layer (PBL) where humans breathe (see Chapters 4 and 5 for more on fuel mapping and consumption).

If we think in terms of our ability to take action on fires, an effective classification scheme is pre-, active-, and post-fire actions.

1. Pre-fire. Pre-fire actions could include quantifying the amount and structure of fuel contained in ecosystems (aboveground, surface, and ground fuel), assessing the health of ecosystems, prescribed fire or mechanical treatments, building fire safe communities, and predicting potential fire danger. Because our landscapes are large, characterizing fuels typically relies on a combination of data from ground-based field campaigns, modeling, and remote sensing. Vegetation indices are a type of satellite-data product that can distinguish the "greenness" or health of an ecosystem from space. An unhealthy ecosystem, due to a drought or insect damage, is more likely to burn, especially when fire danger conditions are high. Before a fire, communities have an opportunity to take action by supporting safe building practices and reducing fuels. Fire danger and fire weather predictions are critical to fire management planning both before and during an active fire. Fire danger data consist of local- and/or satellite-based meteorology to predict the potential for a dangerous fire, given an ignition and available fuel (discussed below).

2. Active fire. During active wildland fires, organizations and agencies manage human, data, and information resources across localities, states, territories, and international borders. Ground-based intelligence is critical as are satellite data and modeling to monitor fire progression, rate of spread, severity, intensity, and expected fire direction to determine response and potential evacuations.

In terms of visibility and air quality (AQ), smoke from active fires can be as hazardous as the fires themselves, thus, predicting smoke detrainment and transport is a significant objective. There are numerous global polar-orbiting (two views per instrument per day of the same geographic region at moderate-to-high resolution) and geostationary (lower resolution view of same geographic region every 5–30 min) satellites that can detect active fire, fire radiative power (FRP), visible smoke transport, and proxies for AQ, such as aerosol optical depth (AOD) (see Chapters 3 and 7 for more on characterizing biomass burning with satellite data). There was also a space-based lidar, on Cloud-Aerosol Lidar and Infrared Pathfinder Satellite Observations (CALIPSO), that could view the vertical extent of smoke through the atmosphere. Vertical profiles are important for understanding smoke aerosol transport and for detecting smoke detrained at the surface (nose level) where aerosols trigger poor AQ and affect health.

3. Post-fire. Depending on the severity of fires, post-fire hazards can be substantial, and these include the risk of debris flows that can result in the loss of life, infrastructure, and damage to critical water resources. Access to postfire landscapes is important to assessing risk, but standing dead tree stags represent a significant threat to forest workers and anyone in the forest. Satellite data products can highlight the areas most severely burned, to enable enhanced decisions and to provide a basis for constraining access. For example, in the United States, Burned Area Emergency Response (BAER) teams are responsible for providing a rapid assessment of infrastructure and the environment within seven days of fire containment. One product that emergency response agencies produce that enhances decision-making capabilities is a product that classifies relative burn severity, Burned Area Reflectance Classifications (BARC) maps, which are difference maps generated using pre- and post-burn imagery.

Fire regimes are defined as the overarching state of a fire configuration in a specific ecosystem or ecozone. Based on a generalized fire history of ecosystems and/or ecozones, measures can include fire frequency or the mean return interval of fire, seasonality, severity and intensity, burned area and patchiness (percentage of burned area), depth of burn, rate of spread, structural fuel classification (crown, surface, ladder fuels, soil organic matter (SOM)), ecosystem health, and long-term impacts on soil, flora, and fauna, including microbial communities (Bond & Keeley, 2005; Clark, 1988; Heinselman, 1978; Knelman et al., 2015; Pausas et al., 2004; Peterson & Reich, 2001; Shen et al., 2016; Wein & MacLean, 1983; Whitman et al., 2019; Williams et al., 2003).

A modified version of the Countryman (1972) and Moritz (2005) "fire triangle" is shown in Figure 2.2 to

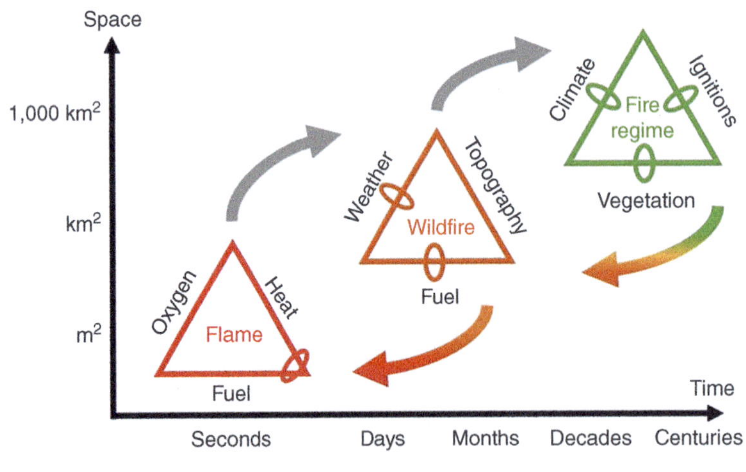

Figure 2.2 Interaction between physical constraints and biomass burning across temporal and spatial scales. Dominant physical processes operate within specific boundaries with the amount of fuel (biomass/vegetation) available (dry enough) to burn, under specific weather and climate constraints, acting to control biomass burning at broad temporal scales. The arrows show the feedbacks from biomass burning across scales and the feedbacks from larger-scale processes on biomass burning. The small circles show interactive controls within that temporal and spatial scale. At small time and space scales, heat from biomass burning serves to dry fuels, resulting in an increasing amount of fuel available to burn. At moderate scales, weather and fuels continually interact (e.g., relative humidity and temperature influence fuel availability, and biomass burning can generate lightning and smokey clouds). Over large geographic areas and long time frames, climate controls ignitions, ecosystem cohorts, and vegetative fuel and structure, while biomass burning feeds back to climate systems (e.g., aerosols on snow/ice/clouds, fire begins and ends succession). This figure is a modification of the "fire triangle" concept conceived by Countryman (1972) and modified by Moritz et al. (2005).

provide a guide to the interactions between fire and the environment at a variety of temporal and spatial scales. For instance, at a fine time and space scale, a flame requires oxygen, fuel, and heat (ignition source) to burn. At times, that heat can serve to dry fuels, generating a feedback that makes additional fuel available (dry enough) to burn (e.g., SOM, peatlands, permafrost). Then at a moderate temporal and spatial scale (days-months, ~1–1,000 km^2), weather acts to dry the fuels necessary to sustain a fire, aided by topography (slope, aspect, elevation), which could provide access to additional vegetative-ladder structure, leading to a rapidly spreading/running uphill fire. At this scale, topography, weather, and fire can act in unison to dry fuels. The severity of a fire is largely controlled by available fuels and weather, while topography is typically stationary. For instance, a persistent high-pressure system, that stalls over a region, is a weather phenomenon that can act to dry fuels resulting in severe fire(s), often at a regional scale. Also, wildland fires can generate their own weather to include lightning, pyrocumulus (pyroCu) and/or pyrocumulonimbus (pyroCb) clouds that inject smoke throughout the troposphere and into the stratosphere (Arreola Amaya & Clements, 2020; Clements et al., 2007; Fromm et al., 2005; Fromm & Servranckx, 2003; Peterson et al., 2015; Rodriguez et al., 2020; Rosenfeld et al., 2007). Then, at seasonal-to-century timescales over large geographic regions, weather and the frequency of lightning ignitions are under the control of climate. At a regional-scale, climate controls the mosaic pattern of unique vegetation cohorts that are under a range of post-disturbance ages and stages of succession. (Chudnikov, 1931; Heinselman, 1978; Shugart et al., 1992; Tumel, 1939; Zackrisson, 1977). Redundant and/or long-term droughts are examples of phenomena that are under the large-scale control of climate that can act to dry fuels resulting in regional-scale extreme fire seasons (Holden et al., 2018; Kitzberger et al., 2007; Rodrigues et al., 2021; Swetnam & Betancourt, 1990; Tosca et al., 2015; Tosca et al., 2013; Yao et al., 2017).

2.2. BIOMASS BURNING, WEATHER, BIOSPHERE, AND CLIMATE INTERACTIONS

Even though biomass burning can be devastating, wildland fire is a natural and integral process on our landscapes (Fig. 2.1). Under the control of weather and climate, wildfire maintains the zonal distribution of biomes and ecosystems, species composition, age structure, and the floristic diversity of ecosystems. Ecosystems evolved with biomass burning, and every system has a natural fire-return interval that can serve to maintain ecosystems or resolve the beginning or end of ecosystem succession (Antonovski et al., 1992; Heinselman, 1981; Van Wagner, 1978; Zackrisson, 1977). Wildland fire can serve to reduce fuels to mitigate wildfire risk, maintain healthy ecosystem functions (e.g., cycle nutrients, open seed cones), reduce total fire emissions, restore and promote native species, maintain natural diversity, decrease invasive species, and reduce tick-borne illnesses. Under controlled conditions, biomass burning provides value by decreasing fuels at the wildland urban interface (WUI) and the wildland urban intermix, which creates defensible space and promotes safe communities for citizens and firefighters. Fuel burning also reduces competitive understory species, which can support the commercial timber industry. Wildland fire is also beneficial for promoting environments for desired and/or endangered flora and fauna (e.g., quail, sage grouse, red-cockaded and black-backed woodpecker, Santa Cruz cypress, natural prairie grasses).

2.2.1. Ecosystems, Fire Return Intervals (FRI), and Carbon Balance

Biomass burning is one of the processes that connects our Earth's interacting system of systems (Fig. 2.1). Climate exerts the largest influence on the species assemblages that are native to a given location, such as the zonal distribution of ecozones: tropical, savanna, desert, temperate, steppe, forest-steppe, boreal-taiga, and arctic. Each ecozone and ecosystem evolved with a typical mean fire return interval (FRI) or a fire interval, which is the number of years between two successive fire events for a given area. Because ecosystems evolved with biomass burning, we can use an ecosystem's mean FRI, along with other ecosystem characteristics (vegetation (fuel) amount and health, fire intensity, depth of burn) to assess potential climate-related change over time.

Even within a biome, ecosystem FRIs can vary, dependent on the evolution of species cohorts, weather, and climate. For instance, the mean FRI in the boreal-taiga ecozone is ~150 yr. However, in the Siberian taiga, in surface-fire dominated *Pinus sylvestris* forests at the southern extent of their range, the FRI is about 10 yr. At the northern extent of this forest, the mean FRI is about 50 yr (Ivanova et al., 2010; McRae et al., 2006). Also in the same Siberian taiga biome, light-coniferous-deciduous *Larix* forests have a mean FRI of 36 yr (range 17–133) in comparison dark-coniferous forest (*Pinus sibirica*, *Picea obovata* (spruce), and *Abies sibirica* (fir)) that have a mean FRI of 196 yr (range 75–725) (Furyaev, 1996a, 1996b; Shvidenko & Nilsson, 2000; Soja et al., 2006). *Larix* spp., and *Pinus sylvestris* are fire tolerant (fire adapted) species that typically survive fires, and dark-coniferous species are shade-tolerant secondary-succession cohorts that do not typically survive a fire (Shugart et al., 1992). At evolutionary time scales, species have adapted unique strategies to resist fire (e.g., thick protective bark) and many species require fire for seeds to germinate (e.g., chemical cues, cones require heat, seedlings require full sun; de Groot et al., 2013; Dyrness & Norum, 1983; Habeck & Mutch, 1973; Keeley, 1993; Nikolov & Helmisaari, 1992; Rogers et al., 2015; Viereck & Johnston, 1990). These boreal-taiga species and FRIs are consistent with Melekhov's Russian fire hazard scale (Melekhov, 1947; Zhdanko, 1965), which serves to provide support for the integral and evolutionary relationship between climate, vegetation, fire, and weather.

Until recently, tropical rain forest (infrequent low-intensity fire) and Arctic tundra (mean FRI 686 yr) rarely burned (Bowman et al., 2011; Habeck & Mutch, 1973; Heinselman, 1981; Payette et al., 1989; Soja et al., 2006; Wein, 1976), but this has changed in recent years, primarily due to human land management and human-induced climate change, respectively. Tropical forests consist of fire-sensitive species, and natural fires are small and rare; one can typically step over the flames (Meggers, 1994, 2017; Saldarriaga & West, 2017). In tropical peatlands that have been drained to support agriculture, like those in Indonesia, the presence of biomass burning on these fire-intolerant ecosystems can have detrimental effects on the global carbon and methane budgets, as well as on local and distant AQ (Wiggins et al., 2018). Similarly, the loss of carbon stored in Alaska has increased as fire severity and fire season length increases (Turetsky et al., 2011). In North American boreal peatlands, a climate-driven increase in burn severity is causing a shift in dominant tree species from coniferous trees that enable long-term carbon storage in underlying peat to deciduous trees that do not provide for peat development (Johnstone & Chapin, 2006). Moreover, in permafrost regions, Abbott led a team of 98 permafrost experts (Abbott et al., 2016) who concluded organic carbon loss from permafrost could increase by 75%, while carbon loss due to burning could increase fourfold by 2100. This contrasts with the previously prevailing notion that permafrost regions would become a carbon sink. Moreover, McCarty and colleagues (2021) recently published a review, aimed at mitigation and adaptation, that documents climate-induced circumboreal change in Arctic and boreal-taiga systems that include ecosystem transformations, increases in fire regimes, and an increase in wildfire emissions, including black carbon (BC), particularly at latitudes above 60°N.

2.2.2. Ecosystems and Feedbacks With Weather and Climate Systems

In terms of climate, biomass burning and smoke exert a huge forcing on direct and indirect climatic processes that influence biospheric, hydrologic, and atmospheric climatic feedbacks (Fig. 2.1). Scarred landscapes directly alter the albedo (relative reflectiveness) of landscapes by altering forest structure and reducing SOM to char, resulting in both positive and negative climate forcings (Betts, 2000; Bonan, 2008; French et al., 2016). A burn scar is dark, consequently it absorbs more solar radiation at the surface, much like dark pavement, resulting in a positive feedback (more warning). However, as vegetation recovers, the younger early-successionary species are more reflective, which is a negative feedback to the climate system (a cooling; Randerson et al., 2006; Rogers et al., 2013). These responses can be at continental scales, as Rogers et al. (2015) demonstrated using satellite data to show the postfire spring albedo increase is larger in boreal North America, which is dominated by crown fire, in comparison with boreal northern Eurasia, where surface fire dominates and often burns less intense and less severe, although a larger area is affected in northern Eurasia.

After a fire has removed vegetation from a landscape, the nutrient, microbial, and hydrologic cycle can remain altered for years, depending on the fire severity and the ecosystem (low FRI systems typically recover quickly, e.g., grasslands). Burn severity can accentuate harsh abiotic conditions, leading to changes in the soil bacterial community structure, enzyme activity, and bacterial abundance and diversity, affecting plant-microbe interactions in secondary succession (Knelman et al., 2015; Shen et al., 2016). Additionally, a severely burned landscape is more likely to be affected by debris flows, similar to the floods and debris flows that followed the Thomas fire that burned in California in 2017, destroying over 100 homes and resulting in the deaths of 21 people. Postfire overland water flow can also release arsenic and metals, which had previously been stored and filtered by landscapes, to severely damage downstream water supplies (Fig. 2.1; Murphy et al., 2020).

Severely burned Arctic and boreal-taiga systems can remain altered for many decades. These systems store the largest pool of terrestrial carbon on Earth, largely in peatlands, permafrost-lain and deep organic soils (Bradshaw & Warkentin, 2015; Scharlemann et al., 2014; Tarnocai et al., 2009; Zinke et al., 1986), thus these systems have the potential to amplify carbon released to the atmosphere and rivers (positive climate forcing). Wang and colleagues (2021) showed the sensitivity of permafrost thaw on streamflow increases in the three great Siberian watersheds (Ob, Yenisei, and Lena) from 1936 to 2019. This increasing freshwater flush to the Arctic Ocean could influence thermohaline circulation (saltwater is heavier, so freshwater could inhibit ocean circulation), which is an important modulator of global climate (positive climate feedback) (Eldevik & Nilsen, 2013; Knies et al., 2007; Rennermalm et al., 2006). Indeed, extreme fires burning in the permafrost-lain ecosystems in the Republic of

Sakha (formerly Republic of Yakutia), Siberia, in 2019, 2020, and 2021 have likely released carbon that has been stored for millennia (McCarty et al., 2021, 2020; York et al., 2021). This is due to climate-induced increases in extreme temperatures and dryness, manifested by high-pressure weather systems that stalled over Siberia, as predicted (IPCC, 2021; Witze, 2020; York et al., 2021).

2.2.3. Fire Weather and Fire Danger

Climate also controls the synoptic-scale and meso-scale weather patterns that result in storms, droughts, lightning, and the stalled high-pressure systems that generate the conditions that precede extreme fire danger (Balzter et al., 2005; IPCC, 2021). Globally, fire danger, fire weather, and fire season length has increased and/or is predicted to increase with medium to high confidence, depending on the geographic region (Flannigan et al., 2013; IPCC, 2021). Fire weather models are used to predict fire danger and the potential for extreme fires, and these data are publicly available and rapidly provided to fire managers to enable the allocation of resources and provide situational awareness. For example, the Canadian Wildland Fire Information System (CWFIS) is widely used in many countries to predict daily fire danger, based on cumulative meteorological information (temperature (T), wind speed (WS), relative humidity (RH), and noon-to-noon precipitation (P); Amiro et al., 2004; CFDG, 1992; Lawson & Armitage, 2008; Van Wagner, 1987). The CWFIS program is freely available and can be modified for specific ecosystem characteristics. Output variables include the Fire Weather Index (FWI), Fine Fuel Moisture Code (FFMC), Duff Moisture Code (DMC), Drought Code (DC), Initial Spread Index (ISI), and a Buildup Index (BUI). In Canada, the CWFIS products also include fire behavior and additional active-fire intelligence, based on fuel databases and other regional information.

In the United States, the National Interagency Coordination Center's (NICC) Predictive Services and National Oceanic Atmospheric Administration's (NOAA) National Weather Service (NWS) work together to produce numerous climate, drought, and fire danger indices for decision-support intelligence to anticipate fire activity and allocate resources. Examples include the National Fire Danger Rating System (NFDRS) and Wildland Fire Assessment System (WFAS) (Giuseppe et al., 2016). In general, these systems provide intelligence on the dryness or availability of fuels to burn, from fine fuels (e.g., leaves, pine needles, twigs) that could dry in 1 hr or less (1 hr fuels) to larger or less accessible fuels (duff, small branches; 10 hr), to fuels that would take even longer to dry (peat, larger branches; 100 hr, 1,000 hr fuels).

2.3. STATE OF GLOBAL FIRE REGIMES

Recent research shows there is already evidence of climate-induced increases in extreme fires, globally (Bowman et al., 2017; IPCC, 2021; Littell et al., 2009; Soja et al., 2007; Westerling et al., 2006; York et al., 2020), which exacerbates our ability to maintain healthy ecosystems, AQ, economies, and societies (Cassella, 2022; Cresswell et al., 2022; Normile, 2022). Jolly and colleagues (2015) used several fire weather indices from 1979 to 2013 to show a global increase in the frequency of long fire-weather seasons, which coincide with actual long and large fire seasons, and they showed a doubling of the area affected. In a regional investigation, despite the influence of suppression across the western United States, several studies have concluded increases in burned area across the west in the twentieth century were largely controlled by climate (Littell et al., 2009; Westerling et al., 2006). In the United States, the mean 10 yr wildland fire burned area has steadily increased, nearly tripling decadal means, from a mean of 1.18 Mha (2.9 M acres) in the 1980s to 3.03 Mha (7.5 M acres) since 2011 (data source: National Interagency Fire Center (NIFC)). Also, since 1960, burned area has exceeded an annual estimate of 4 Mha (9.9 M acres) only four times, and all these extreme years have been recorded since 2006. Every year, there are reports, globally, of the worst fire seasons, in terms of burned area, economic cost, and lives lost. In recent decades, annual records have been continually broken. For example, in 2017, record breaking fires burned in California (Thomas fire) and British Columbia (previous worst 1958), and these records were both surpassed by fires burning in 2018. This is a global phenomenon, with many recent prominent examples of wildfire disasters: Australia in 2009 (Black Saturday) and 2019–2020 (Black Summer), Russia (2010, 2015, 2018, 2019, 2021), Canada (2014, 2018), Greece (2018, 2021), Portugal (2016–2018), Mediterranean fires (2009), Chile (2017), Bolivia (2010), India (2020), Sweden (2018), Brazil (2019, Amazon), Indonesia (1998, 2019), and there is even burning in Greenland (2017, 2019). Almost as an exclamation, again in 2022, extreme wildfires are burning across Europe, the United States, boreal-arctic (Canada, Siberia, Alaska), Africa (Morocco), South America (Chile, Argentina), and Australia. Extreme biomass burning events are increasingly in the news, because they are record breaking, not sensationalized.

The Arctic tundra experienced the earliest fire seasons in the MODIS satellite record in both 2019 and 2020, surpassing the previous record by a whopping 40 days. Also, 2019 and 2020 were the largest Arctic fire seasons, with 2019 being twice the size of the next largest year, 2013, and 2020 surpassed 2019 by about 25% (York et al., 2021). Then, in 2021, the Republic of Sakha, Siberia, which is almost twice the size of the state of Alaska, experienced the most burned area in the satellite record (>8.4 Mha; 20.8 M acres). This is significant, not only due to the size of the burned area, but also because this region holds some of the largest terrestrial carbon reservoirs on Earth. Sakha is underlain with permafrost, so burning deeply into the SOM releases carbon and methane that has been stored for millennia, thus releasing major greenhouse gas stores, amplifying climate change (positive feedback). This has strong implications for AQ at nose level for two reasons: smoke from fuels burning in permafrost is not fully combusted, therefore it is inefficient burning that is high in pollutants that are toxic to human health (Kim et al., 2018b); and this smoke is not injected high in the atmosphere, consequently the smoke tends to hug the surface where humans and other animals breathe. Not all burning in the tundra and taiga is severe, but as climate changes, there is an increased amount of severe burning, which has climate implications. These include a deeper active-layer depth thaw, thermokarst formation, increased freshwater flow to the Arctic oceans, alterations in hydrothermal dynamics, and changes in biogeochemical processes. These changes could inhibit the ability of mammals to migrate to food sources, and result in substantially damaged infrastructure and crops, hindering the ability of wildlife and humans to live sustainably in these remote northern regions. At the same time, these challenging landscapes are understudied, particularly in Siberia, where these massive landscapes are critically significant to regional and global climate.

From a global perspective, burned area and active fires are decreasing, which is consistent with earlier work (Fig. 2.3; Andela et al., 2017; Sokolik et al., 2019), and these decreases are largely in human-dominated systems, under the control of human management. The analysis shown in Figure 2.3 is based on MODIS data, and these data are consistent with the Visible Infrared Imaging Radiometer Suite (VIIRS) instrument data record. We consider two distinct products: the MODIS burned area product is analyzed from 2001 through 2020, and the thermal anomaly (active-fire) data are valid back to 2003 when both the Aqua and Terra instruments were available the entire year (Giglio et al., 2003, 2016; Justice et al., 2002; Roy et al., 1999, 2008). Decreases in burned area are largely driven by Africa (contributes ~55.2% to the decrease). This result is consistent with Andela and colleagues (2017), who reported decreases in burned area from 1998 to 2013, particularly in savannas, cerrados,

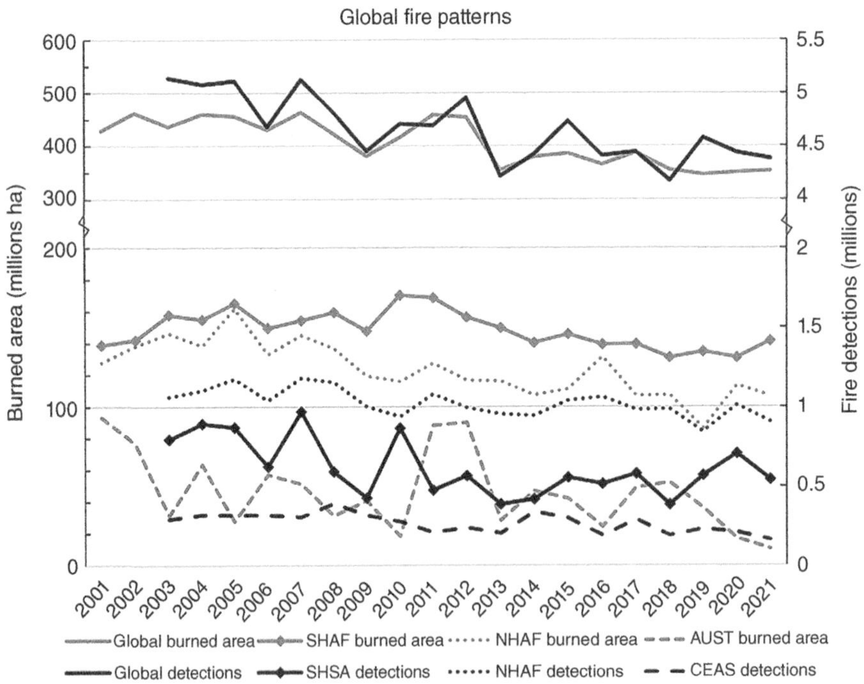

Figure 2.3 Satellite-derived global and regional burned area, showing a global decrease in total burned area and active-fire detection. The strongest contributors to the negative relationship are Southern Hemisphere (SH) and Northern Hemisphere (NH) Africa (AF), on human-dominated landscapes. South America (SA) is another large contributor to the decreasing trend in detections, also on human-managed landscapes. Australia (AUST) and Central Asia (CEAS) also show decreasing trends. Decreasing trends in South America and Australia are likely due to large burned areas early in the fire record and the brief extent of the satellite record in comparison with mean fire return intervals (Bowman et al., 2009; Sokolik et al., 2019).

and grasslands, where the demand for agricultural expansion and intensification is the dominant driver of the decrease. Personal communication with coauthors support the argument for this continued decline. Australia contributes about 28% to the global decline in burned area, which is largely due to large burned areas in 2001, 2011, and 2012, not an actual decrease, but rather related to the minimal temporal extent of the satellite-data record in comparison with FRIs and large fire years early in the satellite record.

In terms of active-fire detections, the Southern Hemisphere South America contributes about 42.5% to the global decline, largely owing the massive areas burned before 2011. The rate of deforestation and fire is still significant and disconcerting, considering that deforestation of 2% of the forest per year will result in the annihilation of the old-growth primary rainforest flora and fauna in just 50 yr. Unlike temperate and boreal systems, which have evolved with massive fires and experienced glaciations, the nutrients stored in these tropical forests are stored in the aboveground vegetation, so these forests will not fully recover in human lifetimes. The Amazonian and Atlantic basin tropical forests evolved with negligible fires over millennia, and the biodiversity contained in these systems will return only in geologic time (Ruiz Pessenda et al., 2009). In 2011, there were actions taken at federal, state, and local levels that resulted in a decrease in the rate of deforestation in the Amazon forest in Brazil, which demonstrates that human management and conservation can reduce deforestation. In some systems, ignitions were reduced by 69% (Aragão & Shimabukuro, 2010; Schmidt et al., 2018). Unfortunately, in 2019, burning in South America increased, following President Bolsonaro's encouragement to loggers and farmers to clear the land. When the Brazilian National Institute for Space Research (INPE) reported satellite data showed an 84% increase in fires in 2019 in comparison with the same period in 2018, Bolsonaro dismissed the head of INPE. Forest fires in closed canopy tropical systems initiate positive fire feedbacks, promoting an increase in the number of fires, due to open access (e.g., roads), accidental spread from nearby pastures, and the increased susceptibility of partially logged or damaged forests (Cochrane et al., 1999).

2.3.1. Fire Regime Change and Future Predictions

Fire regimes are changing, as predicted, with climate and human factors. Flannigan and colleagues (2013) used cumulative fire weather severity and Global Climate Models (GCMs) to estimate future fire regimes, and they concluded cumulative fire severity and fire-season length would increase. The largest increases are predicted in Northern Hemisphere temperate, boreal-taiga, and arctic systems, as well as Peru and southern China (severity >3 times, fire season 3 to >20 days). In another study, Soja and colleagues (2007) used three decades of model predictions, with published evidence of change, to show that ecosystems (treeline migrations, moisture-stress dieback) are changing, as models predicted. They also concluded fire regimes are changing (Soja et al., 2006, 2007) and these increases in extreme and severe fires often act as a catalyst for ecosystem change. Based on the rapidity of change in some regions, Soja and colleagues (2007) suggested a potential nonlinear rapid response to changes in climate. These results are in agreement with current evidence, nonetheless there are hopeful examples of potential mitigation and adaptation. In one example, in an investigation of U.S. western grasses, Abatzoglou and Kolden (2011) identified several climate-related variables that favor invasive grasses, and suggested land managers use models to identify and locate the native species most at risk to mitigate climate stressors. Tchebakova and colleagues (2011) demonstrated the potential for increased crop range and yield across central Siberia, and Parfenova and colleagues (2019) modeled the potential for an increase in human sustainability and attractiveness across Asian Russia.

Across landscapes, predictions include increases in lightning (Chen et al., 2021; Price & Rind, 1994), extreme wildfires (Di Virgilio et al., 2019; Ruffault et al., 2020), and fire severity (Keyser & Westerling, 2019). Veraverbeke and colleagues (2017) found 55% of the variability in burned area at the northern treeline in boreal North America can be described by lightning ignitions in large fire years, which is correlated with temperature and precipitation. Cattau and colleagues (2020) found more intense and larger fires are associated with lightning ignitions in the United States, so lightning ignitions may be more of a concern than had been originally considered. Then, Hurteau and colleagues (2019) analyzed the effect of reduced fuels due to current extreme burning, as opposed to constant fuels, and they found that even after accounting for a 14.3% decrease in fuels, continued significant wildfires with associated AQ challenges will continue in the future. In a recent literature review of predicted future fire regimes (9 global-, 2 biome-, and 22 regional-scale), Bowman and colleagues (2020) found 79% of the studies predicted increases in fire weather, burned area or fire activity, while 18% showed mixed results and one study predicted no change. Indeed, fire regimes are predicted to change and there is ample evidence that fires are becoming larger and more severe across our landscapes, in sync with changes in weather and climate. These fire regime changes will continue in the future with associated increases in unhealthy poor air quality (Reid & Maestas, 2019).

2.4. SMOKE TRANSPORT, CLOUDS, CLIMATE, AIR QUALITY (AQ), AND HEALTH

Biomass burning directly injects smoke emissions into the atmosphere, which can result in poor AQ and altered atmospheric processes. These include changes in the radiation balance, patterns of precipitation, and atmospheric circulation (see Fig. 2.1). Smoke is a complicated mixture of aerosols and gases that are often chemically reactive and interact with the atmosphere in complex ways that are not adequately understood (Sokolik et al., 2019; see Chapter 6 for more on biomass burning emissions). Smoke can travel over long distances to affect climate and AQ in distant communities (Baker et al., 2018; Damoah et al., 2004; Wotawa & Trainer, 2000).

These small particles or smoke aerosols can act as cloud condensation nuclei (CNN) and/or ice-nucleating particles (INP) to impact the formation of clouds, cloud radiative forcing, cloud lifetime, and inhibit near-field precipitation, all of which feed back to climate systems (Andreae et al., 2004; Carslaw et al., 2010; Hobbs & Radke, 1969; Rosenfeld, 1999; Tosca et al., 2013; Twomey, 1974; Zamora et al., 2016). While in the atmosphere, smoke can cause significant impacts to solar and ultraviolet fluxes, thus affecting the radiation balance at the top of the atmosphere, at the surface, and throughout the atmospheric profile (Natarajan et al., 2012; Ward et al., 2018). Zamora and colleagues (2016) estimated that the indirect smoke-driven cloud albedo effect could decrease local summertime shortwave radiation flux by 2 to 4 Wm^2 in the Arctic and subarctic.

Smoke-burdened clouds have an impact on regional lower-altitude cumulus clouds and climate. Smoke-contaminated clouds have a higher cloud droplet number and smaller droplet sizes, which can inhibit precipitation (Twomey, 1974), because the small droplets are less likely to grow and fall out as rain (Twohy et al., 2021), exacerbating already dry fuel conditions. Holden (2018) suggested that precipitation during the western fire season exerts the strongest control on burned area either directly through its wetting effect or indirectly through feedbacks to the vapor pressure deficit. Tosca and colleagues (2015) found aerosols from human-lit fires in the African tropics inhibit convection, especially when aerosol optical depths are high, which served to dry the surface and enable increased fire activity.

Smoke from biomass burning is detrained in the PBL, where the pollutants can cause hazardous AQ, and at the same time, wildland fire smoke is often injected throughout the troposphere and/or to the stratosphere, where smoke can be transported aloft, then descend to cause poor AQ across continents (Baker et al., 2016; Fishman et al., 1986; Jaffe et al., 2013; Wilkins et al., 2020 and 2018; Wotawa & Trainer, 2000). Smoke can be transported over long distances, extensively affecting atmospheric albedo, and can be deposited on snow and ice, forcing freshwater to melt more quickly than it would otherwise (Flanner, 2013; Natarajan et al., 2012; Skiles et al., 2018; Sokolik et al., 2019; Thomas et al., 2017; Ward et al., 2018; Whitlow et al., 1994). Additionally, smoke injected beyond the troposphere to the stratosphere can affect longer term climate forcings that can be similar in magnitude and extent to volcanic eruptions (Peterson et al., 2015, 2018). In 2019–2020, smoke from severe fires burning in Australia were injected into the stratosphere, and this smoke circumnavigated the Earth for more than a year, where the smoke aerosols substantially cooled oceanic cloud-free regions (Hirsch & Koren, 2021).

Despite the importance of the vertical distribution of smoke in the atmosphere for AQ and climate, our ability to monitor and model smoke injection and detrainment height is limited. Lidar instruments can be rapidly deployed to specific fires to provide multilayered smoke plume injection and detrainment heights, as these evolve over time, so that meteorological data are integrated with fire energy and convective plume entrainment processes. These data are incredibly valuable for science and modeling, but these data are limited (Clements et al., 2018). Additionally, there are two satellite instruments that can provide smoke-plume injection and/or detrainment heights: the Cloud-Aerosol Lidar with Orthogonal Polarization (CALIOP) instrument on the CALIPSO satellite; and the Multi-Angle Imaging SpectroRadiometer (MISR) on Terra. MISR has a well-developed global database of smoke plume injection heights, and in comparison with CALIOP, MISR is better able to characterize a larger number of smoke plumes near fires due to the instrument's wider swath (Kahn et al., 2007; Val Martin et al., 2012). CALIOP has an enhanced capability to identify thin, tenuous smoke layers and the vertical structure of smoke through the entire atmospheric column to the surface (Kim et al., 2018a; Omar et al., 2009). Initial CALIOP analysis finds that most smoke is detrained above the PBL (~78%; Soja et al., 2019; Thomas et al., 2017; Wilkins et al., 2022). In contrast, MISR concludes that most smoke (88%–96%) is detrained in the PBL, which is likely because MISR is a morning overpass, and wildfires peak in the late afternoon (Val Martin et al., 2010).

2.4.1. Smoke and Health

Smoke from biomass burning is an unhealthy pollutant that statistically increases hospital visits that are directly associated with increases in respiratory (asthma, pneumonia, acute bronchitis, decreased lung function, painful breathing) and cardiovascular symptoms, myocardial infarction, ischemic heart disease, heart failure, dysrhythmia, pulmonary embolism, ischemic stroke, transient ischemic attack, and death (DeFlorio-Barker et al., 2019; Kondo et al., 2019; Magzamen et al., 2021; Rappold et al., 2011; Reid et al., 2016; Wettstein et al., 2018; see Chapters 6, 10–12). To a novice, it may seem apparent that smoke is harmful, but statistical confirmation, attribution, and validation are challenging (Reid et al., 2016). These evidence-based approaches involve teasing apart the effects of smoke from confounding variables, such as temperature, relative humidity, and other sources of pollution (Cleland et al., 2021). The transported smoke is coupled to affected communities through models, where the health effects are statistically related to smoke event(s) (Liu et al., 2015). In one investigation, Thelen and colleagues (2013) used statistical and process-based models to demonstrate that during peak smoke emissions, a person is 50% more likely to require emergency care than in nonfire conditions. An increase of 10 μg/m^3 daily wildfire $PM_{2.5}$ was associated with a 2.7% increase in Intensive Care Unit admissions 5 days later (Sorensen et al., 2021). This type of research suggests that health care systems should forecast critical care resources. The annual economic estimate of short-term smoke exposure was estimated between $11 billion and $20 billion, with a long-term estimate between $76 billion and $130 billion, which surpasses firefighting cost by approximately an order of magnitude (Fann et al., 2018).

Every fire season, the impacts of smoke on human health receives considerable attention from both the public and policymakers. Wildland fire smoke exposure is rapidly being defined as an environmental justice (EJ) issue, with increasing attention on outdoor workers and socially disadvantaged groups that have limited adaptive capacity and can be disproportionally exposed (D'Evelyn et al., 2022). Rappold and colleagues (2017) identified vulnerable communities and concluded vulnerable communities experience more smoke exposure and health impacts from wildland fires. Davies and colleagues (2018) found wildfire vulnerability is spread unequally across race and ethnicity, with census tracts that were majority Black, Hispanic, or Native American experiencing about 50% greater vulnerability to wildfire compared with other census tracts. This attention provides an opportunity to advance collaborative planning with conservation, economics, EJ, and public health stakeholders, to work towards a balanced blueprint with planners, forest and fire managers, air quality, and public health interests.

In recent work, emissions from wildland fires have been found to contain microbial aerosols that hold viable microbial cells and biological INP. The source of these cells and culturable bacteria is smoke emitted from dead vegetation (Moore et al., 2021). As the health effects from human exposure to the small $PM_{2.5}$ aerosols in smoke is becoming increasingly well researched, the potential health repercussions from microbes and bacteria have not yet been scrutinized (Kobziar & Thompson, 2020). In one investigation, Landguth and colleagues (2020) found high daily average $PM_{2.5}$ concentrations during the wildfire season were associated with increased influenza the following winter (16%–22% increase in rate per 1 μg/m^3 increase in average daily summer $PM_{2.5}$). Moreover, burning landscapes will volatilize what resides on the landscape, which can include mercury (Hg), lead (Pb), zinc (Zn), cadmium (Cd), tin (Sn), copper (Cu), and other toxins (CARB, 2021; Kristensen & Taylor, 2012; Sparks & Wagner, 2021; Turetsky et al., 2006) (see Fig. 2.1).

2.4.2. Small Biomass Burning Events, Smoke, and Health on Human-Dominated Landscapes

Small, short-lived, human-ignited biomass burning events are notably underestimated, unnoticed, and under-investigated. Small biomass burning events can include croplands, orchards, silviculture, piles, slash, land clearing, understory burning, and prescription burns on public and private lands. Research has shown that small biomass burning events and burned areas are vastly underestimated (Al-Saadi et al., 2008; Huang et al., 2018; Larkin et al., 2014; Randerson et al., 2012; Soja et al., 2009). In fact, Nowell and colleagues (2018) found 92% of the biomass burned in Florida is from prescribed burning but only 25% of the burned area is detected by satellite, which distorts air pollution emission inventories and our perception of fire regimes, especially in human-dominated landscapes. Also, the relationship to drought is distorted (Nowell et al., 2018), because in human-dominated landscapes, landowners and local governments choose not to burn during drought conditions.

Globally, fire is used as a tool by farmers and agricultural managers to control pests, weeds and weed diseases, and to fertilize soils, as well as an inexpensive, timely, and effective method to remove excess crop residue to facilitate planting and increase crop yields (Chen et al., 2005; Hari et al., 2003; McCarty et al., 2007; Pyne, 1993). Although there are justifications for cropland burning, cropland burning is a source of aerosols that results in air pollution and regional haze (Chameides et al., 1999; McCarty, 2011; Streets et al., 2001). Additionally, recent research shows that cropland burning leaves open bare soils that promote increased water loss through evapotranspiration and produces hydrophobic ash that reduces water infiltration, as well as increasing susceptibility to wind and water erosion (Pikul & Zuzel, 1994; Verhulst et al., 2011; Wuest et al., 2005). Residue burning has also been found to increase carbon loss and decrease SOM (Norman & Bry, 2016; Smith, 2013). Recently, there have been movements to educate and lead farmers and agricultural managers away from traditional crop residue burning through education and extension agency research and science-exchange information (e.g., Fire Science Exchange Network; Chen et al., 2005).

Agricultural burning is prominent across the Earth, in every populated country, nonetheless burning of these lands is not well quantified, so neither are the health effects from these smaller but seasonally persistent biomass burning events. The global spatial distribution of agricultural fire was found to vary annually in a similar manner, with two peak global agricultural seasons, April to May, and August (Korontzi et al., 2006). The majority of burning was in Eastern European croplands (includes Russian), although there was a decrease in cropland and rangeland burning from 2002 to 2012 (McCarty et al., 2017). The second largest cropland areas burned were in central Asia and Asiatic Russia. In China, farmers burn about one quarter of their crop residue, despite the government providing financial incentives to retain crop residue (Chen et al., 2019). Eastern European Russia is a main contributor of black carbon (BC) to the Arctic, acting as a short-lived climate forcer, strongly influencing snow-ice albedo and Arctic haze (McCarty et al., 2012). In the United States, even though wildfire receives the most notoriety, prescribed and small biomass burning events often exceed or are equivalent to the area burned by western wildland fires in normal fire years (EPA, 2016; Melvin, 2015; Nowell et al., 2018). The Environmental Protection Agency (EPA) has initiated efforts to accurately quantify crop residue and rangeland burning in the National Fire Emissions Inventory (Pouliot et al., 2017), and this effort is continuing.

These small, persistent fires are not extreme, so these biomass burning events do not typically receive attention. However, these small fires persist for months, every year, and are certainly worthy of AQ and health analyses. Accurately detecting and quantifying these biomass burning events is challenging, because information is often not recorded in localities. Historically, we did not have the capability to accurately quantify these biomass burning events from space (Ramo et al., 2021), and satellite data are critical to defining and quantifying fire regimes. Yet, satellite detection and quantification is inhibited for several reasons on these landscapes: persistent cloud cover, the small size of these fires relative to the detection limits of space-based instruments, lack of fire intensity, and timing (Huang et al., 2018; Nowell et al., 2018; Soja et al., 2009). Humans set these biomass burning events at times when they are easiest to control, which is often under conditions when (1) visibility is limited and (2) it is early morning or late afternoon, when polar-orbiting satellites are not overhead. Many of these fires are less intense, understory burning, and/or short-lived small burns (area <20 ha; <50 acres) that challenge satellite detection capabilities.

In many populated regions of the world, small biomass burning events are the dominant fire and smoke source, which is where the detrimental health effects from smoke could be most prominent. However, because these small burns are not well defined, the ability to conduct cogent health impact assessments is inadequate. The need for investigating the health effects of small biomass burning events is important in these understudied regions, which are often located in Black, Indigenous, and people of color (BIPOC) communities (Tessum et al., 2021). In one example, the impact of smoke from sugarcane burning was highlighted in a recent *Palm Beach Post* article (Ramadan et al., 2021), which shows the communities effected disproportionally live in poverty. The authors analyzed PurpleAir sensor data across western Palm Beach County, Florida, where the residents refer to the ash that falls for 4–6 months of the year as "black snow." PM emitted from these agricultural fires is reported to be greater in west Palm Beach County than any other U.S. and Florida counties (Nowell et al., 2022). Health impact assessments could serve to inform future epidemiological studies that refine smoke exposure impacts to chemical constituents and/or aerosol size on susceptible downwind communities. Aerosols from small biomass burning events may not meet the daily EPA 24 hr $PM_{2.5\mu m}$ standards of 35 μg/m^3, but because this biomass burning is seasonally persistent, annual average standards of 12.0 μg/m^3 could be exceeded. Regardless, health impact assessments and epidemiological studies are fundamental to understanding the human health impact of smoke, and the prerequisite for these studies is accurate biomass burning accounting,

2.5. ACTIONS FOR MITIGATION: TOOLS IN THE FIRE COMMUNITY SANDBOX

Although we must accept that climate is changing fire regimes, there are actions we can take to minimize fire and smoke risk to communities and our landscapes. Climate change is increasing the flammability of many of our wildlands and the wildland urban interface (WUI). These changes are not simple linear increases towards warmer and drier conditions. Rather, climate change is manifesting itself as extreme weather events, longer fire seasons, and unpredictable fire behavior. In many ecosystems, due to past fire exclusion, we have far too much built-up fuel on our landscapes. Historic logging often removed larger fire-resistant trees, which is also partly responsible for our current fire and forest management challenge. Human development is continuing to encroach in the WUI, complicating the already challenging conditions for wildland firefighters. In the United States, the number of new homes in the WUI grew by 41% from

1990 to 2010, and the land area in the WUI increased by 33% (Radeloff et al., 2018). When combined with extreme fire weather, fires are larger, less predictable, and more intense.

Creating a new and more resilient path forward begins by acknowledging the ecological and social changes that have taken place on landscapes over the last century and developing a plan to change the way we think about and fight fires on climate-dominated landscapes. The impacts of increased fire activity can be mitigated through effective stewardship of fire regimes. This action can be achieved through evidence-based fire management that incorporates the necessity for prescribed, controlled, Indigenous, and cultural burning, as well as wildfire, based on both qualitative and quantitative ecological- and cost-benefit analyses.

The International Association of Wildland Fire (IAWF) recently authored an international "Position Statement on Climate Change and Wildland Fire." It identified actions to reduce risk and highlighted the impacts and implications of climate change for wildfire and land management services, and the communities they protect. It also provided a vision statement, which supports actions to (1) safely and effectively extinguish wildfires when necessary, (2) use prescribed and wildland fire when possible to meet protection and land and resource management objectives, (3) manage natural resources through progressive fuel reduction to increase landscape resilience, and (4) create fire-adapted communities that can accept shared responsibility for addressing how to co-exist with smoke and wildland fire (see https://www.iawfonline.org/wp-content/uploads/2021/11/IAWF-Position-statement-on-climate-change-_-FINAL.pdf).

On the environmental justice forefront, a California bill clears the path for ex-inmates to become firefighters (Assembly Bill No. 2147). In the past, prisoners received valuable training and took part in California's most extreme fires, but they were not permitted to become firefighters due to their convictions, which can now be expunged. Additionally, IAWF hosted a series of ignite talks focused on diversity and inclusivity that are publicly available to support the community (see https://www.iawfonline.org/events/webinars/). These include Mission Critical Teams; Constraining the Blaze: Transgender Stories; Cultural Safety of Indigenous Wildland Firefighters; Inclusive Terminology; Living with PTSD; Fitness and Mental Health Well-Being. Globally, we are struggling to find qualified personnel as the demand for controlled prescription fire increases and wildfire risk escalates. An inclusive community is a stronger community.

The ecosystem and fire challenges before us create conditions that put firefighters, communities, infrastructure, and historic ecosystems at risk, which compels us to use all of the viable, safe, and effective fire management tools available. Critical management tools in wildland firefighters' operational toolbox include increasing the pace and scale of restoration thinning, prescription burns, and allowing some wildfires to burn. Both scientific literature and fire management experience strongly support the use of ecologically beneficial, or good, fire on landscapes. Restoring fire to landscapes through controlled prescribed and managed wildfires can lead to reductions in fuels, long-term reductions in suppression costs, and reduction in the societal and environmental impacts of severe wildfires. Allowing more good fire on landscapes will help limit the spread and severity of future wildfires, promote long-term ecological resilience, and provide firefighters another tool to manage landscapes.

This translates to conducting more large-scale, multijurisdictional, multiagency prescribed burns to restore the beneficial role of fire in resilient ecosystems. Managing ecosystems with controlled-fire prescriptions on landscapes is not without risk, but the cumulative fire risks far outweigh the prescription-fire risk. Ecosystems are going to burn. The question is, do we burn ecosystems when we are more able to control the fires under safe prescription parameters, guided by local fire managers or wait until the fire conditions are extreme and the possibility of suppression and control are minimal? Full-suppression strategies, especially for the most severe fires, are not a safe or viable option, even for the most experienced and best-equipped teams of firefighters. In many cases, their only option is to plan exit strategies and wait for the weather to change.

Increased prescription burning is not a sustained viable option without recognized public support. Developing public understanding for the overarching long-term benefits of fire on our landscapes to mitigate potential risk is critical to continued support and the restoration of resilience on our landscapes. Communities can be enabled to take the lead in maintaining defensible spaces and reducing hazardous fuels near their homes. Social scientists are essential to understanding perceived and/or actual barriers to communicating the value of fire-safe communities, as are practical examples from similar communities (Miller, 2013). In the case of Miller (2013), firefighters and a high-end community were initially strongly resistant to prescribed fire. Through extensive planning and convincing the community of the value of fire, this community now argues for fire. There are a growing number of communities that are convinced of the value of prescribed fire in their communities to reduce fuels and create defensible space, which substantially enhances a community's ability to sustain a wildfire. There are articles from Greece, North Carolina, Oregon, Kansas, Ghana, and the Basque Country in a recent *Wildfire* publication (IAWF, 2022), along with an article that imagines a balanced future with prescribed fire.

Additionally, fire-prone communities could strongly advocate for revised fire-safe building and construction codes, with direction from organizations like the National Fire Protection Association (NFPA). Increasingly, communities are forced by fire to evacuate their homes and businesses, because full-suppression strategies are not safe or viable possibilities. Home and business losses are minimized in fire-safe communities, where fire fighters know the community and know the space is defensible. Working with first responders, communities can learn to recognize where and when suppression can be used safely and effectively, while at the same time setting priorities for community protection and identifying trigger points for community evacuations, in advance of an emergency.

2.6. CHALLENGES, CRITICAL ENGAGEMENT, AND MOVING FORWARD

Challenges and potential actions are described throughout this chapter, and the chapters that follow will layout a blueprint for understanding AQ and health in our interconnected communities. Fire regimes are adversely changing, compelling us to improve our understanding of the connections between biomass burning, fuels, smoke transport, atmospheric chemistry, AQ, and health, which are ultimately regulated through weather and climate. It is clear that we need to move toward interdisciplinary works, similar in spirit to this book. Interdisciplinary research can provide a comprehensive background, upon which our decisions will impact our potential to diminish, modulate, or amplify biomass burning, climate, and AQ for health. Until we understand the data, models, research, and practices of our disparate communities, we will be unable to develop sound biomass burning-AQ-health conclusions that will lead to science-based decisions and actions. For instance, while horizontal data (e.g., aerosol optical depth, visual smoke) are geographically useful, they are not an indicator of smoke at the surface where we breath. This book provides the basis to connect and understand communities that have not typically interacted.

As you read through this work, which contains the details necessary to understand the current state of biomass burning and well-being, we hope you will view each chapter within the context of a larger arrangement, one that could serve to move us forward towards action and balance.

Central topics critical for current biomass burning to AQ and health research and engagement include: enhance understanding of climate feedbacks; connect fire weather with fuels; quantify fuels across landscapes as they change daily to annually; increase good fire on landscape; restore resilient ecosystem; adaptation and mitigation; include short and long-term impacts of seasonally persistent small and extreme smoke on health; increase comprehension of biosphere-atmosphere-chemical processes on AQ; enhance small- to large-scale predictive and global models; increase and enhance fire-safe, healthy, and informed communities; and integrate health impact assessments and epidemiological studies. We must learn to live with fire on our landscapes and that requires a balanced evidence-based approach that incorporates agencies and local knowledge in making well-informed decisions for fire and landscape management, with air quality specialists and health professionals.

2.7. RECENT DEVELOPMENTS

Biomass burning has been a topical and newsworthy subject in the last decades, which has drawn public attention to the importance of biomass burning on our landscapes and the health of firefighters and communities that are inundated with smoke. Sustained droughts and reoccurring extreme fire seasons have served to highlight the connections between biomass burning, AQ, weather, and climate. To further the evidence, a devastating Australian *State of the Environment* report (Australia, 2022) was recently released that found bushfire risk has increased by 30%; the 2019 and 2020 bushfires killed or displaced millions of animals and threatened human health with smoke; 18 ecosystems are deemed at risk of collapse; and Australia has already lost more mammal species than any other continent and has one of the highest rates of species decline in the world. Australian scientists hope the ominous environmental assessment, which details the devastating impacts of climate change, will spur public and governmental action toward accelerated decreases in emissions (Cassella, 2022; Normile, 2022). In another damaging article, published in the United Kingdom, unions criticized the government over losses of resources and firefighter jobs (11,500 since 2010), while demands are increasing under climate change (Mahmood, 2022).

On a positive note, the response to climate-driven escalations in droughts and wildfires, is the largest investment in the resilience of physical and natural systems in American history, the Infrastructure Investment and Jobs Act, passed in November 2021 (https://www.congress.gov/bill/117th-congress/house-bill/3684). This historic law ensures investments in climate resiliency that are crucial to ensuring local, state, and Tribal communities have the resources they need to bolster climate resilience and protect natural areas from intensifying droughts, wildfires, flooding, and legacy pollution. Specifically relevant to biomass burning and health are (1) a $1.5 billion investment in wildfire resilience (e.g., risk reduction; salary increases; maintenance of fuel breaks, and prohibition of new roads; Joint Fire Science Program research, thinning and prescribed fire); (2) $1.4 billion investment in ecosystem restoration and resilience (e.g., invasive pest detection, prevention, and eradication; reforestation on both public and private lands; abandoned mine land restoration); (3) a $466 million investment in Tribal climate resilience and infrastructure (total $3.3 billion for wildland fire); and an $8.3 billion investment in water and drought resilience. Two requirements are (1) the U.S. Department of Agriculture (USDA) Forest Service (USFS), and the Department of the Interior (DOI) are required to report accomplishments annually and develop 5-yr treatment, monitoring, and maintenance plans for fuels reductions; and (2) collaborative landscape-scale natural area restorations on public and private lands and collaborative restoration programs to restore water quality and/or fish passage on federal lands. States will apply for funding through a competitive process requiring proposals and plans for the projects' scope, anticipated improvements, and benefits. These investments are expected to aid communities to tackle the climate crisis, while creating well-paying jobs, advancing environmental justice, and boosting local economies.

As part of the Bipartisan Infrastructure Law (BIL), on 21 June 2022, the White House announced wildland firefighters would receive a minimum of $15 per hour (cost $24.3 million) and directed no fewer than 1,000 seasonal wildland firefighters be assigned full-time, permanent, year-round federal employment. Also, on 3 July 2022, firefighters should have received temporary bridge salaries (lesser of $20,000/year or 50% base salary), with retroactive pay (back to 1 October 2021). Additionally, under the BIL, the Office of Personnel Management (OPM) established the new GS-0456, Wildland Firefighter series, which replaces the Forestry Technician or Range Technician series and provides a clear path for career advancement through the federal government. The BIL establishes a joint DOI and USDA program to increase focus on wildland firefighters' physical and mental well-being in a targeted interagency effort to provide trauma-informed mental health care. This joint effort supports the development of mitigation strategies to minimize wildland firefighter exposure to line-of-duty environmental hazards and adds staffing capacity specifically focused on mental health and employee support efforts for firefighters.

In accordance, organizations, such as the National Science Foundation (NSF), National Aeronautics and Space Administration (NASA), USFS, and National Oceanic and Atmospheric Administration (NOAA), are preparing strategic interagency plans that include a future with increasing fire regimes. The numerous interagency interactions are often suggested and led by the Office of Science and Technology Policy (OSTP) (e.g., USDA Office of the Secretary (OSEC) Wildfire Smoke Impacts Subgroup). NSF hosted the Wildfire and the Biosphere Innovation Lab that included leading interdisciplinary fire scientists to formulate innovative strategies and campaigns, and to set priorities to advance "wildfire and living system" science in 2021 (Shuman et al., 2022). Also in 2021, NASA and the National Institute of Environmental Health Sciences (NIEHS) met to establish a new Science for Disaster Reduction (SDR) Interagency Working Group (WG) to facilitate data integration between geoscience and health agencies to improve our response to disasters.

NASA hosted stakeholder workshops to identify agency priorities and gaps to align potential future contributions to biomass burning science and applications from remote sensing, field campaigns, instrumentation, predictions, models, technology, and data fusion to the priorities, gaps, and requirements reported by stakeholders. The ultimate goal is to support decision-support research, tools, and operational agencies. NASA Applied Sciences established a new program, Wildland Fire, and the FireSense project was announced, which spans disciplines to include research, applied sciences, instruments, and technologies. NASA Health and Air Quality Applied Sciences Team (HAQAST) is a collaborative team that works in partnership with public health and air quality agencies across the globe to use NASA data and tools for the public benefit. Increasingly, through these teams, interagency partnerships are producing interdisciplinary work that connects biomass burning with AQ to move science and application forward (O'Neill et al., 2021). NOAA and the OSTP are currently leading the Interagency Council for Advancing Meteorological Services (ICAMS) efforts to coordinate federal agencies to foster meteorological services to provide societal benefits, which includes fire weather, smoke transport, and AQ (e.g., Working Group (WG) on Fire Weather). The ICAMS Subcommittee for Atmospheric Composition Information and Services (SC-ACIS) established two pertinent teams: WG for Atmospheric Transport, Dispersion, and Volcanic Ash (WG-ATDVA), and the WG for Wildfire, Controlled Burns, and Sources Services (WG-Fire).

After decades of work, on 1 July 2022, the International Agency for Research on Cancer (IARC), the cancer agency of the World Health Organization (WHO), announced an evaluation of the carcinogenicity of occupational exposure on firefighters. They concluded there was sufficient evidence for cancer in humans for mesothelioma and bladder cancer, and there was limited evidence for cancer in humans for colon cancer, prostate cancer, testicular cancer, melanoma of the skin, and Non-Hodgkin's lymphoma (Demers et al., 2022; Smith et al., 2020). At the time of this publication, the U.S. Federal Firefighters Fairness Act of 2022 (passed House 11 May 2022) was making its way through congress (https://www.congress.gov/bill/117th-congress/house-bill/2499). This bill would make it easier for federally employed firefighters who contract chronic obstructive pulmonary disease, mesothelioma, and some cancers to qualify for federal workers' compensation.

Recent field campaigns include Fire and Smoke Model Evaluation Experiment (FASMEE 2019–2021) and Fire Influence on Regional to Global Environments and Air Quality (FIREX-AQ 2019). FASMEE is a multiagency, interdisciplinary collaborative effort to measure fuels, fire behavior, fire energy, meteorology, smoke, and fire effects to advance the understanding of wildland fire to promote better prediction of the spread of fire and smoke to support model development and enhance decision-support capabilities. The objective of the interagency NOAA/NASA campaign, FIREX-AQ, was to provide measurements of trace gas and aerosol emissions for wild-land and small fires and to relate these emissions to fuels and fire conditions at the point of emission, characterize the conditions relating to plume rise, follow plumes down-wind to understand chemical transformation and AQ impacts, and assess the efficacy of satellite data to define fire and to estimate emissions from the fires sampled. This unique work resulted in research that connects the amount of fuel burning on the ground to the smoke transects sampled during the campaign and to modeled data, in an effort to more accurately quantify biomass burning emissions to enhance chemical transport models for AQ and health (Wiggins et al., 2021). This mission begins to bridge the gap between fuels, remote sensing, fire behavior, and in-situ aircraft measurements to ultimately enhance our understanding of and ability to quantify the short-term variability in fire emissions for AQ and health.

The world's first Geostationary Environment Monitoring Spectrometer (GEMS) instrument was launched on the Korean satellite, Cheollian, in February 2020 by the Republic of Korea. This momentous AQ satellite enables hourly monitoring of air pollution for almost 20 Asian countries. Two additional pollution-monitoring satellites are expected in orbit, one focused on Europe (Sentinel 4, expected to launch in 2024) and the other focused on North America, Tropospheric Emissions: Monitoring of Pollution (TEMPO). TEMPO was launched 7 April 2023 and is the first space-based instrument to monitor major air pollutants (e.g., ozone, nitrogen dioxide) every daylight hour at a high-spatial resolution across North America. These sensors will monitor their regions and provide the data necessary to monitor and measure intercontinental pollution transport. Together these satellites will form a global GEO constellation for AQ, which will revolutionize AQ forecasts and create groundbreaking new datasets of atmospheric chemistry measurements.

Global satellite missions will continue to provide information and data to support biomass burning, fuels quantification, AQ, and health. VIIRS instruments will continue to be hosted on the Joint Polar Satellite System (JPSS) series of polar-orbiting satellites, enabling visible views of Earth, and providing meteorological and active-fire detection data. Geostationary Meteorological Satellites (GMS) are a series of global satellites that will continue to be launched to provide high-temporal, low-spatial resolution imagery for monitoring weather, fire, and smoke (Milrad, 2018). WildFireSat is a Canadian mission that is expected to launch in 2028. The mission intends to support wildfire management by providing more precise information on fire, smoke, AQ, and carbon accounting. The Multi-Angle Imager for Aerosols (MAIA) is an instrument that is currently under development that will derive abundance and characteristics of ground-level PM over several major cities to explicitly target epidemiologists and health organizations. The NASA- Indian Space Research Organization (ISRO) Synthetic Aperture Radar (NISAR) mission will provide radar data to observe the vertical and horizontal distribution of vegetation (biomass) to enhance understanding of ecosystem change, trends, responses to disturbance and recovery.

Several satellite missions are in the planning stages that can include international partners (e.g., CNES, JAXA, CSA). These missions are based on the Decadal Survey, which is driven by input from the science community and policy experts and sponsored by NASA, NOAA, and the USGS. The National Research Council identifies and prioritizes leading-edge scientific questions and the observations required to answer these principal science questions (NRC, 2007). Two of these are particularly relevant to this community: (1) Aerosol, Cloud, Convection and Precipitation (ACCP, currently the Atmospheric Observing System (AOS) mission); and (2) Surface Biology and Geology (SBG). These are the first NASA missions that have explicitly included input from stakeholder communities from conception, as well as the strong foundational science core.

1. The goal of AOS is to provide essential data for improved forecasts of weather, AQ, and climate. AOS promises to provide unprecedented insights into processes and the vertical structure of our atmosphere to (a) determine how aerosols affect AQ and the global energy balance and (b) target processes that influence the prediction of severe weather and climate.

2. Surface Biology and Geology (SBG) is focused on providing data products that have the ability to improve our understanding of terrestrial vegetation physiology, function, and health; inland and coastal ecosystem physiology, function, and health; active surface change (e.g., landslides); and the effects of change in land use (e.g., energy, water, carbon fluxes). Overarching applied goals are to provide the data products that will enhance our ability to manage agriculture, natural habitats, water use and water quality, and urban development.

2.8. PERTINENT AND REFERENCED WEBSITES

AFAC, Australian National Council for Fire and Emergency Services. https://www.afac.com.au/

AFIS, South African Advanced Fire Information System, a near real-time satellite-based fire monitoring system for Africa and the globe. https://www.afis.co.za/

AirNow, Reports air quality using the official Air Quality Index (AQI) to communicate healthy-to-unhealthy AQ to the public in a effort to protect health; provides air quality information at local, state, national, and world views through Facebook, Twitter, news media and website. https://www.airnow.gov/ https://www.airnow.gov/fires/using-airnow-during-wildfires/

ARSET, NASA trainings for beginners and advanced practitioners for air quality, fire, agriculture, climate, disasters, land, and water resources management. https://appliedsciences.nasa.gov/what-we-do/capacity-building/arset/about-arset

AOS, Atmosphere Observing System (formerly ACCP, Aerosol and Cloud, Convection and Precipitation), mission to optimize comprehension of interactive aerosols, clouds, atmospheric convection, and precipitation for improved forecasts of weather, air quality and climate by providing insight into the vertical structure of the atmosphere. https://aos.gsfc.nasa.gov/

Australia Forest Fire, up-to-date information on Australia's forests including forest-related publications and maps, and links to forest management organizations and institutions; hosted by in the Australian Government Department of Agriculture, Fisheries and Forestry (ABARES). https://www.agriculture.gov.au/abares/forestsaustralia/forest-data-maps-and-tools/fire-data

Australia My FireWatch, interactively browse satellite observed hotspots https://myfirewatch.landgate.wa.gov.au/

BAER, Burned Area Emergency Response, BAER teams rapidly assess the potential for postfire hazards. https://burnseverity.cr.usgs.gov/baer/

BlueSky modeling framework, links a variety of independent fire information, fuel loading, fire consumption, fire emissions, and smoke dispersion models to enable easy simulations. https://www.fs.usda.gov/pnw/tools/bluesky-framework

CALIPSO, Cloud-Aerosol Lidar and Infrared Pathfinder Satellite Observations, a partnership between NASA and the French Space Agency, CNES. High-vertical resolution data shows the vertical extent or profile of smoke through atmosphere. Provides aerosol profile and aerosol layer data. https://asdc.larc.nasa.gov/project/CALIPSO

CAWFE, Coupled Atmosphere-Wildland Fire Environment, a coupled weather prediction and fire behavior model by National Center for Atmospheric Research. https://ral.ucar.edu/solutions/products/coupled-atmosphere-wildland-fire-environment-cawfe

CEOS, Committee on Earth Observation Satellites, coordinates Earth observations for ease of data access for the user communities. https://ceos.org/

Climate and Clean Air Coalition, a global coalition, hosted by the United Nations Environment Programme, that supports high-priority initiatives (https://breathelife2030.org/), host a free library, and provides relative global news. https://www.ccacoalition.org/en

CNES, Centre National d'Etudes Spatiales, government agency responsible for shaping and implementing France's space policy in Europe. https://cnes.fr/en

Conabio, Mexican fire alert system that provides detections, vegetation and conditions. http://incendios.conabio.gob.mx/#

CONAFOR, Comisión Nacional Forestal, a Mexican agency tasked with developing, supporting and promoting the conservation, sustainability and restoration of Mexico's forest. https://www.gob.mx/conafor

Manejo del Fuego, Fire management. https://snif.cnf.gob.mx/incendios/

Copernicus, European Union's Earth observation programme. https://www.copernicus.eu/en

CSA, Canadian Space Agency, advance the knowledge of space through science to ensure space science and technology provide social and economic benefits for Canadians. https://www.asc-csa.gc.ca/eng/

CWFIS, Canadian Wildland Fire Information System, daily fire weather and fire behavior maps. https://cwfis.cfs.nrcan.gc.ca/home

Decadal Survey for Earth Science and Applications from Space. https://www.nationalacademies.org/our-work-decadal-survey-for-earth-science-and-applications-from-space

DOI, Department of the Interior, protects and manages U.S. natural resources and cultural heritage; provides scientific and other information about resources; and honors its trust responsibilities or special commitments to American Indians, Alaska Natives, and affiliated Island Communities. https://www.doi.gov/

EFFIS, European Forest Fire Information System, supports the protection of forests against fires in the EU and neighboring countries. https://effis.jrc.ec.europa.eu/

EOSDIS Worldview, Interactive interface for browsing full-resolution, global, daily satellite images of wildfire smoke and observing fire detections. https://worldview.earthdata.nasa.gov/

EPA, Environmental Protection Agency, National Ambient Air Quality Standards (NAAQS) for Particulate Matter. https://www.epa.gov/pm-pollution/national-ambient-air-quality-standards-naaqs-pm#rule-summary

EPA Air Data, Air Quality data collected at outdoor monitors across the US https://www.epa.gov/outdoor-air-quality-data

EPA EJScreen, Environmental Justice Screening and Mapping Tool, a new mapping and screening for tool environmental justice (EJ), based on nationally consistent data and an approach that combines environmental and demographic indicators in maps and reports. https://www.epa.gov/ejscreen

FASMEE, Fire and Smoke Model Evaluation Experiment. https://www.fasmee.net/

FEMA U.S. Fire Administration, Federal Emergency Management Agency, entity of U.S. Department of

Homeland Security; the mission is to support and strengthen fire and emergency medical services (EMS) and stakeholders to prepare for, prevent, mitigate and respond to all hazards. https://www.usfa.fema.gov/about/

FIRECAST, a fully automated analysis and alert system that delivers a range of near real-time monitoring products tailored to a user's specific needs, particularly in under-served remote communities (e.g., Amazonian indigenous communities, Madagascar), supports strategic land-use policies and conservation interventions, hosted by Conservation International. https://firecast.conservation.org/

FireMARS, Fire Monitoring, Accounting and Reporting System for Canada, provides burned area polygons, carbon monitoring and accounting, and Greenhouse Gas inventories. https://www.nrcan.gc.ca/our-natural-resources/forests/wildland-fires-insects-disturbances/fire-monitoring-reporting-tool/13159

Fire Science Exchange Network, U.S,.national collaboration of 15 fire science exchanges that provide relevant, current, wildland fire science and information to federal, state, local, tribal, and private stakeholders within ecologically similar regions. https://www.firescience.gov/JFSP_exchanges.cfm

FIREX-AQ 2019, Fire Influence on Regional to Global Environments and Air Quality Campaign, NOAA and NASA, provides comprehensive observations to investigate the impact on air quality and climate from wildfires and agricultural fires across in theUnited States. https://www-air.larc.nasa.gov/missions/firex-aq/index.html

FIRMS, Fire Information for Resource Management System, Global data provided from VIIRS and MODIS instruments that includes burned area, Thermal Anomalies or active -fire detections and FRP. https://firms.modaps.eosdis.nasa.gov/

GEMS, Geostationary Environmental Monitoring Spectrometer, GEMS launched by the Republic of Korea in February 2020, enables the hourly monitoring of air pollution levels for almost 20 countries in Asia. https://www.ball.com/aerospace/programs/gems

GEOS-Chem, Goddard Earth Observing System Chemistry Model. https://geoschem.github.io/

GFMC, Global Fire Monitoring Center, an umbrella of 14 independent Regional Wildland Fire Networks whose mission is to reduce negative impacts of landscape fires and advance ecological role of natural fire in fire-dependent ecosystems. https://gfmc.online/

GFWED, Global Fire WEather Database, integrates weather factors influencing the likelihood of a vegetation fire starting and spreading, provides a globally consistent fire weather dataset. https://svs.gsfc.nasa.gov/4634

GMS, Geostationary Meteorological Satellites, a series of satellites that will continue to be launched to monitor global weather, fire, and smoke. These include: Japan Meteorological Agency (JMA) Himawari, EUMETSAT Meteosat, Russian Meteor, China Fengyun, U.S. GOES-East and GOES-West.

GOES, Geostationary Operational Environmental Satellite program, a joint NOAA and NASA effort that consist of an eastward- and westward-viewing satellite that provides meteorological, atmospheric, and fire data and information - operational series lifetime through December 2036.

GOFC-GOLD, Global Observation of Forest Cover-Global Observation of Land Dynamics, a coordinated international program working to provide ongoing space-based and in situ observations of the land surface to support sustainable management of terrestrial resources at different scales. https://gofcgold.org/

GTOS, Global Terrestrial Observing System, a program for observation, modelling, and analysis of terrestrial ecosystems to support sustainable development, Food and Agriculture Organization of the United Nations. https://www.fao.org/geospatial/projects/detail/en/c/1035185/

GWIS, Global Wildfire Information System. https://gwis.jrc.ec.europa.eu/

HAQAST, NASA Health and Air Quality Applied Sciences Team. https://haqast.org/

IAWF, International Association of Wildland Fire, a professional membership association dedicated to uniting the global wildland fire community. https://www.iawfonline.org/

iNAP, Situation reports from wildfires detailed fire behavior and fuels (formerly FAMWEB), https://nap.nwcg.gov/NAP/

INPE, Instituto Nacional de Pesquisas Espaciais, Brazilian National Institute for Space Research. http://www.inpe.br/amazonia1/en/ https://www.gov.br/inpe/pt-br

International Charter Space and Major Disasters activation, a method through which agencies freely share data, information, and resources for disaster response worldwide, working through a country-level agency designee https://disasterscharter.org/web/guest/home

ISRO, Indian Space Research Organization, objectives are develop space technology, enable private space technology development, establish major space systems for communications, television broadcasting, and meteorological services; https://new.isro.gov.in/

JAXA, Japan Aerospace Exploration Agency, agency designed to support the Japanese overall aerospace development and utilization through integrated operations from basic research and development, to utilization. https://global.jaxa.jp/

JFSP, Joint Fire Science Program, jointly funded by the Department of the Interior (DOI) and USFS, identifies and funds high-priority fire science research needs that will enhance the decision-making ability of fire and fuels managers, natural resource managers, and others to meet their management objectives. https://www.firescience.gov/index.cfm

JPSS, Joint Polar Satellite System, a collaborative effort between NOAA, NASA and industry partners to develop an advanced series of polar-orbiting environmental satellites that includes the VIIRS instruments.

MAIA, Multi-Angle Imager for Aerosols. https://maia.jpl.nasa.gov/

MISR, Multi-angle Imaging SpectroRadiometer (MISR), Stereographically analyzed smoke plume height. https://misr.jpl.nasa.gov/

NASA Decadal Survey. https://science.nasa.gov/earth-science/decadal-surveys

NFPA, National Fire Protection Associate, Leading organization providing information and knowledge resources on fire, electrical and related hazards; developed a comprehensive strategy that lays out key policy changes to reduce the wildfire devastation. https://www.nfpa.org/; https://nfpa.org/wildfirepolicy

NICC Fire Danger. https://www.predictiveservices.nifc.gov/fuels_fire-danger/fuels_fire-danger.htm

NICC, National Interagency Coordination Center, Predictive services for climate, drought, and fire danger indices. https://www.nifc.gov/nicc/index.htm

NICC Weather. https://www.predictiveservices.nifc.gov/weather/weather.htm

NIFC, National Interagency Fire Center, manages national fire programs for each federal fire agency, along with partners including the National Association of State Foresters, the U.S. Fire Administration, and the National Weather Service to provide leadership, policy oversight and coordination to manage the nation's wildland fire programs. https://www.nifc.gov/

NISAR, NASA-ISRO SAR, NASA- Indian Space Research Organisation (ISRO) Synthetic Aperture Radar a U.S. and Indian InSAR mission. https://nisar.jpl.nasa.gov/

NOAA HMS, Hazard Mapping System Smoke Product, a satellite-based human-drawn smoke product for North America that includes shapefiles, maintained by 24/7 image analysts and updated multiple times a day. https://www.ospo.noaa.gov/Products/land/hms.html

NOAA NWS, National Oceanic and Atmospheric Administration National Weather Service, Fire weather predictive services. https://www.weather.gov/fire/

NWCG, National Wildfire Coordination Group, national interagency wildland fire information and operations. https://www.nwcg.gov/

Planet, transforming the earth observation industry with the highest frequency satellite data commercially available. https://www.planet.com/

PurpleAir, commercial publicly-available low-cost air quality monitors that collect data that can be shared with the public, more recently through AirNow; When corrected, sensors accurately report NowCast AQI categories 90% of the time; https://www2.purpleair.com/

QUIC-Fire, A physics-based simulation tool that simulates the interactive effects of fuel structure, local winds and fire behavior for fire planning. https://doi.org/10.1016/j.envsoft.2019.104616

SBG, Surface Biology and Geology. https://sbg.jpl.nasa.gov/

Sentinel, a series of Earth observation missions, on behalf of the European Space Agency (ESA) and European Commission initiative Copernicus. https://sentinel.esa.int/web/sentinel/home

Sentinel-4. https://sentinel.esa.int/web/sentinel/missions/sentinel-4

TEMPO, Tropospheric Emissions: Monitoring of Pollution. http://tempo.si.edu/

USGS, U.S. Geological Survey, Geospatial data of the United States. https://www.usgs.gov/

WFAS, Wildland Fire Assessment System, Continental United States forecasted and observed daily percentiles and thresholds from the National Fire Danger Rating System (NFDRS) including Fire Danger Rating, Energy Release Component (ERC), Burning Index, and Severe Fire Danger Index. https://www.wfas.net/

WHO, World Health Organization, Uses science to champion global efforts to achieve better health by connecting countries, people and partners to give everyone, everywhere an equal chance at a safe and healthy life. https://www.who.int/

WRF-SFIRE, Weather Research and Forecasting Model, a coupled fire-atmosphere model from National Center for Atmospheric Research. https://ral.ucar.edu/solutions/products/wrf-fire-wildland-fire-modeling

ACKNOWLEDGMENTS

We greatly appreciate NASA support from John Hayne's Applied Sciences Health and Air Quality Applied Sciences Team (HAQAST), Award Number 80NSSC21K0514, and Barry Lefer's NASA Tropospheric Chemistry Program, Award Number 80NSSC18K0685. We are also appreciative of Katherine Froiland for developing Figure 2.1 for this chapter.

REFERENCES

Abatzoglou, J. T., & Kolden, C. A. (2011). Climate change in western US deserts: Potential for increased wildfire and invasive annual grasses. *Rangeland Ecology and Management*, *64*(5), 471–478. https://doi.org/10.2111/REM-D-09-00151.1

Abbott, B. W., Jones, J. B., Schuur, E. A. G., Chapin, F. S., III, Bowden, W. B., Bret-Harte, M. S., et al. (2016). Biomass offsets little or none of permafrost carbon release from soils, streams, and wildfire: an expert assessment. *Environmental Research Letters*, *11*(3), 034014. https://doi.org/10.1088/1748-9326/11/3/034014

Al-Saadi, J., Soja, A. J., Pierce, R. B., Szykman, J. J., Wiedinmyer, C., Emmons, L., et al. (2008). Intercomparison of near-real-time biomass burning emissions estimates constrained by satellite fire data. *Journal of Applied Remote Sensing*, *2*, 021504, https://doi.org/10.1117/1111.2948785

Altshuler, S. L., Zhang, Q., Kleinman, M. T., Garcia-Menendez, F., Moore, C. T., Hough, M. L., et al. (2020). Wildfire and prescribed burning impacts on air quality in the United States. *Journal of the Air and Waste Management Association*, *70*(10), 961–970. https://doi.org/10.1080/10962247.2020.1813217

Amiro, B., Logan, K., Wotton, B., Flannigan, M., Todd, J., Stocks, B., & Martell, D. (2004). Fire weather index system components for large fires in the Canadian boreal forest. *International Journal of Wildland Fire*, *13*, 391–400. https://doi.org/10.1071/WF03066

Andela, N., Morton, D. C., Giglio, L., Chen, Y., van der Werf, G. R., Kasibhatla, P. S., et al. (2017). A human-driven decline in global burned area. *Science*, *356* (6345), 1356–1362. https://doi.org/10.1126/science.aal4108

Andreae, M. O., Rosenfeld, D., Artaxo, P., Costa, A. A., Frank, G. P., Longo, K. M., et al. (2004). Smoking rain clouds over the Amazon. *Science*, *303* (5662), 1337–1342. https://doi.org/10.1126/science.1092779

Antonovski, M. Y., Ter-Mikaelian, M. T., & Furyaev, V. V. (1992). A spatial model of long-tem forest fire dynamics and its applications to forests in western Siberia. In H. H. Shugart, R. Leemans, & G. B. Bonan (Eds.), *A systems analysis of the global boreal forest* (pp. 373–403). New York: Cambridge University Press.

Aragão, L. E., & Shimabukuro, Y. E. (2010). The incidence of fire in Amazonian forests with implications for REDD. *Science*, *328* (5983), 1275–1278. https://doi.org/10.1126/science.1186925

Arreola Amaya, M., & Clements, C. B. (2020). Evolution of plume core structures and turbulence during a wildland fire experiment. *Atmosphere*, *11* (8). https://doi.org/10.3390/atmos11080842

Australia (2022). *Australia state of the environment 2021*. https://soe.dcceew.gov.au/

Baker, K. R., Woody, M. C., Tonnesen, G. S., Hutzell, W., Pye, H. O. T., Beaver, M. R., et al. (2016). Contribution of regional-scale fire events to ozone and PM2.5 air quality estimated by photochemical modeling approaches. *Atmospheric Environment*, *140*, 539–554. https://doi.org/10.1016/j.atmosenv.2016.06.032

Baker, K. R., Woody, M. C., Valin, L., Szykman, J., Yates, E. L., Iraci, L. T., et al. (2018). Photochemical model evaluation of 2013 California wild fire air quality impacts using surface, aircraft, and satellite data. *Science of the Total Environment*, *637–638*, 1137–1149. https://doi.org/10.1016/j.scitotenv.2018.05.048

Balzter, H., Gerard, F. F., George, C. T., Rowland, C. S., Jupp, T. E., McCallum, I., et al. (2005). Impact of the Arctic oscillation pattern on interannual forest fire variability in Central Siberia. *Geophysical Research Letters*, *32*(14). https://doi.org/10.1029/2005GL022526

Bar-Yosef, O. (2002). The Upper Paleolithic revolution. *Annual Review of Anthropology*, *31*, 363–393. https://doi.org/10.1146/annurev.anthro.31.040402.085416

Betts, R. A. (2000). Offset of the potential carbon sink from boreal forestation by decreases in surface albedo. *Nature*, *408* (6809), 187–190. https://doi.org/10.1038/35041545

Bonan, G. B. (2008). Forests and climate change: Forcings, feedbacks, and the climate benefits of forests. *Science*, *320* (5882), 1444–1449. https://doi.org/10.1126/science.1155121

Bond, W. J., & Keeley, J. E. (2005). Fire as a global "herbivore": The ecology and evolution of flammable ecosystems. *Trends in Ecology and Evolution*, *20*(7), 387–394. https://doi.org/10.1016/j.tree.2005.04.025

Bowman, D. M., Balch, J. K., Artaxo, P., Bond, W. J., Carlson, J. M., Cochrane, M. A., et al. (2009). Fire in the Earth system. *Science*, *324* (5926), 481–484. https://doi.org/10.1126/science.1163886

Bowman, D. M. J. S., Balch, J., Artaxo, P., Bond, W. J., Cochrane, M. A., D'Antonio, C. M., et al. (2011). The human dimension of fire regimes on Earth. *Journal of Biogeography*, *38* (12), 2223–2236. https://doi.org/https://doi.org/10.1111/j.1365-2699.2011.02595.x

Bowman, D. M. J. S., Kolden, C. A., Abatzoglou, J. T., Johnston, F. H., van der Werf, G. R., & Flannigan, M. (2020). Vegetation fires in the Anthropocene. *Nature Reviews Earth and Environment*, *1*(10), 500–515. https://doi.org/10.1038/s43017-020-0085-3

Bowman, D. M. J. S., Williamson, G. J., Abatzoglou, J. T., Kolden, C. A., Cochrane, M. A., & Smith, A. M. S. (2017). Human exposure and sensitivity to globally extreme wildfire events. *Nature Ecology and Evolution*, *1*(3), 0058. https://doi.org/10.1038/s41559-016-0058

Bradshaw, C. J. A., & Warkentin, I. G. (2015). Global estimates of boreal forest carbon stocks and flux. *Global and Planetary Change*, *128*, 24–30. https://doi.org/10.1016/j.gloplacha.2015.02.004

Byram, George M. (1959). Combustion of forest fuels. In K. P. Davis (Eds.), *Forest fire: control and use* (pp. 61–89). New York, NY: McGraw-Hill.

CARB (2021). New analysis shows spikes of metal contaminants, including lead, in 2018 Camp Fire wildfire smoke. https://ww2arb.ca.gov/sites/default/files/2021–07/Camp_Fire_report_July2021.pdf

Carslaw, K. S., Boucher, O., Spracklen, D. V., Mann, G. W., Rae, J. G. L., Woodward, S., & Kulmala, M. (2010). A review of natural aerosol interactions and feedbacks within the Earth system. *Atmospheric Chemistry and Physics*, *10* (4), 1701–1737. https://doi.org/10.5194/acp-10-1701-2010

Cassella, C. (2022). *Australia drops a bombshell: An environment report card that nobody should ignore*. https://www.sciencealert.com/australia-s-2021-environment-report-is-everything-voters-feared-it-would-be

Cattau, M. E., Wessman, C., Mahood, A., & Balch, J. K. (2020). Anthropogenic and lightning-started fires are becoming larger and more frequent over a longer season length in the U.S.A. *Global Ecology and Biogeography*, *29*(4), 668–681. https://doi.org/10.1111/geb.13058

CFDG (1992). Canada fire danger group: Development and structure of the Canadian Forest Fire Behavior Prediction System (Info. Rep. ST-X-3). https://cfs.nrcan.gc.ca/publications?id=10068

Chameides, W. L., Yu, H., Liu, S. C., Bergin, M., Zhou, X., Mearns, L., et al. (1999). Case study of the effects of atmospheric aerosols and regional haze on agriculture: An opportunity to enhance crop yields in China through emission controls? *Proceedings of the National Academy of Sciences*, *96*(24), 13626–13633. https://doi.org/10.1073/pnas.96.24.13626

Chen, J., Gong, Y., Wang, S., Guan, B., Balkovic, J., & Kraxner, F. (2019). To burn or retain crop residues on croplands? An integrated analysis of crop residue management in China. *Science of the Total Environment*, *662*, 141–150. https://doi.org/10.1016/j.scitotenv.2019.01.150

Chen, Y., Romps, D. M., Seeley, J. T., Veraverbeke, S., Riley, W. J., Mekonnen, Z. A., et al. (2021). Future increases in Arctic lightning and fire risk for permafrost carbon. *Nature Climate Change*, *11* (5), 404–410. https://doi.org/10.1038/s41558-021-01011-y

Chen, Y., Tessier, S., Cavers, C., Xu, X., & Monero, F. (2005). A survey of crop residue burning practices in Manitoba. *Applied Engineering in Agriculture*, *21*(3), 317–323. https://doi.org/10.13031/2013.18446

Chudnikov, P. I. (1931). *Impact of fire on regeneration of Urals forests*. Moscow: Selkhozgiz (Agricultural State Publishing House).

Clark, J. S. (1988). Effect of climate change on fire regimes in northwestern Minnesota. *Nature*, *334* (6179), 233–235. https://doi.org/10.1038/334233a0

Cleland, S. E., Serre, M. L., Rappold, A. G., & West, J. J. (2021). Estimating the acute health impacts of fire-originated PM2.5 exposure during the 2017 California wildfires: Sensitivity to choices of inputs. *GeoHealth*, *5*(7), e2021GH000414. https://doi.org/10.1029/2021GH000414

Clements, C. B., Lareau, N. P., Kingsmill, D. E., Bowers, C. L., Camacho, C. P., Bagley, R., et al. (2018). The rapid deployments to wildfires experiment (RaDFIRE): Observations from the fire zone. *Bulletin of the American Meteorological Society*, *99* (12), 2539–2559. https://doi.org/10.1175/BAMS-D-17-0230.1

Clements, C. B., Zhong, S., Goodrick, S., Li, J., Potter, B. E., Bian, X., et al. (2007). Observing the dynamics of wildland grass fires: FireFlux, A field validation experiment. *Bulletin of the American Meteorological Society*, *88*(9), 1369–1382. https://doi.org/10.1175/BAMS-88-9-1369

Cochrane, M. A., Alencar, A., Schulze, M. D., Souza Jr, C. M., Nepstad, D. C., Lefebvre, P., & Davidson, E. A. (1999). Positive feedbacks in the fire dynamic of closed canopy tropical forests. *Science*, *284*(5421), 1832–1835. https://doi.org/10.1126/science.284.5421.1832

Countryman, C. M. (1972). *The fire environment concept*. Berkeley, CA: USDA Forest Service, Pacific Southwest Forest and Range Experiment Station; reprinted as National Fire Equipment System Publication NFES 2166. National Wildfire Coordinating Group, Boise, Idaho.

Cresswell, I., Janke, T., & Johnston, E. (2022). *Australia state of the environment 2021: Overview*. https://doi.org/APO-318651

Damoah, R., Spichtinger, N., Forster, C., James, P., Mattis, I., Wandinger, U., et al. (2004). Around the world in 17 days: Hemispheric-scale transport of forest fire smoke from Russia in May 2003. *Atmospheric Chemistry and Physics*, *4*(5), 1311–1321. https://doi.org/10.5194/acp-4-1311-2004

Davies, I. P., Haugo, R. D., Robertson, J. C., & Levin, P. S. (2018). The unequal vulnerability of communities of color to wildfire. *PloS One*, *13*(11), e0205825. https://doi.org/10.1371/journal.pone.0205825

DeFlorio-Barker, S., Crooks, J., Reyes, J., & Rappold, A. G. (2019). Cardiopulmonary effects of fine particulate matter exposure among older adults, during wildfire and non-wildfire periods, in the United States 2008–2010. *Environmental Health Perspectives*, *127*(3), 037006. https://doi.org/10.1289/EHP3860

de Groot, W. J., Cantin, A. S., Flannigan, M. D., Soja, A. J., Gowman, L. M., & Newbery, A. (2013). A comparison of Canadian and Russian boreal forest fire regimes. *Forest Ecology and Management*, *294*, 23–34. https://doi.org/https://doi.org/10.1016/j.foreco.2012.07.033

Demers, P. A., DeMarini, D. M., Fent, K. W., Glass, D. C., Hansen, J., Adetona, O., et al. (2022). Carcinogenicity of occupational exposure as a firefighter. *The Lancet Oncology*. https://doi.org/10.1016/S1470-2045(22)00390-4

D'Evelyn, S. M., Jung, J., Alvarado, E., Baumgartner, J., Caligiuri, P., Hagmann, R. K., et al. (2022). Wildfire, smoke exposure, human health, and environmental justice need to be integrated into forest restoration and management. *Current Environmental Health Reports*. https://doi.org/10.1007/s40572-022-00355-7

Di Virgilio, G., Evans, J. P., Blake, S. A., Armstrong, M., Dowdy, A. J., Sharples, J., et al. (2019). Climate change increases the potential for extreme wildfires. *Geophysical Research Letters*, *46* (14), 8517–8526. https://doi.org/10.1029/2019GL083699

Dyrness, C. T., & Norum, R. A. (1983). The effects of experimental fires on black spruce forest floors in interior Alaska. *Canadian Journal of Forestry*, *13*, 879–893.

Eldevik, T., & Nilsen, J. E. Ø. (2013). The Arctic-Atlantic thermohaline circulation. *Journal of Climate*, *26*(21), 8698–8705. https://doi.org/10.1175/JCLI-D-13-00305.1

EPA (2016). *2014 national emissions inventory, version 1*. Technical support document. https://www.epa.gov/sites/production/files/2016-12/documents/nei2014v1_tsd.pdf

Fann, N., Alman, B., Broome, R. A., Morgan, G. G., Johnston, F. H., Pouliot, G., et al. (2018). The health impacts and economic value of wildland fire episodes in the U.S.: 2008–2012. *Science of the Total Environment*, *610–611*, 802–809. https://doi.org/10.1016/j.scitotenv.2017.08.024

Fishman, J., Minnis, P., & Reichle, H. G., Jr. (1986). Use of satellite data to study tropospheric ozone in the tropics. *Journal of Geophysical Research: Atmospheres*, *91*(D13), 14451–14465. https://doi.org/10.1029/JD091iD13p14451

Flanner, M. G. (2013). Arctic climate sensitivity to local black carbon. *Journal of Geophysical Research: Atmospheres*, *118*(4), 1840–1851. https://doi.org/10.1002/jgrd.50176

Flannigan, M., Cantin, A. S., de Groot, W. J., Wotton, M., Newbery, A., & Gowman, L. M. (2013). Global wildland fire season severity in the 21st century. *Forest Ecology and Management*, *294*, 54–61. https://doi.org/10.1016/j.foreco.2012.10.022

French, N. H. F., Whitley, M. A., & Jenkins, L. K. (2016). Fire disturbance effects on land surface albedo in Alaskan tundra. *Journal of Geophysical Research: Biogeosciences*, *121*(3), 841–854. https://doi.org/10.1002/2015JG003177

Fromm, M., Bevilacqua, R., Servranckx, R., Rosen, J., Thayer, J. P., Herman, J., et al. (2005). Pyro-cumulonimbus injection of smoke to the stratosphere: Observations and impact of a super blowup in northwestern Canada on 3–4 August 1998. *Journal of Geophysical Research: Atmospheres*, *110* (D8). https://doi.org/10.1029/2004JD005350

Fromm, M. D., & Servranckx, R. (2003). Transport of forest fire smoke above the tropopause by supercell convection. *Geophysical Research Letters*, *30* (10). https://doi.org/10.1029/2002GL016820

Furyaev, V. V. (1996a). Pyrological regimes and dynamics of the southern Taiga forests in Siberia. In J. G. Goldammer & V. V. Furyaev (Eds.), *Fire in ecosystems of boreal Eurasia* (pp. 168–185). Dordrecht: Kluwer Academic Publishers.

Furyaev, V. V. (1996b). *Role of fires in the forest regeneration process*. Novosibirsk, Russia: Nauka.

Giglio, L., Descloitres, J., Justice, C. O., & Kaufman, Y. J. (2003). An enhanced contextual fire detection algorithm for MODIS. *Remote Sensing of Environment*, *87*(2), 273–282. https://doi.org/10.1016/S0034-4257(03)00184-6

Giglio, L., Schroeder, W., & Justice, C. O. (2016). The collection 6 MODIS active fire detection algorithm and fire products. *Remote Sensing of Environment*, *178*, 31–41. https://doi.org/10.1016/j.rse.2016.02.054

Giuseppe, F. D., Pappenberger, F., Wetterhall, F., Krzeminski, B., Camia, A., Libertá, G., et al. (2016). The potential predictability of fire danger provided by numerical weather prediction. *Journal of Applied Meteorology and Climatology*, *55*(11), 2469–2491. https://doi.org/10.1175/jamc-d-15-0297.1

Habeck, J. R., & Mutch, R. W. (1973). Fire-dependent forests in the northern Rocky Mountains. *Quarternary Research*, *3*, 408–424.

Hari, O. M., Dhiman, S. D., Hemant, K., & Sajjan, K. (2003). Biology and management of *Phalaris minor* in wheat under a rice/wheat system. *Weed Research*, *43*(1), 59–67. https://doi.org/https://doi.org/10.1046/j.1365-3180.2003.00319.x

Heinselman, M. L. (1978). *Fire intensity and frequency as factors in the disturbance and structure of northern ecosystems*. Paper presented at the Fire Regimes and Ecosystem Properties, Honolulu.

Heinselman, M. L. (1981). Fire and succession in the conifer forests of northern North America. In D. C. West, H. H. Shugart, & D. B. Botkin (Eds.), *Forest succession: Concepts and application* (pp. 374–405). New York: Springer-Verlag.

Hirsch, E., & Koren, I. (2021). Record-breaking aerosol levels explained by smoke injection into the stratosphere. *Science*, *371*(6535), 1269–1274. https://doi.org/10.1126/science.abe1415

Hobbs, P. V., & Radke, L. F. (1969). Cloud condensation nuclei from a simulated forest fire. *Science*, *163* (3864), 279–280. https://doi.org/https://doi.org/10.1126/science.163.3864.279

Holden, Z. A., Swanson, A., Luce, C. H., Jolly, W. M., Maneta, M., Oyler, J. W., et al. (2018). Decreasing fire season precipitation increased recent western US forest wildfire activity. *Proceedings of the National Academy of Sciences*, *115*(36), E8349–E8357. https://doi.org/10.1073/pnas.1802316115

Huang, R., Zhang, X., Chan, D., Kondragunta, S., Russell, A. G., & Odman, M. T. (2018). Burned area comparisons between prescribed burning permits in southeastern United States and two satellite-derived products. *Journal of Geophysical Research: Atmospheres*, *123*(9), 4746–4757. https://doi.org/10.1029/2017JD028217

Hurteau, M. D., Liang, S., Westerling, A. L., & Wiedinmyer, C. (2019). Vegetation-fire feedback reduces projected area burned under climate change. *Scientific Reports*, *9* (1), 2838. https://doi.org/10.1038/s41598-019-39284-1

IAWF (2022). WILDFIRE: Uniting the global wildland fire community. *Wildfire Magazine*, *31*(2). https://www.iawfonline.org/issue/2022-quarter-2/

IPCC (2021). *Climate change 2021: The physical science basis*. Contribution of Working Group I to the Sixth Assessment Report of the Intergovernmental Panel on Climate Change. Cambridge University Press. https://www.ipcc.ch/report/ar6/wg1/

Ivanova, G. A., Ivanov, V. A., Kukavskaya, E. A., & Soja, A. J. (2010). The frequency of forest fires in Scots pine stands of Tuva, Russia. *Environmental Research Letters*, *5* (1), 015002. https://doi.org/10.1088/1748-9326/5/1/015002

Jaffe, D. A., Wigder, N., Downey, N., Pfister, G., Boynard, A., & Reid, S. B. (2013). Impact of wildfires on ozone exceptional events in the western U.S. *Environmental Science and Technology*, *47* (19), 11065–11072. https://doi.org/10.1021/es402164f

Johnstone, J. F., & Chapin, F. S. (2006). Effects of soil burn severity on postfire tree recruitment in boreal forest. *Ecosystems*, *9* (1), 14–31. https://doi.org/10.1007/s10021-004-0042-x

Jolly, W. M., Cochrane, M. A., Freeborn, P. H., Holden, Z. A. Brown, T. J., Williamson, G. J., et al. (2015). Climate-induced variations in global wildfire danger from 1979 to 2013. *Nature Communications*, *6*, 7537. https://doi.org/10.1038/ncomms8537

Justice, C. O., Giglio, L., Korontzi, S., Owens, J., Morisette, J. T., Roy, D., et al. (2002). The MODIS fire products. *Remote Sensing of Environment*, *83*(1), 244–262. https://doi.org/10.1016/S0034-4257(02)00076-7

Kahn, R. A., Li, W. H., Moroney, C., Diner, D. J., Martonchik, J. V., & Fishbein, E. (2007). Aerosol source plume physical characteristics from space-based multiangle imaging. *Journal of Geophysical Research: Atmospheres*, *112*(D11). https://doi.org/10.1029/2006JD007647

Karkanas, P., Shahack-Gross, R., Ayalon, A., Bar-Matthews, M., Barkai, R., Frumkin, A., et al. (2007). Evidence for habitual use of fire at the end of the Lower Paleolithic: Site-formation processes at Qesem Cave, *Israel. Journal of Human Evolution*, *53*(2), 197–212. https://doi.org/10.1016/j.jhevol.2007.04.002

Keeley, J. E. (1993). Smoke-induced flowering in the fire-lily *Cyrtanthus ventricosus. South African Journal of Botany*, *59*(6), 638.

Keeley, J. E. (2008). Fire. In: S. E. Jørgensen and B. D. Fath (Eds), *Encyclopedia of Ecology*. (pp. 1557–1564). Elsevier Science.

Keyser, A. R., & Westerling, A. L. (2019). Predicting increasing high severity area burned for three forested regions in the western United States using extreme value theory. *Forest Ecology and Management*, *432*, 694–706. https://doi.org/10.1016/j.foreco.2018.09.027

Kim, M. H., Omar, A. H., Tackett, J. L., Vaughan, M. A., Winker, D. M., Trepte, C. R., et al. (2018a). The CALIPSO version 4 automated aerosol classification and lidar ratio selection algorithm. *Atmospheric Measurement Techniques*, *11*(11), 6107–6135. https://doi.org/10.5194/amt-11-6107-2018

Kim, Y. H., Warren, S. H., Krantz, Q. T., King, C., Jaskot, R., Preston, W. T., et al. (2018b). Mutagenicity and lung toxicity of smoldering vs. flaming emissions from various biomass fuels: implications for health effects from wildland fires. *Environmental Health Perspectives, 126* (1), 017011. https://doi.org/10.1289/EHP2200

Kitzberger, T., Brown, P. M., Heyerdahl, E. K., Swetnam, T. W., & Veblen, T. T. (2007). Contingent Pacific-Atlantic Ocean influence on multicentury wildfire synchrony over western North America. *Proceedings of the National Academy of Sciences, 104* (2), 543–548. https://doi.org/10.1073/pnas.0606078104

Knelman, J. E., Graham, E. B., Trahan, N. A., Schmidt, S. K., & Nemergut, D. R. (2015). Fire severity shapes plant colonization effects on bacterial community structure, microbial biomass, and soil enzyme activity in secondary succession of a burned forest. *Soil Biology and Biochemistry, 90,* 161–168. https://doi.org/10.1016/j.soilbio.2015.08.004

Knies, J., Matthiessen, J., Mackensen, A., Stein, R., Vogt, C., Frederichs, T., et al. (2007). Effects of Arctic freshwater forcing on thermohaline circulation during the Pleistocene. *Geology, 35*(12), 1075–1078. https://doi.org/10.1130/G23966A.1

Kobziar, L. N., & Thompson, G. R., III,(2020). Wildfire smoke, a potential infectious agent. *Science, 370* (6523), 1408–1410. https://doi.org/10.1126/science.abe8116

Kondo, M. C., De Roos, A. J., White, L. S., Heilman, W. E., Mockrin, M. H., Gross-Davis, C. A., et al. (2019). Meta-analysis of heterogeneity in the effects of wildfire smoke exposure on respiratory health in North America. *International Journal of Environmental Research and Public Health, 16* (6). https://doi.org/10.3390/ijerph16060960

Korontzi, S., McCarty, J., Loboda, T., Kumar, S., & Justice, C. (2006). Global distribution of agricultural fires in croplands from 3 years of Moderate Resolution Imaging Spectroradiometer (MODIS) data. *Global Biogeochemical Cycles, 20* (2). https://doi.org/10.1029/2005GB002529

Kristensen, L. J., & Taylor, M. P. (2012). Fields and forests in flames: Lead and mercury emissions from wildfire pyrogenic activity. *Environmental Health Perspectives, 120* (2), a56–a57. https://doi.org/10.1289/ehp.1104672

Kukavskaya, E. A., Soja, A. J., Petkov, A. P., Ponomarev, E. I., Ivanova, G. A., & Conard, S. G. (2013). Fire emissions estimates in Siberia: evaluation of uncertainties in area burned, land cover, and fuel consumption. *Canadian Journal of Forest Research-Revue Canadienne De Recherche Forestiere, 43* (5), 493–506. https://doi.org/DOI 10.1139/cjfr-2012-0367

Landguth, E. L., Holden, Z. A., Graham, J., Stark, B., Mokhtari, E. B., Kaleczyc, E., et al. (2020). The delayed effect of wildfire season particulate matter on subsequent influenza season in a mountain west region of the USA. *Environment International, 139,* 105668. https://doi.org/10.1016/j.envint.2020.105668

Larkin, N. K., Raffuse, S. M., & Strand, T. M. (2014). Wildland fire emissions, carbon, and climate: U.S. emissions inventories. *Forest Ecology and Management, 317,* 61–69. https://doi.org/10.1016/j.foreco.2013.09.012

Lawson, B. D., & Armitage, O. B. (2008). *Weather guide for the Canadian Forest Fire Danger Rating System.* Edmonton AB: Natural Resources Canada, Canadian Forest Service, Northern Forestry Centre.

Littell, J. S., McKenzie, D., Peterson, D. L., & Westerling, A. L. (2009). Climate and wildfire area burned in western US ecoprovinces, 1916–2003. *Ecological Applications, 19* (4), 1003–1021. https://doi.org/10.1890/07-1183.1

Liu, J. C., Pereira, G., Uhl, S. A., Bravo, M. A., & Bell, M. L. (2015). A systematic review of the physical health impacts from non-occupational exposure to wildfire smoke. *Environmental Research, 136,* 120–132. https://doi.org/10.1016/j.envres.2014.10.015

Loboda, T. V., Hoy, E. E., Giglio, L., & Kasischke, E. S. (2011). Mapping burned area in Alaska using MODIS data: A data limitations-driven modification to the regional burned area algorithm. *International Journal of Wildland Fire, 20* (4), 487–496. https://doi.org/10.1071/WF10017

Magzamen, S., Gan, R. W., Liu, J., O'Dell, K., Ford, B., Berg, K., et al. (2021). Differential cardiopulmonary health impacts of local and long-range transport of wildfire smoke. *GeoHealth, 5* (3), e2020GH000330. https://doi.org/10.1029/2020GH000330

Mahmood, B. (2022). Firefighters' union blasts government policy of cutting resources amid heatwave crisis. *Left Foot Forward.*

McCarty, J. L. (2011). Remote sensing-based estimates of annual and seasonal emissions from crop residue burning in the contiguous United States. *Journal of the Air and Waste Management Association, 61*(1), 22–34. https://doi.org/10.3155/1047-3289.61.1.22

McCarty, J. L., Aalto, J., Paunu, V. V., Arnold, S. R., Eckhardt, S., Klimont, Z., et al. (2021). Reviews and syntheses: Arctic fire regimes and emissions in the 21st century. *Biogeosciences, 18*(18), 5053–5083. https://doi.org/10.5194/bg-18-5053-2021

McCarty, J. L., Ellicott, E. A., Romanenkov, V., Rukhovitch, D., & Koroleva, P. (2012). Multi-year black carbon emissions from cropland burning in the Russian Federation. *Atmospheric Environment, 63,* 223–238. https://doi.org/10.1016/j.atmosenv.2012.08.053

McCarty, J. L., Justice, C. O., & Korontzi, S. (2007). Agricultural burning in the southeastern United States detected by MODIS. *Remote Sensing of Environment, 108* (2), 151–162, [https://doi.org/110.1016/j.rse.2006.1003.1020].

McCarty, J. L., Krylov, A., Prishchepov, A. V., Banach, D. M., Tyukavina, A., Potapov, P., et al. (2017). Agricultural fires in European Russia, Belarus, and Lithuania and their impact on air quality, 2002–2012. In G. Gutman & V. Radeloff (Eds.), *Land-cover and land-use changes in Eastern Europe after the collapse of the Soviet Union in 1991* (pp. 193–221). Cham: Springer International Publishing.

McCarty, J. L., Smith, T. E. L., & Turetsky, M. R. (2020). Arctic fires re-emerging. *Nature Geoscience, 13* (10), 658–660. https://doi.org/10.1038/s41561-020-00645-5

McRae, D. J., Conard, S. G., Ivanova, G. A., Sukhinin, A. I., Baker, S. P., Samsonov, Y. N., et al. (2006). Variability of fire behavior, fire effects, and emissions in Scotch Pine forests of central Siberia. *Mitigation and Adaptation Strategies for Global Change, 11*(1), 45–74. https://doi.org/10.1007/s11027-006-1008-4

Meggers, B. J. (1994). Archeological evidence for the impact of mega-Niño events on Amazonia during the past two millennia. *Climatic Change, 28*(4), 321–338. https://doi.org/10.1007/BF01104077

Meggers, B. J. (2017). Climatic oscillation as a factor in the prehistory of Amazonia. *American Antiquity*, *44*(2), 252–266. https://doi.org/10.2307/279075

Melekhov, I. S. (1947). *The forest nature and forest fire*. Arkhangelsk, Russia: OGIZ Publishers.

Melvin, M. A. (2015). *2015 National prescribed fire use survey report*. Coalition of Prescribed Fire Councils, Inc. https://doi.org/Technical Report 02-15. http://www.stateforesters.org/2015-national-prescribed-fire-use-survey-report

Miller, S. (2013). *Burning in their backyards and having them say "Thank you."* Saint Johns River Water Management District, Florida. Media Webinar, Wildland Fire Lessons Learned Center. https://www.frames.gov/catalog/14914

Milrad, S. (2018). 11: Satellite imagery. In S. Milrad (Ed.), *Synoptic analysis and forecasting* (pp. 145–161). Elsevier.

Moore, R. A., Bomar, C., Kobziar, L. N., & Christner, B. C. (2021). Wildland fire as an atmospheric source of viable microbial aerosols and biological ice nucleating particles. *The ISME Journal*, *15* (2), 461–472. https://doi.org/10.1038/s41396-020-00788-8

Moritz, M. A., Morais, M. E., Summerell, L. A., Carlson, J. M., & Doyle, J. (2005). Wildfires, complexity, and highly optimized tolerance. *Proceedings of the National Academy of Sciences of the United States of America*, *102* (50), 17912–17917. https://doi.org/10.1073/pnas.0508985102

Murphy, S. F., McCleskey, R. B., Martin, D. A., Holloway, J. M., & Writer, J. H. (2020). Wildfire-driven changes in hydrology mobilize arsenic and metals from legacy mine waste. *Science of the Total Environment*, *743*, 140635. https://doi.org/10.1016/j.scitotenv.2020.140635

Natarajan, M., Pierce, R. B., Schaack, T. K., Lenzen, A. J., Al-Saadi, J. A., Soja, A. J., et al. (2012). Radiative forcing due to enhancements in tropospheric ozone and carbonaceous aerosols caused by Asian fires during spring 2008. *Journal of Geophysical Research: Atmospheres*, *117* (D6). https://doi.org/10.1029/2011JD016584

National Wildfire Coordinating Group (2022). *NWCG glossary of wildland fire, PMS 205*. https://www.nwcg.gov/publications/pms205

Nikolov, N., & Helmisaari, H. (1992). Silvics of the circumpolar boreal forest tree species. In H. H. Shugart, R. Leemans, & G. B. Bonan (Eds.), *A systems analysis of the global boreal forest* (pp. 13–84). New York: Cambridge University Press.

Norman, C. R., & Bry, K. R. (2016). Alternative residue management practice effects over time on near-surface soil properties in a wheat-soybean, double-crop system in Eastern Arkansas. In J. Ross (Ed.), *Arkansas soybean research studies 2014*. Arkansas Agricultural Experiment Station Research Series. Fayetteville.

Normile, D. (2022). Australian scientists hope grim new environmental assessment will spur government action. *ScienceInsider*. https://doi.org/https://doi.org/10.1126/science.ade0184

Nowell, H. K., Holmes, C. D., Robertson, K., Teske, C., & Hiers, J. K. (2018). A new picture of fire extent, variability, and drought interaction in prescribed fire landscapes: Insights from Florida government records. *Geophysical Research Letters*, *45*(15), 7874–7884. https://doi.org/10.1029/2018GL078679

Nowell, H. K., Wirks, C., Val Martin, M., van Donkelaar, A., Martin, R. V., Uejio, C. K., & Holmes, C. D. (2022). Impacts of Sugarcane Fires on Air Quality and Public Health in South Florida. *Environmental Health Perspectives*, 130(8), CID: 087004-14. https://doi.org/10.1289/EHP9957

NRC (2007). *Earth science and applications from space: National imperatives for the next decade and beyond*. National Research Council. Washington, DC: The National Academies Press.

Omar, A. H., Winker, D. M., Vaughan, M. A., Hu, Y., Trepte, C. R., Ferrare, R. A., et al. (2009). The CALIPSO automated aerosol classification and lidar ratio selection algorithm. *Journal of Atmospheric and Oceanic Technology*, *26*(10), 1994–2014. https://doi.org/10.1175/2009JTECHA1231.1

O'Neill, S. M., Diao, M., Raffuse, S., Al-Hamdan, M., Barik, M., Jia, Y., et al. (2021). A multi-analysis approach for estimating regional health impacts from the 2017 Northern California wildfires. *Journal of the Air and Waste Management Association*, *71*(7), 791–814. https://doi.org/10.1080/10962247.2021.1891994

Parfenova, E., Tchebakova, N., & Soja, A. (2019). Assessing landscape potential for human sustainability and "attractiveness" across Asian Russia in a warmer 21st century. *Environmental Research Letters*, *14*(6), 065004. https://doi.org/10.1088/1748-9326/ab10a8

Pausas, J. G., Bradstock, R. A., Keith, D. A., & Keeley, J. E. (2004). Plant functional traits in relation to fire in crown-fire ecosystems. *Ecology*, *85*(4), 1085–1100. https://doi.org/10.1890/02-4094

Payette, S., Morneau, C., Sirois, L., & Desponts, M. (1989). Recent fire history of the northern Quebec biomes. *Ecology*, *70*, 656–673.

Peterson, D. A., Campbell, J. R., Hyer, E. J., Fromm, M. D., Kablick, G. P., Cossuth, J. H., et al. (2018). Wildfire-driven thunderstorms cause a volcano-like stratospheric injection of smoke. *Climate and Atmospheric Science*, *1* (1), 30. https://doi.org/10.1038/s41612-018-0039-3

Peterson, D. A., Hyer, E. J., Campbell, J. R., Fromm, M. D., Hair, J. W., Butler, C. F., et al. (2015). The 2013 rim fire: Implications for predicting extreme fire spread, pyroconvection, and smoke emissions. *Bulletin of the American Meteorological Society*, *96*(2), 229–247. https://doi.org/10.1175/BAMS-D-14-00060.1

Peterson, D. W., & Reich, P. B. (2001). Prescribed fire in oak savanna: Fire frequency effects on stand structure and dynamics. *Ecological Applications*, *11*(3), 914–927. https://doi.org/https://doi.org/10.1890/1051-0761(2001)011[0914:PFIOSF]2.0.CO;2

Pikul, J. L., Jr., & Zuzel, J. F. (1994). Soil crusting and water infiltration affected by long-term tillage and residue management. *Soil Science Society of America Journal*, *58* (5), 1524–1530. https://doi.org/10.2136/sssaj1994.03615995005800050036x

Ponomarev, E. I., & Shvetsov, E. G. (2015). Satellite detection of forest fires and geoinformation methods of calibration of the results. *Studies in Earth Remote Sensing*, *1*, 84–91. https://doi.org/10.7868/S0205961415010054

Pouliot, G., Rao, V., McCarty, J. L., & Soja, A. (2017). Development of the crop residue and rangeland burning in the 2014 National Emissions Inventory using information from multiple sources. *Journal of the Air and Waste Management Association*, *67*(5), 613–622. https://doi.org/10.1080/10962247.2016.1268982

Price, C., & Rind, D. (1994). Possible implications of global climate change on global lightning distributions and frequencies. *Journal of Geophysical Research*, *99*(D5), 10823–10831.

Pyne, S. J. (1993). Keeper of the flame: A survey of anthropogenic fire. In J. Sons (Ed.), *Fire in the environment: its ecological, climatic, and atmospheric chemical importance* (pp. 245–266). New York.

Pyne, S. J. (2019). *Fire: A brief history*. University of Washington Press.

Radeloff, V. C., Helmers, D. P., Kramer, H. A., Mockrin, M. H., Alexandre, P. M., Bar-Massada, A., et al. (2018). Rapid growth of the US wildland-urban interface raises wildfire risk. *Proceedings of the National Academy of Sciences*, *115* (13), 3314–3319. https://doi.org/10.1073/pnas.1718850115

Ramadan, L., Ngu, A., & Miller, M. (2021). The smoke comes every year. Sugar companies say the air is safe. ProPublica, *The Palm Beach Post*. https://projects.propublica.org/black-snow/

Ramo, R., Roteta, E., Bistinas, I., van Wees, D., Bastarrika, A., Chuvieco, E., et al. (2021). African burned area and fire carbon emissions are strongly impacted by small fires undetected by coarse resolution satellite data. *Proceedings of the National Academy of Sciences*, *118* (9), e2011160118. https://doi.org/10.1073/pnas.2011160118

Randerson, J. T., Chen, Y., van der Werf, G. R., Rogers, B. M., & Morton, D. C. (2012). Global burned area and biomass burning emissions from small fires. *Journal of Geophysical Research: Biogeosciences*, *117* (G4). https://doi.org/10.1029/2012JG002128

Randerson, J. T., Liu, H., Flanner, M. G., Chambers, S. D., Jin, Y., Hess, P. G., et al. (2006). The impact of boreal forest fire on climate warming. *Science*, *314*(5802), 1130–1132. https://doi.org/10.1126/science.1132075

Rappold, A. G., Reyes, J., Pouliot, G., Cascio, W. E., & Diaz-Sanchez, D. (2017). Community vulnerability to health impacts of wildland fire smoke exposure. *Environmental Science and Technology*, *51*(12), 6674–6682. https://doi.org/10.1021/acs.est.6b06200

Rappold, A. G., Stone Susan, L., Cascio Wayne, E., Neas Lucas, M., Kilaru Vasu, J., Carraway Martha, S., et al. (2011). Peat bog wildfire smoke exposure in rural North Carolina is associated with cardiopulmonary emergency department visits assessed through syndromic surveillance. *Environmental Health Perspectives*, *119* (10), 1415–1420. https://doi.org/10.1289/ehp.1003206

Reid, C. E., & Maestas, M. M. (2019). Wildfire smoke exposure under climate change: Impact on respiratory health of affected communities. *Current Opinion in Pulmonary Medicine*, *25* (2), 179. https://doi.org/10.1097%2FMCP.0000000000000552

Reid, C. E., Brauer, M., Johnston Fay, H., Jerrett, M., Balmes John, R., & Elliott Catherine, T. (2016). Critical review of health impacts of wildfire smoke exposure. *Environmental Health Perspectives*, *124*(9), 1334–1343. https://doi.org/10.1289/ehp.1409277

Rennermalm, A. K., Wood, E. F., Déry, S. J., Weaver, A. J., & Eby, M. (2006). Sensitivity of the thermohaline circulation to Arctic Ocean runoff. *Geophysical Research Letters*, *33*(12). https://doi.org/10.1029/2006GL026124

Rodrigues, M., Peña-Angulo, D., Russo, A., Zúñiga-Antón, M., & Cardil, A. (2021). Do climate teleconnections modulate wildfire-prone conditions over the Iberian Peninsula? *Environmental Research Letters*, *16* (4), 044050. https://doi.org/10.1088/1748-9326/abe25d

Rodriguez, B., Lareau, N. P., Kingsmill, D. E., & Clements, C. B. (2020). Extreme pyroconvective updrafts during a megafire. *Geophysical Research Letters*, *47*(18), e2020GL089001. https://doi.org/10.1029/2020GL089001

Rogers, B. M., Randerson, J. T., & Bonan, G. B. (2013). High-latitude cooling associated with landscape changes from North American boreal forest fires. *Biogeosciences*, *10* (2), 699–718. https://doi.org/10.5194/bg-10-699-2013

Rogers, B. M., Soja, A. J., Goulden, M. L., & Randerson, J. T. (2015). Influence of tree species on continental differences in boreal fires and climate feedbacks. *Nature Geoscience*, *8*(3), 228–234. https://doi.org/https://doi.org/10.1038/ngeo2352

Rosenfeld, D. (1999). TRMM observed first direct evidence of smoke from forest fires inhibiting rainfall. *Geophysical Research Letters*, *26* (20), 3105–3108. https://doi.org/10.1029/1999GL006066

Rosenfeld, D., Fromm, M., Trentmann, J., Luderer, G., Andreae, M. O., & Servranckx, R. (2007). The Chisholm firestorm: Observed microstructure, precipitation and lightning activity of a pyro-cumulonimbus. *Atmospheric Chemistry and Physics*, *7*(3), 645–659. https://doi.org/10.5194/acp-7-645-2007

Roy, D. P., Boschetti, L., Justice, C. O., & Ju, J. (2008). The collection 5 MODIS burned area product: Global evaluation by comparison with the MODIS active fire product. *Remote Sensing of Environment*, *112* (9), 3690–3707. https://doi.org/10.1016/j.rse.2008.05.013

Roy, D. P., Giglio, L., Kendall, J. D., & Justice, C. O. (1999). Multi-temporal active-fire based burn scar detection algorithm. *International Journal of Remote Sensing*, *20*(5), 1031–1038. https://doi.org/10.1080/014311699213073

Ruffault, J., Curt, T., Moron, V., Trigo, R. M., Mouillot, F., Koutsias, N., et al. (2020). Increased likelihood of heat-induced large wildfires in the Mediterranean Basin. *Scientific Reports*, *10*(1), 13790. https://doi.org/10.1038/s41598-020-70069-z

Ruiz Pessenda, L. C., De Oliveira, P. E., Mofatto, M., de Medeiros, V. B., Francischetti Garcia, R. J., Aravena, R., et al. (2009). The evolution of a tropical rainforest/grassland mosaic in southeastern Brazil since 28,000 14C yr BP based on carbon isotopes and pollen records. *Quaternary Research*, *71*(3), 437–452. https://doi.org/10.1016/j.yqres.2009.01.008

Saldarriaga, J. G., & West, D. C. (2017). Holocene fires in the northern Amazon Basin. *Quaternary Research*, *26*(3), 358–366. https://doi.org/10.1016/0033-5894(86)90095-5

Scharlemann, J. P. W., Tanner, E. V. J., Hiederer, R., & Kapos, V. (2014). Global soil carbon: Understanding and managing the largest terrestrial carbon pool. *Carbon Management*, *5*(1), 81–91. https://doi.org/10.4155/cmt.13.77

Schmidt, I. B., Moura, L. C., Ferreira, M. C., Eloy, L., Sampaio, A. B., Dias, P. A., et al. (2018). Fire management in the Brazilian savanna: First steps and the way forward. *Journal of Applied Ecology*, *55*(5), 2094–2101. https://doi.org/10.1111/1365-2664.13118

Shen, J.-p., Chen, C. R., & Lewis, T. (2016). Long term repeated fire disturbance alters soil bacterial diversity but not the abundance in an Australian wet sclerophyll forest. *Scientific Reports*, *6*, 19639–19639. https://doi.org/https://doi.org/10.1038/srep19639

Shugart, H. H., Leemans, R., & Bonan, G. B. (1992). *A systems analysis of the global boreal forest*. New York: Cambridge University Press.

Shuman, J. K., Balch, J. K., Barnes, R. T., Higuera, P. E., Roos, C. I., Schwilk, D. W., et al. (2022). Reimagine fire science for the anthropocene. *PNAS Nexus*, *1*, 1–14. https://doi.org/10.1093/pnasnexus/pgac115

Shvidenko, A. Z., & Nilsson, S. (2000). Extent, distribution, and ecological role of fire in Russian forests. In E. S. Kasischke & B. J. Stocks (Eds.), *Fire, climate change, and carbon cycling in the boreal forest* (pp. 132–150). New York: Springer-Verlag.

Skiles, S. M., Flanner, M., Cook, J. M., Dumont, M., & Painter, T. H. (2018). Radiative forcing by light-absorbing particles in snow. *Nature Climate Change*, *8*(11), 964–971. https://doi.org/10.1038/s41558-018-0296-5

Smith, M. T., Guyton, K. Z., Kleinstreuer, N., Borrel, A., Cardenas, A., Chiu, W. A., et al. (2020). The key characteristics of carcinogens: relationship to the hallmarks of cancer, relevant biomarkers, and assays to measure them. *Cancer Epidemiology, Biomarkers & Prevention*, *29*(10), 1887–1903. https://doi.org/10.1158/1055-9965.EPI-19-1346

Smith, S. F. (2013). *Long-term residue and water management effects on soil respiration and soil aggregate stability in a wheat-soybean, double-crop system in eastern Arkansas*. University of Arkansas, Fayetteville University

Soja, A. J., Al-Saadi, J., Giglio, L., Randall, D., Kittaka, C., Pouliot, G., et al. (2009). Assessing satellite-based fire data for use in the National Emissions Inventory. *Journal of Applied Remote Sensing*, *3*(031504), 29. https://doi.org/https://doi.org/10.1117/1.3148859

Soja, A. J., Gargulinski, E. M., & Choi, H. (2019). *CALIOP-based biomass burning smoke plume detrainment height*. Paper presented at the Proceedings for the 6th International Fire Behavior and Fuels Conference, Albuquerque, New Mexico, USA.

Soja, A. J., Loboda, T. V., & Jandt, R. (2020). Fire regimes: An agent of rapid ABZ change. In B. N. Duncan, L. E., et al. (Eds.), Space-based observations for understanding changes in the Arctic-boreal zone, *Reviews of Geophysics*, *58*, e2019RG000652.

Soja, A. J., Shugart, H. H., Sukhinin, A., Conard, S., & Stackhouse, P. W. (2006). Satellite-derived mean fire return intervals as indicators of change in Siberia (1995–2002). *Mitigation and Adaptation Strategies for Global Change*, *11*(1), 75–96. https://doi.org/10.1007/s11027-006-1009-3

Soja, A. J., Tchebakova, N. M., French, N. H. F., Flannigan, M. D., Shugart, H. H., Stocks, B. J., et al. (2007). Climate-induced boreal forest change: Predictions versus current observations. *Global and Planetary Change*, *56*(3), 274–296. https://doi.org/10.1016/j.gloplacha.2006.07.028

Sokolik, I. N., Soja, A. J., DeMott, P. J., & Winker, D. (2019). Progress and challenges in quantifying wildfire smoke emissions, their properties, transport, and atmospheric impacts. *Journal of Geophysical Research: Atmospheres*, *124*(23), 13005–13025. https://doi.org/10.1029/2018JD029878

Sorensen, C., House, J. A., O'Dell, K., Brey, S. J., Ford, B., Pierce, J. R., et al. (2021). Associations between wildfire-related PM2.5 and intensive care unit admissions in the United States, 2006–2015. *GeoHealth*, *5*(5), e2021GH000385.

Sparks, T. L., & Wagner, J. (2021). Composition of particulate matter during a wildfire smoke episode in an urban area. *Aerosol Science and Technology*, *55*(6), 734–747. https://doi.org/10.1080/02786826.2021.1895429

Streets, D. G., Gupta, S., Waldhoff, S. T., Wang, M. Q., Bond, T. C., & Yiyun, B. (2001). Black carbon emissions in China. *Atmospheric Environment*, *35*(25), 4281–4296. https://doi.org/10.1016/S1352-2310(01)00179-0

Swetnam, T. W., & Betancourt, J. L. (1990). Fire-southern oscillation relations in the southwestern United States. *Science*, *249*(4972), 1017–1020. https://doi.org/10.1126/science.249.4972.1017

Tarnocai, C., Canadell, J. G., Schuur, E. A. G., Kuhry, P., Mazhitova, G., & Zimov, S. (2009). Soil organic carbon pools in the northern circumpolar permafrost region. *Global Biogeochemical Cycles*, *23*(2). https://doi.org/10.1029/2008GB003327

Tchebakova, N. M., Parfenova, E. I., Lysanova, G. I., & Soja, A. J. (2011). Agro-climatic potential across central Siberia in an altered twenty-first century. *Environmental Research Letters*, *6*(045207). https://doi.org/10.1088/1748-9326/1086/1084/045207

Tessum, C. W., Paolella, D. A., Chambliss, S. E., Apte, J. S., Hill, J. D., & Marshall, J. D. (2021). PM2.5 polluters disproportionately and systemically affect people of color in the United States. *Science Advances*, *7*(18), eabf4491. https://doi.org/10.1126/sciadv.abf4491

Thelen, B., French, N. H. F., Koziol, B. W., Billmire, M., Owen, R. C., Johnson, J., et al. (2013). Modeling acute respiratory illness during the 2007 San Diego wildland fires using a coupled emissions-transport system and generalized additive modeling. *Environmental Health*, *12*, Article number: 94. https://doi.org/https://doi.org/10.1186/1476-069X-12-94

Thomas, J. L., Polashenski, C. M., Soja, A. J., Marelle, L., Casey, K. A., Choi, H. D., et al. (2017). Quantifying black carbon deposition over the Greenland ice sheet from forest fires in Canada. *Geophysical Research Letters*, *44*(15), 7965–7974. https://doi.org/10.1002/2017GL073701

Tosca, M. G., Diner, D. J., Garay, M. J., & Kalashnikova, O. V. (2015). Human-caused fires limit convection in tropical Africa: First temporal observations and attribution. *Geophysical Research Letters*, *42*(15), 6492–6501. https://doi.org/10.1002/2015GL065063

Tosca, M. G., Randerson, J. T., & Zender, C. S. (2013). Global impact of smoke aerosols from landscape fires on climate and the Hadley circulation. *Atmospheric Chemistry and Physics*, *13*(10), 5227–5241. https://doi.org/10.5194/acp-13-5227-2013

Tumel, V. F. (1939). *Changes in the soils and permafrost regimes following vegetation cover burning*. Reports of the Permafrost Commission, Issue VIII. Moscow-Leningrad: Academy of Sciences of the USSR.

Turetsky, M. R., Harden, J. W., Friedli, H. R., Flannigan, M., Payne, N., Crock, J., et al. (2006). Wildfires threaten mercury stocks in northern soils. *Geophysical Research Letters*, *33*(16). https://doi.org/10.1029/2005GL025595

Turetsky, M. R., Kane, E. S., Harden, J. W., Ottmar, R. D., Manies, K. L., Hoy, E., et al. (2011). Recent acceleration of biomass burning and carbon losses in Alaskan forests and peatlands. *Nature Geoscience*, *4*(1), 27–31. https://doi.org/10.1038/ngeo1027

Twohy, C. H., Toohey, D. W., Levin, E. J. T., DeMott, P. J., Rainwater, B., Garofalo, L. A., et al. (2021). Biomass burning smoke and its influence on clouds over the western U. S. *Geophysical Research Letters*, *48*(15), e2021GL094224. https://doi.org/10.1029/2021GL094224

Twomey, S. (1974). Pollution and the planetary albedo. *Atmospheric Environment (1967)*, *8*(12), 1251–1256. https://doi.org/10.1016/0004-6981(74)90004-3

Val Martin, M., Kahn, R. A., Logan, J. A., Paugam, R., Wooster, M., & Ichoku, C. (2012). Space-based observational constraints for 1-D fire smoke plume-rise models. *Journal of Geophysical Research: Atmospheres*, *117*(D22). https://doi.org/10.1029/2012JD018370

Val Martin, M., Logan, J. A., Kahn, R. A., Leung, F. Y., Nelson, D. L., & Diner, D. J. (2010). Smoke injection heights from fires in North America: Analysis of 5 years of satellite observations. *Atmospheric Chemistry and Physics*, *10*(4), 1491–1510. https://doi.org/10.5194/acp-10-1491-2010

Van Wagner, C. E. (1978). Age class distribution and the forest fire cycle. *Canadian Journal of Forest Research*, *8*, 220–227.

Van Wagner, C. E. (1987). *Development and structure of the Canadian Forest Fire Weather Index System*. Canadian Forestry Service, Headquarters, Ottawa, Forestry Technical Report 35. http://cfs.nrcan.gc.ca/publications?id=19927

Veraverbeke, S., Rogers, B. M., Goulden, M. L., Jandt, R. R., Miller, C. E., Wiggins, E. B., et al. (2017). Lightning as a major driver of recent large fire years in North American boreal forests. *Nature Climate Change*, *7*(7), 529–534. https://doi.org/10.1038/nclimate3329

Verhulst, N., Nelissen, V., Jespers, N., Haven, H., Sayre, K. D., Raes, D., et al. (2011). Soil water content, maize yield and its stability as affected by tillage and crop residue management in rainfed semi-arid highlands. *Plant and Soil*, *344*(1), 73–85. https://doi.org/10.1007/s11104-011-0728-8

Viereck, L. A., & Johnston, W. F. (1990). Black spruce. In *Silvics of North America* (pp. 227–237). Washington DC: Forest Service.

Wang, P., Huang, Q., Pozdniakov, S. P., Liu, S., Ma, N., Wang, T., et al. (2021). Potential role of permafrost thaw on increasing Siberian river discharge. *Environmental Research Letters*, *16*(3), 034046. https://doi.org/10.1088/1748-9326/abe326

Ward, J. L., Flanner, M. G., Bergin, M., Dibb, J. E., Polashenski, C. M., Soja, A. J., et al. (2018). Modeled response of Greenland snowmelt to the presence of biomass burning-based absorbing aerosols in the atmosphere and snow. *Journal of Geophysical Research: Atmospheres*, *123*(11), 6122–6141. https://doi.org/10.1029/2017JD027878

Wein, R. W. (1976). Frequency and characteristics of Arctic tundra fires. *Arctic*, *29*, 213–222.

Wein, R. W., & MacLean, D. A. (1983). *The role of fire in northern circumpolar ecosystems*. Chichester (UK), New York: John Wiley & Sons.

Westerling, A. L., Hidalgo, H. G., Cayan, D. R., & Swetnam, T. W. (2006). Warming and earlier spring increase western U.S. forest wildfire activity. *Science*, *313*(5789), 940–943. https://doi.org/10.1126/science.1128834

Wettstein, Z. S., Hoshiko, S., Fahimi, J., Harrison, R. J., Cascio, W. E., & Rappold, A. G. (2018). Cardiovascular and cerebrovascular emergency department visits associated with wildfire smoke exposure in California in 2015. *Journal of the American Heart Association*, *7*(8), e007492. https://doi.org/10.1161/JAHA.117.007492

Whitlow, S., Mayewski, P., Dibb, J., Holdsworth, G., & Twickler, M. (1994). An ice-core-based record of biomass burning in the Arctic and Subarctic, 1750–1980. *Tellus B: Chemical and Physical Meteorology*, *46*(3), 234–342. https://doi.org/10.3402/tellusb.v46i3.15794

Whitman, T., Whitman, E., Woolet, J., Flannigan, M. D., Thompson, D. K., & Parisien, M.-A. (2019). Soil bacterial and fungal response to wildfires in the Canadian boreal forest across a burn severity gradient. *Soil Biology and Biochemistry*, *138*, 107571. https://doi.org/10.1016/j.soilbio.2019.107571

Wiggins, E. B., Anderson, B. E., Brown, M. D., Campuzano-Jost, P., Chen, G., Crawford, J., et al. (2021). Reconciling assumptions in bottom-up and top-down approaches for estimating aerosol emission rates from wildland fires using observations from FIREX-AQ. *Journal of Geophysical Research: Atmospheres*, *126*(24), e2021JD035692. https://doi.org/10.1029/2021JD035692

Wiggins, E. B., Czimczik, C. I., Santos, G. M., Chen, Y., Xu, X., Holden, S. R., et al. (2018). Smoke radiocarbon measurements from Indonesian fires provide evidence for burning of millennia-aged peat. *Proceedings of the National Academy of Sciences*, *115*(49), 12419–12424. https://doi.org/10.1073/pnas.1806003115

Wilkins, J. L., de Foy, B., Thompson, A. M., Peterson, D. A., Hyer, E. J., Graves, C., et al. (2020). Evaluation of stratospheric intrusions and biomass burning plumes on the vertical distribution of tropospheric ozone over the midwestern United States. *Journal of Geophysical Research: Atmospheres*, *125*(18), e2020JD032454. https://doi.org/10.1029/2020JD032454

Wilkins, J. L., Pouliot, G., Foley, K., Appel, W., & Pierce, T. (2018). The impact of US wildland fires on ozone and particulate matter: a comparison of measurements and CMAQ model predictions from 2008 to 2012. *International Journal of Wildland Fire*, *27*(10), 684–698. https://doi.org/10.1071/WF18053

Wilkins, J. L., Pouliot, G., Pierce, T., Soja, A., Choi, H., Gargulinski, E., et al. (2022). An evaluation of empirical and statistically based smoke plume injection height parametrisations used within air quality models. *International Journal of Wildland Fire*, *31*(2), 193–211. https://doi.org/10.1071/WF20140

Williams, P. R., Congdon, R. A., Grice, A. C., & Clarke, P. J. (2003). Effect of fire regime on plant abundance in a tropical eucalypt savanna of north-eastern Australia. *Austral Ecology*, *28*(3), 327–338. https://doi.org/10.1046/j.1442-9993.2003.01292.x

Witze, A. (2020). The Arctic is burning like never before, and that's bad news for climate change. *Nature*, *585*, 336+. Retrieved from https://link.gale.com/apps/doc/A635608962/AONE?u=anon~802e63a1&sid=googleScholar&xid=db21925d

Wotawa, G., & Trainer, M. (2000). The influence of Canadian forest fires on pollutant concentrations in the United States. *Science*, *288*(5464), 324–328. https://doi.org/10.1126/science.288.5464.324

Wuest, S. B., Caesar-TonThat, T. C., Wright, S. F., & Williams, J. D. (2005). Organic matter addition, N, and residue burning effects on infiltration, biological, and physical properties of an intensively tilled silt-loam soil. *Soil and Tillage Research*, *84*(2), 154–167. https://doi.org/10.1016/j.still.2004.11.008

Yao, Q., Brown, P. M., Liu, S., Rocca, M. E., Trouet, V., Zheng, B., et al. (2017). Pacific-Atlantic Ocean influence on wildfires in northeast China (1774 to 2010). *Geophysical Research Letters*, *44*(2), 1025–1033. https://doi.org/10.1002/2016GL071821

York, A., Bhatt, U., Gargulinski, E., Grabinski, Z., Jain, P., Soja, A., et al. (2021). Sidebar 5.1: Wildland fires in the high northern latitudes. State of the Climate in 2020. *Bulletin of the American Meteorological Society*, *102*(8), S274–S276, https://doi.org/210.1175/BAMS-D-1121-0086.1171.

York, A., Bhatt, U. S., Gargulinski, E., Grabinski, Z., Jain, P., Soja, A., et al. (2020). Wildland fire in high northern latitudes. *NOAA Arctic Report Card 2020*. https://doi.org/10.25923/2gef-3964

Zackrisson, O. (1977). Influence of forest fires on the North Swedish boreal forest. *Oikos*, *29*, 22–32. https://doi.org/https://doi.org/10.2307/3543289

Zamora, L. M., Kahn, R. A., Cubison, M. J., Diskin, G. S., Jimenez, J. L., Kondo, Y., et al. (2016). Aircraft-measured indirect cloud effects from biomass burning smoke in the Arctic and subarctic. *Atmospheric Chemistry and Physics*, *16*(2), 715–738. https://doi.org/10.5194/acp-16-715-2016

Zhdanko, V. A. (1965). Scientific basis of development of regional scales and their importance for forest fire management. In I. S. Melekhov (Ed.), *Contemporary problems of forest protection form fire and firefighting*. Moscow: Lesnaya Promyshlennost.

Zinke, P. J., Millemann, R. E., & Boden, T. A. (1986). *Worldwide organic soil carbon and nitrogen data*. Carbon Dioxide Information Center, Environmental Sciences Division, Oakridge National Laboratory. https://www.osti.gov/servlets/purl/543663

3

Mapping and Characterizing Fire

Louis Giglio[1], David P. Roy[2], Michael L. Humber[1], Evan Ellicott[1], Maria Zubkova[1], and Christopher O. Justice[1]

ABSTRACT

In this chapter we focus on the mapping and characterization of biomass burning as a precursor to the fire emissions modeling that is the main topic of this book. Our discussion will necessarily emphasize satellite-based fire monitoring given the widespread use of remotely sensed fire data and its uniquely comprehensive spatial and temporal coverage.

3.1. OVERVIEW OF MAPPING APPROACHES

3.1.1. Ground-Based Approaches

We initially and briefly discuss ground-based fire mapping for the sake of completeness. Within the scope of this book, ground-based mapping and characterization of fire, while providing detailed information, is only practical in the context of small, prescribed fires. For example, a considerable body of research on fire spread, fuel consumption, and post-fire effects have been underpinned by field measurements of prescribed fires (e.g., Ward et al., 1996; Cofer et al., 1998; Kasischke et al., 2000; Korontzi et al., 2003; McCaw et al., 2012; Johnston et al., 2018). Assessment of wildfires is typically not planned and so has limited pre-fire field data available, which can present considerable interpretative challenges, for example, in inferring whether observed effects were definitively caused by fire and not by other confounding factors (Smith et al., 2010).

The primary limitation of ground-based mapping and characterization is that the sample of fires is small and unsuitable for large-scale fire assessment. Moreover, field-based empirical models may not be applicable to different ecosystems or fuel characteristics or to locations that burned under different meteorological conditions. Nevertheless, a number of studies have used ground-based fire records for regional analyses (e.g., Tekeli et al., 2009).

3.1.2. Airborne Approaches

Modern remote sensing arose from photogrammetry and analog airborne camera photographs used for surveying and reconnaissance. The potential of airborne camera photography for monitoring fire was recognized in the 1940s and 1950s (Garver, 1948; Spurr, 1952). Airborne sensors are now used extensively in support of fire management in some countries and are in addition sometimes employed in regional fire science studies (e.g., Kaufman et al., 1998b; King et al., 2003; Oertel et al., 2003; Riggan et al., 2004; Schroeder et al., 2014a). As with satellites, the use of airborne platforms for fire monitoring generally requires specialized sensors that

[1] *Department of Geographical Sciences, University of Maryland, College Park, Maryland, USA*

[2] *Department of Geography, Environment and Spatial Sciences, and Center for Global Change and Earth Observations, University of Michigan, East Lansing, Michigan, USA*

Landscape Fire, Smoke, and Health: Linking Biomass Burning Emissions to Human Well-Being, Geophysical Monograph 280, First Edition.
Edited by Tatiana V. Loboda, Nancy H. F. French, and Robin C. Puett.

DOI: 10.1002/9781119757030.ch03

operate in the middle-infrared (~4 μm) spectral region (we defer a discussion of this important point to section 3.4). In the last decade, unmanned aircraft systems, now often referred to as *drones*, have become of interest for fire research and for localized tactical fire assessment. These are considered briefly below.

In the 1960s, the U.S. Forest Service initiated research into the application of thermal infrared technology to support detection and monitoring of wildfires, particularly from fixed-wing aircraft. Since 1969, the National Infrared Operations, or NIROPS, has been flying thermal infrared detection and mapping missions providing critical fire intelligence for incident management teams. During the peak fire season, NIROPS flies nightly sorties using two aircraft, each equipped with an infrared sensor that sees a 6 mi swath of land and can map 300,000 ac (121,000 ha) per hour. From an altitude of 3,000 m, a NIROPS sensor can detect an actively burning fire just 15 cm across and place it within 4 m on a map. During intensive fire seasons with multiple wildfires in various states, the demand for NIROPS coverage far exceeds the aircraft availability.

Airborne infrared missions help fill gaps in satellite observations and provide a greater understanding of scaling observations from the ground to space. For example, NASA's Autonomous Modular Sensor-Wildfire (AMS) has been flown aboard aircraft in support of fire science missions (Ambrosia & Hinkley, 2008; Schroeder et al., 2014a), and the NASA-NOAA FIREX-AQ, provides comprehensive observations to investigate the impact on air quality and climate from wildfires and agricultural fires across the continental United States.

Unmanned aircraft systems (UAS), often termed *drones*, can support response and recovery operations, gather situational awareness in a short time, and enable fast mapping for incident response as well as for postincident recovery, at lower cost than a traditional aircraft. In unfamiliar territory, rugged terrain, or shifting fire conditions, deploying a small UAS to provide situational awareness is becoming commonplace. The fact that UAS can fly lower than helicopters means they can provide more detailed information, while also being capable of navigating tight or dangerous spaces. In addition, UAS have been deployed to administer prescribed burns. While controlled burns are traditionally lit with handheld drip torches, unique payloads built specifically for fire ignition allow a UAS pilot to remotely initiate a controlled fire from a safe distance. Rapid research and development are underway for the use of UAS in firefighting, in particular for autonomous fire suppression, and for assessment of fuel loads and condition, and fire rate of spread and behavior (e.g., Aydin et al., 2019; Burke et al., 2019; Innocente & Grasso, 2019; Filkov et al., 2020).

3.1.3. Satellite Approaches

Overview

Satellite-based remote sensing has long been recognized as the only practical way to monitor fire activity over large areas and in particular at the global scale. For this discussion, it is helpful to explicitly distinguish between *active fire detection*, whereby the instantaneous state of an actively burning fire is mapped and possibly characterized, and *burned* (or *burnt*) *area mapping*, in which the cumulative spatial extent affected by a fire (or fires) during a particular time period is mapped (depending on the application, this time period can vary from days to years). Figure 3.1 illustrates this distinction using the 2016 Soberanes fire in the western United States as an example. We defer a discussion of both aspects of satellite-based fire mapping to section 3.3.

Important also in the present discussion is the distinction between polar-orbiting and geostationary satellites since, crucially, the type of satellite orbit strongly constrains the sampling in time when observations can be made. Polar-orbiting satellites pass over the polar regions and accumulate daily coverage of the Earth's surface as the planet rotates (Fig. 3.2). Environmental polar-orbiting satellites typically have a 600–900 km altitude and are in sun-synchronous orbits such that the satellite passes over a given point on the Earth's surface at approximately the same local time(s) each day. This scheme makes it simpler to monitor change over time. The temporal overpass frequency of polar-orbiting satellites is typically twice per day at the equator, once in daytime followed by a nighttime overpass 12 hr later, rising to >10 times per day at the poles. Combining satellite data from different polar-orbiting satellites that carry duplicate instrumentation provides increased temporal observation frequency.

Geostationary satellites trade spatial coverage and spatial resolution to achieve greater temporal sampling than polar-orbiting systems. Geostationary systems orbit at an altitude of 35,785 km over or close to the plane of the equator, and consequently the satellite continuously sits over the same point on the Earth's surface (Fig. 3.2). Sensors on board geostationary platforms can observe any point on the visible full-disk hemisphere, and recent geostationary systems can take observations typically in the range of every 1 to 30 min. However, because the geostationary orbit is considerably higher than polar orbits, sensors on board geostationary satellites acquire imagery at coarser spatial resolution than their polar-orbiting counterparts.

Commercial small satellites, sometimes referred to as *cubesats*, are increasingly being launched into polar orbits. From 2011 to 2020, nearly 4,000 spacecraft were

Figure 3.1 Sentinel-2 MSI L1C composite of the Soberanes Fire near Monterrey, California, in the United States on 23 August 2016. In this false color image, MSI bands 12 (2.2 μm), 8a (0.865 μm), and 4 (0.665 μm) are encoded as the red, green, and blue channels, respectively. Note the smoke and bright red areas in the southeast side of the fire, where the flaming front is located.

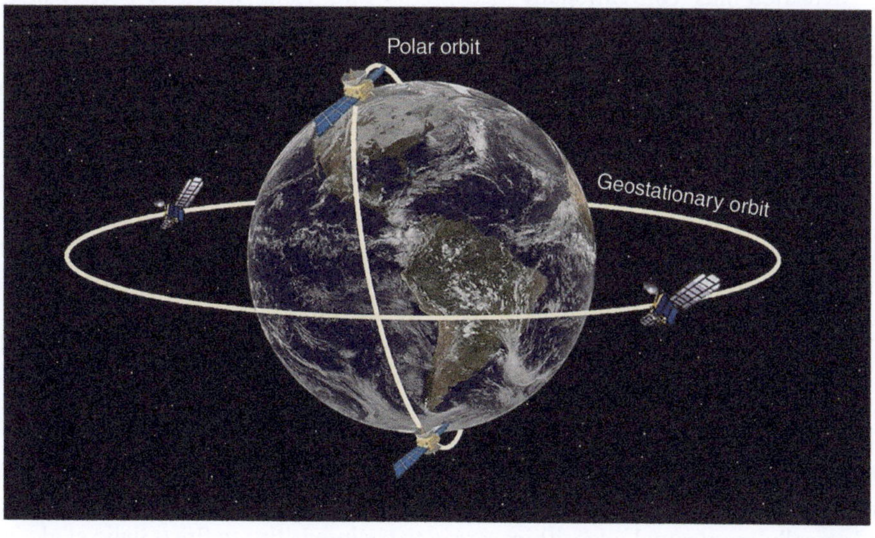

Figure 3.2 Illustration of the two main satellite orbit configurations: geostationary and sun-synchronous polar satellite orbits. NASA / NOAA GOES-13 satellite image showing Earth on 2 March 2010 at 8:45 UTC.

launched globally, of which 75% were small and weighed < 600 kg (Bryce Space and Technology, 2021). Only a minority are dedicated to science missions, but their capabilities show potential to assist in discoveries across numerous and diverse scientific disciplines beyond those typically attainable with higher-cost single satellite approaches (NAS, 2022). For example, Planet Labs has deployed a constellation of PlanetScope small satellites into sun-synchronous polar orbits that provide multispectral data at 3–4 m spatial resolution (Planet Team, 2021) with a 30 hr global median average revisit interval (Roy et al., 2021). In the future, our expectation is, that pending appropriate detector technology development, constellations with fire monitoring capabilities will be developed.

3.2. SATELLITE-BASED FIRE MAPPING AND CHARACTERIZATION PRINCIPLES

3.2.1. Active Fire Mapping and Characterization

We distinguish between active fire detection, in which the presence of one or more actively burning fires within the sensor footprint is noted, and active fire characterization, whereby an attempt is made to retrieve one or more attributes of the actively burning fire(s). Not all sensors that are capable of detecting active fires are also capable of characterizing them. Regardless, for detection as well as characterization the resulting information reflects the instantaneous state of the fire at the time the sensor observation is acquired.

Detection

The temperature and also size of an actively burning fire determine the emitted radiance that is sensed at specific wavelengths by the remote sensing system. Satellite-based fire detection has conventionally been undertaken at moderate resolution (~1 km) using a band located near 4 μm to exploit the high levels of blackbody radiation emitted in this region of the electromagnetic spectrum at typical vegetation fire temperatures. In fact, this middle infrared region is near-optimal for daytime satellite-based fire detection (Fig. 3.3); for nighttime fire detection the region is not necessarily optimal but more than capable. In all but the most rudimentary detection algorithms, a second long-wave infrared channel, usually near 11 μm, is used to help remove the component of the integrated mid-infrared signal that is contributed by the cooler ~300 K nonfire land surface that simultaneously resides within the sensor footprint. Additional specialized tests, often employing reflective channels, are used to eliminate false detections caused by sun glint and other surface features that might be mistaken for a fire (e.g., bright coastline). Typically, contextual algorithms are used, which consider iteratively the surrounding pixels to enable more sensitive active fire detection. Comprehensive reviews of satellite-based active fire detection algorithms can be found in Robinson (1991), Cahoon et al. (2000), Li et al. (2001), and Wooster et al. (2021).

A key advantage of the conventional approach is that actively burning fires that occupy an area considerably smaller than the observation footprint (e.g., as little as 0.1%) can be detected. This is because smoldering (~600 K) and flaming (~1,000 K) fires emit in excess of 100 and 1,000 times, respectively, as much blackbody radiation in the 3–5 μm middle-infrared portion of the electromagnetic spectrum (Fig. 3.3) compared with that emitted from the typical ~300 K nonburning land surface (Giglio et al., 2008).

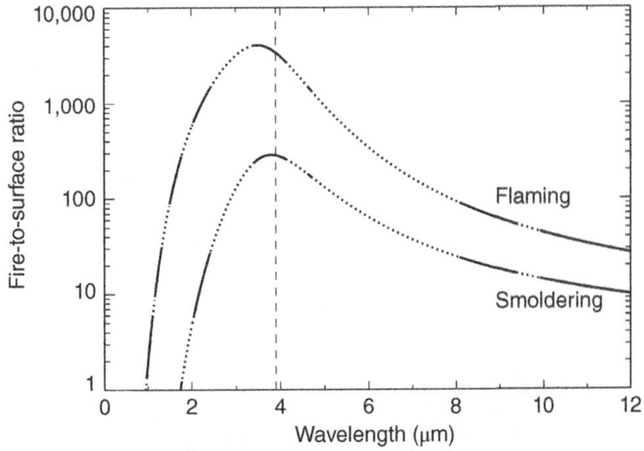

Figure 3.3 Ratio of typical fire radiance to typical daytime land surface radiance, as a function of wavelength, for 1,000 K flaming and 600 K smoldering fires. The solid portions of both curves denote the locations of atmospheric windows through which the surface is visible from space with comparatively little atmospheric absorption. This is a highly idealized case with the land surface assumed to be a grey body of 15% reflectance, and no atmospheric extinction. Vertical dashed line indicates the approximate location of a generic ~3.9 μm middle-infrared fire channel suitable for active fire detection (adapted from Giglio et al., 2008).

A major advantage of the ~4 μm channel detection approach is the ability to detect small fires. However, with medium and high spatial resolution sensors (10 m–100 m), this advantage wanes, as a fire will occupy a larger proportion of the sensor footprint. However, often the fire is sufficiently hot that it emits radiance that can be detected both in the shortwave infrared (SWIR) as well as the long-wave infrared (LWIR). In practice, SWIR channels are typically used as they are present on medium and high spatial resolution sensors. Daytime fire detection using a SWIR channel requires one or more additional reflective channels, often a red band, that is comparatively insensitive to fire (Giglio et al., 2008). For nighttime fire detection, when reflected sunlight is not present, simple SWIR radiance thresholds can often be employed.

An alternative, specialized, active fire detection approach that has been proposed and tested employs narrow, dedicated fire channels to look for signature emission lines associated with vegetation combustion from potassium at 766.5 nm and 769.9 nm (Vodacek et al., 1997) as well as carbon monoxide near 2 μm

(Worden et al., 1997; Dennison, 2006). However, to date, this has not been implemented systematically.

Characterization via Fire Radiative Power

Fire radiative power (FRP) (units: W) may be retrieved from middle-infrared wavelength remotely sensed data and provides an assessment of the radiative strength of a fire at the time of satellite overpass (Kaufman et al., 1998a). The FRP is related to the fire rate of fuel consumption and smoke emission as heat yields are relatively constant among vegetation types (Stott, 2000; Ichoku et al., 2008). The FRP has in turn been used to characterize fire types, for example, heading and backing fires (Smith & Wooster, 2005), crown and ground fires (Wooster & Zhang, 2004), or land use related based fire type categories (Roy & Kumar, 2017). Of interest also is the maximum FRP, as the maximum fire intensity affects vegetation processes like grass and tree response to fires (Archibald et al., 2010; Heward et al., 2013).

Given sufficient temporal sampling, FRP can provide reasonably accurate estimates of the fuel consumption rate and, ultimately, the total fuel consumed (Kaufman et al., 1998a; Wooster et al., 2012). The key to this route is to temporally integrate the FRP over the lifetime of a fire (T), yielding the fire radiative energy (FRE) (units: J) within the sensor footprint:

$$\mathrm{FRE} = \int_0^T \mathrm{FRP}(t)\,\mathrm{d}t. \tag{3.1}$$

The utility of the FRE is that the FRE is linearly related to the total biomass burned (Freeborn et al., 2008; Wooster et al., 2005), and the energy content of the greater majority of vegetation spans a comparatively narrow range. Consequently, it is possible to directly estimate the mass M of combusted vegetation solely from FRE using a simple multiplicative conversion factor (0.368 kg/MJ). In practice, few satellite-based sensors acquire observations sufficiently often to reliably estimate FRE via equation 3.1, and as a result FRE retrievals typically rely on modeled diurnal variability, possibly in conjunction with data fusion, to operate (e.g., Boschetti & Roy, 2009; Kumar et al., 2011).

The FRP for N different fire components within a pixel, each having its own temperature and area, can be defined (Wooster et al., 2003) as:

$$\mathrm{FRP} = A_{\mathrm{pix}}\varepsilon\sigma\sum_{i=1}^{N} p_i T_i^4 \tag{3.2}$$

where p_i is the fraction of the pixel occupied by the ith fire component with temperature T_i, A_{pix} is the area of the pixel, ε is the weighted mean emissivity of the fire components, and σ is the Stefan-Boltzmann constant. For the special case of a MWIR wavelength near 4 μm, Wooster et al. (2003) showed that FRP can be remotely sensed using a single middle-infrared band via the relationship

$$\mathrm{FRP} \approx \frac{A_{\mathrm{pix}}\varepsilon\sigma}{a\varepsilon_{\mathrm{MWIR}}}\left(L - L_{\mathrm{b}}\right), \tag{3.3}$$

where L is the MWIR radiance of a fire pixel, L_{b} is an independent estimate of the radiance contributed by the nonfire portion of the pixel, $\varepsilon_{\mathrm{MWIR}}$ is the middle-infrared fire emissivity, and a is a sensor-specific empirical constant. This unique property of the middle-infrared was first noted by Kaufman et al. (1998a), who derived an empirical relationship relating the FRP to brightness temperature but was later refined to the above radiance-based form.

Other Characterization

A recurring goal of fire characterization has been to supplement active fire detection with an estimate of the instantaneous temperature and subpixel area of the fire (or fires) within the sensor footprint. Most approaches for estimating this information (e.g., Prins & Menzel, 1992; Zhang & Kondragunta, 2008; Peterson et al., 2013) are based on the bispectral retrieval proposed by Dozier (1981) using mid-infrared and long-wave infrared channels. However, the retrieval relies upon a number of assumptions that are rarely satisfied for typical vegetation fires, and as a result cannot be performed with sufficient accuracy for the majority of detectable fires (Giglio & Kendall, 2001; Shephard & Kennelly, 2003; Giglio & Schroeder, 2014). Useful estimates of these quantities are consequently largely limited to high-resolution (and often hyperspectral) sensors for which the majority or entirety of the footprint is occupied by the fire (e.g., Dennison & Roberts, 2009; Waigl et al., 2019). A comprehensive review of hyperspectral fire remote sensing can be found in Veraverbeke et al. (2018).

3.2.2. Burned Area Mapping and Characterization

Burned Area Mapping

While active fire detection records the instantaneous state of a fire at a particular instant in time, burned area mapping seeks to map the cumulative spatial extent over which fires occurred. An approximate date of the burn is typically reported as well. Comprehensive reviews of satellite-based burned area mapping can be found in Arino et al. (2001), Roy et al. (2013), and Chuvieco et al. (2019).

Vegetation fires transform living and dead leaf and woody material from the overstory and/or understory into char and ash, and may also expose the underlying soil. Black char is typically produced rather than white mineral ash that is only produced under complete or near complete combustion (Smith & Hudak, 2005). All of these changes are nonpermanent. Post-fire forest regrowth may take several decades whereas savanna grasses regrow rapidly (typically weeks to months), and a recent global analysis with MODIS satellite data found that the median burned area persistence time was 29 days (Melchiorre & Boschetti, 2018).

Satellite-based burned area mapping methods generally attempt to exploit the change in surface reflectance post-fire (Fig. 3.4). Typically, post-fire reflectance decreases and is associated with black char deposition, which has a relatively flat spectra from visible to middle-infrared wavelengths (0.3–3.5 μm) and contrasts strongly with the reflectance of vegetation (Pereira, 2003; Roy & Landmann 2005). Spectral band ratios are often used as they reduce sensitivity to solar illumination and topographic variations found at reflective wavelengths. Ratios based on near-infrared (~0.8 μm) and shortwave infrared (2.2 μm) reflectance have long been known to be particularly suited for discrimination between unburned and burned vegetation (García & Caselles, 1991). More information regarding band selection and the use of spectral indices in the context of burned area mapping may be found in Pereira et al. (1999), Key & Benson (2006), and Roy et al. (2006).

While passive optical sensors compose the majority of satellite-based systems that have been used for burned area mapping, active sensors, in particular synthetic aperture radar (SAR), have also been used to map burned areas on a regional basis (e.g., French et al., 1999; Tanase et al., 2010; Belenguer-Plomer et al., 2021). SAR, which operates in the microwave portion of the electromagnetic spectrum (i.e., wavelengths of several mm to ~1 m) and can penetrate clouds and smoke and is thus particularly attractive for use in regions prone to persistent cloud cover (e.g., much of the tropics). However, SAR data are sensitive to many sensing and environmental factors and, for example, the change in backscatter pre- and post-fire can be comparable to changes in moisture content and seasonal phenology in savannas that may reduce burned area mapping capability (Mathieu et al., 2019). More information regarding the use of active sensors for burned area mapping may be found in Chuvieco et al. (2019).

There is no consensus burned area mapping algorithm. However, most algorithms use multitemporal observations under a change detection scheme to take advantage of the temporal persistency of fire effects. The algorithms are thus applied on a single pixel basis to reflectance time series. In these approaches, the definition of the magnitude of spectral change associated with the conversion of vegetation to burned vegetation is critical. A subset of burned area mapping methods use active fire detections to determine the threshold and/or supplement the reflectance and spectral band ratio time series.

Burned Area Characterization

Once the cumulative spatial extent of a fire has been mapped, additional analyses may be used to characterize selected aspects of the combustion process within the burned (or fire-affected) area. Perhaps, most commonly, this information consists of an assessment of burn severity, a variably defined measure of "the degree to which an ecosystem has changed owing to [a] fire" that "incorporates both short- and long-term post-fire effects on the local and regional environment" (Lentile et al, 2006, p. 322). A comprehensive discussion of burn severity and the related concept of fire severity, which

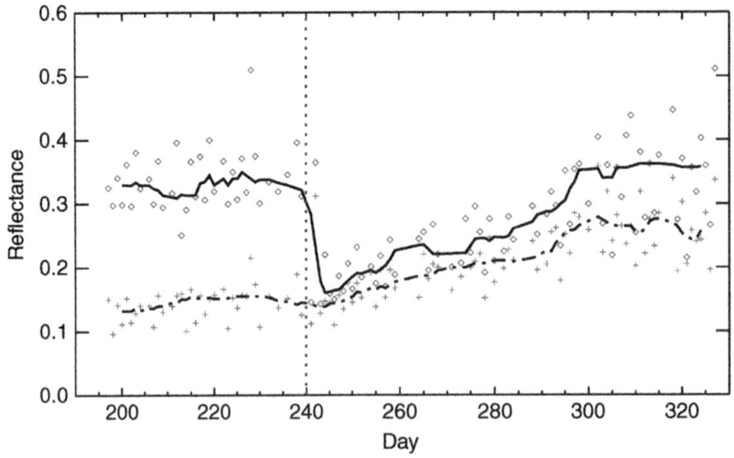

Figure 3.4 MODIS band-5 (diamonds) and band-7 (plus signs) daily surface reflectance time series from mid-July through late November 2001 for a 500 m grid cell in northern Australia. The cell is located within an area that burned on ordinal day 240 (red dotted vertical line). Solid (band 5) and dashed (band 7) lines show the time series after smoothing (adapted from Giglio et al., 2009).

operates on shorter time scales, and remote sensing techniques for assessing these metrics may be found in Lentile et al. (2006). The change in a pre- and post-fire spectral index, such as the Normalized Burn Ratio (Key & Benson, 2006), is often used as an indicator of severity. However, linkages between remotely sensed observations and ground assessments of severity are often imperfect and inherently indirect (Roy et al., 2006; French et al., 2008). Broadly speaking, the target of burned area characterization may consist of any number of post-fire vegetation or land-surface attributes (e.g., tree mortality, soil moisture, etc.), but common to each is to initially map the spatial extent of the fire-affected area.

3.3. SATELLITE-BASED MAPPING AND CHARACTERIZATION RESEARCH AND PRODUCTS

3.3.1. Early History

The history of satellite-based fire monitoring began with NOAA's polar-orbiting TIROS-N satellite, launched in 1978, on which resided the first of Advanced Very High Resolution Radiometer (AVHRR). Dozier (1981) and Matson and Dozier (1981) showed that the 3.8 μm middle-infrared channel of this sensor was extremely sensitive to subpixel, high temperature sources, such as fires, within its field of view. Numerous AVHRR fire-detection studies followed (e.g., Malingreau et al., 1985; Muirhead & Cracknell, 1985; Flannigan & Vonder Haar, 1986), which, in conjunction with the growing availability of satellite data over the next decade, demonstrated the potential for spaceborne sensors to supply large-scale information about biomass burning (Malingreau, 1990; Justice et al., 1993). A key element facilitating this progression was the unique sensitivity of the AVHRR middle-infrared channels to actively burning fires that were considerably smaller than the 1 km sensor footprint.

Many refinements followed this early work, yielding a large collection of AVHRR fire detection algorithms (e.g., Kaufman et al., 1990; Lee & Tag, 1990; Setzer & Pereira, 1991; Flasse & Ceccato, 1996; Justice et al., 1996, Giglio et al., 1999; Boles & Verbyla, 2000; Li et al., 2000; Lasaponara et al., 2003; Soja et al., 2004), all of which were constrained by limitations of both the AVHRR instrument itself and the NOAA platforms on which the family of AVHRR instruments was flown (Li et al., 2001; Giglio & Roy, 2020). A comprehensive survey of the AVHRR sensor and its capabilities and limitations can be found in Cracknell (1997).

Relevant work during this formative period was not confined to the AVHRR. Prins and Menzel (1992, 1994) similarly exploited the Visible Infrared Spin Scan Radiometer (VAS) on board NOAA's Geostationary Operational Environmental Satellites (GOES). The nominal spatial resolution of the middle-infrared VAS channel used for fire detection was 8 km. Prins et al. (1998) later extended this capability to the (then) new GOES Imager, which had a nominal spatial resolution of 4 km in the channels useful for fire detection. The second Along-Track Scanning Radiometer (ATSR-2), on board ESA's second European Remote Sensing (ERS-2) spacecraft, and its Advanced Along-Track Scanning Radiometer (AATSR) successor, which flew on ESA's ENVISAT satellite, were used by Arino and Rosaz (1999) to produce the 1 km ATSR World Fire Atlas, which in the pre-MODIS era provided the longest time series of global active fire data. The low dynamic range of the ATSR-2 and AATSR mid-infrared channel restricted active fire detection to nighttime imagery. Giglio et al. (2000) used 2.5 km imagery from the Visible and Infrared Scanner (VIRS) on board NASA's Tropical Rainfall Measuring Mission (TRMM) satellite for fire detection. The TRMM platform remains unique among satellites that have been used for fire monitoring (although it was not intended for this purpose) because its low inclination orbit meant that the platform passed over the surface at different times of day and demonstrated that many regions have different diurnal fire variations. Notably, Elvidge et al. (1996) pursued a different approach with the Operational Linescan System (OLS) on board the suite of Defense Meteorological Satellite Program (DMSP) polar orbiting satellites. The OLS had no mid-infrared imaging capability but included a visible/near-infrared channel that permitted fires (and other light sources) to be detected at night.

3.3.2. The MODIS Era

A characteristic of the early history of satellite-based fire monitoring was that the focus was almost exclusively on active fire detection using sensors that were not designed for fire monitoring. This situation changed markedly with the launch of NASA's Terra and Aqua satellites in 1999 and 2001, respectively, as part of NASA's Earth Observing System (EOS) with respective 10:30 and 13:30 equatorial overpasses. In addition to other sensors, these satellites carried the 36-band Moderate Resolution Imaging Spectroradiometer (MODIS), which included two high-saturation 4 and 11 μm channels designed specifically for fire detection and characterization at a spatial resolution of 1 km. The MODIS sensor in addition has two 250 m and five 500 m reflective channels, which for the first time permitted large-scale retrospective mapping of burned area. A suite of MODIS fire products, both active fire and burned area, were subsequently produced by NASA (Justice et al., 2002; Giglio et al., 2016; Giglio et al., 2018) and have been used extensively over the past two decades. The demand for active fire information for operational fire management led NASA to develop a near real-time capability for EOS (LANCE). This system provides MODIS fire data within 2 hr of satellite acquisition (Davies et al., 2009).

3.3.3. Geostationary Satellite Developments

In 2004, the European Organization for the Exploitation of Meteorological Satellites (EUMETSAT) launched the first of multiple geostationary Meteosat satellites carrying the Spinning Enhanced Visible and Infrared Imager (SEVIRI). With a nominal spatial resolution of 4.8 km, SEVIRI provides comparatively high frequency active-fire observations that have been used in numerous studies (e.g., Roberts et al., 2005; Roberts & Wooster, 2007). SEVIRI active fire products remain in operational production to this day.

Somewhat later, the Advanced Baseline Imager (ABI; Schmit et al., 2017) was launched on the NOAA GOES-16 and GOES-17 satellites in November 2016 and March 2018, respectively. With a dedicated 3.9 μm fire-imaging channel and nominal spatial resolution of 2 km, the ABI is a greatly enhanced successor to the earlier GOES VAS and Imager instruments. The Japan Meteorological Agency (JMA) Advanced Himawari Imager (AHI) is an ABI-clone system respectively launched on board the Himawari-8 and Himawari-9 satellites (Bessho et al., 2016) in October 2014 and November 2016. Located at a longitude of 140°E, the Himawari satellites provide coverage of major biomass burning areas in Asia and Oceania.

3.3.4. Beyond MODIS

Based in part on experience with MODIS data, improved fire-monitoring capabilities were incorporated into newer sensors deployed on polar-orbiting satellites. In October 2011, the first Visible Infrared Imaging Radiometer Suite (VIIRS) instrument was launched on board the joint NASA-NOAA Suomi National Polar-Orbiting Partnership (S-NPP) satellite, a preparatory mission for the NOAA Joint Polar Satellite System (JPSS). Among its numerous 750 m moderate resolution bands, VIIRS includes two 4 μm and 11 μm channels designed for fire detection and characterization. In addition to this advantage in spatial resolution over the 1 km MODIS sensor, VIIRS employs an onboard aggregation scheme that greatly constrains pixel growth over each scan; for traditional polar-orbiting sensors (e.g., MODIS and AVHRR), this growth is much larger and significantly degrades the sensitivity to small fires toward the edge of the scan. The third advantage of VIIRS lies within its five 375 m imagery channels, which include a 3.8 μm channel not intended for fire monitoring but which nevertheless has been successfully exploited by Schroeder et al. (2014b) to produce a high-fidelity global active fire product that is an order of magnitude more sensitive to small fires than MODIS. The S-NPP VIIRS was followed by the JPSS-1 VIIRS in late 2017. Additional JPSS platforms, each carrying a VIIRS sensor, are planned for launch over the next decade.

Launched also in 2017 was the Sentinel-3A satellite by ESA as part of the European Commission's comprehensive Earth Observation Programme (now referred to as *Copernicus*), followed a year later with Sentinel-3B. Present on both Sentinel-3 platforms are the Sea and Land Surface Temperature Radiometer (SLSTR), which has 500 m dedicated fire channels (Wooster et al., 2012; Xu et al., 2020), as well as the 300 m Ocean and Land Color Instrument (OLCI) that has since been exploited for burned area mapping (Lizundia-Loiola et al., 2021).

3.3.5. Coarse Resolution Burned Area Mapping

Early examples of satellite-based burned area mapping include Kasischke and French (1995), who used 15 day AVHRR NDVI composite imagery to detect burns in Alaskan boreal forests, Roy et al. (1999), who used 1 km AVHRR image composites to map savanna burns in southern Africa, Fraser et al. (2000), who mapped burns in Canada using AVHRR imagery, Stroppiana et al. (2003), who used Satellite Pour l'Observation de la Terre (SPOT) VEGETATION data to map burns in Australian savanna, and George et al. (2006), who used two 16 day MODIS image composites to map burn scars in central Russia. These efforts were almost exclusively regional but, in a few instances, yielded short-term global burned area data sets such as GLOBSCAR (Simon et al., 2004) and GBA2000 (Tansey et al., 2004).

Advances in sensor, computing, and algorithm capabilities have led to the production of a variety of long-term, global, moderate resolution burned area data sets over the past decade, beginning with NASA's MCD45A1 MODIS 500 m monthly burned area product (Roy et al., 2002, 2005), which was superseded by the current MCD64A1 500 m monthly burned area product (Giglio et al., 2009, 2018). An example of the MCD64A1 product, which is representative of moderate-resolution global burned area products in general, is shown in Figure 3.5. Complementing these NASA-sponsored data sets are several remotely sensed global burned area data sets produced under the auspices of ESA's CCI program, including the 300 m MERIS FireCCI41 (Chuvieco et al., 2016), 250 m MODIS FireCCI50/FireCCI51 (Chuvieco et al., 2018), and 300 m Sentinel-3 C3SBA10 (Lizundia-Loiola et al., 2021) burned area products.

To date, geostationary data have not been used to systematically generate burned area products. This is because

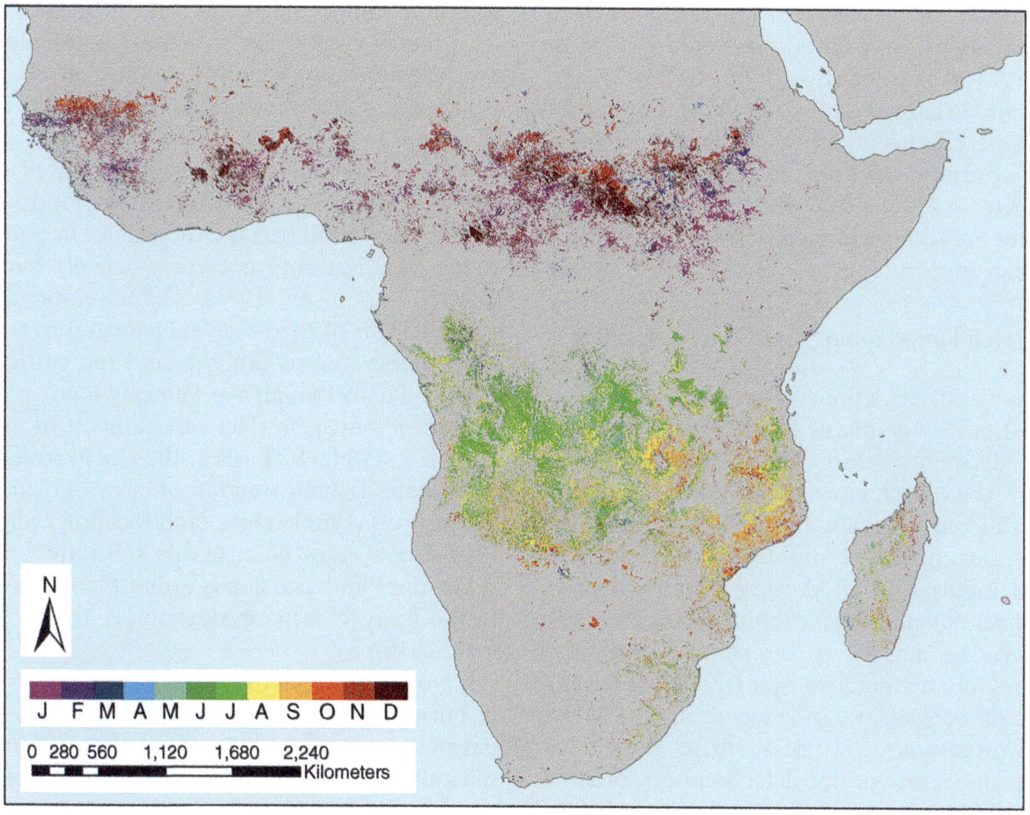

Figure 3.5 Cumulative 2019 annual burned area for Africa as mapped with NASA's MODIS Collection-6 MCD64A1 monthly burned area product. The calendar month of the burn date is indicated by a rainbow color scale. The grey and light blue areas correspond to unburned land and water, respectively.

geostationary data have traditionally had coarse spatial resolution, poor registration, and a small number of spectral bands unsuitably placed spectrally for burned area mapping (Boschetti et al., 2003). However, the new generation of geostationary satellites has potential for burned area mapping due to its significantly improved spatial resolution and typically NIR and SWIR bands that are well suited for burned area mapping (Roy et al., 2021).

Recent burned area mapping efforts (e.g., Boschetti et al., 2015; Hawbaker et al., 2017; Roteta et al., 2019; Roy et al., 2019) are understandably almost universally focused on exploiting medium spatial resolution 10 m to 30 m satellite sensors that are able to capture the large population of small burns that cannot be mapped using older, coarser resolution sensors (Laris, 2005). Notably, the Landsat program spans a nearly 50 yr record of terrestrial observations. The first Landsat satellite, Landsat 1, was launched in 1972 and an additional six functional Landsat satellites have been deployed since then (Wulder et al., 2019). With a series of increasingly capable multispectral sensors (the Multispectral Scanner System (MSS), with a nominal spatial resolution of ~60 m, followed by the 30 m Thematic Mapper (TM), Enhanced Thematic Mapper (ETM+), and, most recently, the Operational Land Imager (OLI) on board the Landsat 8 satellite launched in 2013), the ability to map burned areas is well established. However, because Landsat data were not free prior to 2008 and because of early computer processing limitations, most research was on using only one or a small number of Landsat images (Hall et al., 1980; Richards, 1984; Chuvieco & Congalton, 1988).

Until recently, the relatively infrequent Landsat repeat cycle of 16 days (18 days for Landsat 1–3) generally limited the platform to regional fire studies (e.g., García & Caselles, 1991) and in the context of coarse-resolution burned area product validation, where global, wall-to-wall reference maps are not required (e.g., Roy & Boschetti, 2009). This situation changed with the launch of ESA's Sentinel-2A and Sentinel-2B satellites in 2015 and 2017, respectively, which carry the MultiSpectral Instrument (MSI) with 10 m and 20 m reflective channels that are well placed for burned area mapping (Huang et al., 2016; Roteta et al., 2019). Used in combination, the two MSIs and OLI provide a global median average revisit interval of 2.9 days (Li & Roy, 2017), a temporal resolution that for the first time enabled "the development of burned area mapping algorithms at medium resolution using time series approaches" that could identify the approximate day of burning (Roy et al., 2019, p. 2).

The OLI and MSI are also capable of detecting actively burning fires despite having no bands designed specifically for this purpose (Murphy et al., 2016; Schroeder et al., 2016; Kumar & Roy, 2017), but applications of the resulting maps are somewhat more limited as a result of their comparatively low temporal resolution.

3.3.6. Integrated Fire Mapping and Characterization

As satellite-based active fire and burned area products have matured, a new emphasis has been placed on the creation of derived products that combine multiple sources of fire data. Such integrated products draw from data across different thematic types, such as active fire and burned area maps, or different sensors such as MODIS and Landsat 7's ETM+ (e.g., Boschetti et al., 2015). Approaches that combine multiple sources of fire data are useful for improving the characterization of individual fires and fire patterns over large areas because they facilitate a convergence-of-evidence approach that can be used to parameterize fire attributes more accurately. For example, active fire detections can often be considered as better estimators of fire timing than burned area products (Boschetti et al., 2010; Humber et al., 2019), and high spatial resolution sensors can capture details that are lost at coarse spatial resolutions (Roteta et al., 2019; Roy et al., 2019).

In recent efforts, combinations of active fire and burned area products have been used to identify individual fire events from the global datasets at coarse spatial resolutions. These integrated sources can provide more information than the individual datasets because they summarize characteristics on a per-fire basis rather than for individual cells or detections. One such example is the Fire Atlas based on VIIRS active fire and MODIS burned area detections, which estimates fire sizes, durations, directions, and speed based on a fire progression tracking approach (Andela et al., 2019). The FRY database is a similar data set based only on burned area data but from two different sources, MODIS and MERIS, that includes "functional traits" of extracted fire patches similar to the parameters identified by the Fire Atlas (Laurent et al., 2018).

3.4. CAVEATS AND LIMITATIONS OF REMOTE SENSING DATA

Note: This section is an updated and expanded version of material from Giglio (2006).

Given the growing reliance on satellite data for an ever-expanding range of applications, it is important to recognize and understand the limitations of such data. This is especially important in the context of biomass burning since there is at present a plethora of satellites and sensors used for fire monitoring, not all of which were designed for fire monitoring. Among some of the most important issues are the following:

1. Due to limitations of the sensor as well as the detection algorithm, and obscuration by clouds, smoke, and forest canopy, and because the surface may not be sensed at the time burning, not every actively burning fire is detected (Robinson, 1991). The envelope of detectable fires (in terms of their size and temperature) is dependent upon observation conditions and the particular instrument and detection algorithm being used.

2. Burned area products are difficult to generate reliably on a systematic basis at the global scale. In general, pixels with a small fraction of area burned and/or low combustion completeness, and locations where the persistence of spectral changes due to fire are short, typically in savannas and grasslands rather than in boreal forests, are less likely to be mapped reliably (Laris, 2005; Boschetti et al., 2019).

3. Not every fire detection, and not every pixel reported as burned is real. The frequency at which these false alarms arise is dependent upon observation conditions and the particular instrument and detection algorithm being used. Examples of active-fire false alarms and their impact on downstream applications can be found in Schroeder et al. (2008) and Zhang et al. (2018).

4. The suite of satellites and sensors used for fire monitoring provide very different spatial and temporal sampling strategies. This can lead to inconsistencies in the respective fire data sets.

5. Although the growing availability of satellite data, coupled with the relative ease with which those data may be processed (e.g., via machine learning), has greatly expanded the generation of remotely sensed fire data sets, relatively little attention given to product quality control and robust product accuracy assessment.

6. Although a less common occurrence than in the past, the fire information provided by spaceborne sensors often does not exactly match the fire information desired by the broad community of users that now routinely makes use of remotely sensed fire data. For example, data latency may be longer than desired for the emergency response community, and the comparatively coarse spatial resolution of long-term fire data sets can be an impediment for land managers.

REFERENCES

Ambrosia, V. G., & Hinkley, E. (2008). NASA science serving society: Improving capabilities for fire characterization to effect reduction in disaster losses. *Proceedings of the IEEE Geoscience and Remote Sensing Symposium*, *4*. https://doi.org/10.1109/IGARSS.2008.4779800

Andela, N., Morton, D. C., Giglio, L., Paugam, R., Chen, Y., Hantson, S., et al. (2019). The Global Fire Atlas of individual fire size, duration, speed and direction. *Earth System Science Data*, *11*(2), 529–552. https://doi.org/10.5194/essd-11-529-2019

Archibald, S., Scholes, R. J., Roy, D. P., Roberts, G., & Boschetti, L. (2010). Southern African fire regimes as revealed by remote sensing. *International Journal of Wildland Fire*, *19*, 861–878.

Arino, O., & Rosaz, J.-M. (1999). 1997 and 1998 world ATSR fire atlas using ERS-2 ATSR-2 data. In L. F. Neuenschwander, K. C. Ryan, & G. E. Golberg (Eds.), *Proceedings of the Joint Fire Science Conference* (pp.177–182). Boise: University of Idaho and the International Association of Wildland Fire.

Arino, O., Piccolini, I., Kasischke, E., Siegert, F., Chuvieco, E., Martin, P., et al. (2001). Methods of mapping surfaces burned in vegetation fires. In F. J. Ahern et al. (Eds.), *Global and regional wildfire monitoring from space: Planning a coordinated international effort* (pp. 227–255).

Aydin, B., Selvi, E., Tao, J., & Starek, M. J. (2019). Use of fire-extinguishing balls for a conceptual system of drone-assisted wildfire fighting. *Drones*, *3*(1), 17.

Belenguer-Plomer, M. A., Tanase, M. A., Chuvieco, E., & Bovolo, F. (2021). CNN-based burned area mapping using radar and optical data. *Remote Sensing of Environment*, *260*, 112468.

Bessho, K., Date, K., Hayashi, M., Ikeda, A., Imai, T., Inoue, H., et al. (2016). An introduction to Himawari-8/9: Japan's new-generation geostationary meteorological satellites. *Journal of the Meteorological Society of Japan*. Ser. II, *94*(2), 151–183.

Boles, S. H., & Verbyla, D. L. (2000). Comparison of three AVHRR based fire detection algorithms for interior Alaska. *Remote Sensing of Environment*, *72*, 1–16.

Boschetti, L., & Roy, D. P. (2009). Strategies for the fusion of satellite fire radiative power with burned area data for fire radiative energy derivation. *Journal of Geophysical Research*, *114*, D20302. https://doi.org/10.1029/2008JD011645

Boschetti, L., Brivio, P. A., & Gregoire, J. M. (2003). The use of Meteosat and GMS imagery to detect burned areas in tropical environments. *Remote Sensing of Environment*, *85*(1), 78–91.

Boschetti, L., Roy, D. P., Giglio, L. Huang, H., Zubkova, M., & Humber, M. L. (2019). Global validation of the collection 6 MODIS burned area product. *Remote Sensing of Environment*, *235*, 11490.

Boschetti, L., Roy, D. P., Justice, C. O., & Giglio, L. (2010). Global assessment of the temporal reporting accuracy and precision of the MODIS burned area product. *International Journal of Wildland Fire*, *19*(6), 705, https://doi.org/10.1071/wf09138.

Boschetti, L., Roy, D. P., Justice, C. O., & Humber, M. L. (2015). MODIS-Landsat fusion for large area 30 m burned area mapping. *Remote Sensing of Environment*, *161*, 27–42, https://doi.org/10.1016/j.rse.2015.01.022

Bryce Space and Technology (2021). *SmallSats by the numbers 2021*. BryceTech, Alexandria, VA. Brycetech.com/reports

Burke, C., Wich, S., Kusin, K., McAree, O., Harrison, M. E., Ripoll, B., et al. (2019). Thermal-drones as a safe and reliable method for detecting Subterranean peat fires. *Drones*, *3*(1), 23.

Cahoon D. R., Stocks, B. J., Alexander, M. E., Baum, B. A., & Goldammer, J. G. (2000). Wildland fire detection from space: Theory and application. In J. L. Innes et al. (Eds.), *Biomass burning and its inter-relationships with the climate system*. Advances in Global Change Research, vol. 3. Springer, Dordrecht. https://doi.org/10.1007/0-306-47959-1_9

Chuvieco, E., & Congalton, R. G. (1988). Mapping and inventory of forest fires from digital processing of TM data. *Geocarto International*, *3*(4), 41–53.

Chuvieco, E., Lizundia-Loiola, J., Pettinari, M. L., Ramo, R., Padilla, M., Tansey, K., et al. (2018). Generation and analysis of a new global burned area product based on MODIS 250 m reflectance bands and thermal anomalies. *Earth System Science Data*, *10*, 2015–2031. https://doi.org/ 10.5194/essd-10-2015-2018

Chuvieco, E., Mouillot, F., van der Werf, G. R., San Miguel, J., Tanasse, M., Koutsias, N., et al. (2019). Historical background and current developments for mapping burned area from satellite earth observation. *Remote Sensing of Environment*, *225*, 45–64.

Chuvieco, E., Yue, C., Heil, A., Mouillot, F., Alonso-Canas, I., Padilla, M., et al. (2016). A new global burned area product for climate assessment of fire impacts. *Global Ecology and Biogeography*, *25*, 619–629.

Cofer, W. R., III, Winstead, E. L., Stocks, B. J., Goldammer, J. G., & Cahoon, D. R. (1998). Crown fire emissions of CO_2, CO, H_2, CH_4, and TNMHC from a dense jack pine boreal forest fire. *Geophysical Research Letters*, *25*, 3919 – 3922.

Cracknell, A. P. (1997). *The advanced very high resolution radiometer*. London: Taylor and Francis Ltd.

Davies, D. K., Ilavajhala, S., Wong, M. M., & Justice, C. O. (2009). Fire information for resource management system: Archiving and distributing MODIS active fire data. *IEEE Transactions on Geoscience and Remote Sensing*, *47*, 72–79.

Dennison, P. E. (2006). Fire detection in imaging spectrometer data using atmospheric carbon dioxide absorption. *International Journal of Remote Sensing*, *27*, 3049–3055.

Dennison, P. E., & Roberts, D. A. (2009). Daytime fire detection using airborne hyperspectral data. *Remote Sensing of Environment*, *113*, 1646–1657.

Dozier, J. (1981). A method for satellite identification of surface temperature fields of subpixel resolution. *Remote Sensing of Environment*, *11*, 221–229.

Elvidge, C. D., Kroehl, H. W., Kihn, E. A., Baugh, K. E., Davis, E. R., & Hao, W. M. (1996). Algorithm for the retrieval of fire pixels from DMSP operational linescan system data. In J. S. Levine (Ed.), *Biomass burning and global change: Remote sensing, modeling and inventory development, and biomass burning in Africa*, vol. *1* (pp. 73–85). Cambridge: MIT Press.

Filkov, A., Cirulis, B., & Penman, T. (2020). Quantifying merging fire behavior phenomena using unmanned aerial vehicle technology. *International Journal of Wildland Fire*, *30*(3), 197–214, https://doi.org/ 10.1071/WF20088.

Flannigan, M. D., & Vonder Haar, T. H. (1986). Forest fire monitoring using NOAA satellite AVHRR. *Canadian Journal of Forest Research*, *16*, 975–982.

Flasse, S. P., & Ceccato, P. (1996). A contextual algorithm for AVHRR fire detection. *International Journal of Remote Sensing*, *17*, 419–424.

Fraser, R. H., Li, Z., & Cihlar, J. (2000). Hotspot and NDVI differencing synergy (HANDS): A new technique for burned area mapping over boreal forest. *Remote Sensing of Environment*, *74*, 362–376.

Freeborn, P. H., Wooster, M. J., Hao, W. M., Ryan, C. A., Nordgren, B. L., Baker, S. P., et al. (2008). Relationships between energy release, fuel mass loss, and trace gas and aerosol emissions during laboratory biomass fires. *Journal of Geophysical Research, 113*(D1), D01102. https://doi.org/10.1029/2007JD008489.

French, N. H., Bourgeau-Chavez, L. L., Wang, Y., & Kasischke, E. S. (1999). Initial observations of Radarsat imagery at fire-disturbed sites in interior Alaska. *Remote Sensing of Environment, 68*, 89–94.

French, N. H., Kasischke, E. S., Hall, R. J., Murphy, K. A., Verbyla, D. L., Hoy, E. E., et al. (2008). Using Landsat data to assess fire and burn severity in the North American boreal forest region: An overview and summary of results. *International Journal of Wildland Fire, 17*(4), 443–462.

García, M. L., & Caselles, V. (1991). Mapping burns and natural reforestation using Thematic Mapper data. *Geocarto Int, 6* (1), 31–37.

Garver, R. D. (1948). Aerial photographs in forest surveys. *Journal of Forestry, 46*, 104–106.

George, C., Rowland, C., Gerard, F., & Balzter, H. (2006). Retrospective mapping of burnt areas in Central Siberia using a modification of the normalized difference water index. *Remote Sensing of Environment, 104*, 346–359.

Giglio, L. (2006). Detection, evaluation, and analysis of global fire activity using MODIS data. Doctoral dissertation, University of Maryland, UMI/ProQuest document number 3212607.Giglio, L., & Kendall, J. D. (2001). Application of the Dozier retrieval to wildfire characterization: A sensitivity analysis. *Remote Sensing of Environment, 77*, 34–49.

Giglio, L., & Kendall, J. D. (2001). Application of the Dozier retrieval to wildfire characterization: A sensitivity analysis. *Remote Sensing of Environment, 77*, 34–49.

Giglio, L., & Roy, D. P. (2020). On the outstanding need for a long-term, multi-decadal, validated and quality assessed record of global burned area: caution in the use of Advanced Very High Resolution Radiometer data. *Science of Remote Sensing, 2*, 100007.

Giglio, L., & Schroeder, W. (2014). A global feasibility assessment of the bi-spectral fire temperature and area retrieval using MODIS data. *Remote Sensing of Environment, 152*, 166–173.

Giglio, L., Boschetti, L., Roy, D. P., Humber, M. L., & Justice, C. O. (2018). The collection 6 MODIS burned area mapping algorithm and product. *Remote Sensing of Environment, 217*, 72–85.

Giglio, L., Csiszar, I., Restás, Á., Morisette, J. T., Schroeder, W., Morton, D., et al. (2008). Active fire detection and characterization with the Advanced Spaceborne Thermal Emission and Reflection Radiometer (ASTER). *Remote Sensing of Environment, 112*, 3055–3063.

Giglio, L., Kendall, J. D., & Justice, C. O. (1999). Evaluation of global fire detection algorithms using simulated AVHRR infrared data. *International Journal of Remote Sensing, 20*, 1947–1985.

Giglio, L., Kendall, J. D., & Tucker, C. J. (2000). Remote sensing of fires with the TRMM VIRS. *International Journal of Remote Sensing, 21*, 203–207.

Giglio, L., Loboda, T., Roy, D. P., Quayle, B., & Justice, C. O. (2009). An active-fire based burned area mapping algorithm for the MODIS sensor. *Remote Sensing of Environment, 113*, 408–420.

Giglio, L., Schroeder, W., & Justice, C. O. (2016). The collection 6 MODIS active fire detection algorithm and fire products. *Remote Sensing of Environment, 178*, 31–41.

Hall, D. K., Ormsby, J. P., Johnson, L., & Brown, J. (1980). Landsat digital analysis of the initial recovery of burned tundra at Kokolik river, Alaska. *Remote Sensing of Environment, 10*, 263–272.

Hawbaker, T. J., Vanderhoof, M. K., Beal, Y. -J., Takacs, J. D., Schmidt, G. L., Falgout, J. T., et al. (2017). Mapping burned areas using dense time-series of Landsat data. *Remote Sensing of Environment, 198*, 504–522.

Heward, H., Smith, A. M. S., Roy, D. P., Hoffman, C. M., Tinkham, W. T., & Morgan, P. (2013). Is burn severity related to fire intensity? Observations from landscape scale remote sensing. *International Journal of Wildland Fire, 22*, 910–918.

Huang, H., Roy, D. P., Boschetti, B., Zhang, H. K., Yan, L., Kumar, S. S., et al. (2016). Separability analysis of Sentinel-2A multi-spectral instrument (MSI) data for burned area discrimination. *Remote Sensing, 8* (10), 873.

Humber, M. L., Boschetti, L., Giglio, L., & Justice, C. O. (2019). Spatial and temporal intercomparison of four global burned area products. *International Journal of Digital Earth, 12*(4). https://doi.org/ 10.1080/17538947.2018.1433727.

Ichoku, C., Giglio, L., Wooster, M. J., & Remer, L. A. (2008). Global characterization of biomass-burning patterns using satellite measurements of fire radiative energy. *Remote Sensing of Environment, 112*, 2950–2962.

Innocente, M. S., & Grasso, P. (2019). Self-organizing swarms of firefighting drones: Harnessing the power of collective intelligence in decentralised multi-robot systems. *Journal of Computational Science, 34*, 80–101.

Johnston, J. M., Wheatley, M. J., Wooster, M. J., Paugam, R., Davies, G. M., & DeBoer, K. A. (2018). Flame-front rate of spread estimates for moderate scale experimental fires are strongly influenced by measurement approach. *Fire, 1* (1), 16. https://doi.org/10.3390/fire1010016

Justice, C. O., Giglio, L., Korontzi, S., Owens, J., Morisette, J., Roy, D., et al. (2002). The MODIS fire products. *Remote Sensing of Environment, 83*, 244–262.

Justice, C. O., Kendall, J. D., Dowty, P. R., & Scholes, R. J. (1996). Satellite remote sensing of fires during the SAFARI campaign using NOAA AVHRR data. *Journal of Geophysical Research, 101*(D23):23851–23863.

Justice, C. O., Malingreau, J.-P., & Setzer, A. W. (1993). Satellite remote sensing of fires: Potential and limitations. In P. J. Crutzen & J. G. Goldammer (Eds.), *Fire in the Environment* (pp. 77–88). New York: Wiley.

Kasischke, E. S., & French, N. H. F. (1995). Locating and estimating the areal extent wildfires in Alaskan boreal forest using multiple-season AVHRR NDVI composite data. *Remote Sensing of Environment, 51*, 263–275.

Kasischke, E. S., O'Neill, K. P, French, N. H. F., & Bourgeau-Chavez, L. L. (2000). Controls on patterns of biomass burning in Alaskan boreal forests. In E. S. Kasischke & B. J. Stocks (Eds.), *Fire, climate change, and carbon cycling in the North American boreal forest* (pp. 173–196). New York: Springer-Verlag.

Kaufman, Y. J., Justice, C. O., Flynn, L. P., Kendall, J. D., Prins, E. M., Giglio, L., et al. (1998a). Potential global fire monitoring from EOS-MODIS. *Journal of Geophysical Research, 103*(D24), 32215–32238.

Kaufman, Y. J., Kleidman, R. G., & King, M. D. (1998b). SCAR-B fires in the tropics: Properties and remote sensing from EOS-MODIS. *Journal of Geophysical Research, 103*(D24), 3,955–31968.

Kaufman, Y. J., Setzer, A., Justice, C. O., Tucker, C. J., Pereira, M. C., & Fung, I. (1990). Remote sensing of biomass burning in the tropics. In J. G. Goldammer (Ed.), *Fire in the tropical biota: Ecosystem processes and global challenges* (pp. 371–399). Berlin: Springer-Verlag.

Key, C. H., & Benson, N. C. (2006). Landscape assessment (LA): Sampling and analysis methods. In D. C. Lutes et al. (Eds.), *FIREMON: Fire effects monitoring and inventory system. General technical report RMRS-GTR-164-CD* (pp. LA1–LA51). Fort Collins, CO: Rocky Mountain Research Station, U.S. Department of Agriculture, Forest Service. http://www.fs.fed.us/rm/pubs/rmrs_gtr164.pdf

King, M. D., Platnick, S., Moeller, C. C., Revercomb, H. E., & Chu, D. E. (2003). Remote sensing of smoke, land, and clouds from the NASA ER-2 during SAFARI 2000. *Journal of Geophysical Research, 108*(D13). http://dx.doi.org/10.1029/2002JD003207

Korontzi S., Ward, D. E., Susott, R. A., et al. (2003). Seasonal variation and ecosystem dependence of emission factors for selected trace gases and PM2.5 for southern African savanna fires. *Journal of Geophysical Research, 108*, 4758. https://doi.org/10.1029/ 2003JD003730

Kumar, S., & Roy, D. P. (2017). Global operational land imager Landsat-8 reflectance-based active fire detection algorithm. *International Journal of Digital Earth*. https://doi.org/10.1080/17538947.2017.1391341

Kumar, S. S., Roy, D. P., Boschetti, L., & Kremens, R. (2011). Exploiting the power law distribution properties of satellite fire radiative power retrievals: A method to estimate fire radiative energy and biomass burned from sparse satellite observations. *Journal of Geophysical Research, 116*, D19303. https://doi.org/10.1029/2011JD015676

Laris, P. S. (2005). Spatiotemporal problems with detecting and mapping mosaic fire regimes with coarse-resolution satellite data in savanna environments. *Remote Sensing of Environment, 99*, 412–424.

Lasaponara, R., Cuomo, V., Macchiato, M. F., & Simoniello, T. (2003). A self-adaptive algorithm based on AVHRR multitemporal data analysis for small active fire detection. *International Journal of Remote Sensing, 24*(8), 1723–1749.

Laurent, P., Mouillot, F., Yue, C., Ciais, P., Moreno, M. V., & Nogueira, J. M. P. (2018). FRY, a global database of fire patch functional traits derived from space-borne burned area products. *Scientific Data, 5*, 180132. https://doi.org/ 10.1038/sdata.2018.132

Lee, T. F., & Tag, P. M. (1990). Improved detection of hotspots using AVHRR 3.7 µm channel. *Bulletin of the American Meteorological Society, 71*, 1722–1730.

Lentile, L. B., Holden, Z., Smith, A. M. S., Falkowski, M. J., Hudak, A. T., Morgan, P., et al. (2006). Remote sensing techniques to assess active fire and post-fire effects. *International Journal of Wildland Fire, 15*(3), 319–345.

Li, J., & Roy, D. P. (2017). A global analysis of sentinel-2A, sentinel-2B and Landsat-8 data revisit intervals and implications for terrestrial monitoring. *Remote Sensing, 9* (9), 902.

Li, Z., Kaufman, Y. J., Ichoku, C., Fraser, R., Trishchenko, A., Giglio, L., et al. (2001). A review of AVHRR-based active fire detection algorithms: Principles, limitations, and recommendations. In F. J. Ahern, J. G. Goldammer, & C. O. Justice (Eds.), *Global and regional vegetation monitoring from space: Planning a coordinated international effort* (pp. 199–225). The Hague: SPB Academic Publishing.

Li, Z., Nadon, S., & Cihlar, J. (2000). Satellite-based detection of Canadian boreal forest fire: development and application of the algorithm. *International Journal of Remote Sensing, 21*, 3057–3069.

Lizundia-Loiola, J., Franquesa, M., Boettcher, M., Kirches, G., Pettinari, M. L., & Chuvieco, E. (2021). Operational implementation of the burned area component of the Copernicus Climate Change Service: from MODIS 250 m to OLCI 300 m data, *Remote Sensing, 13*, 4295.

Mathieu, R., Main, R., Roy, D. P., Naidoo, L., & Yang, H. (2019). The effect of surface fire in Savannah Systems in the Kruger National Park, South Africa, on the backscatter of C-Band Sentinel-1 Images. *Fire, 2*(3), 37.

Matson, M., & Dozier, J. (1981). Identification of subresolution high temperature sources using a thermal IR sensor. *Photogrammetric Engineering and Remote Sensing, 47*, 1311–1318.

Malingreau, J. -P. (1990). The contribution of remote sensing to the global monitoring of fires in tropical and subtropical ecosystems. In J. G. Goldammer(Ed.), *Fire in the tropical biota: Ecosystem processes and global challenges* (pp. 337–370). Berlin: Springer-Verlag.

Malingreau, J. -P., Stephens, G., & Fellows, L. (1985). Remote sensing of forest fires: Kalimantan and North Borneo in 1982–3. *Ambio, 14*, 314–315.

McCaw, W. L., Gould, J. S., Cheney, N. P., Ellis, P. F. M., & Anderson, W. R. (2012). Changes in behavior of fire in dry eucalypt forest as fuel increases with age. *Forest Ecology and Management, 271*, 170–181.

Melchiorre, A., & Boschetti, L. (2018). Global analysis of burned area persistence time with MODIS data. *Remote Sensing, 10*(5), 750.

Muirhead, K., & Cracknell, A. (1985). Straw burning over Great Britain detected by AVHRR. *International Journal of Remote Sensing, 65*, 827–833.

Murphy, S. W., de Souza Filho, C. R., Wright, R., Sabatino, G., & Pabon, R. C. (2016). HOTMAP: Global hot target detection at moderate spatial resolution. *Remote Sensing of Environment, 177*, 78–88. https://doi.org/10.1016/j.rse.2016.02.027

NAS (2022). *Leveraging commercial space for earth and ocean remote sensing*. Washington, DC: The National Academies Press. https://doi.org/10.17226/26380

Oertel, D., Briess, K., Halle, W., Neidhardt, M., Lorenz, E., Sandau, R., et al. (2003). Airborne forest fire mapping with an adaptive infrared sensor. *International Journal of Remote Sensing, 24*(18), 3663–3682.

Pereira, J. M. C. (2003). Remote sensing of burned areas in tropical savannas. *International Journal of Wildland Fire, 12*, 259–270. https://doi.org/ 10.1071/Wf03028

Pereira, J. M. C., Sá, A. C. L., Sousa, A. M. O., Silva, J. M. N., Santos, T. N., & Carreiras, J. M. B. (1999). Spectral characterization and discrimination of burnt areas. In E. Chuvieco (Ed.), *Remote sensing of large wildfires* (pp.123–138). Berlin: Springer-Verlag.

Peterson, D., Wang, J., Ichoku, C., Hyer, E., & Ambrosia, V. (2013). A sub-pixel-based calculation of the fire radiative power from MODIS observations: 1. Algorithm development and initial assessment. *Remote Sensing of Environment*, *129*, 262–279. https://doi.org/ 10.1016/j.rse.2012.10.036

Planet Team (2021). *Planet imagery product specifications*. https://assets.planet.com/docs/Planet_Combined_Imagery_Product_Specs_letter_screen.pdf

Prins, E. M., & Menzel, W. P. (1992). Geostationary satellite detection of biomass burning in South America. *International Journal of Remote Sensing*, *13*, 2783–2799.

Prins, E. M., & Menzel, W. P. (1994). Trends in South American biomass burning detected with the GOES VAS from 1983–1991. *Journal of Geophysical Research*, *99*(D8), 16719–16735.

Prins, E. M., Feltz, J. M., Menzel, W. P., & Ward, D. E. (1998). An overview of GOES-8 diurnal fire and smoke results for SCAR-B and the 1995 fire season in South America. *Journal of Geophysical Research*, *103*(D24), 31821–31836.

Richards, J. A. (1984). Thematic mapping from multitemporal image data using the principal components transformation. *Remote Sensing of Environment*, *16*, 34–36.

Riggan, P. J., Tissell, R. G., Lockwood, R. N., Brass, J. A., Pereira, J. A. R., Miranda, H., et al. (2004). Remote measurement of energy and carbon flux from wildfires in Brazil. *Ecological Applications*, *14*, 855–872.

Roberts, G., & Wooster, M. J. (2007). New perspectives on African biomass burning dynamics. *EOS*, *88*(38), 369–370.

Roberts, G., Wooster, M. J., Perry, G. L.W., Drake, N., Rebelo, L. -M., & Dipotso, F. (2005). Retrieval of biomass combustion rates and totals from fire radiative power observations: Application to southern Africa using geostationary SEVIRI imagery. *International Journal of Remote Sensing*, *110*, D21111. https://doi.org/10.1029/2005JD006018

Robinson, J. M. (1991). Problems in global fire evaluation: Is remote sensing the solution? In J. Levine (Ed.), *Global biomass burning* (pp. 67–73). Cambridge: MIT Press.

Roteta, E., Bastarrika, A., Padilla, M., Storm, T., & Chuvieco, E. (2019). Development of a Sentinel-2 burned area algorithm: Generation of a small fire database for sub-Saharan Africa. *Remote Sensing of Environment*, *222*, 1–17. https://doi.org/ 10.1016/j.rse.2018.12.011.

Roy, D. P., & Boschetti, L. (2009). Southern Africa validation of the MODIS, L3JRC, and GlobCarbon burned-area products. *IEEE Transactions on Geoscience and Remote Sensing*, *47*, 1032–1044.

Roy, D. P., & Kumar, S. S. (2017). Multi-year MODIS active fire type classification over the Brazilian tropical moist forest biome. *International Journal of Digital Earth*, *10*, 54–84.

Roy, D. P., & Landmann, T. (2005). Characterizing the surface heterogeneity of fire effects using multi-temporal reflective wavelength data. *International Journal of Remote Sensing*, *26*, 4197–4218.

Roy, D. P., Boschetti, L., & Smith, A. M. (2013). Satellite remote sensing of fires. *Fire Phenomena and the Earth System: An Interdisciplinary Guide to Fire Science*, 77–96.

Roy, D. P., Boschetti, L., & Trigg, S. N. (2006). Remote sensing of fire severity: Assessing the performance of the Normalized Burn Ratio. *IEEE Geoscience and Remote Sensing Letters*, *3*, 112–116.

Roy, D. P., Giglio, L., Kendall, J. D., & Justice, C. O.(1999). Multi-temporal active-fire based burn scar detection algorithm. *International Journal of Remote Sensing*, *20*, 1031–1038.

Roy, D. P., Huang, H., Boschetti, L., Giglio, L., Yan, L., Zhang, H. H., et al. (2019). Landsat-8 and Sentinel-2 burned area mapping: A combined sensor multi-temporal change detection approach. *Remote Sensing of Environment*, *231*, 111254. https://doi.org/10.1016/j.rse.2019.111254

Roy, D. P., Huang, H., Houborg, R., & Martins, V. S. (2021). A global analysis of the temporal availability of PlanetScope high spatial resolution multi-spectral imagery. *Remote Sensing of Environment*, *264*, 112586.

Roy, D. P., Jin, Y., Lewis, P. E., & Justice, C. O. (2005). Prototyping a global algorithm for systematic fire-affected area mapping using MODIS time series data. *Remote Sensing of Environment*, *97*, 137–162.

Roy, D. P., Lewis, P., & Justice, C. O. (2002). Burned area mapping using multi- temporal moderate spatial resolution data: A bi-directional reflectance model-based expectation approach, *Remote Sensing of Environment*, *83*, 263–286.

Roy, D. P., Li, Z., Giglio, L., Boschetti, L., & Huang, H. (2021). Spectral and diurnal temporal suitability of GOES Advanced Baseline Imager (ABI) reflectance for burned area mapping. *International Journal of Applied Earth Observations and Geoinformation*, *96*, 102271.

Schmit, T. J., Griffith, P., Gunshor, M. M., Daniels, J. M., Goodman, S. J., & Lebair, W. J. (2017). A closer look at the ABI on the GOES-R series. *Bulletin of the American Meteorological Society*, *98*(4), 681–698.

Schroeder, W., Ellicott, E., Ichoku, C., Ellison, L., Dickinson, M. B., Ottmar, R. D., et al. (2014a). Integrated active fire retrievals and biomass burning emissions using complementary near-coincident ground, airborne and spaceborne sensor data. *Remote Sensing of Environment*, *140*: 719–730. https://doi.org/10.1016/j.rse.2013.10.010

Schroeder, W., Oliva, P., Giglio, L., & Csiszar, I. (2014b). The new VIIRS 375 m active fire detection data product: Algorithm description and initial assessment. *Remote Sensing of Environment*, *143*, 85–96.

Schroeder, W., Oliva, P., Giglio, L., Quayle, B., Lorenz, E., & Morelli, F. (2016). Active fire detection using Landsat-8/OLI data. *Remote Sensing of Environment*, *185*, 210–220, https://doi.org/10.1016/j.rse.2015.08.032.

Schroeder, W., Ruminski, M., Csiszar, I., Giglio, L., Prins, E., Schmidt, C., et al. (2008). Validation analyses of an operational fire monitoring product: The hazard mapping system. *International Journal of Remote Sensing*, *29*, 6059–6066.

Setzer, A. W., & Pereira, M. C. (1991). Amazonia biomass burnings in 1987 and an estimate of their tropospheric emissions. *Ambio*, *20*, 19–22.

Shephard, M. W., & Kennelly, E. J. (2003). Effect of band-to-band coregistration on fire property retrievals. *IEEE Transactions on Geoscience and Remote Sensing*, *41*, 2648–2661.

Simon, M., Plummer, S., Fierens, F., Hoelzemann, J. J., & Arino, O. (2004). Burnt area detection at global scale using ATSR-2: The GLOBSCAR products and their qualification, *Journal of Geophysical Research: Atmospheres*, *109*, D14S02. https://doi.org/10.1029/2003JD003622

Smith, A. M. S., & Hudak, A. T. (2005). Estimating combustion of large downed woody debris from residual white ash. *International Journal of Wildland Fire*, *14*, 245–248. https://doi.org/ 10.1071/WF05011

Smith, A. M. S., & Wooster, M. J. (2005). Remote classification of head and backfire types from MODIS fire radiative power and smoke plume observations. *International Journal of Wildland Fire*, *14*, 249–254.

Smith, A. M. S., Eitel, J. U. H., & Hudak, A. T. (2010). Spectral analysis of charcoal on soils: implications for wildland fire severity mapping methods. *International Journal of Wildland Fire*, *19*, 976–983. https://doi.org/ 10.1071/WF09057

Soja, A. J., Sukhinin, A. I., Cahoon, D. R., Jr., Shugart, H. H., & Stackhouse, P. W., Jr. (2004). AVHRR-derived fire frequency, distribution and area burned in Siberia. *International Journal of Remote Sensing*, *25*, 1939–1960.

Spurr, S. H. (1952). Aerial photographs in forest management. *Photogrammetria*, *9*, 33–41.

Stott, P. (2000). Combustion in tropical biomass fires: a critical review. *Progress in Physical Geography*, *24*, 355–377.

Stroppiana, D., Grégoire, J. -M., & Pereira, J. M. C. (2003). The use of SPOT VEGETATION data in a classification tree approach for burnt area mapping in Australian savanna. *International Journal of Remote Sensing*, *24*, 2131–2151.

Tanase, M. A., Santoro, M., de la Riva, J., Perez-Cabello, F., & Le Toan, T. (2010). Sensitivity of X-, C-, and L-band SAR backscatter to burn severity in Mediterranean pine forests. *IEEE Transactions on Geoscience and Remote Sensing*, *48*, 3663–3675.

Tansey, K., Grégoire, J. -M., Stroppiana, D., Sousa, A., Silva, J., Pereira, J. M. C., Boschetti, L., et al. 2004). Vegetation burning in the year 2000: Global burned area estimates from SPOT VEGETATION data. *Journal of Geophysical Research: Atmospheres*, *109*, D14S03. https://doi.org/10.1029/2003JD003598

Tekeli, A. E., Sönmez, A., Erdi, E., & Demir, F. (2009). Validation studies of EUMETSAT's active fire monitoring product over Turkey. *International Journal of Wildland Fire*, *18*, 517–526. https://doi.org/10.1071/WF07088

Veraverbeke, S., Dennison, P., Gitas, I., Hulley, G., Kalashnikova, O., Katagis, T., et al. (2018). Hyperspectral remote sensing of fire: state-of-the-art and future perspectives. *Remote Sensing of Environment*, *215*, 2015–2121, https://doi.org/ 10.1016/j.rse.2018.06.020

Vodacek, A., Kremens, R. L., Fordham, A. J., Vangorden, S. C., Luisi, D., & Schott, J. R. (2002). Remote optical detection of biomass burning using a potassium emission signature. *International Journal of Remote Sensing*, *23*, 2721–2726.

Waigl, C. F., Prakash, A., Stuefer, M., Verbyla, D., & Dennison, P. (2019). Fire detection and temperature retrieval using EO-1 Hyperion data over selected Alaskan boreal forest fires. *International Journal of Applied Earth Observation and Geoinformation*, *81*, 72–84.

Ward, D. E., Hao, W. M., Susott, R. A., et al. (1996). Effect of fuel composition on combustion efficiency and emission factors for African savanna ecosystems. *Journal of Geophysical Research*, *101*, 23569–23576.

Wooster, M., et al. (2012). Sentinel-3 SLSTR active fire detection and FRP product: Pre-launch algorithm development and performance evaluation using MODIS and ASTER datasets. *Remote Sensing of Environment*, *120*, 236–254.

Wooster, M. J., & Zhang, Y. H. (2004). Boreal forest fires burn less intensely in Russia than in North America. *Geophysical Research Letters*, *31*(20).

Wooster, M. J., Roberts, G., Giglio, L., Roy, D. P., Freeborn, P., Boschetti, L., et al. (2021). Satellite remote sensing of active fires: History and current status, applications and future requirements. *Remote Sensing of Environment*, *267*, 112694. https://doi.org/10.1016/j.rse.2021.112694

Wooster, M. J., Roberts, G., Perry, G., & Kaufman, Y. J. (2005). Retrieval of biomass combustion rates and totals from fire radiative power observations: Calibration relationships between biomass consumption and fire radiative energy release. *Journal of Geophysical Research*, *110*, D21111. https://doi.org/10.1029/2005JD006318

Wooster, M. J., Zhukov, B., & Oertel, D. (2003). Fire radiative energy for quantitative study of biomass burning: derivation from the BIRD experimental satellite and comparison to MODIS fire products. *Remote Sensing of Environment*, *86*, 83–107.

Worden, H., Beer, R., & Rinsland, C. P. (1997). Airborne infrared spectroscopy of 1994 western wildfires. *Journal of Geophysical Research*, *102*, 1287–1299.

Wulder, M. A., Loveland, T. R., Roy, D. P., Crawford, C. J., Masek, J. G., Woodcock, C. E., et al. (2019). Current status of Landsat program, science, and applications. *Remote Sensing of Environment*, *225*, 127–147.

Xu, W., Wooster, M. J., He, J., & Zhang, T. (2020). First study of Sentinel-3 SLSTR active fire detection and FRP retrieval: Night-time algorithm enhancements and global intercomparison to MODIS and VIIRS AF products. *Remote Sensing of Environment*, *248*, 111947.

Zhang, T., Wooster, M. J., De Jong, M. C., & Xu, W. (2018). How well does the 'small fire boost' methodology used within the GFED4.1s fire emissions database represent the timing, location and magnitude of agricultural burning? *Remote Sensing*, *10*, 823. https://doi.org/10.3390/rs10060823

Zhang, X., & Kondragunta, S. (2008). Temporal and spatial variability in biomass burned areas across the USA derived from the GOES fire product. *Remote Sensing of Environment*, *112*, 2886–2897. https://doi.org/ 10.1016/j.rse.2008.02.006

4

Wildland Fuel Characterization Across Space and Time

Susan J. Prichard[1], Eric Rowell[2], Robert E. Keane[3], Andrew T. Hudak[4], Duncan Lutes[3], and E. Louise Loudermilk[5]

ABSTRACT

Wildland fuels are foundational to smoke prediction and generally contribute to high uncertainty in smoke production and dispersion modeling. Until recently, wildland fuels and fuel consumption have been mapped using traditional methods to estimate the cover, height, and biomass (kg m^{-2}). Over the past decade, significant progress has been made in describing and quantifying fuels more accurately with 3D characterization and quantification across large spatial scales. This chapter introduces wildland fuels and approaches to mapping them at scales relevant to smoke management, smoke impacts to communities, and future research needs. Topics covered include (1) an introduction to wildland fuels, (2) traditional approaches to mapping wildland fuels. (3) next-generation approaches to wildland fuel mapping, and (4) research and development needs. Because source characterization of wildland fuels is critical to predicting smoke impacts, reviewing how to measure and map wildland fuel biomass and consumption provides important context for fire and fuels managers, smoke scientists, and policy makers. Although recent developments in 3D fuel mapping are promising, there are still many challenges associated in scaling-measured values to the scale of wildland fire events and assessing wildland fuels such as coarse wood and organic soils (duff, peat lands) that can contribute to long-duration smoke events.

4.1. INTRODUCTION

As smoke from wildland fires becomes an increasing concern to communities, improved predictions of smoke impacts to communities is essential (Rappold et al., 2011; Williamson et al., 2016; Liu et al., 2019; Holm et al., 2021). The amount, composition, and structure of fuels that burn in wildland fires play a central role in the type and amount of air pollutants emitted by wildland fire events, and how long they are produced (Byram, 1959; Ottmar et al., 2007, Weise & Wright, 2014). Wildland fuels are defined as the combustible biomass of live and dead vegetation (Keane, 2015). In their most basic expression, fuels are stored potential energy that when released by fire, generate heat and emit air pollutants. However, as we explore in this chapter, the structure, composition, and chemical composition of wildland fuels are highly variable and are anything but basic (Cruz et al., 2018). Along with weather and topography, the amount, type, structure, spatial continuity, and condition of fuels that

[1] *School of Environmental and Forest Sciences, University of Washington, Seattle, Washington, USA*

[2] *Tall Timbers Research Station, Tallahassee, Florida, USA*

[3] *Rocky Mountain Research Station, United States Forest Service, Missoula, Montana, USA*

[4] *Forestry Sciences Laboratory, Rocky Mountain Research Station, United States Forest Service, Moscow, Idaho, USA*

[5] *Southern Research Station, United States Forest Service, Athens, Georgia, USA*

Landscape Fire, Smoke, and Health: Linking Biomass Burning Emissions to Human Well-Being, Geophysical Monograph 280, First Edition.
Edited by Tatiana V. Loboda, Nancy H. F. French, and Robin C. Puett.

DOI: 10.1002/9781119757030.ch04

burn in a wildland fire event drive fire spread, energy release, fuel consumption, and smoke production (Fig. 4.1; Larkin et al., 2010).

Characterizing and mapping the structure and variability of wildland fuels are key elements of smoke modeling and can be difficult to accurately represent without site-specific information. Geography, climate, seasonality, and past disturbances all factor into the type and quantity of wildland fuels that are available for combustion and how much smoke is produced from wildland fire events. For example, a fully cured grassland at the end of a growing season generally burns as a fast-moving fire with short-duration smoke production and relatively minimal $PM_{2.5}$ emissions compared with sites with higher available biomass for burning such as shrublands and forests (Cook et al., 2016). Following extensive fire exclusion, mixed-conifer forests generally have high biomass, accumulated understory fuels, and deep organic soils (Hessburg et al., 2019; Hagmann et al., 2021). In the absence of frequent fire, many mixed conifer forests have multilayered canopies that can support crown fires with large plume development and long-duration smoke emissions from the smoldering of coarse wood and organic soil layers (De Groot et al., 2007; Urbanski, 2014; Miller et al., 2019). However, forests that have been regularly prescribed burned often have light understory fuels that contribute to low-intensity surface fires that cannot reach the forest canopy; much like grasslands, these low-intensity surface fires can produce short-duration emissions compared with fire-excluded forests (Haikerwal et al., 2015; Liu et al., 2017; Jaffe et al., 2020).

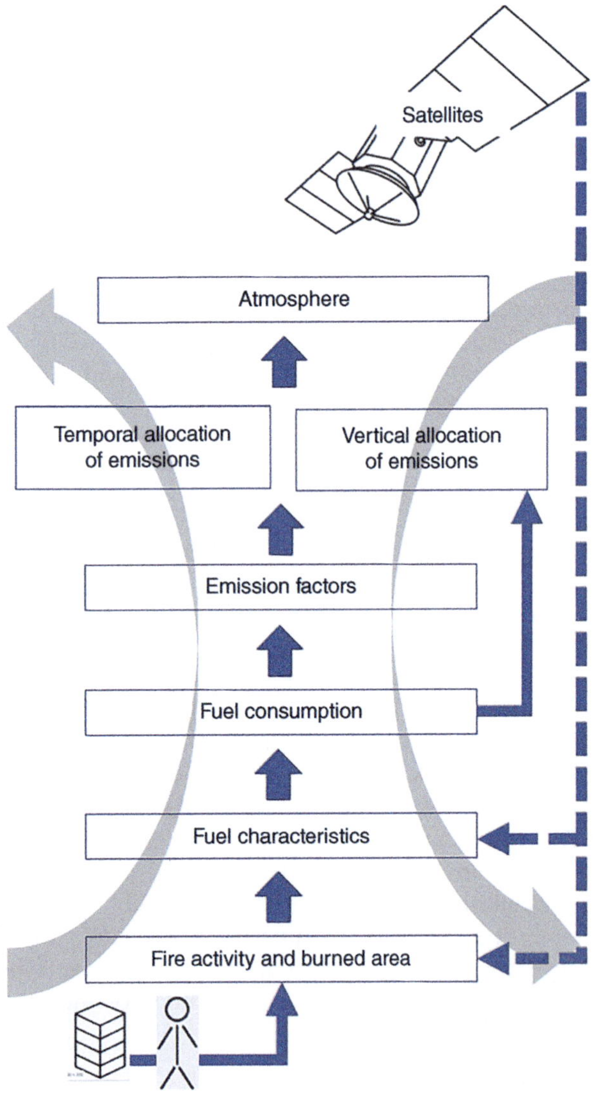

Figure 4.1 Interrelated steps to estimating wildland fire emissions from burned area, fuel characterization, and consumption to emissions and dispersion modeling (after Ottmar et al., 2009).

Because the structure, continuity, and composition of wildland fuels matter so much for fire behavior and smoke prediction, wildland fuelbeds often need to be characterized into the types of fuels that are available for combustion, often organized in vertical layers, including tree canopies, dead tree boles (termed snags), shrub stems and leaves, grass and herbaceous vegetation, sound and rotten wood, needle and leaf litter, moss, lichen, and ground fuels, which are defined as partially to fully decomposed organic soils (Ottmar et al., 2007). Smoke emissions are then estimated based on type and mass of fuel consumed (g or kg), which is used to determine smoke composition by multiplying mass by emission factors for specific pollutants and fuel types (Urbanski, 2014; Prichard et al., 2020).

Of all variables involved in estimating smoke emissions, the amount of available fuel and proportion that is consumed often represent the greatest sources of uncertainty (Ottmar, 2014; Drury et al., 2014). Errors in estimates of pre-burn fuel biomass can contribute as much as 80% when calculating emissions, and errors in estimating fuel consumption can contribute another 30% (Peterson, 1987). Fuels, topography, and weather compose the classic fire environment triangle (Countryman, 1972). However, wildland fuels are the only one of the three variables that can be managed to influence fire behavior before an ignition occurs (see Box 4.1).

Although maps of wildland fuel biomass are widely available (e.g., LANDFIRE; Rollins, 2009), they generally have high uncertainty because they are derived from vegetation layers classified from remotely sensed imagery and are not based on direct observations. Reliable estimates of fuels generally require more detailed information

Box 4.1 Fuel Reduction Treatments and Emissions Scenarios

Fuel reduction treatments are used to reduce surface and canopy fuels and to mitigate future wildfire hazard (Omi, 2015; Kalies & Kent, 2016). In the following example, we evaluate potential consumption and emissions in managed dry, mixed conifer forests of the eastern Cascade Mountains (northwestern United States). The three fuelbeds represent an unmanaged forest dominated by Douglas fir (*Pseudotsuga menziesii*), ponderosa pine (*Pinus ponderosa*), and grand fir (*Abies grandis*) that have large accumulations of surface and canopy fuels from an extended period of fire exclusion. The Thin Rx scenario represents a restored forest that was thinned from below, removing midstory trees and thinning overstory trees. This treatment was followed by a prescribed understory burn, which reduced accumulated surface fuels. The thin-only scenario represents postthinning fuels that are elevated from the accumulated posttreatment logging slash and litter (see Table B4.1).

Consumption and $PM_{2.5}$ emissions scenarios were estimated in Consume (Prichard et al., 2007) under two scenarios. A summer wildfire scenario was evaluated to compare potential $PM_{2.5}$ emissions under dry fuels typical of summer wildfire seasons in dry mixed conifer forests of the eastern Cascade Mountains. Fuel moisture and other burn day inputs are listed in Table B4.2. Because thinned forests lack continuity between surface and canopy fuels, canopy consumption is often minimal. We therefore assumed 10% canopy consumption under the summer wildfire scenario and no consumption under the prescribed fire scenario.

Predicted $PM_{2.5}$ emissions differ markedly across scenarios. Under summer wildfires, the thinned and prescribed burn unit has sparse understory fuels that contribute little to $PM_{2.5}$ emissions. In contrast, downed woody debris and ground fuels in the thin only unit pose a substantial emissions potential. The unmanaged forest (control) has high crown fire potential, modeled at 75% canopy consumption. Although it might seem counterintuitive that thin-only treatments pose comparable risk to high $PM_{2.5}$ emissions as unmanaged forests, Figure B4.1 demonstrates that the majority of fuels that contribute to wildfire emissions are surface fuels. If thinning results in a rearrangement of canopy fuels to surface fuels, it would be expected that high emissions could result from a summer wildfire.

The prescribed burn scenario shows a similar contrast between treatments. Maintenance burning is common in restored dry forests, and, as presented in this scenario, actual $PM_{2.5}$ emissions can be quite low once fuels have been reduced from the first prescribed burn. In contrast, the thin-only unit has substantial emissions potential in ground fuels, litter, and especially downed wood. Where possible, additional biomass removal including whole-tree yarding and biomass utilization can reduce the emissions potential of recently thinned sites. In this case, the unmanaged control has less emissions potential than the thinned unit due to the elevated downed wood in the thin-only unit.

Table B4.1 Input fuel loadings (Mg/Ha)

	Canopy	Shrub	Herb	Wood	LLM	Ground	Total
Control	120.94	1.49	0.22	41.62	3.18	17.75	188.59
Thin only	62.55	1.31	1.12	87.96	10.41	17.75	232.45
Thin Rx	62.55	0	1.12	10.94	0.48	2.11	128.55

Note: Canopy, shrub, herb, downed wood (wood), litter-lichen-moss (LLM), and ground fuels (ground) of the three comparison fuelbeds including an unmanaged control, thin only and thin and prescribed underburn (Thin Rx)

Table B4.2 Environmental variable inputs to consume for summer wildfire and prescribed burn scenarios

	1,000 h FM (%)	Duff FM (%)	Litter FM (%)	Shrub consumption (%)	Canopy consumption (%)
Summer wildfire					
Control	10	10	6	100	75
Thin and thin Rx	10	10	6	100	10
Prescribed burn					
Control	20	40	15	100	10
Thin and thin Rx	20	40	15	100	0

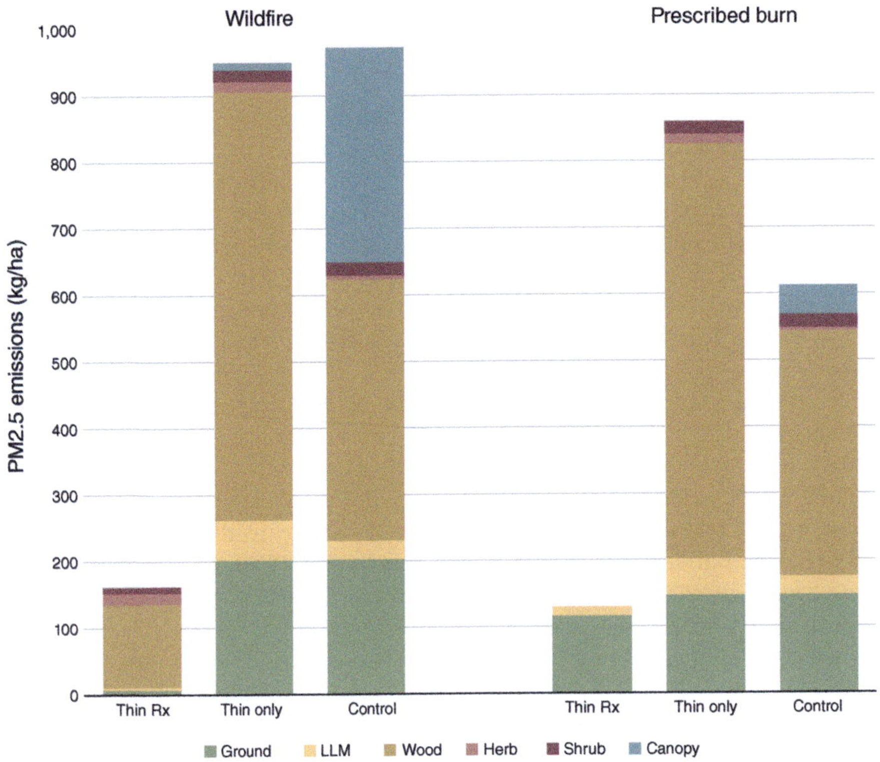

Figure B4.1 Comparison of predicted $PM_{2.5}$ emissions (kg/ha) under wildfire and prescribed fire scenarios for mixed conifer forests under three comparative management strategies including thinning and prescribed burning, thin only and prolonged fire exclusion (control).

than is provided by remotely sensed imagery and classified vegetation. For example, fuels that most often burn in a forest fire are obscured by forest canopies and are strongly dictated by past disturbances or management activities (Keane, 2015; Prichard et al., 2019). Passive remote sensing imagery may provide operational maps of forest vegetation type and cover but do not reflect the amount, structure, or condition of subcanopy fuels the drive fire behavior and consumption (Keane et al., 2001).

With increased availability of light detection and ranging (lidar) and structure-from-motion (SfM) photogrammetry, remotely sensed data sets are beginning to facilitate 3D fuel mapping (Loudermilk et al., 2009; Wang & Glenn, 2009; Hoff et al., 2019; Hudak et al., 2020). As such, methods for next-generation fuel characterization are being developed at scales and resolutions appropriate for physics-based computational fluid dynamics (CFD) models that are capable of resolving fire-atmosphere interactions, heat release, and smoke production (Linn et al., 2002; Mell et al., 2007; Loudermilk et al., 2009; Hudak et al., 2016a, 2020). However, even these new approaches to mapping fuels in 3D have limitations. In particular, distribution of downed logs and stumps may vary at fine spatial scales and be obscured from airborne remote sensing platforms (Brown, 1974; Keane, 2015), and there are currently no practical approaches for measuring forest floor layers (litter and duff depth, cover and bulk density) using remote sensing techniques. Reliable estimates of these fuel categories across burn units often are critical to anticipating long-term smoke impacts and residual smoldering following the flaming fire front.

This chapter introduces wildland fuels and approaches to mapping them at scales relevant to smoke management, smoke impacts to communities, and future research needs. Topics covered include (1) an introduction to wildland fuels, (2) traditional approaches to mapping wildland fuels, (3) next-generation approaches to wildland fuel mapping, and (4) research and development needs. Because source characterization of wildland fuels is critical to predicting smoke impacts, reviewing how to measure and map wildland fuel biomass and consumption provides important context for fire and fuels managers, smoke scientists, and policy makers. This chapter provides annotated content of a more detailed review recently published within the National Smoke Assessment with an emphasis on wildland fuel mapping applications relative to smoke modeling and forest health (Prichard et al., 2022).

4.2. WHAT ARE WILDLAND FUELS AND HOW DO THEY CONTRIBUTE TO SMOKE?

In wildland fire science and management, wildland fuels are often characterized in three vertical fuel layers: canopy, surface, and ground fuels (Keane, 2015; Gale et al., 2021). Canopy fuels are the biomass above the surface fuel layer (approximately >2 m). Surface fuels generally include biomass within 2 m above the ground surface. Ground fuels are all organic matter below the ground line, where the ground line is usually just below the litter (Oi soil horizon) and include the Oe and Oa soil horizons (collectively, "duff"), defined as partly to fully decomposed litter. Fuelbed layers are composed of finer-scale elements called *fuel strata* and *categories* (Fig. 4.2).

Fuel strata characterize the vertical profile of wildland fuels, and fuel categories are defined for specific applications, such as fire behavior prediction. For example, the downed wood stratum often contains fuel categories including fine wood (<8 cm diameter), coarse wood (>8 cm diameter), stumps, and piles (Riccardi et al., 2007). Fine wood is mostly consumed during the flaming phase of combustion (Ottmar, 2014) and is important for surface fire behavior prediction and flaming emissions, whereas coarse wood partially burns during the flaming phase of combustion and then often smolders long after the passage of the flaming front (Fig. 4.3), contributing to long-duration heat release and smoke (Albini, 1976; Hollis et al., 2010; Hyde et al., 2011).

Fuel strata and categories are also characterized by their physical and chemical properties, including bulk density, biomass (mass per area, kg m^{-2}), surface area (m^2), and heat content (J kg^{-1}), which are used as inputs to fire behavior and effects models (Keane, 2015). The finest scale of fuel description is the fuel particle, defined as a specific piece of fuel that is part of a fuel category (Gale et al., 2021). A particle can be an intact or fragmented woody stick, grass blade, shrub leaf, or pine needle. Fuel particles have a broad range of physical properties, including density (kg m^{-3}), volume (m^3), and shape. The properties of fuel categories, strata, and fuelbeds are often quantified from statistical summaries of properties of the particles that compose them. For example, quadratic mean diameter and mean surface area-to-volume ratio (cm^2 cm^{-3}) are often applied to size classes of wood particles for fire and smoke modeling (e.g., Brown, 1974).

Because live and dead fuels generally have very different fuel moistures and availability for burning, wildland fuels are also defined by their status (live/dead) and condition (fuel moisture). Dead fuel is suspended or

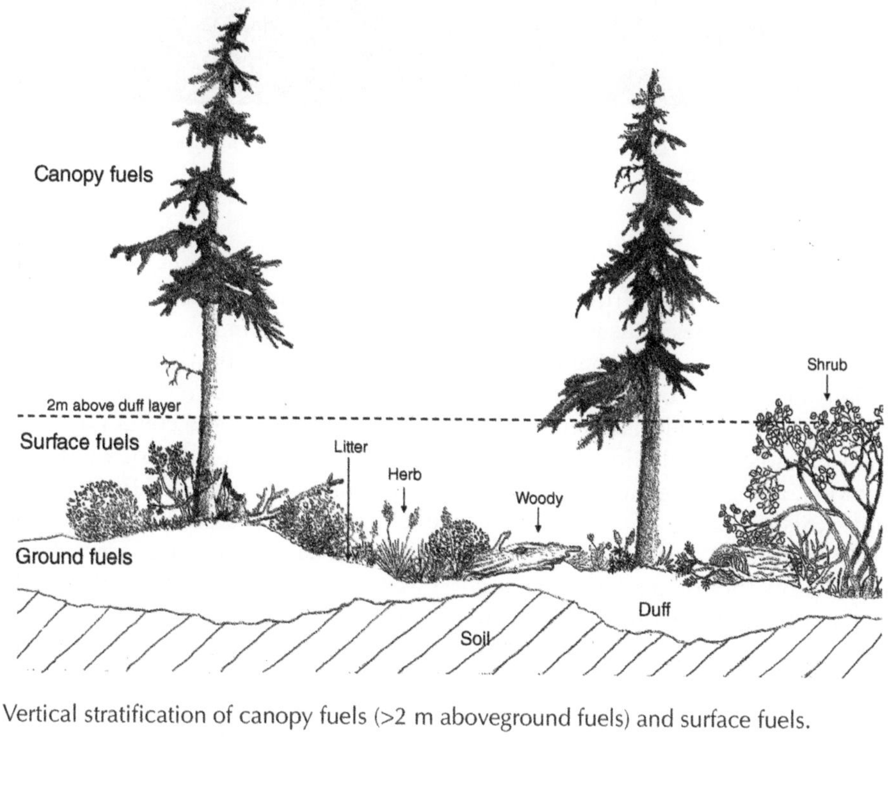

Figure 4.2 Vertical stratification of canopy fuels (>2 m aboveground fuels) and surface fuels.

(a) (b) (c)

Figure 4.3 Representative photos of (a) flaming, (b) short-term smoldering, and (c) long-term (residual) smoldering combustion (courtesy of Roger Ottmar).

downed dead biomass, and live fuel is the biomass of living organisms including vascular plants (trees, shrubs, herbs) and nonvascular plants (mosses, lichens, and liverworts). Live and dead fuel properties are both influenced by short-term variations in weather, but live fuel moistures are primarily controlled by chemical content (Fares et al., 2017) and plant physiology, including phenology, transpiration, evaporation, and soil moisture, which differ among taxa and across regional climate (Nelson, 2001; Jolly et al., 2014). In contrast, dead fuel moisture is dictated by the physical properties of the fuel (e.g., size, density, surface area) and their interaction with site climate, weather dynamics (wind, solar radiation, vapor pressure deficit), and available soil moisture (Fosberg et al., 1970; Viney, 1991; Banwell et al., 2013; Byram & Nelson, 2015). Live foliar moisture content influences when and where grasses, shrubs, and tree crowns are available to burn and can be highly variable across space and time. For example, green grass cover generally presents a barrier to fire spread whereas partially to fully cured grass facilitates rapid fire spread.

Because the type and condition of fuels vary over time, characterizing seasonality of wildland fuelbeds is critical for refining predictions of fire behavior and smoke productions. Differences in foliar moisture content between green and cured grasslands are well understood, but forests can also have marked differences in available fuels between growing and nongrowing seasons. For example, mixed pine and hardwood forests of southeastern North America are less available to burn during warm, humid, summer months with generally high monthly precipitation (Mitchell et al., 2009). However, dormant seasons are often marked by lower monthly precipitation with highly flammable litter fuelbeds that are generally composed of forest litter, grasses, and shrubs. Similarly, during the spring dip in boreal and subboreal landscapes, conifer tree species experience a pronounced decline in foliar moisture content prior to bud burst (Little, 1970; Chrosciewicz, 1986). During this short period of time, these forests can be particularly conducive to crown fires (Jolly et al., 2014).

The structure of wildland fuels is also an important characteristic that can strongly influence how fires spread and produce smoke. At fine spatial scales, the porosity of fuels influences how much available oxygen and air flow there is to support combustion (Fons, 1946). Complexes of dry fine wood, grass, and shrubs often burn efficiently and contribute most to flaming combustion. In contrast, duff is densely packed with low combustion efficiency and mostly contributes to smoldering combustion. Across coarser spatial scales, gaps in fuel structure influence fire spread, including whether a forest can support transitions from surface to crown fires (Weise et al., 2018) and how readily fires can spread from tree crown to tree crown (Parsons et al., 2017; Ziegler et al., 2017). For example, vertical gaps between understory fuels (shrubs, grasses, and litter) and canopy fuels can prevent surface fires from burning into forest canopies (Fig. 4.4). Similarly, the spatial continuity of surface fuels often strongly drives or constrains fire behavior (Atchley et al., 2021). Depending on fire intensity and wind speed, discontinuities in surface fuels such as patches of bare mineral soil or green vegetation with high foliar moisture content can create impediments to fire spread. Under mild weather and low wind speeds, surface fuel heterogeneity can strongly influence wildland fire behavior. However, vertical and horizontal gaps in fuel structure matter far less under more extreme fire weather and behavior. Desert ecosystems offer the most extreme example of fuel discontinuity; although desert vegetation is often dry enough to ignite a fire, fire spread is limited by the lack of continuity in sparse fuels (Gill & Allan, 2008; Swetnam et al., 2016).

Geospatial information on past fuel reduction treatments including thinning, prescribed burning, and past wildfires, is important for wildland fuel mapping because it can refine estimates of wildland fuel biomass and continuity. Fuel reduction treatments such as forest thinning and prescribed burning are generally designed to create gaps in fuel structure and, for a time, reduce the continuity and availability of fuels to burn in subsequent wildfire events (Kalies & Kent, 2016; Prichard et al., 2017). Following fuel reduction treatments or wildland fire events in fire-prone forests, understory fuels may either present barriers to fire spread through lack of continuity or support low intensity fire with concomitantly low emissions (Fulé et al., 2012; Omi, 2015). Specific information about the type of treatment and time since treatment are therefore critical to updating maps of wildland fuels and are used in wildland fuels mapping projects described in section 4.3.

Where organic soil layers including litter and ground fuels exist, they often comprise a substantial proportion of total available biomass and contribute disproportionately to smoke emissions including long-term smoldering events (Urbanski et al., 2014; Mickler et al., 2017). However, approaches to characterize peat land and forest floor layers have not advanced much in recent decades. To date, airborne remote sensing techniques are unable to

Figure 4.4 Conceptual diagram of mixed severity fire with a depiction of crown fire (middle) and surface fire (left and right) (courtesy of Bob Van Pelt).

penetrate organic soils and cannot resolve their density or depth. Instead, field observations in addition to models of organic soil accumulation (Zakem et al., 2021), decomposition, and changing moisture characteristics are needed to complement 3D fuel measurement techniques. Recent research on spatial distributions of ground fuel depth, biomass, and other characteristics in long-unburned forests of the southeastern United States emphasizes fine-scale spatial and temporal variability in ground fuels and the potential problems of sampling across forest stands or burn units (Kreye et al., 2014).

4.2.1. Temporal and Spatial Dynamics of Wildland Fuels

Much of the inherent complexity of wildland fuels is driven by how vegetation dynamics varies over space and time. Numerous ecological processes influence live and dead fuel dynamics, but the following four are particularly important in governing spatial and temporal distributions of wildland fuels (Keane, 2015).

1. Development. Wildland fuels accumulate from the establishment, growth, and mortality of vegetation (development). The rate of biomass accumulation, or productivity of vegetation, is dictated by interactions of the plant species available to occupy a site and the physical environment (climate, soils, and topography).
2. Deposition. Over time, portions of living biomass shed or die and are deposited on the ground to become dead surface fuels.
3. Decomposition. Belowground and aboveground dead organic matter is eventually decomposed by microbes and soil macrofauna.
4. Disturbance. Fire, insects, and disease and other disturbance agents act on living and dead biomass to change the magnitude, trend, and direction of fuel accumulation in space and time.

Wildland fuel properties and their distributions are a cumulative result of interactions of the above processes that create shifting mosaics of fuel conditions on fire-prone landscapes. The four processes listed above create such heterogeneous fuel distributions that any property of fuel often exhibits high variability. For example, the biomass per unit area, kg m^{-2} of fine woody debris, often termed *fine wood loading,* can vary by two to three times its mean over a small (<10 ha) prescribed burn unit (Keane et al., 2012a).

These four processes interact to influence fuel development, and the interactions depend on the ecosystem and corresponding climate and disturbance regimes. For example, live and dead vegetation characteristics often correlate to development and deposition, whereas climate drives decomposition and disturbance (Keane, 2008). Vegetation composition is sometimes used as a surrogate for fuels (Keane et al., 1998; Menakis et al., 2000), but often does not adequately represent the role of decomposition and disturbance on fuelbed development.

Understanding spatial and temporal variability of wildland fuels provides insight into how fuel consumption influences smoke emissions (Anderson, 1976) and how fuel management influences fuel properties. Processes that drive fuel dynamics also inform novel fire behaviors and effects (Parsons et al., 2010) and can contribute to new fuel classifications, sampling methods, and geospatial data sets (Keane, 2015). Spatial configuration of fuel characteristics is needed for next-generation fire effects and behavior models that rely on 3D fuel inputs and represent fire with CFD modeling (Linn et al., 2002; Mell et al., 2007; King et al., 2008; Parsons et al., 2010). It is this variability, combined with the inherent sampling error of various fuel sampling techniques, that makes estimating biomass for smoke prediction particularly challenging.

4.3. TRADITIONAL APPROACHES TO MAPPING WILDLAND FUELS FOR SMOKE MODELING AND EMISSIONS INVENTORIES

Traditional approaches to mapping wildland fuels fuel biomass (kg m^{-2}) and other characteristics (e.g., vegetation height and cover) are based on modeled relationships between classified vegetation and fuels (Keane, 2013). For example, LANDFIRE supports 30 m raster maps of wildland fuels and other data layers important to fire and smoke modeling. Wildland fuel characteristics are modeled from LANDSAT Thematic Mapper (TM) imagery, which is used to estimate vegetation cover and existing vegetation type (https://landfire.gov/evt.php), and airborne lidar data sets that inform vegetation height and cover metrics for 30 m pixels. Disturbances, including past fires, forest insects, windstorms, and forest management, are also used to update map layers and represent vegetation and fuel succession. Supported LANDFIRE fuel layers include surface fire behavior fuel models, canopy fuel, fuel loading models (Lutes et al., 2006), and fuelbeds for the Fuel Characteristics Classification System (Prichard et al., 2013).

Many biomass maps are also available at coarser spatial scales, based on remotely sensed optical, lidar, and radar data, and are used for global carbon modeling and emissions mapping. Examples include the U.S. Environmental Protection Agency's emission inventory (U.S. EPA, 2017), global carbon and biomass mapping (e.g., ESA Biomass Climate Change Initiative map by Santoro & Cartus, 2019), and wildland fuel maps based on major vegetation types (e.g., Pettinari & Chuvieco, 2016).

Although widely used, associating fuel characteristics with remotely sensed products has limitations due to imagery resolution and canopies that obscure surface and ground fuels (Keane et al., 2013). Because fuel categories vary independently of each other, high variability of fuel characteristics within a site may overwhelm unique fuelbed identification across sites (Prichard et al., 2019). A common approach to wildland fuel mapping is to assign fire behavior fuel models based on vegetation type and disturbance history for fire simulations (Burgan et al., 1998; Scott & Burgan, 2005). However, fuel models generally are too simplistic to represent the complexity of wildland fuels and do not support categories that are important to quantify for smoke modeling such as coarse wood and ground fuels (Sikkink & Keane, 2008; Keane, 2015).

Because mapping products are often unreliable at the scale of prescribed burning, field-based fuel characterization is required to further refine fuels for site-specific applications. Field surveys often include visual assessments, including use of photo load plots to assess live and dead surface fuels (Keane & Dickinson, 2007), photo series to visually estimate fuel loadings from similar sites (https://depts.washington.edu/nwfire/dps), and direct measurement (see Keane, 2015, for more detail on this topic). The most common direct measurement methods include planar intercept for sampling downed woody fuels and fixed-area plots for sampling the biomass per unit area of fuels (Brown, 1974; Catchpole & Wheeler 1992; Bate et al., 2004). Litter and ground fuel characterization generally involves destructive sampling and estimations based on depth measurements and has remained virtually unchanged over the last four decades (Brown et al., 1982; DeBano et al., 1998). For destructive sampling, plots are sampled to develop local bulk density values from dry-weight biomass measurements of a known volume (Keane, 2015). Some ecosystems may have patchy soil organic matter coverage (deserts, woodlands, sagebrush grasslands) making sampling difficult and often requiring a field measurement of ground fuel and litter cover. A common approach to estimate biomass (kg m^{-2}) of ground fuels is from measured depths of organic soil horizons (m), which are multiplied by bulk density (kg m^{-3}) to estimate biomass (kg m^{-2}; (Brown et al., 1982; Deluca & Boisvenue, 2012).

Until recently, a major gap in our understanding of wildland fuels has been a lack of spatial dimensionality in fuel characterization, which is necessary to reduce uncertainty and increase precision of inputs to fire behavior, fuel consumption, and smoke modeling. Due to the high spatial variability of wildland fuels and lack of correlation between fuel strata and categories, estimations based on traditional fuel measurement techniques often result in high variance and lack of precision (Brown et al., 1982). For example, planar intersect sampling of woody fuel biomass incorporates only one dimension of point intersections with a transect line (Brown, 1971). Given that fine and coarse wood can vary differently across space, linear sampling may not capture spatial variability of fine and coarse wood (Keane & Gray, 2013). Other conventional fuel inventory techniques, such as photo series, may also be inappropriate because fuels vary at spatial scales that might be different from the scales represented by the photo.

Whether collected in the field or through remote sensing, traditional wildland fuel mapping often provides a single estimate of biomass or depth for a given pixel. As such, the inherent uncertainty in wildland fuel estimation cannot be used to inform confidence intervals or probabilistic ranges of actual values. This is an important consideration for smoke modeling. With better knowledge of actual ranges in fuel distributions for mapped vegetation, we can provide better source characterization for wildland fire events and estimated uncertainty in smoke production. For example, in a recent study, Prichard et al. (2019) compiled biomass estimates by major fuel category based on plot data collected throughout the United States. They were able to estimate (based on the well-studied fuels, mostly forests) probability distributions of plot estimates. These values can be used to inform whether there are sufficient observations to fit distributions, and, if so, fitted distributions can be used to provide statistically derived estimates of means, ranges, and most likely values (e.g., 50th vs 95th quantile).

4.4. NEXT-GENERATION WILDLAND FUEL MAPPING

Advances in remote sensing techniques offer promising approaches to characterizing the structure and biomass of wildland fuels. These include airborne and ground-based lidar and structure-from-motion photogrammetry (SfM) (Loudermilk et al., 2009; Hudak et al., 2016a; Cooper et al., 2017). Airborne lidar scanning (ALS) is used operationally for precision forest inventory of tree stems and crowns and is useful for mapping vegetation height and cover and biomass (Hudak et al., 2008; García et al., 2011; Silva et al., 2016, 2017). However, its coarser resolution (3–5 m) limits its ability to adequately characterize understory and surface fuels, especially through an overstory forest canopy (Skowronski et al., 2011; Hudak et al., 2016a, b). Neither lidar nor other remote sensing systems can penetrate the forest floor to measure litter and ground fuel biomass, although recent work suggests that robust estimates of litter biomass may be possible in some forest types (Rowell et al., 2020).

Terrestrial laser scanning (TLS) is an emerging technology used to estimate the biomass and structure of surface, subcanopy, and canopy fuels (Loudermilk et al., 2009; Seielstad et al., 2011; Rowell et al., 2020). Mounted on a tripod or vehicle, TLS units obtain scan distances at sub-cm scales from the instrument location to vegetation, surface fuel, and other object surfaces and can penetrate through foliage layers. The lidar signal, which amounts to a 3D cloud of X, Y, and Z points, can then be related to biomass by constructing statistical models where destructively sampled bulk density for various categories is correlated to statistical metrics derived from the lidar point cloud data (Fig. 4.5). It can be difficult to differentiate between fuel categories using TLS in heterogeneous fuelbeds, and integration with multispectral imagery is sometimes necessary for image interpretation (e.g., Hudak et al., 2020).

Structure-from-motion (SfM) technology uses high-resolution images, often collected from cameras mounted on drones, to create 3D multispectral images of vegetation and fuels (Zarco-Tejada et al., 2014). Although photogrammetric point clouds have inferior vegetation penetration compared with lidar, the multispectral capabilities of digital cameras make assignment of plant functional type or live/dead status more feasible than from the single near-infrared or green channel data in most lidar sensors (Bright et al., 2016; Hudak et al., 2020). Integrating short-range SfM using digital cameras, mobile phones, or high definition (4K) digital video allows for fine-scale, 3D representations of wildland fuels in true color or multispectral images (Bright et al., 2016; Wallace et al., 2019). Once calibrated with field-based measurements (Fig. 4.6), these data sets can provide 3D mapping of live and dead canopy and surface fuel biomass and structure with applications for biomass mapping, fire behavior modeling, and fuel consumption measurements.

Because TLS and SfM images are sampled at high resolution, they can be merged into 3D point clouds for fine-scale mapping and quantification of live and dead surface and canopy fuels. Highly resolved spatial data from TLS and SfM expands sampling beyond the domains of traditional destructive plots and planar intersect fuel surveys. In particular, TLS excels at capturing detailed pre- and post-fire 3D data that represent continuous changes in estimates of bulk density at fine scales (Hudak et al., 2020). These technologies offer promising ways to characterize the vertical and horizontal distribution of wildland fuels with applications for fire behavior and smoke modeling (e.g., Rowell et al., 2016). High resolution 3D mapping is making pre- and post-burn fuel monitoring of spatially resolved fuel consumption possible. These advances have the potential to refine estimated fuel consumption and smoke emissions; they have implications for post-fire monitoring of vegetation response to fire.

Physics-based models of fire behavior and smoke production that rely on 3D fuel characterization are under evaluation for operational use (e.g., Linn et al., 2020). These rapid developments in fire behavior and smoke modeling (e.g., Atchley et al., 2021) are in close coordination with 3D fuel characterization and mapping approaches. When integrated with fire behavior monitoring (Moran et al., 2019), 3D consumption data can be used to evaluate fuel and fire interactions that produce smoke from a range of fire types and behavior.

Although recent developments in 3D fuel mapping are promising, there are still no established methods to scale 3D fuels data from fine-scale field measurements to the larger spatial scales (e.g., burn units or watersheds). Before such mapping applications can be developed, modelers need to identify how fuel metrics (e.g., biomass, bulk density, surface-area-to-volume ratio, and heat content) and characteristics (e.g., fuel type, status, and condition) can be assigned from sampled values to large spatial scales, across fire types (e.g., prescribed fire, wildfire, surface fire versus canopy fire) and across managed landscapes.

Figure 4.5 Example of a point cloud generated from terrestrial lidar scanning with denser fuels represented in blue and more porous fuels in red.

Figure 4.6 Example image of a forest understory generated from close-range structure-from-motion photogrammetry.

Appropriate scales for fuels mapping depend on the application and type of wildland fire event. For example, coarse-scale grid cells (e.g., 5 × 5 × 5 m) may be sufficient to represent crown fuels during extreme fire spread events, where fire weather and topography dominate fire behavior and smoke production patterns. In contrast, fine-scale fuel heterogeneity is often critical for accurate fire behavior predictions in a low-intensity surface fire such as a prescribed burn in light surface fuels. A forest that has been recently thinned and prescribed burned contains available fuels but in a structure that is unavailable to burn in most wildland fire events. However, column-driven fire spread combined with strong winds could exceed the threshold to burning for that site. Similarly, sites with high live fuel moisture in grass and shrub fuels may present barriers to fire spread under normal fire-weather conditions, but thresholds to burning can be exceeded by exceptional fire weather and spread events.

Mapping 3D fuels at the scale of wildland fire events, including wildfires and prescribed fires, presents several challenges to hierarchically scaled observations across spatial scales. Field and remote sensing measurements may be taken at similar scales but are inherently difficult to integrate due to intermixed fuels (e.g., complexes of live and dead shrubs, grasses, and leaf litter) and challenges in colocating and relating field and remote sensing measurements. However, the field of 3D fuel characterization is rapidly advancing with several promising approaches. Calibrated with field data sets, TLS and SfM photogrammetry point clouds offer a novel and scalable advancement in fuel characterization with highly resolved bulk density estimates for known volumes (Rowell et al., 2020). Robust coupling involves colocating techniques between individual 3D field plots and point-cloud data sets. For example, a recent approach to 3D field sampling (Hawley et al., 2018) was designed specifically to link 3D fuel types and fuel mass, collected within 10 × 10 × 10 cm cubes to the same resolution of volume TLS point clouds of vegetation structure. Additional work is needed to create machine-learning algorithms to classify 3D point-cloud data sets generated from TLS and/or SfM photogrammetry, then apply rule-based assignments of metrics such as bulk density, surface-area-to-volume ratios, and fuel moisture content to each classified object or volume for broader scale mapping applications.

4.5. RESEARCH NEEDS AND FUTURE DIRECTIONS

With rapid advancement in remote sensing and machine learning algorithms coupled with cloud computing, we anticipate that lidar and SfM photogrammetry will become readily accessible and provide highly resolved mapping of wildland fuels for fire behavior and smoke modeling. However, point-cloud imagery has its limitations, and future maps will likely involve an integration between remotely sensed and modeled fuel characterizations. In particular, research is needed to develop operational methods for coordinated image acquisition, interpretation, and mapping.

Given this reality, the largest gap in our knowledge of wildland fuels remains how to model fuel dynamics over space and time to create up-to-date and realistic characterizations for smoke modeling. Spatiotemporal interactions of living and dead vegetation are complex yet critical for providing temporally resolved estimates of fuel production and distribution. The 3D spatial complexity of fuels, and their dynamics over time, translates to similar complexity and variability in the availability of fuels to burn and their contribution to fire behavior and effects. However, the life cycle of fuels as it relates to vegetation dynamics and feedbacks with fire has not been fully defined. Furthermore, the temporal dynamics of fuels can be distinct between fine-scale changes in fine fuel moisture and the coarse-scale changes (e.g., vegetation structure, productivity, and climate).

Because long-duration smoke events create significant smoke impacts to communities, future research is needed to identify potential sources of smoldering consumption that can contribute to the severity and duration of smoke impacts to communities from prescribed burning. Organic soils can compose a substantial proportion of total site biomass and contribute disproportionately to smoke emissions including long-term smoldering events (Rappold et al., 2011; Urbanski, 2014). Because some fuels, including forest floor and peat-land soils, cannot be remotely sensed, future approaches to fuel characterization will involve a combination of traditional methods and new technologies that are currently under development. For example, in boreal ecosystems where the majority of biomass is stored in peat-land soils, Chasmer et al. (2017) showed that variation in forest floor depth could be quantified by differencing the ground-surface elevation models derived from separate pre- and post-fire lidar collections. However, in most fuelbeds, ground fuel layers are too shallow relative to the vertical precision of airborne lidar to detect changes in ground fuel depth due to consumption. Realistically, updating maps of wildland fuels will require linkages between remotely sensed data sets and ecological process models to account for the diversity of wildland fuels and challenges with remotely sensing forest floor layers. Ecological models that simulate development, deposition, decomposition, and disturbance can capture multiscaled fuel dynamics and translate them to fire behavior and smoke modeling at relevant spatial scales (Keane, 2015).

4.6. CONCLUSIONS

Wildland fuels are foundational to smoke prediction and generally contribute to high uncertainty in smoke production and dispersion modeling (Weise & Wright, 2014). Until recently, fuels and fuel consumption have been mapped using traditional methods to estimate the cover, height, and biomass (kg m^{-2}) of wildland fuels across dominant ecosystems of North America. Over the past decade, significant progress has been made in describing and quantifying fuels more accurately with 3D characterization and quantification across large spatial scales.

One of the biggest challenges in characterizing fuels is the high spatial and temporal variability that is present in wildland fuels in nearly all types of ecosystems. Most fuels managers do not have routine access to high-tech tools or high-resolution data (e.g., 3D characterization of fuels). Therefore, practical approaches are needed to improve field-based fuel characterization, fire behavior modeling, and consumption modeling, which will in turn inform the potential contribution of specific fuels (coarse wood, rotten stumps, basal accumulations, deep organic soil layers) to fire emissions and smoke dispersion.

Given the spatial and temporal complexity of wildland fuel dynamics, a better understanding is needed of the ecology of vegetation and fuels. In future decades, we anticipate that climate change will drive substantial changes in vegetation and fire dynamics, with concomitant changes in fuelbeds and their contribution to fuel consumption and emissions. Developing or revising ecological process models to ensure compatibility with next-generation fire behavior and smoke models will improve characterization of wildland fuel dynamics and both reduce and inform uncertainty in smoke predictions.

REFERENCES

Albini, F. A. (1976). *Estimating wildfire behavior and effects*. General Technical Report INT-GTR-30. Ogden: U.S. Forest Service, Intermountain Forest and Range Research Station.

Anderson, H. E. (1976). Fuels: The source of the matter. In *Air Quality and Smoke from Urban and Forest Fires: Proceedings*. Fort Collins: Colorado State University.

Atchley, A. L., Linn, R., Jonko, A., Hoffman, C., Hyman, J. D., Pimont, F., et al. (2021). Effects of fuel spatial distribution on wildland fire behavior. *International Journal of Wildland Fire*, *30*, 179–189. https://doi.org/10.1071/WF20096

Banwell, F. M., Varner, J. M., Knapp, E. E., & Van Kirk, R. W. (2013). Spatial, seasonal, and diel forest floor moisture dynamics in Jeffery pine-white fir forests of the Lake Tahoe Basin, USA. *Forest Ecology and Management*, *305*, 11–20. https://doi.org/10.1016/j.foreco.2013.05.005

Bate, L. J., Torgersen, T. R., Wisdom, M. J., & Garton, E. O. (2004). Performance of sampling methods to estimate log characteristics for wildlife. *Forest Ecology and Management*, *199*, 83–102. https://doi.org/10.1016/j.foreco.2004.04.021

Bright, B. C., Loudermilk, E. L., Pokswinski, S. M., Hudak, A. T., & O'Brien, J. J. (2016). Introducing close-range photogrammetry for characterizing forest understory plant diversity and surface fuel structure at fine scales. *Canadian Journal of Remote Sensing*, *42*, 460–472. https://doi.org/10.1080/07038992.2016.1229598

Brown, J. K. (1971). A planar intersect method for sampling fuel volume and surface area. *Forest Science*, *17*, 96–102. https://doi.org/10.1093/forestscience/17.1.96

Brown, J. K. (1974). *Handbook for inventorying downed woody material*. General Technical Report INT-16. Ogden: U.S. Forest Service, Intermountain Forest Experiment Station.

Brown, J. K., Oberheu, R. D., & Johnston, C. M. (1982). *Handbook for inventorying surface fuels and biomass in the interior west*. GTR-INT-129. Ogden: USDA Forest Service, Intermountain Forest and Range Experiment Station. https://www.fs.usda.gov/treesearch/pubs/29401

Burgan, R. E., Klaver, R. W., & Klaver, J. M. (1998). Fuel models and fire potential from satellite and surface observations. *International Journal of Wildland Fire*, *8*(3), 159–170. https://doi.org/10.1071/WF9980159

Byram, G. M. (1959). Combustion of forest fuels. In K. P. Brown (Ed.), *Forest fire: Control and use*. New York: McGraw-Hill.

Byram, G. M., & Nelson, R. M., Jr. (2015). *An analysis of the drying process in forest fuel material*. E-General Technical Report SRS-200. Asheville, NC: USDA Forest Service, Southern Research Station. https://www.treesearch.fs.fed.us/pubs/47925

Catchpole, W. R., & Wheeler, C. J. (1992). Estimating plant biomass: A review of techniques. *Australian Journal of Ecology*, *17*, 121–131. https://doi.org/10.1111/j.1442-9993.1992.tb00790.x

Chasmer, L. E., Hopkinson, C. D., Petrone, R. M., & Sitar, M. (2017). Using multitemporal and multispectral airborne lidar to assess depth of peat loss and correspondence with a new active normalized burn ratio for wildfires. *Geophysical Research Letters*, *44*, 11,851–11,859. https://doi.org/10.1002/2017GL075488

Chrosciewicz, Z. (1986). Foliar moisture content variations in four coniferous tree species of central Alberta. *Canadian Journal of Forest Research*, *16*, 157–162. https://doi.org/10.1139/x86-029

Cook, G. D., Meyer, C. P., Muepu, M., & Liedloff, A. C. (2016). Dead organic matter and the dynamics of carbon and greenhouse gas emissions in frequently burnt savannas. *International Journal of Wildland Fire*, *25*, 1252–1263. https://doi.org/10.1071/WF15218

Cooper, S. D., Roy, D. P., Schaaf, C. B., & Paynter, I. (2017). Examination of the potential of terrestrial laser scanning and structure-from-motion photogrammetry for rapid nondestructive field measurement of grass biomass. *Remote Sensing*, *9*, 531. https://doi.org/10.3390/rs9060531

Countryman, C. M. (1972). *The fire environment concept*. Berkeley, CA: Pacific Southwest Forest and Range Experiment Station. https://www.frames.gov/documents/behaveplus/publications/Countryman_1972_TheFireEnvironmentConcept_ocr.pdf

Cruz, M. G., Gould, J. S., Hollis, J. J., & McCaw, W. L. (2018). A hierarchical classification of wildland fire fuels for

Australian vegetation types. *Fire 1*(13), 1–39. https://doi.org/10.3390/fire1010013

DeBano, L. F., Neary, D. G., & Folliott, P. F. (1998). *Fire's effect on ecosystems*. New York: John Wiley and Sons.

De Groot, W. J., Landry, R., Kurz, W. A., et al., (2007). Estimating direct carbon emissions from Canadian wildland fires. *International Journal of Wildland Fire*, *16*, 593–606. https://doi.org/10.1071/WF06150\

Deluca, T. H., & Boisvenue, C. (2012). Boreal forest soil carbon: Distribution, function and modeling. *Forestry: An international journal of forest research*, *85*(2), 161–184. https://doi.org/10.1093/forestry/cps003

Drury, S. A., Larkin, N. S., Strand, T. T., Huang, S., Strenfel, S. J., Banwell, E. M., et al. (2014). Intercomparison of fire size, fuel loading, fuel consumption, and smoke emissions estimates on the 2006 Tripod Fire, Washington, USA. *Fire Ecology 10*, 56–83. https://doi.org/10.4996/fireecology.1001056

Fares, S., Bajocco, S., Salvati, L., Camarretta, N., Dupuy, J. -L., Xanthopoulos, G., et al. (2017). Characterizing potential wildland fire fuel in live vegetation in the Mediterranean region. *Annals of Forest Science*, *74*(1). https://doi.org/10.1007/s13595-016-0599-5

Fons, W. L. (1946). Analysis of fire spread in light forest fuels. *Journal of Agricultural Research 72*(3), 92–121. https://www.fs.fed.us/psw/publications/fons/psw_1946_fons001.pdf

Fosberg, M. A., Lancaster, J. W., & Schroeder, M. J. (1970). Fuel moisture response-drying relationships under standard and field conditions. *Forest Science*, *16*, 121–128. https://doi.org/10.1093/forestscience/16.1.121

Fulé, P. Z., Crouse, J. E., Roccaforte, J. P., & Kalies, E. L. (2012). Do thinning and/or burning treatments in western USA ponderosa or Jeffrey pine dominated forests help restore natural fire behavior? *Forest Ecology and Management*, *269*, 68–81. https://doi.org/10.1016/j.foreco.2011.12.025

Gale, M. G., Cary, G. J., Van Dijk, A. I. J. M., & Yebra, M. (2021). Forest fire fuel through the lens of remote sensing: Review of approaches, challenges and future directions in the remote sensing of biotic determinants of fire behavior. *Remote Sensing of Environment*, *255*, 112282. https://doi.org/10.1016/j.rse.2020.112282

García, M., Danson, F. M., Riano, D., Chuvieco, E., Ramirez, F. A., & Bandugula, V. (2011). Terrestrial laser scanning to estimate plot-level forest canopy fuel properties. *International Journal of Applied Earth Observation and Geoinformation*, *13*, 636–645. https://doi.org/10.1016/j.jag.2011.03.006

Gill, A. M., & Allan, G. (2008). Large fires, fire effects and the fire-regime concept. *International Journal of Wildland Fire*, *17*, 688–695. https://doi.org/10.1071/WF07145

Hagmann, R. K., Hessburg, P. F., Prichard, S. J., Povak, N. A., Brown, P. M., Fulé, P. Z., et al. (2021). Evidence for widespread changes in the structure, composition, and fire regimes of western North American forests. Invited feature. *Ecological Applications*, *e02431*. https://doi.org/10.1002/eap.2431

Haikerwal, A., Reisen, F., Sim, M. R., Abramson, M. J., Meyer, C. P., Johnston, F. H., et al. (2015). Impact of smoke from prescribed burning: Is it a public health concern? *Journal of the Air and Waste Management Association*, *65*(5), 592–598. https://doi.org/10.1080/10962247.2015.1032445

Hawley, C. M., Loudermilk, E. L., Rowell, E. M., & Pokswinski, S. (2018). A novel approach to fuel biomass sampling for 3D fuel characterization. *MethodsX*, *5*, 1597–1604. https://doi.org/10.1016/j.mex.2018.11.006

Hessburg, P. F., Miller, C. L., Parks, S. A., Povak, N. A., Taylor, A. H., Higuera, P. E., et al. (2019). Climate, environment, and disturbance history govern resilience of western North American forests. *Frontiers in Ecology and Evolution*, *7*, 239. https://doi.org/10.3389/fevo.2019.00239

Hoff, V., Rowell, E., Teske, C., Queen, L., & Wallace, T. (2019). Assessing the relationship between forest structure and fire severity on the North Rim of the Grand Canyon. *Fire*, *2*, 10. https://doi.org/10.3390/fire2010010

Hollis, J., Matthews, S., Ottmar, R. D., & Prichard, S. J. (2010). Testing woody fuel consumption models for application in Australian southern eucalypt forest fires. *Forest Ecology and Management 260*, 6. https://doi.org/10.1016/j.foreco.2010.06.007

Holm, S. M., Miller, M. D., & Balmes, J. R. (2021). Health effects of wildfire smoke in children and public health tools: a narrative review. *Journal of Exposure Science and Environmental Epidemiology*, *31*, 1–20. https://doi.org/10.1038/s41370-020-00267-4

Hudak, A. T., Bright, B. C., Pokswindki, S., Loudermilk, E. L., O'Brien, J. J., Hornsby, B. J., et al. (2016a). Mapping forest structure and composition from low-density LiDAR for informed forest, fuel, and fire management at Eglin Air Force Base, Florida, USA. *Canadian Journal of Remote Sensing*, *42*, 411–427. https://doi.org/10.1080/07038992.2016.1217482

Hudak, A. T., Crookston, N. L., Evans, J. S., Hall, D. E., & Falkowski, M. J. (2008). Nearest neighbor imputation of species-level, plot-scale forest structure attributes from LiDAR data. *Remote Sensing of Environment*, *112*, 2232–2245. https://doi.org/10.1016/j.rse.2007.10.009

Hudak, A. T., Dickinson, M. B., Bright, B. C., Kremens, R. L, Loudermilk, E. L., O'Brien, J. J., et al. (2016b). Measurements relating fire radiative energy density and surface fuel consumption: RxCADRE 2011 and 2012. *International Journal of Wildland Fire*, *25*, 25–37. https://doi.org/10.1071/WF14159

Hudak. A. T., Kato, A., Bright, B. C., Loudermilk, E. L., Hawley, C., Restaino, J. C., et al. (2020). Toward spatially explicit quantification of pre- and post-fire fuels and fuel consumption from traditional and point cloud measurements. *Forest Science*, *fxz085*. https://doi.org/10.1093/forsci/fxz085

Hyde, J. C., Smith, A. M. S., Ottmar, R. D., Alvarado, E. C., & Morgan, P. (2011). The combustion of sound and rotten coarse woody debris: A review. *International Journal of Wildland Fire*, *20*, 163–174. https://doi.org/10.1071IWF09113

Jaffe, D. A., O'Neill, S. M., Larkin, N. K., Holder, A. L., Peterson, D. L., Halofsky, J. E., et al. (2020). Wildfire and prescribed burning impacts on air quality in the United States. *Journal of the Air and Waste Management Association*, *70*(6), 583–615. https://doi.org/10.1080/10962247.2020.1749731

Jolly, W. M., Hadlow, A. M., & Huguet, K. (2014). De-coupling seasonal changes in water content and dry matter to predict live conifer foliar moisture content. *International Journal of Wildland Fire*, *23*, 480–489. https://doi.org/10.1071/WF13127

Kalies, E. L., & Yocom Kent, L. (2016). Tamm review: Are fuel treatments effective at achieving ecological and social objectives? A systematic review. *Forest Ecology and Management 375*, 84–95.

Keane, R., Gray, K., Bacciu, V., & Leirfallom, S. (2012a). Spatial scaling of wildland fuels for six forest and rangeland ecosystems of the northern Rocky Mountains, USA. *Landscape Ecology*, *27*, 1213–1234. https://doi.org/10.1007/s10980-012-9773-9

Keane, R. E. (2008). Biophysical controls on surface fuel litterfall and decomposition in the northern Rocky Mountains, USA. *Canadian Journal of Forest Research*, *38*, 1431–1445.

Keane, R. E. (2013). Describing wildland surface fuel loading for fire management: A review of approaches, methods and systems. *International Journal of Wildland Fire*, *22*, 51–62. https://doi.org/10.1071/WF11139

Keane, R. E. (2015). *Wildland fuel fundamentals and applications*. New York: Springer.

Keane, R. E., & Dickinson, L. J. (2007). *The photoload sampling technique: Estimating surface fuel loadings using downward looking photographs*. General Technical Report RMRS-190. Fort Collins: U.S. Forest Service, Rocky Mountain Research Station.

Keane, R. E., & Gray, K. (2013). Comparing three sampling techniques for estimating fine woody down dead biomass. *International Journal of Wildland Fire*, *22*, 1093–1107. https://doi.org/10.1071/WF13038

Keane, R. E., Burgan, R. E., & Wagtendonk, J. V. (2001). Mapping wildland fuels for fire management across multiple scales: Integrating remote sensing, GIS, and biophysical modeling. *International Journal of Wildland Fire*, *10*, 301–319. https://doi.org/10.1071/WF01028

Keane, R. E., Garner, J. L., Schmidt, K. M., Long, D. G., Menakis, J. P., & Finney, M. A. (1998) *Development of input spatial data layers for the FARSITE fire growth model for the Selway-Bitterroot Wilderness Complex, USA*. General Technical Report RMRS-3. Ogden: U.S. Forest Service, Rocky Mountain Research Station.

Keane, R. E., Herynk, J. M., Toney, C., Urbanski, S. P., Lutes, D. C., & Ottmar, R. D. (2013). Evaluating the performance and mapping of three fuel classification systems using Forest Inventory and Analysis surface fuel measurements. *Forest Ecology and Management*, *305*, 248–263. https://doi.org/10.1016/j.foreco.2013.06.001

King, K. J., Bradstock, R. A., Cary, G. J., Chapman, J., & Marsden-Smedley, J. B. (2008). The relative importance of fine-scale fuel mosaics on reducing fire risk in south-west Tasmania, Australia. *International Journal of Wildland Fire*, *17*, 421–430. https://doi.org/10.1071/WF07052

Kreye, J. K., Varner, J. M., & Dugaw, C. J. (2014). Spatial and temporal variability of forest floor duff characteristics in long-unburned *Pinus palustris* forests. *Canadian Journal of Forest Research*, *44*, 1477–1486. https://doi.org/10.1139/cjfr-2014-0223

Larkin, N. K., O'Neill, S. M., Solomon, R., Raffuse, S., Strand, T., Sullivan, D. C., et al. (2010). The BlueSky smoke modeling framework. *International Journal of Wildland Fire*, *18*, 906–920. https://doi.org/10.1071/WF07086

Linn, R., Reisner, J., Colman, J. J., & Winterkamp, J. (2002). Studying wildfire behavior using FIRETEC. *International Journal of Wildland Fire*, *11*, 233–246. https://doi.org/10.1071/WF02007

Linn, R. R., Goodrick, S., Brambilla, S., Brown, M. J., Middleton, R. S., O'Brien, J. J., et al. (2020). QUIC-fire: A fast-running simulation tool for prescribed fire planning. *Environmental Modelling and Software*, *125*, 104616. https://doi.org/10.1016/j.envsoft.2019.104616

Little, C. H. A. (1970). Seasonal changes in carbohydrate and moisture content in needles of balsam fire (*Abies balsamea*). *Canadian Journal of Botany*, *48*, 2021–2028. https://doi.org/10.1139/b70-295

Liu, X., Huey, L. G., Yokelson, R. J., et al. (2017). Airborne measurements of western U.S. wildfire emissions: comparison with prescribed burning and air quality implications. *JGR Atmospheres*, *122*(11), 6108–6129. https://doi.org/10.1002/2016JD026315

Liu, Y., Kochanski, A., Baker, K. R., Mell, W., et al. (2019). Fire behavior and smoke modeling: Model improvement and measurement needs for next-generation smoke research and forecasting systems. *International Journal of Wildland Fire*, *28*, 570–588. https://doi.org/10.1071/WF18204

Loudermilk, E. L., Hiers, J. K., O'Brien, J. J., Mitchell, R. J., Singhania, J. C., & Fernandez, W. P. (2009). Ground-based LIDAR: A novel approach to quantify fine-scale fuelbed characteristics. *International Journal of Wildland Fire*, *18*, 676–685. https://doi.org/10.1071/WF07138

Lutes, D. C., Keane, R. E., & Caratti, J. F. (2006). A surface fuel classification for estimating fire effects. *International Journal of Wildland Fire*, *18*, 802–814. https://doi.org/10.1071/WF08062

Mell, W., Jenkins, M. A., Gould, J., & Cheney, P. (2007). A physics-based approach to modeling grassland fires. *International Journal of Wildland Fire 16*, 1–22. https://doi.org/10.1071/WF06002

Menakis, J. P., Keane, R. E., & Long, D. G. (2000). Mapping ecological attributes using an integrated vegetation classification system approach. *Journal of Sustainable Forestry*, *11*, 245–265. https://doi.org/10.1300/J091v11n01_10

Mickler, R. A., Welch, D. P., & Bailey, A. D. (2017). Carbon emissions during wildland fire on a North American temperate peat land. *Fire Ecology*, *13*, 34–57. https://doi.org/10.4996/fireecology.1301034

Miller, C., O'Neill, S., Rorig, M., & Alvarado, E. (2019). Air-quality challenges of prescribed fire in the complex terrain and wildland urban interface surrounding Bend, *Oregon. Atmosphere*, *10*, 515. doi:10.3390/atmos10090515

Mitchell, R. J., Hiers, J. K., O'Brien, J., & Starr, G. (2009). Ecological forestry in the Southeast: understanding the ecology of fuels. *Journal of Forestry*, *107*, 391–397. doi: 10.1093/JOF/107.8.391

Moran, C. J., Seielstad, C. A., Cunningham, M. R., Hoff, V., Parsons, R. A., Queen, L., et al. (2019). Deriving fire behavior metrics from UAS imagery. *Fire*, *2*, 36. https://doi.org/10.3390/fire2020036

Nelson, R. M., Jr. (2001). Water relations of forest fuels. In E.A. Johnson & K. Miyanishi (Eds.), *Forest fires: Behavior and ecological effects* (pp. 79–149). San Diego, CA: Academic Press. https://doi.org/10.1016/B978-012386660-8/50006-4

Omi, P. N. (2015). Theory and practice of wildland fuels management. *Current Forestry Reports 1*, 100–117. https://doi.org/10.1007/s40725-015-0013-9

Ottmar, R. D. (2014). Wildland fire emissions, carbon, and climate: modeling fuel consumption. *Forest Ecology and Management*, *317*, 41–50. https://doi.org/10.1016/j.foreco.2013.06.010

Ottmar, R. D., Isabel, M. A., & Sandberg, D. V. (2009). Characterizing sources of emissions from wildland fires. In A. Bytnerowicz, et al. (Eds.), *Wildland fires and air pollution*. Amsterdam, The Netherlands: Developments in Environmental Science 8.

Ottmar, R. D., Sandberg, D. V., Riccardi, C. L., & Prichard, S. J. (2007). An overview of the fuel characteristic classification system: Quantifying, classifying, and creating fuelbeds for resource planning. *Canadian Journal of Forest Research*, *37*, 2383–2393. https://doi.org/10.1139/X07-077

Parsons, R. A., Linn, R. R., Pimont, F., Hoffman, C., Sauer, J., Winterkamp, J., et al. (2017). Numerical investigation of aggregated fuel spatial pattern impacts on fire behavior. *Land*, *6*, 43. https://doi.org/10.3390/land6020043

Parsons, R. A., Mell, W. E., & McCauley, P. (2010). Linking 3D spatial models of fuels and fire: effects of spatial heterogeneity on fire behavior. *Ecological Modeling*, *222*, 679–691. https://doi.org/10.1016/j.ecolmodel.2010.10.023

Peterson, J. L. (1987). *Analysis and reduction of the errors of predicting prescribed burn emissions*. PhD thesis. Seattle: University of Washington.

Pettinari, M. L., & Chuvieco, E. (2016). Generation of a global fuel data set using the Fuel Characteristic Classification System. *Biogeosciences*, *13*, 2061–2076. https://doi.org/10.5194/bg-13-2061-2016

Prichard, S., Ottmar, R., & Anderson, G. (2007). CONSUME user's guide and scientific documentation. https://www.fs.fed.us/pnw/fera/research/smoke/consume/consume30_users_guide.pdf

Prichard, S. J., Keane, R., Rowell, E., Hudak, A., Loudermilk, E. A., Lutes, D., et al. (2022). *U.S. Forest Service national smoke assessment: Fuels and consumption*. Springer.

Prichard, S. J., Kennedy, M. C., Andreu, A. G., Eagle, P. C., French, N. H., & Billmire, M. (2019). Next-generation biomass mapping for regional emissions and carbon inventories: incorporating uncertainty in wildland fuel characterization. *Journal of Geophysical Research: Biogeosciences*, *124*, 3699–3716. https://doi.org/10.1029/2019JG005083

Prichard, S. J., O'Neill, S. M., Eagle, P., Andreu, A. G., Drye, B., Dubowy, J., et al. (2020). Wildland fire emission factors in North America: Synthesis of existing data, measurement needs and management applications. *International Journal of Wildland Fire*, *29*(2), 132–147. https://doi.org/10.1071/WF19066

Prichard, S. J., Sandberg, D. V., Ottmar, R. D., Eberhardt, E., Andreu, A., Eagle, P., et al. (2013). *Fuel Characteristic Classification System Version 3.0: Technical documentation*. PNW-GTR-887. U.S. Forest Service Pacific Northwest Research Station, Portland, OR.

Prichard, S. J., Stevens-Rumann, C. S., & Hessburg, P. F. (2017). Tamm Review. Shifting global fire regimes: Lessons from reburns and research needs. *Forest Ecology and Management*, *396*, 217–233. http://dx.doi.org/10.1016/j.foreco.2017.03.035

Rappold, A. G., Stone, S. L., Cascio, W. E., Neas, L. M., et al., (2011). Peat bog wildfire smoke exposure in rural North Carolina is associated with cardiopulmonary emergency department visits assessed through syndromic surveillance. *Environmental Health Perspectives*, *119*, 1415–1420.

Riccardi, C. L., Prichard, S. J., Sandberg, D. V., & Ottmar, R. D. (2007). Quantifying physical characteristics of wildland fuels using the Fuel Characteristic Classification System. *Canadian Journal of Forest Research*, *37*, 2413–2420.

Rollins, M. G. (2009). LANDFIRE: A nationally consistent vegetation, wildland fire, and fuel assessment. *International Journal of Wildland Fire*, *18*, 235–249.

Rowell, E., Loudermilk, E. L., Hawley, C., Pokswinski, S., Seielstad, C., Queen, L., et al. (2020). Coupling terrestrial laser scanning with 3D fuel biomass sampling for advancing wildland fuels characterization. *Forest Ecology and Management*, *462*, 117945. https://doi.org/10.1016/j.foreco.2020.117945

Rowell, E., Loudermilk, E. L., Seielstad, C., & O'Brien, J. J. (2016). Using simulated 3D surface fuelbeds and terrestrial laser scan data to develop inputs to fire behavior models. *Canadian Journal of Remote Sensing*, *42*, 443–459. https://doi.org/10.1080/07038992.2016.1220827

Santoro, M., & Cartus, O. (2019). *ESA Biomass Climate Change Initiative (Biomass_cci): Global data sets of forest above-ground biomass for the year 2017, vol.1*. Centre for Environmental Data Analysis. https://doi.org/10.5285/bedc59f37c9545c981a839eb552e4084

Scott, J. H., & Burgan, R. E. (2005). *Standard fire behavior fuel models: A comprehensive set for use with Rothermel's surface fire spread model*. General Technical Report RMRS-153. Fort Collins, CO: Forest Service, Rocky Mountain Research Station.

Seielstad, C., Stonesifer, C., Rowell, E., & Queen, L. (2011). Deriving fuel mass by size class in Douglas-fir (*Pseudotsuga menziesii*) using terrestrial laser scanning. *Remote Sensing*, *3*, 1691–1709. https://doi.org/10.3390/rs3081691

Sikkink, P. G., & Keane, R. E. (2008). A comparison of five sampling techniques to estimate surface fuel loading in montane forests. *International Journal of Wildland Fire*, *17*, 363–379. https://doi.org/10.1071/WF07003

Silva, C. A., Hudak, A. T., Vierling, L. A., Klauberg, C., Garcia, M., Ferraz, A., et al. (2017). Impacts of airborne lidar pulse density on estimating biomass stocks and changes in a selectively logged tropical forest. *Remote Sensing*, *9*, 1068. https://doi.org/10.3390/rs9101068

Silva, C. A., Hudak, A. T., Vierling, L. A., Loudermilk, E. L., O'Brien, J. J., Hiers, J. K., et al. (2016). Imputation of individual longleaf pine (*Pinus palustris* Mill.) tree attributes from field and LiDAR data. *Canadian Journal of Remote Sensing*, *42*, 554–573. https://doi.org/10.1080/07038992.2016.1196582

Skowronski, N. S., Clark, K. L., Duveneck, M., & Hom, J. (2011). Three-dimensional canopy fuel loading predicted using upward and downward sensing LIDAR systems. *Remote Sensing of Environment*, *115*, 703–714. https://doi.org/10.1016/j.rse.2010.10.012

Swetnam, T. W., Farella, J., Roos, C. I., Liebmann, M. J., Falk, D. A., & Allen, C. D. (2016). Multiscale perspectives of fire, climate and humans in western North America and the Jemez Mountains, USA. *Philosophical Transactions of the Royal*

Society B: Biological Sciences, *371*, 20150168. https://doi.org/10.1098/rstb.2015.0168

Urbanski, S. (2014). Wildland fire emissions, carbon, and climate: emission factors. *Forest Ecology and Management*, *317*, 51–60. https://doi.org/10.1016/j.foreco.2013.05.045

U.S. EPA (2017). *Profile of version 1 of the 2014 National Emissions Inventory*. https://www.epa.gov/sites/production/files/2017-04/documents/2014neiv1_profile_final_april182017.pdf

Viney, N. R. (1991). A review of fine fuel moisture modeling. *International Journal of Wildland Fire*, *1*, 215–234. https://doi.org/10.1071/WF9910215

Wallace, L., Saldias, D. S., Reinke, K., Hillman, S., Hally, B., & Jones, S. (2019). Using orthoimages generated from oblique terrestrial photography to estimate and monitor vegetation cover. *Ecological Indicators*, *101*, 91–101.

Wang, C., & Glenn, N. F. (2009). Estimation of fire severity using pre- and post-fire LiDAR data in sagebrush steppe rangelands. *International Journal of Wildland Fire*, *18*, 848–856. https://doi.org/10.1071/WF08173

Weise, D. R., & Wright, C. S. (2014). Wildland fire emissions, carbon and climate: Characterizing wildland fuels. *Forest Ecology and Management*, *317*, 26–40.

Weise, D. R., Cobian–Iñiguez, J., & Princevac, M. (2018). Surface to crown transition. In L. Manzello (Ed.), *Encyclopedia of wildfires and wildland-urban interface (WUI) fires*. Cham: Springer International Publishing. https://doi.org/10.1007/978-3-319-51727-8_24-1

Williamson, G. J., Bowman, D. M. J. S., Price, O. F., Henderson, S. B., & Johnston, F. H. (2016). A transdisciplinary approach to understanding the health effects of wildfire and prescribed fire smoke regimes. *Environmental Research Letters*, *11*, 125009. https://doi.org/10.1088/1748-9326/11/12/125009

Zakem, E. J., Cael, B. B., & Levine, N. M. (2021). A unified theory for organic matter accumulation. *PNAS*, *118*,(6), e2016896118. https://doi.org/10.1073/pnas.2016896118

Zarco-Tejada, P. J., Diaz-Varela, R., Angileri, V., & Loudjani, P. (2014). Tree height quantification using very high resolution imagery acquired from an unmanned aerial vehicle (UAV) and automatic 3D photo-reconstruction methods. *European Journal of Agronomy*, *55*, 89–99. https://doi.org/10.1016/j.eja.2014.01.004

Ziegler, J. P., Hoffman, C., Battaglia, M., & Mell, W. (2017). Spatially explicit measurements of forest structure and fire behavior following restoration treatments in dry forests. *Forest Ecology and Management*, *386*, 1–12. https://doi.org/10.1016/j.foreco.2016.12.002

5

Biomass Burning Fuel Consumption and Emissions for Air Quality

Nancy H. F. French[1] and Andrew T. Hudak[2]

ABSTRACT

Biomass burning emissions estimates of gas and aerosol air pollutants are used in modeling and forecasting smoke and air quality. These emissions originate from combustion of the fuels found in the landscape and vary based on the terrain, weather, fuel type, and condition of the fuelbed, including the arrangement of fuels, during the fire event. Combustion describes the flaming and smoldering burning phases of a fire that, summed over the duration of the event, equals the fuel consumed. This chapter provides a review of approaches to quantify fuel consumption from biomass burning relevant for air quality assessment. The indirect approach estimates consumption based on the difference between pre- and post-fire fuel loads within the burn area or models it by applying a burning efficiency to the pre-fire fuel load. The direct approach uses a combustion factor to translate fire radiative power (FRP) measured from thermal infrared instruments into the consumption rate; temporally integrating FRP over the duration of the fire provides fire radiative energy (FRE), which the same combustion factor translates to total consumption. Methods to map the location, strength, and composition of fire emissions, which are a function of fuel characteristics and fire environment, are provided. This chapter provides a discussion on the uncertainty of biomass burning emissions inventory products and considers new approaches for improving fuel consumption mapping.

5.1. INTRODUCTION

Emissions from biomass burning are the direct product of the consumption of fuels during fire, made up of the live and dead vegetation found in the landscape (Ottmar, 2014). The process is fuels driven and mediated by weather and topography to create spatial-temporal complexity in how and where fuels burn and the resulting emissions. The combustion process produces energy (heat and light), gases, and solids of various sizes that are either left as residue at the site (ash and char) or lofted into the air as particulate matter (PM), an important component of smoke. A variety of different gaseous and particulate combustion products are emitted into the atmosphere depending on the fuel type and conditions during burning, including the fuel moisture, arrangement of fuels, terrain, and fire weather, which influence the intensity and efficiency of burning (Prichard et al., 2020).

Global carbon emissions from biomass burning have been estimated to average 2,200 TgC yr^{-1} from 1997 to 2016, and extend as high as 3,000 TgC in 1 yr (1997) (van der Werf et al., 2017), with high interannual variation. Additionally, differences are found between studies, depending on the approach (Hoelzemann et al., 2004; Ito & Penner, 2004; van der Werf et al., 2006). In years since

[1] *Michigan Tech Research Institute, Michigan Technological University, Ann Arbor, Michigan, USA*

[2] *Forestry Sciences Laboratory, Rocky Mountain Research Station, United States Forest Service, Moscow, Idaho, USA*

Landscape Fire, Smoke, and Health: Linking Biomass Burning Emissions to Human Well-Being, Geophysical Monograph 280, First Edition.
Edited by Tatiana V. Loboda, Nancy H. F. French, and Robin C. Puett.

DOI: 10.1002/9781119757030.ch05

these studies, several regions have experienced unprecedented levels of fire, so these data from the late 1990s and early 2000s should be considered an illustration of the complexity of the question "How much pollution does biomass burning contribute to the atmosphere?" Improving estimates of the magnitude and composition of biomass burning emissions has gained attention since 2000, pointing to the growing concern to better understand biomass burning emissions for both climate and air quality (Larkin et al., 2020).

An emissions inventory is a periodically generated accounting (e.g., hourly, daily, annual) of the amount of pollutants released into the atmosphere from a source. Emissions inventories can be used as input to atmospheric models to map the concentrations of smoke in the atmosphere (see Chapters 7–9), or by land and air quality managers to assess expected emissions impacts on air quality, visibility, and health, or in carbon cycle modeling and management (e.g., Gurney et al., 2020; Pouliot et al., 2008). Within the context of this book, the focus of this chapter is on emissions of pollutants from biomass burning relevant to air quality and human health; however, biomass burning emissions inventories are vital for calculating the impact of fire on atmospheric carbon and climate, but are not covered in this chapter.

The general approach to an emissions inventory for any pollution source is a product of two factors: activity and emission factor (EF) (IPCC, 2007). Activity is defined as the amount of fuel consumption, and the EF converts consumption into emissions. Mass of emissions of a given gas or PM category, M_x (g), is determined with:

$$M_x = M_c \times EF_x, \tag{5.1}$$

where M_c is the mass of fuel consumed (kg) and EF_x (g kg^{-1}) is the emissions factor in mass of gas or PM (x) emitted per mass of fuel consumed (IPCC, 2007; Urbanski, 2014).

Emissions from fuels that burn under controlled conditions, such as gasoline in an internal combustion engine, are well characterized via controlled laboratory experiments. In this case, fuels are standardized and EFs are consistent for a given fuel type, making fuel consumption (activity, in the general context of emissions inventory) the primary variable responsible for determining source emissions strength (the amount of emissions from a source or event). EFs for biomass burning are more varied and complex due to the variety of fuel and combustion conditions possible in open burning. They have been determined through both controlled combustion laboratory measurements and field-based measurements from hundreds of fuel types and various burning conditions. Compilation of EFs for biomass fuels by Prichard et al. (2020) includes more than 12,000 records on 267 air pollutants defined for various fuel types, burn types, locations, and combustion phase (flaming or smoldering combustion).

In the case of biomass burning, in general, the amount of area burned dictates the amount of fuel consumed and is the primary driver of biomass burning emissions magnitude; more burned area means more emissions, with some important qualifiers. Specific attention to the fuel type, density, structure, size, and condition, as well as combustion phase provides more refined estimates of emissions from landscape fire than can be achieved by tallying the area burned alone (French et al., 2002; Prichard et al., 2020). This is because emission contributions greatly differ between different vegetation types (e.g., forest, shrublands, or grasslands) and fuel components, which include aboveground components of tree foliage, fine branches, and shrubs; fine woody fuels (live or dead) and coarse woody debris (decay status); and surface organic material (litter and duff) that compose the fuelbed (see Chapter 4). The relative contributions of these different materials vary greatly within and between wildland fuelbed types, introducing great complexity that drives both fire behavior and fire emissions.

At the same time, the fire environment introduces complexity. Variability in space and time in solar insolation, precipitation, humidity, and other weather variables in the weeks to days prior to the event will influence fuel moisture. Fuel moisture can influence combustion type, namely the flaming and smoldering phases of a fire, which dictates the type of emissions based on combustion efficiency. Combustion efficiency is a measure of how much energy results from a given amount of fuel, with higher combustion efficiency resulting in more complete burning of fuels than low-efficiency combustion. For biomass burning, combustion efficiency is defined as the amount of excess in CO_2 in comparison to total emitted carbon (Yokelson et al., 1999). The fire environment before and during the event, therefore, is an important factor in determining biomass burning emissions across space and time.

Biomass burning emissions inventory and mapping is required as a primary input to smoke concentration mapping, numerical smoke prediction, coupled fire weather-smoke models, and air quality models, as described in Chapters 7, 8, and 9. In this chapter, we review methods to model biomass burning emissions by explaining approaches used to determine fuel consumption and the uncertainties associated with consumption estimates. In section 5.2, we present a general overview of emission factors used in biomass burning emissions inventories, as well as the concept of combustion efficiency, both of which are more completely reviewed in the literature (Prichard et al., 2020; Urbanski, 2014; Urbanski et al., 2021). Approaches to map the location, strength, and composition of fire emissions, considering needs for both retrospective and real-time smoke modeling and forecasting, are considered in section 5.3. Retrospective mapping of emissions has value for greenhouse-gas and aerosol emissions assessments for carbon accounting and climate change modeling as well as for comparing to health outcomes from smoke pollution exposure, which is reviewed later in this book (Chapters 12 and 13). Real-time emissions inventories are used in forecasting potential smoke concentrations downwind from an event (Chapters 7, 8, and 9) or for taking action to avoid exposure (Chapter 10). Finally, in section 5.4, new approaches under development using remote sensing observations for improving estimates of fuel consumption and biomass burning emissions are discussed.

5.2. MODELING BIOMASS BURNING EMISSIONS

Emissions from a fire event are calculated (equation 5.1) from the total fuel consumption for an event (activity, in the general context of emissions inventory) and emission factors (EF) for combustion products for the variety of fuels and conditions found on the landscape. Determination of both fuel consumption and emission factors in the case of biomass burning must take into consideration many factors. The characteristics of the fuels and the conditions of the burn are very important in determining both the amount of fuel consumption achieved during the burning and combustion efficiency, which drives EF. Fuel consumption is determined in various ways in emissions inventory approaches. Two general approaches for determining biomass consumption for wildland fire are presented here followed by a section reviewing the role of emissions factors and combustion efficiency in determining biomass burning emissions.

5.2.1. Fuel Consumption

Conceptually, the most straightforward method to estimate consumption is to difference estimates of the pre- and post-fire fuel loading (biomass density) measured on the ground. At the scale of fire events, maps of pre- and post-fire fuel loading can be derived from pre- and post-fire remotely sensed data trained from field plot data. McCarley et al. (2020) differenced pre- and post-fire fuel maps derived from pre- and post-fire lidar collections to map consumption estimates. Michalek et al. (2000) employed 30 m resolution Landsat data to extend field-measured consumption across a burn site, stratified by fuel type. However, data available from measurement of field-derived consumption are limited due to the onerous task of collecting sufficient quantitative field data to represent the area of interest. Because of this limitation, a field measurement approach cannot be accomplished for quantifying emissions on anything but a local scale.

For comprehensive estimation and mapping of fuel consumption (M_c) two general approaches are commonly used (Wooster et al., 2013; Fig. 5.1). Either of these methods provide the estimate for M_c in equation 5.1 to derive pollutant emissions. The indirect model-based approach uses an estimate of consumption based on the difference between pre- and post-fire fuel loading estimates (McCarley et al., 2020; Ottmar, 2014); alternatively, it can be modeled by multiplying the pre-fire fuel load by a burning efficiency factor. The direct approach estimates consumption based on fire radiative power (FRP) observations, which integrated over time provide an estimate of fire radiative energy (FRE) (Wooster et al., 2005). The terms *indirect* and *direct* refer to how field or remotely sensed observations relate temporally to combustion as a biophysical process, with the indirect approach made from a post-fire assessment, and the direct approach conducted at the time of the event based on active fire data; both approaches can be applied at multiple spatial scales. By the indirect approach, fuel consumption can be estimated at the small scale of paired pre- and post-fire field plots, either destructively from harvest samples (Ottmar, 2014) or remotely from pre- and post-fire terrestrial lidar (Hudak et al., 2020), at the landscape level of an entire fire event from airborne lidar collected pre- and post-fire (McCarley et al., 2020), and at regional-global scales from satellite image passes acquired before and after the burn (Pettinari & Chuvieco, 2016; Rollins, 2009). If post-fire fuel load estimates are unavailable, as is often the case, then a burning efficiency factor (fraction of fuel consumed) can be applied to the pre-fire fuel load estimate to derive consumption. However, it is not possible to directly observe fuel mass loss as it occurs by combustion during a fire event. On the other hand, consumption can be estimated from FRP observations, which provides a suitable surrogate for combustion as it occurs (Wooster et al., 2013). Moreover, FRP and FRE metrics scale readily between ground radiometers and airborne and satellite thermal imaging (Dickinson et al., 2016), making the direct approach a powerful method for near-real-time fuel consumption estimation. These two general approaches are used in the variety of emissions inventory systems described later in this chapter and in Chapter 9.

5.2.2. Mapping Consumption: The Indirect Approach

The indirect approach to mapping consumption is a model-based framework employing generalized information on pre-fire fuel characteristics and the level of consumption based on fuel conditions (French et al., 2014; Ottmar, 2014). Consumption is a product of the

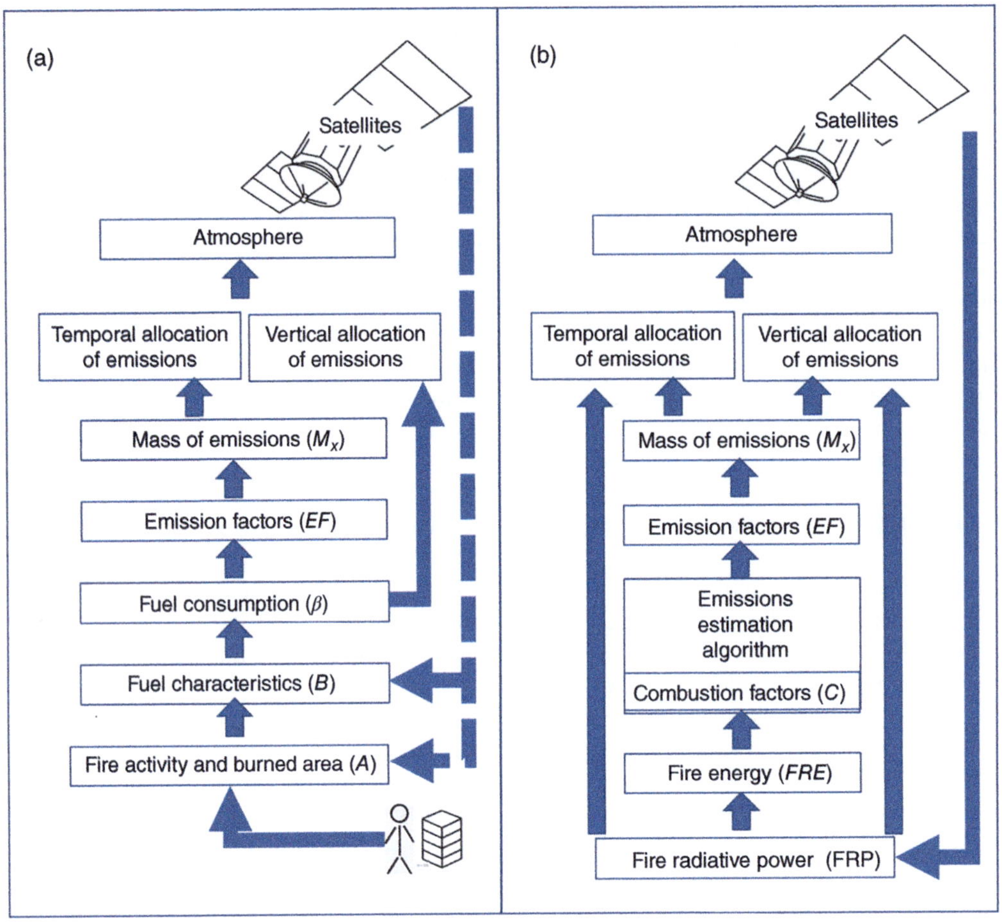

Figure 5.1 Two approaches to determine emissions from biomass burning. Fuel consumption is determined via two methods: (a) The indirect combustion approach considers the pre-fire fuels and proportion of fuel consumed (β). (b) The direct approach uses fire radiative energy to estimate consumption based on fuel-type specific combustion factors (C). Emissions factor (EF) for each product is applied based on combustion conditions. Various methods are used to allocate emissions temporally and vertically for use in atmospheric smoke modeling (see Chapters 8 and 9) (figure courtesy of S. O'Neill).

burn area, the amount of biomass in that area, and the fraction of the biomass consumed (Fig. 5.1), which is also termed *burning efficiency* or *combustion completeness* (van der Werf et al., 2006). The fraction consumed can be defined as a function of several variables, including the moisture conditions of the fuel and combustion conditions in the fire environment. Seiler and Crutzen (1980) used this general framework to make the first global estimates of contemporary carbon emissions from biomass burning, separately considering variation in each term for different biomes and different types of land management. Using this approach, the mass of total fuel consumption (M_c, kg) for an event is calculated using the following general equation:

$$M_c = A \times B \times \beta, \tag{5.2}$$

where A is area burned (m^2) in an event or region, B is biomass density (fuel loading per unit area; kg m^{-2}), and β is the fraction of biomass consumed (0–1, unitless), also termed the *burning efficiency factor*. Circling back to equation 5.1, applying an emissions factor, EF_x, provides the mass of emissions M_x for fuel type x:

$$M_x = A \times B \times \beta \times EF_x. \tag{5.3}$$

Quantification of each of these terms is described in detail next. This general approach has been established as a standard method for fuel consumption from biomass burning. For carbon emissions, the fraction of fuels that is carbon is included as a carbon emission factor, typically 0.45 or 0.5, the nominal carbon content of biomass. For pollutant emissions, used in atmospheric chemistry and air quality modeling, emission factors are applied as shown in equation 5.1 and further discussed later in this chapter.

Active Fire and Burn Area Mapping

In Chapter 2, the drivers of fire occurrence are reviewed explaining that some biomes are naturally fire prone, while other regions have little to no natural fire, and still others are influenced by anthropogenic burning for land clearing and other land use needs (Bowman et al., 2009). The amount of biomass burning varies greatly from region to region and, in some fire-prone ecosystems, from year to year. Determining where and when fire is occurring and the amount of area burned has been a major focus for mapping and monitoring air quality because the magnitude of burning (the amount of burn area (A)) is the dominant factor in many regions driving interannual variability in smoke (van der Werf et al., 2006). Prior to the 1990s, the magnitude of global biomass burning was not well known (Levine, 1991), and there were wide discrepancies in estimates of global fires occurrence (Kasischke & Penner, 2004). The advent of satellite sensing has provided comprehensive observations to better quantify fire occurrence with thermal sensing (Giglio et al., 2003; Kaufman et al., 1998). Multispectral sensing has advanced for mapping land change and burn area products at landscape scales with Landsat (Hawbaker et al., 2020; Key & Benson, 2006) and at regional to global scales with MODIS and similar systems (Giglio et al., 2006). Estimates of biomass burning (burn area and fire counts) for various regions of the world are fairly well constrained due to the advent of satellite-based systems for fire monitoring (Chapters 2 and 3).

For some regions and fuel types, accurate accounting of biomass burning is challenging. Satellite-based methods for determining burn area suffer from a mismatch in resolution between the sensor and the phenomenon; where fire is typically confined to a narrow fire-line of meters to submeter scale, satellite sensors detect energy at several meters to kilometer scale. This requires methods to subsample the information, often leading to bias of too much area burned (Hyer & Reid, 2009). On the other hand, satellite sensors also suffer from characteristics leading to undercounting of burn area by missing fires due to both the sampling rate of the orbiting satellite sensor and the obscuration of the fires under tree canopies or cloudy skies (Cochrane et al., 1999; Johnston et al., 2018; Morton et al., 2011; Souza et al., 2005). Agricultural burning and other small fires, such as pile burning practices, are missed in many emissions inventories for these reasons (Randerson et al., 2012). In addition, smoldering fires that are known to last for weeks to months, and can even reignite after winter hiatus, can consume large areas of organic peat soils, emissions that are undeniably undercounted (McCarty et al., 2021).

Figure 5.2 provides an example of a case in the southeastern United States where prescribed fires are common. The extensive omission errors are caused by the 16 day overpass frequency of the Landsat satellite, coupled with occlusion from cloud cover. Moreover, most fires in the ecosystems at both locations are low-intensity prescribed surface fires burning beneath the tree canopy; hence, canopy occlusion causes further underestimation of the true burn area, as recorded on the ground, causing some fires to be entirely missed (shown as empty blue polygons). In this case, satellite-based accounting misses 80% of the actual area burned recorded from fire records. This is a common problem in accounting for fire in any forests that have dense canopies and support understory burning, including tropical forests, which are typically low-intensity understory fires where dense forest canopies and clouds are common (Morton et al., 2013). Methods that use pre- and post-burn images for mapping burn area, and do not rely on active fire, are employed (Hawbaker et al., 2020); however most burn area products are seeded with locations mapped from active fire sensing (Giglio et al., 2006), because detection of burn locations in some biomes and ecosystems can be subtle, resulting in missed fires. Chapter 3 reviews these and other caveats in relying on satellite data alone for emissions inventory.

Because satellite remote sensing cannot capture all burning relevant to air quality, approaches that include record-keeping techniques are often required. Within the United States, the USEPA relies on a data consolidation effort that combines fire records with satellite detections (SmartFire-2) (Larkin et al., 2020) to account for fire in the triennial National Emissions Inventory (NEI). Therefore, the United States and other national fire activity assessments for emissions reporting rely heavily on satellite-based active fire and burn area data, as reviewed later in this chapter and in Chapter 9. In general, fire accounting will always miss some fires and area burned, as has been documented by Hall et al. (2016).

Biomass and Fraction Consumed

Estimating the amount of consumption within the perimeter of a burn based on the model shown in equation 5.2 can be highly uncertain due to the high variability in both the amount of fuel (loading) at the site (B) and the fraction consumed (β) (French et al., 2002, 2004; Urbanski et al., 2011; van Leeuwen et al., 2014). To estimate emissions where no pre- or post-fire fuel loading measurements are available, consumption must be modeled based on field data that relates consumption to measurable variables (fuel characteristics and moisture status). Examples of consumption models are the Consume model (Prichard et al., 2006), the First Order

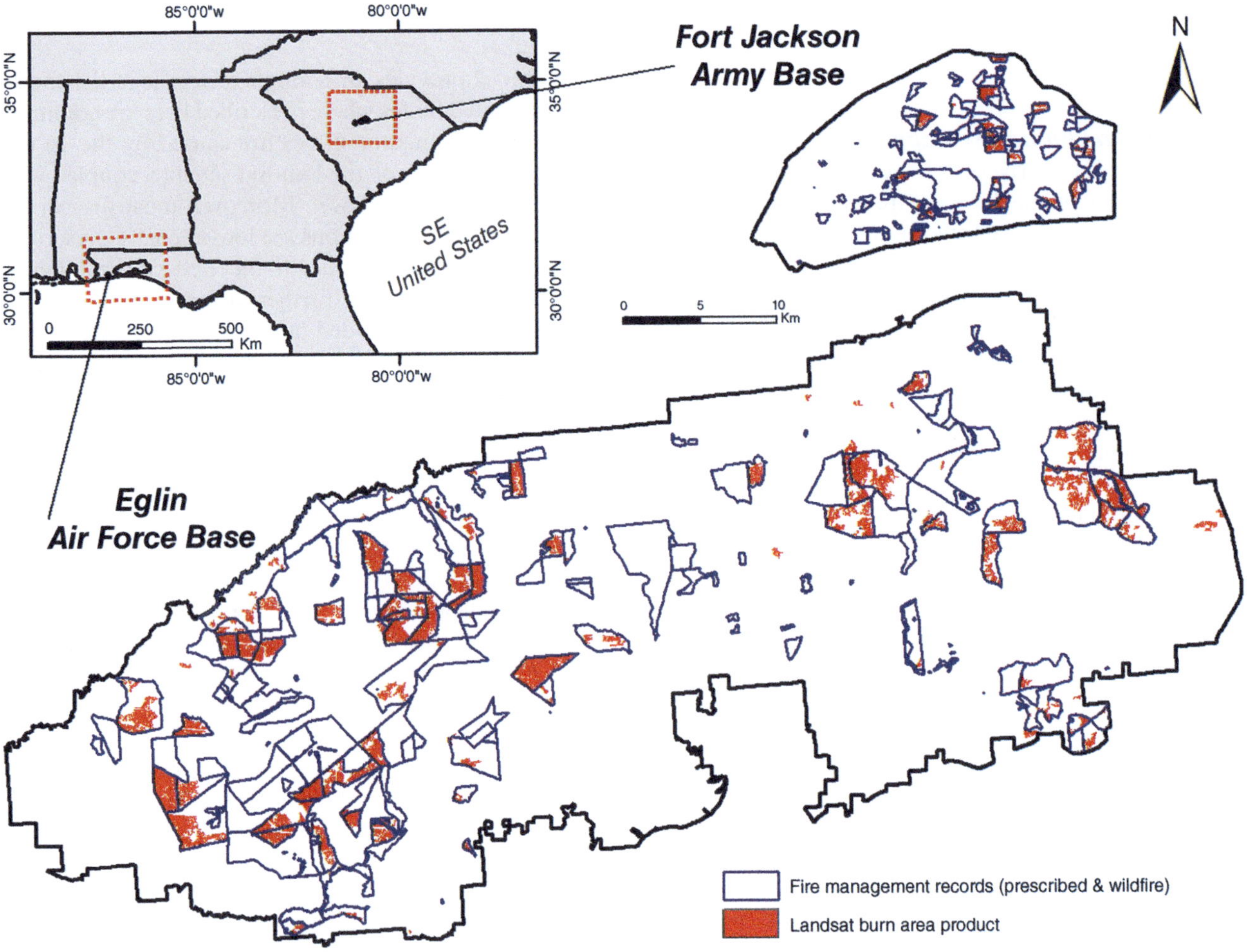

Figure 5.2 Fire management records of 2014 prescribed and unintentional fire perimeters versus remote sensing-derived burn area from the Landsat burn area product (Hawbaker et al., 2020) at Ft. Jackson Army Base and Eglin Air Force Base, located in the southeastern United States, showing extensive omission errors from canopy and cloud obscuration and low satellite repeat frequency (figure courtesy Dr. Nuria Sanchez-Lopez).

Fire Effects (FOFEM) model (Reinhardt et al., 1997), and CanFIRE (de Groot et al., 2009), which are used as the basis of several emissions model systems. All of these models rely on knowledge of the fire environment, including the pre-fire fuels and the fuel moisture, which are a function of weather and climate.

Pre-fire fuel loadings can be estimated from remotely sensed data, typically multispectral (e.g., Landsat) satellite imagery, trained from field plot measures. National LANDFIRE maps are the best example in the United States (Rollins, 2009). The Fuel Characteristics Classification System (FCCS) developed by Ottmar et al. (2007) provides a comprehensive framework for summarizing complex fuelbeds into their component materials (trees, shrubs, herbaceous, coarse and fine woody debris fractions, litter, duff); the proportions of these components vary greatly by vegetation type. Landsat imagery FCCS fuelbeds have been mapped for the United States based on LANDFIRE Existing Vegetation Type (EVT) maps (Rollins, 2009). Globally, FCCS has been mapped from MODIS-derived vegetation type for use in fire emissions modeling in many regions around the world, including South America (Pettinari et al., 2013), and as a global product (Pettinari & Chuvieco, 2016). The original purpose of developing the FCCS was for fire planning and management to provide fire managers with site-level fuel information from which they could estimate fire behavior (e.g., spread rate), consumption, emissions, and fire effects for a given set of fire weather and fuel moisture conditions. The Consume and FOFEM models were developed to use FCCS fuels for site-level consumption estimation but have been used with LANDFIRE FCCS layers to increase their applicability at broad scales (French et al., 2014) and in smoke model environments (e.g., BlueSky Tools; Larkin & O'Neill, 2018; Larkin et al., 2009; see Chapter 9).

Weather and climate factors that influence fuel consumption are best considered as a continuum of constraints on combustion, operating at different temporal scales. Fire weather variables (wind, temperature, and RH) have the most immediate effects, operating at seconds to days, while climate variables, such as growing season length, operate at seasonal to decadal time scales. In the absence of fire, climate operates to constrain vegetation production, senescence, and mortality, and thereby fuel loads, as a balance between accumulation and decomposition (Olson, 1963). Occupying the middle ground between weather and climate at an intermediate temporal scale, ranging from hourly to seasonally, is fuel moisture, constrained by the full continuum of weather-climate factors and landscape physiography. These factors include temperature, precipitation, evapotranspiration, relative humidity, wind, slope, aspect, and soil type. Fuel moisture is a critical control on combustion and is difficult to monitor remotely (Chowdhury & Hassan, 2015; Yebra et al., 2013) given that it is temporally very dynamic for each fuel component, depending on particle size and arrangement (see Chapter 4) (Viegas et al., 1992; Viney, 1991). Despite this complexity, moisture status of dead fuels is easier to parameterize in fuel moisture models by fuel size class than live fuel moisture content (Carlson et al., 2007; Finney, 1998; Nelson, 2000, 2001), because plants in their various functional forms exhibit a variety of strategies in response to water limitation and drought (Viegas et al., 2001). However, fuel moisture is the basis of fuel consumption models, providing the most important constraint on combustion, making it a very important variable in emissions estimation, no matter the approach used.

5.2.3. Mapping Consumption: The Direct Approach

Active fire sensing can provide a direct estimate of biomass consumption derived from the energy emitted (Wooster et al., 2003), unlike the indirect approach, which determines consumption from a model-based accounting of fuel consumption (Fig. 5.1). Fire radiative power (FRP) observations are derived from calibrated brightness temperatures measured by thermal infrared imaging sensors (Dozier, 1981; Kaufman et al., 1998). Fire radiative energy (FRE) is the time integral of instantaneous FRP observations, expressed as follows:

$$FRE = \int_{t1}^{t2} FRP(t)\,dt. \tag{5.4}$$

This integral can be approximated by the following equation (Boschetti & Roy, 2009; Hudak et al., 2015):

$$FRE = \sum_{i}^{n} 0.5\left(FRP_i + FRP_{i-1}\right)\left(t_i - t_{i-1}\right), \tag{5.5}$$

where time t is the time in seconds (s) of each FRP observation, i, in the time series.

FRP and FRE are extremely useful because they are linearly related to combustion rate and total consumption, respectively (Wooster et al., 2005). The direct approach to estimating the consumption mass, M_c, is as follows (Prichard et al., 2021; Wooster et al., 2013):

$$M_c = FRE \times C, \tag{5.6}$$

where C is a combustion factor (kg MJ^{-1}) for a given vegetation (fuel) type, which equates to the slope of the linear best fit relationship between and FRE and the mass of fuel consumed, M_c (Wooster et al., 2013). An advantage of the direct approach to consumption estimation by equation 5.6 is that it replaces the burn area, biomass, and fraction consumed variables in equation 5.2. Thus, the combustion factor C (kg MJ^{-1}) converts energy units (MJ) to mass units (kg) of fuel consumed (M_c) to effectively collapse the three predictors used in the indirect approach of equation 5.2 into a single coefficient. The combustion factor varies considerably as a function of fuel moisture (Smith et al., 2013). Circling back to equation 5.1, applying EF_x provides the mass of emissions M_x for fuel type x:

$$M_x = FRE \times C \times EF_x. \tag{5.7}$$

Inherent underestimation biases confound FRE-based consumption estimation from remote sensing observations. The origin of these biases parallel those for estimating consumption in the indirect method. They include obscuration from clouds, smoke, and tree canopies and limitations in observation due to satellite orbit characteristics. Mathews et al. (2016) found that forest with complete or nearly complete canopy closure can obscure fire and may be below the detection threshold of above-canopy FRP sensors. Global satellite FRP products are derived from polar-orbiting satellites (e.g., MODIS) that image a given location on the Earth twice daily at most, assuming no cloud cover (Roberts et al., 2011); this translates into many missed ignitions and unobserved flame front movement between satellite overpasses and where clouds are present during the overpass. At the regional scale of northern Australia, which has a very active fire regime, Boschetti and Roy (2009) used simple kriging to spatially extrapolate sparse MODIS observations of FRP using the independently derived MODIS burn area product, to derive FRE estimates that were less affected by undersampling. Similarly, at the landscape scale of individual prescribed fires, Klauberg et al. (2018) used ordinary kriging and conditional simulation to extrapolate FRE estimates integrated from temporally undersampled FRP observations collected from an infrared camera mounted on a fixed-wing aircraft flying laps over the fire. In both of these examples, geostatistical modeling (e.g., kriging) served to fill in the spatial gaps caused by undetected burning between the sparse fire observations.

5.2.4. Addressing Complexity in Fuels and Combustion

As described in Chapter 4, biomass fuels are heterogeneous in characteristics relevant to emissions. Even in areas of monoculture forests or grasslands, topographic and hydrologic features can influence the fuel conditions and drive variability in combustion. Fuel components, meaning the fuel type and vegetation strata (see Chapter 4), have very different burning properties, making assignment of combustion factors by vegetation type a gross oversimplification, especially for global-scale assessment when generalized vegetation type is often used in this way (e.g., land cover class or biome; Kaiser et al., 2012; van der Werf et al., 2006; Wiedinmyer et al., 2011).

One strategy for addressing fuel complexity for more accurate consumption estimates via the indirect method is by modeling combustion by fuel component or strata, for example assigning different β values in equation 5.2 to the aboveground and surface fuels (French et al., 2002). In this example, equation 5.2 is expanded to include two fuel components as:

$$M_x = A \times \sum B_a \beta_a + B_s \beta_s, \qquad (5.8)$$

where B_a is the average biomass density of aboveground vegetation and B_s is the loading of the surface fuels, with corresponding consumption fractions β, for each stratum. This can be applied to more than two fuel components or strata, and is one of the strategies used in several consumption approaches, including the Global Fire Emissions Database (GFED) (van der Werf et al., 2017; van Leeuwen et al., 2014) and the Consume model, which uses the FCCS fuelbed paradigm (Ottmar, 2014). In addition, separating combustion models by fuel component allows for explicitly defining combustion phase for assigning emission factors, thereby improving the estimate of emissions of non-CO_2 gases by accounting for combustion efficiency (see section 5.2.5).

5.2.5. Emission Factors and Combustion Efficiency

Smoke from biomass burning is a complex mixture of gases and particulate matter (PM) (Table 5.1; Prichard et al., 2020; Urbanski, 2014). The most abundant gases in smoke from fires burning wood and other biomass are carbon dioxide (CO_2), carbon monoxide (CO), methane (CH_4), nitrogen oxides (NO, NO_2), ammonia (NH_3), and sulfur dioxide (SO_2) (USEPA, 2021a). Nitrogen oxides, grouped together as NO_x, are especially important for their role in the formation of ozone (O_3) through a photochemical process, which is a critical pollutant in downwind smoke. Important for air quality, nonmethane hydrocarbons, organic aerosols, and other organic and inorganic pollutant compounds are found in smaller but sometimes critical levels in smoke, most often in incomplete combustion conditions (Andreae & Merlet, 2001). PM in biomass burning smoke is a mixture of fine to very fine solid particles and liquid droplets. Alternative terminologies for PM are *total particulate matter (TPM)* or *total suspended particles (TSP)*. Particulates that are lofted into the air form aerosols, a collection of suspended particles of varying composition, shapes, and sizes. PM is often designated by size; typical designations are particles

Table 5.1 Emission Factors (EF) for the primary pollutants in biomass burning smoke

	Mean (SD) of all data in SERA	Mean (SD) by combustion phase	
Pollutant	Emission factor (EF) (g/kg)	EF flaming (g/kg)	EF smoldering (g/kg)
Carbon dioxide CO_2	1,595.6 (166.2)	1,671.4 (78.2)	1,521.2 (148.5)
Carbon monoxide CO	99.042 (49.430)	70.363 (22.700)	131.593 (40.013)
Particulate matter <2.5mm - $PM_{2.5}$	22.456 (16.954)	15.957 (9.884)	31.804 (18.880)
Methane CH_4	4.294 (3.387)	2.618 (1.345)	7.628 (3.338)
Nitrogen oxides NO_x	3.021 (2.110)	2.755 (1.143)	2.288 (2.016)
Ammonia NH_3	1.386 (1.445)	0.889 (0.768)	2.576 (1.886)
Sulfur dioxide SO_2	1.113 (0.723)	1.025 (0.552)	1.569 (1.158)

Source: This information is compiled and condensed from the Smoke Emissions Reference Application (SERA), Prichard et al. (2020).

Note: Units are in g of a gas or PM per kg of fuel consumed (Mc). Also shown are EFs for flaming and smoldering combustion conditions. Standard deviation (SD) of measurement mean is shown in parentheses. The number of data samples varies. We are considering fire in areas with few to no houses or human-built infrastructure. Smoke from fires that consume structures contain a myriad of additional compounds, which will not be reviewed in this chapter.

with a diameter <2.5 μm ($PM_{2.5}$) and particles with a diameter <10 μm (PM_{10}), which include the smaller category of particles. Smaller particles ($PM_{2.5}$) are more influential on human health and often the subject of health studies because they can be respired deeper into the lungs than coarse particulates (see Chapter 11). Additionally, these small particles stay aloft longer, so can be transported farther, causing impacts to air quality far from the site of the event (Seinfeld & Pandis, 1998).

Emission Factors

For all sources, emission factors (EFs) define the partitioning of combustion products into various emission products and are variable based on the fuel type and combustion conditions. Biomass burning, EFs are determined by measuring the concentration of a pollutant either in the lab or in a field-measured fresh smoke plume against a background concentration (Urbanski et al., 2021). Seasonal and regional variability in EFs is well known. For global scasle emissions inventories, specific EFs are applied by biome (Kaiser et al., 2012; van Leeuwen et al., 2013). For event-based and regional efforts, EFs are on the fuel type and combustion conditions (Prichard et al., 2020). Studies of African savannah grasslands identified some of the underlying factors driving variability in these ecosystems (Hoffa et al., 1999; Korontzi, 2005); these and similar studies provide some guidance for improving EFs used in biomass burning emissions modeling. Regionally defined variability in biomass burning CO emissions and concentrations using six EF scenarios was assessed in a study by van Leeuwen (2014). The study used both spatial variability defined by biome and temporal variability, as opposed to previous approaches that employed static EFs by biome (Akagi et al., 2011). Using this spatial-temporal complexity in EF, van Leeuwen (2014) found an increase in CO emissions for boreal regions (as much as 50%), but lower estimates for Africa (~3.5%), compared with using static EF values, revealing the importance of refined knowledge of EF over space and time for accurate emissions and smoke concentration mapping.

Recent efforts to better characterize the composition of smoke for emissions modeling have resulted in several updates to EFs from biomass burning and other open burning (Akagi et al., 2011; Andreae, 2019) and include the development of the Smoke Emissions Repository Application (SERA) database of biomass burning EFs for smoke modeling (Prichard et al., 2020). SERA lists over 275 species of gases and particulate categories; the primary gases and aerosols and an average EF as reported in SERA are listed in Table 5.1. A subset of the pollutant list in SERA is the criteria air pollutants (CAP) designated by the USEPA. It should be noted that one of the primary pollutants of interest for air quality assessment, ozone (O_3), is not included in this table because it is not a primary product of combustion, rather is formed when nitrous oxides (NO_x) in smoke interact with sunlight through photochemistry. Similarly, aerosols are known to evolve as they age (Huang et al., 2013; Kleinman et al., 2020), so typically EFs are reported from fresh smoke, unless otherwise noted.

Combustion Efficiency and Emissions

Variations in emissions and EFs are driven by various compounds in the biomass as well as the combustion conditions. Combustion type or combustion phase is determined by the conditions under which the combustion process occurs, which is a result of fuel water content, fuel compaction, and ventilation (wind), all of which influence available oxygen. Low oxygen conditions during a smoldering burn produces a very different set of combustion products than clean burning, where oxygen is not restricted, as shown in the EF differences due to combustion phase (Table 5.1). Smoldering conditions reduce the combustion efficiency thereby decreasing the amount of CO_2 at the same time as increasing the formation of CO and gaseous and particulate organic compounds that can have high toxicity, so are therefore of concern for human health (Rappold et al., 2011; see Chapter 11).

Modified combustion efficiency (MCE) is a way to quantify the level of inefficient combustion found especially during the smoldering phase. MCE is calculated from the concentrations of CO_2 and CO in the plume compared to the nonplume (background) concentration, as shown in Eq. 5.9 with Δ indicating plume to background difference (Yokelson et al., 1999):

$$MCE = \Delta CO_2 / (\Delta CO_2 + \Delta CO). \qquad (5.9)$$

Efficient combustion will have an MCE of 1 indicating that all carbon in the fuel is emitted as CO_2. The MCE indicates the relative amount of flaming and smoldering combustion, with a lower MCE indicating more smoldering. Typical MCE values for wildfire biomass combustion are 0.80 to 0.95. MCE measured in smoke provides a way of defining the relative amount of complex combustion products, such as organic gases and aerosols, critical to air quality. As reviewed in Urbanski et al. (2021), MCE tends to differ with fuel type, with MCE highest for herbaceous (grass and forb) and shrub-type fuels and lower for forest fires (Fig. 5.3). In general, MCEs for prescribed fire tend to be higher than for wildfires, indicating cleaner burning in prescribed versus wildfire, although this generalization is broad, and MCE is highly variable in all open burning situations (Prichard et al., 2020).

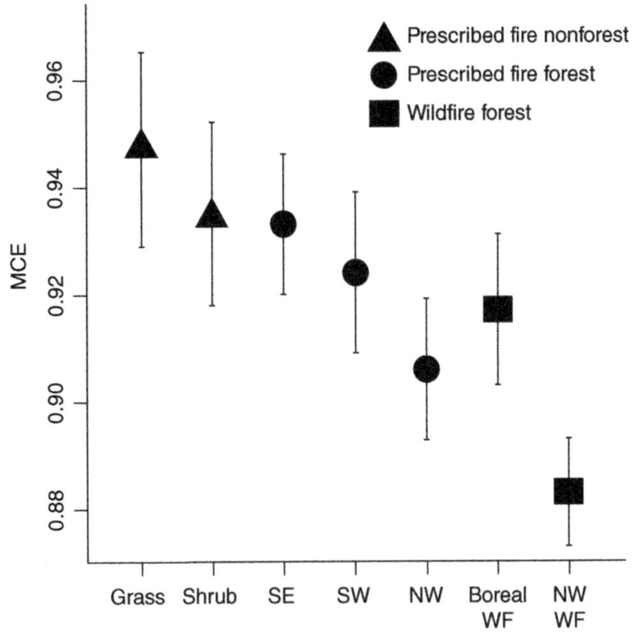

Figure 5.3 Figure demonstrating the variation possible of modified combustion efficiency (MCE) for fire in different vegetation types and biomes of the United States (US). SE = southeast US, SW = southwest US, NW = northwest US, and WF = wildfire. Error bars are 1 standard deviation of the measured data (reproduced from Fig. 6 in Urbanski, 2014).

5.3. BIOMASS BURNING EMISSIONS INVENTORIES AND MAPPING SYSTEMS AND PRODUCTS

The variety of emissions inventory approaches has resulted in disparate estimates of the impacts of fires on pollutant concentrations. Emissions estimates of some chemical species may vary by a factor of 3 to 4 (Urbanski et al., 2011). In this section, we review existing models, systems, and products for emissions inventory and mapping, including approaches used for national fire emissions inventories and modeling systems used for event-scale assessment of smoke plumes to global-scale understanding of atmospheric chemistry. Chapter 9 provides a more specific review of air quality models from several nations that use the methods described in this chapter.

5.3.1. National-Level Emissions Inventories

Monitoring air pollution on a global scale is challenging, requiring multiple governmental and quasi-governmental efforts to bring the science of emissions monitoring to the task (Frost et al., 2013). The work to quantify pollutants has primarily been carried out to inform in-country and transboundary regulatory requirements. The Convention on Long-Range Transboundary Air Pollution (LRTAP) was set up in 1979 by 51 nations to address transboundary air pollution. LRTAP provides a framework for monitoring specific pollutants to address transboundary air pollution in North America, Europe, and Russia and former East Bloc countries (UNECE, 1979). In a separate effort, much of the work on quantifying emissions from biomass burning has been carried out for purposes of climate change monitoring, which is concerned with CO_2 and other climate-relevant gases and aerosols, also called greenhouse gases (GHG). Often GHG assessments present results in total carbon emissions or CO_2 equivalents, rather than for toxic air pollutants that are of concern for human health studies. The initiative to quantify GHG emissions from biomass burning has provided a framework to account for and quantify pollutant source emissions using methods prescribed by the IPCC (2007). Recent efforts to reconcile the different approaches for multiple needs of climate and air quality have provided guidance for including biomass burning in national-scale emissions inventories (Vallack & Rypdal, 2019).

National governments around the world conduct inventories of air pollutants and GHG emissions from biomass burning and other open burning. As an example, the U.S. National Emissions Inventory (NEI) compiles detailed estimates of air pollutants and precursors every 3 yr from state and regional government agencies, reported by county or state, depending on the data source. In addition to general air pollution sources, such as vehicles and power plants, the NEI accounts for fire and agricultural burning with extensive improvements to methods beginning in the 2011 NEI (Fig. 5.4) (Larkin et al., 2020). In a separate effort for climate assessment purposes, the *Inventory of U.S. Greenhouse Gas Emissions and Sinks* (USEPA, 2021b) has been compiled annually since the early 1990s; it is submitted to the United Nations in accordance with the Framework Convention on Climate Change (UNFCCC) reporting requirements (UNFCCC, 2021). Similarly, in the European Union, the European Monitoring and Evaluation Programme (EMEP) started publishing a guidebook on air pollutant emissions in 1992 and it has been maintained under the LRTAP Convention. The guidebook includes emissions sources from forest and grassland fires, as well as agricultural burning and continues to be published by the European Environment Agency (EEA), most recently released in 2019 (European Environment Agency, 2019). Canada and Australia have extensive areas of burning that vary annually. They both conduct detailed accounting of emissions from fire in their annual GHG reports to the UNFCCC using IPCC reporting methodologies, and both have national air quality monitoring programs akin

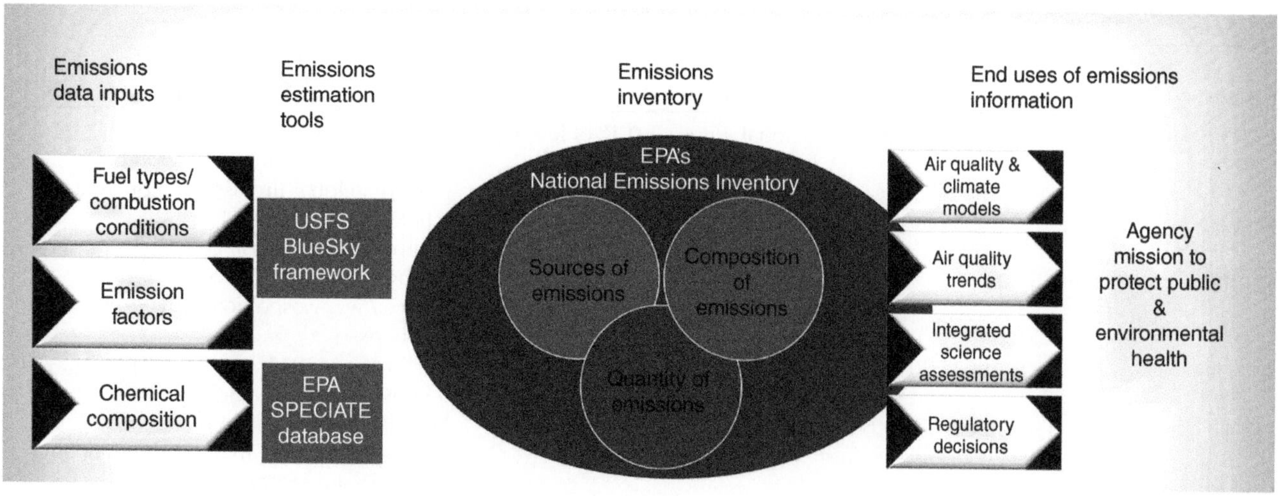

Figure 5.4 Overview of the U.S. Environmental Protection Agency approach to wildland fire emissions inventory. For more information see Holder (2019).

to the U.S. EPA and E.U. EMAP programs. Additional information on country-level efforts to model and track pollution from fire is provided in Chapter 9.

5.3.2. Fire Emissions Inventory Systems

Outside of needs for national and international reporting and regulatory compliance, fire emissions are tracked for a variety of needs and stakeholders. Table 5.2 provides a list of some of the emissions monitoring and inventory systems currently in use. Many of these efforts are focused on GHG emissions tracking, with some having been adopted for integration into air quality models. Global-scale emissions inventory approaches are available for retrospective emissions inventory, such as the Global Fire Emissions Database (GFED) and for direct use in atmospheric chemistry modeling and forecasting (see Chapter 9). Global-scale models provide a comprehensive method to understand fire in the Earth system, and can allow for more complete and accurate modeling of atmospheric processes. The different approaches used incur tradeoffs and assumptions to achieve the goals of the assessment. The Global Fire Assimilation System (GFAS; Kaiser et al., 2012), the NASA Fire Energetics and Emissions Research (FEER) algorithm (Ichoku & Ellison, 2014), and the global biomass burning emission product from geostationary satellites (GBBEP-Geo; Zhang et al., 2019) use the fire energy approach (FRE) described previously, which employs a combustion factor to relate energy and emissions with broad assumptions regarding EFs for an event as well as fuel type and condition. FINN blends indirect and direct modeling approaches with generalized information on fuels and satellite-derived active fire detections to produce emissions on a daily basis (Wiedinmyer et al., 2011). These systems produce results in near-real-time, meaning the estimates are made as fires occur in order to provide data for numerical weather prediction and atmospheric modeling (see Chapter 9).

Finer-scale emissions inventory approaches are used for mapping smoke from individual fires and regions and to gain more specific and refined estimates of emissions for landscape and regional applications. The USDA Forest Service's Wildland Fire Emissions Inventory (WFEI) (Urbanski et al., 2011) was completed to estimate fine-scale emissions for the western United States from 2003 to 2008 in order to assess the reliability and uncertainty of coarse-scale estimates from other global inventories (see Chapter 9). The Canadian Wildland Fire Information System (CWFIS) provides twice-daily input on fire emissions for Canada's Wildfire Smoke Prediction System (FireWork) daily smoke forecast maps. The underlying emissions model for CWFIS modeling is CanFIRE (de Groot et al., 2009), which is also integrated into the national emissions reporting for GHG. The Wildland Fire Emissions Inventory System (WFEIS; French et al., 2014) operates on historical fire data and was developed to provide a spatial mapping capability to nonspatial emissions modeling using Consume (Prichard et al., 2009). WFEIS uses the indirect method, intersecting fire perimeter location and time-stamp data with fuel information from the FCCS (see Chapter 4) mapped by LANDFIRE (Rollins, 2009), and weather-derived fuel moisture, to make an indirect estimate based on equations 5.1 and 5.2. BlueSky is an emissions modeling and smoke forecasting framework that is used to map

Table 5.2 Fire emissions modeling systems currently available showing temporal and spatial domain

System	Approach	Temporal domain	Spatial domain	Spatial scale	Reference(s); websites
GFAS	Direct	2003 pres; NRT	Global	0.1° of lat/long	Kaiser et al., 2012 https://www.ecmwf.int/en/forecasts/dataset/global-fire-assimilation-system
FEER	Direct	2003 pres; NRT	Global	0.1° of lat/long	Ichoku & Ellison, 2014 https://feer.gsfc.nasa.gov
GBBEPx	Direct	NRT	Global	0.1° of lat/long	Zhang et al., 2019 https://www.ospo.noaa.gov/Products/land/gbbepx
GFEDv4	Indirect	1997– 2016	Global	0.25° of lat/long	Giglio et al., 2013; van der Werf et al., 2017 https://www.globalfiredata.org/
FINN	Indirect	2002– 2013 NRT	Global	1 km	Wiedinmyer et al., 2011 https://www2.acom.ucar.edu/modeling/finn-fire-inventory-ncar
BlueSky (emissions)	Indirect	NRT	United States	Various	Larkin et al., 2009 https://www.airfire.org/data/bluesky
WFEIS	Indirect	1983–2020	North America	Various	French et al., 2014 https://wfeis.mtri.org

Note: NRT = near real time, indicating estimates are produced as the events happen.

emissions sources in a way that can be ingested into air quality and other smoke transport models (Larkin et al., 2009). The BlueSky emissions model uses similar inputs as WFEIS and uses the Consume model to make an indirect estimate of consumption and emissions. It has been used for the triennial U.S. NEI to estimate source emissions for input into the Community Model for Air Quality (CMAQ; Byun & Schere, 2006). BlueSky is described in detail in Chapter 9. In addition to the models listed in Table 5.2, Chapter 9 provides additional insight into the emissions modeling and inventory data used in operational smoke mapping and forecasting around the world.

5.4. APPROACHES FOR IMPROVING FUEL CONSUMPTION ESTIMATES FOR EMISSIONS INVENTORIES

Existing approaches for quantifying the magnitude and extent of fire emissions across the landscape are inherently limited by lack of knowledge about conditions within and during the fire event. Most emissions inventories rely on generalized maps or unconfirmed knowledge of fuel loadings and structure. Approaches often use rules of thumb regarding combustion conditions and averages of important parameters that define emission magnitude and type. Furthermore, once emissions are calculated, in order to use the information for smoke exposure, they need to be integrated into a smoke transport model to map air quality concentration (see Chapter 8). To integrate emissions with transport models that operate at minute to hour time steps, a temporal profile must be assigned to allocate the daily results to the required time step of the smoke model, introducing another set of assumptions. Because of these various assumptions and the inherent variability in fuels and the fire environment, large differences are found when simulated smoke composition is compared with remote sensing measurements or in situ observations (e.g., Pfister et al., 2011; Yao et al., 2013). Pfister et al. (2011) modeled sources of CO and found high uncertainty and an underestimate of CO from fires when compared with measurements collected during an airborne sampling campaign aimed at quantifying sources of CO over northern California.

The high variability in emissions from biomass burning has ramifications for air quality modeling and forecasting. Attempts at error analysis for emissions have been elusive, mainly due to the lack of measurements to characterize the uncertainty (French et al., 2004). For this reason, techniques to better describe the system with more complete observations are required. In this section, we review a few avenues under way to help better describe the state of emission variables, including burn area and consumption. Not included here is a review of recent advances in fuels characterization, burn area mapping, and measuring emission factors (EF). Chapter 4 covers advances in fuels characterization relevant for emissions modeling. Burn area improvements are reviewed in Chapter 3. Recent improvements in measuring and quantifying EFs can be found in the literature coming out of the 2018 Western Wildfire Experiment for Cloud Chemistry, Aerosol Absorption and Nitrogen (WE-CAN) (e.g., Selimovic et al., 2019) and the 2019 Fire Influence on Regional to Global Environments and Air Quality (FIREX-AQ) (NOAA, 2021) campaigns, recent intensive studies of North American fire emissions and smoke composition.

5.4.1. Improving Indirect Fuel Consumption Estimations

Estimates of M_c by the indirect approach are highly sensitive to the source of burn area information used (Knorr et al., 2012). Estimates using standard modeling methods (equation 5.2) vary widely, with large uncertainties that are difficult to quantify (Fortin, 2021; French et al., 2004; Urbanski et al., 2011). This inherent variability, and therefore uncertainty, in quantification of fuel consumption is founded in the variability of both the fuel loading (B) and the amount of fuel consumed (β in equation 5.2 and C in equation 5.6). Uncertainties in fuel loading and burn efficiency can exceed 100% (Reid et al., 2009; Wooster et al., 2013), as these predictors are considerably difficult to quantify at any scale (Boschetti & Roy, 2009; French et al., 2004; Schultz et al., 2008; Wooster et al., 2005). It has been empirically demonstrated that fuels in natural fuelbeds fundamentally vary at submeter scales (Keane, 2012, 2015; Keane et al., 2012).

The indirect approach to estimating consumption relies on poorly parameterized empirical models of consumption based on a small sample of consumption measurements by fuel type and strata (e.g., the Consume model; Ottmar, 2014; Prichard et al., 2009). Van Leeuwen et al. (2014) published the first attempt at a comprehensive, global database listing all the available fuel consumption field measurements in the peer-reviewed literature. Estimates range from 4.6 ± 2.2 t ha^{-1} in savannas to 314 ± 196 t ha^{-1} in tropical peat lands, which represents anywhere from 21% to 96% of available fuel consumed, depending on the fuel category and biome (Table 2 in van Leeuwen et al., 2014). Based on van Leeuwen et al. (2014), some biomes and regions are drastically undersampled, including tundra regions, agricultural landscapes, and peat lands. The high level of variability in pre-burn fuel loading (Prichard et al., 2019) and combustion loss makes it very difficult to measure enough samples to fully characterize fuel consumption for this type of model. An improvement would be spatially explicit estimates of consumption that capture fuel heterogeneity, or more samples to inform the empirical models. The challenge is to develop a framework for modeling fuels and fire regimes that accounts for spatiotemporal variability in fuels. This framework will necessarily be informed by both remotely sensed and field measurements collected at multiple spatiotemporal scales. Uncertainties in maps of estimated fuels and consumption are expected to be high at the resolution of pixel-based estimates but improve upon aggregation to entire fire events (Hudak et al., 2015; Klauberg et al., 2018), which is the scale domain of emissions models needed for air quality mapping.

Lidar has been found to be an excellent tool in characterizing forest structure (Wulder et al., 2008), and is now being used experimentally to estimate fuel loading because of its great sensitivity to vegetation structure that correlates to fuel loading (Hudak et al., 2002). Unlike passive optical imagery (e.g., Landsat), which saturates under high biomass conditions, lidar is sensitive to increasing biomass without asymptote (Hudak et al., 2020; Kennedy et al., 2018; Zhang et al., 2014). Airborne lidar collections are becoming widespread, particularly in the United States and particularly in forested lands, where the canopy penetrating capabilities of lidar provide the most accurate ground elevation measurements from which to interpolate a high-resolution (e.g., 1 m) digital terrain models (DTM) from the ground returns. The strategy of the 3D Elevation Program (USGS, 2021) is to collect lidar data wall-to-wall across the coterminous United States within 10 yr, to populate a national elevation data set with unprecedented accuracy and precision at 1 m resolution. Such lidar can also be used to characterize pre-fire fuel loading, providing detailed information not otherwise available in such detail (Popescu et al., 2003; Skowronski et al., 2007). In the case of large wildfires where whole watersheds may be burned, landslides and broad-scale erosion become a critical concern of local watershed managers, especially for municipal water provisions, which has motivated post-fire lidar collections in specific cases to quantify such changes. While relatively rare, these pre- and post-fire lidar collections provide an opportunity to estimate fuel consumption, at unprecedented spatial resolution that no other technology can provide, simply by differencing maps of pre- and post-fire fuel loading, as demonstrated by McCarley et al. (2020). The advent of spaceborne lidar sensors with global coverage (e.g., Global Ecosystem Dynamics Investigation (GEDI); Ice, Cloud, and Land Elevation Satellite (ICESat-2)) provide dramatically expanded monitoring capabilities for estimating consumption using pre- and post-fire lidar. Spaceborne waveform sampling instruments with larger instantaneous coverage (footprint) than typical airborne lidar systems could provide estimates of fuel loading at the footprint level; for

ICESat-2 this is a 25 m diameter footprint, similar in size to a forest inventory plot (Leite et al., 2022). Such estimates can be augmented with pre- and post-fire passive optical imagery (e.g., Sentinel or Landsat) using statistical or geostatistical models (Hudak et al., 2002) to generate pre- and post-fire fuel maps from which to estimate consumption.

5.4.2. Improving the Direct Approach to Fuel Consumption and Emissions Estimation

The biggest contributor to uncertainty in the direct approach to estimating consumption and subsequent emissions, based on FRE, is undersampling of FRP over the course of the fire event. The largest factor contributing to underdetection is the infrequent overpasses of polar orbiting satellites (e.g., twice daily for MODIS) relative to daily active fire at a given location (Boschetti & Roy, 2009). Cloud and smoke cover reduces detection of active fire, as does obscuration of low-intensity surface fires burning in closed canopy forests (Mathews et al., 2016; Morton et al., 2013). Underdetections include small fires that are entirely missed, portions of large fires between potential hot pixel detects that go unobserved, and significant (particularly peak) fire activity that occurs between co-located hot pixel detects in time. These overlapping detections often represent a large proportion of the total energy released but are missed in the subsequent integration of FRE from the undersampled FRP record (Hudak et al., 2015).

A more accurate approach to FRE estimation is to increase fire detections, ideally via high-rate sampling, as is accomplished by geostationary satellites that provide high-rate imaging capability, but at a low spatial resolution (O'Neill et al., 2021; Roberts & Wooster, 2008). O'Neill and Raffuse (2021) demonstrate the use of the U.S. Geostationary Operational Environmental Satellite (GOES) Advanced Baseline Imager (ABI) on the GOES-16 platform to detect fires, compute FRE (see equation 5.3) and estimate emissions using the FEER approach (Ichoku & Ellison, 2014). This approach provides a look at the active fire every 5 min, which allowed modeling of near-surface 1 hr average particulate matter concentrations using an atmospheric dispersion model for the 2020 California Camp wildfire. Limitations include less precise location information and obscuration of the signal by smoke and tree canopies, as previously discussed, but advantages are a more complete accounting of active fire over time. For energetic, fast-moving fires, geostationary-derived information is a major advantage. As techniques to use the high sample-rate data from satellites such as GOES are developed, improvements in smoke tracking and near-real-time forecasting is becoming a reality, as covered in detail in Chapter 9.

5.4.3. Additional Sources of Uncertainty for Mapping Consumption and Emissions.

Developing a realistic estimate of consumption during biomass burning, for both the direct, FRP-based approach and the indirect consumption modeling approach, requires many assumptions due to the complexity of the variables involved. Uncertainty arises from subscale variability in fuels and combustion for both approaches. O'Brien et al. (2016) has shown submeter spatial variation in FRE during fuel combustion. For instance, the FRE of pine cones versus grass fuels in the fuelbed dramatically differs within a 1 m x 1 m microplot, as would the expected emissions. Indeed, inaccurately counting the number of pine cones per meter would dramatically bias emissions estimates, and the same could be said for inaccurate estimations of surface litter loading, duff depth, and other fuel components. Fuel moisture also influences combustion and FRE (Smith et al., 2013), which can change quickly over short timescales as well as spatially. Moisture conditions can be especially impactful in peat-dominated sites, where the water table is an important driver of surface fuel consumption (Bourgeau-Chavez et al., 2020). Most, if not all, fuel components and their moisture content are exceedingly difficult to quantify at the fine scale at which they fundamentally vary, making fuel combustion, even if accounting for vegetation type, one of the largest sources of uncertainty in emissions modeling.

The collective uncertainties in consumption, by either the indirect or direct method, may seem to limit the ability to make reasonable estimates, but factors operating at broader scales serve to constrain fuel loading, consumption, and emissions such that accurate estimates can be derived at higher levels of aggregation. This is not to say that current estimates are necessarily biased, but imprecise at fine scale where fine-scale heterogeneity in fuels and consumption is insufficiently captured. Fuel maps necessarily rely on remote sensing data and empirical models that exploit the statistical relationships between ancillary data provided by remote sensing observables (e.g., spectral reflectance) and the biophysical attributes of interest measured on the ground (e.g., fuels). Such model-based approaches can use bootstrapping of the training data to generate variation in the mapped estimates, which provides a measure of model uncertainty at the pixel level (Urbazaev et al., 2018). Ideally, fuel maps would fully account for all sources of error (sampling, allometric, observation, etc.) that propagate to contribute to overall uncertainty at the pixel level (e.g., LANDFIRE at 30 m resolution). Fully accounting for these cumulative errors remains a challenge (Saatchi et al., 2011); however, in practical terms, much of the fine-scale fuels variation average out upon aggregation (Bell et al., 2018) to the decision space that typically corresponds to entire fire events and resulting emissions.

5.5. SUMMARY AND CONCLUSIONS

In summary, quantifying emissions from biomass burning is complex and inexact, involves observations and modeling tools, and has matured rapidly in the past three decades (Kasischke & Penner, 2004; Larkin et al., 2020; van der Werf et al., 2017). Emission estimates are needed for quantifying the constituents of pollution found in smoke to address air quality concerns. An emissions inventory provides the data needed to run smoke models that can be integrated into air quality maps and used in decision-making and policy development. Many such models are used for both air quality and climate research and applications. Two approaches are in use to estimate consumption. The indirect approach uses a model-based framework for consumption based on knowledge of fuels and conditions, while the direct approach provides real-time assessment of active fire and calculation of emissions given specific assumptions on combustion. All emissions models suffer inaccuracies from lack of complete knowledge of the fire environment; however, new observational methods are under development that will improve this knowledge to better quantify the emissions originating from biomass burning.

ACKNOWLEDGMENTS

This research was supported by the USDA Forest Service, Rocky Mountain Research Station (AH). The findings and conclusions in this publication are those of the authors and should not be construed to represent any official USDA or United States government determination or policy.

REFERENCES

Akagi, S. K., Yokelson, R. J., Wiedinmyer, C., Alvarado, M. J., Reid, J. S., Karl, T., et al. (2011). Emission factors for open and domestic biomass burning for use in atmospheric models. *Atmospheric Chemistry and Physics*, *11*(9), 4039–4072. https://doi.org/10.5194/acp-11-4039-2011

Andreae, M. O. (2019). Emission of trace gases and aerosols from biomass burning – an updated assessment. *Atmospheric Chemistry and Physics*, *19*(13), 8523–8546. https://doi.org/10.5194/acp-19-8523-2019

Andreae, M. O., & Merlet, P. (2001). Emission of trace gases and aerosols from biomass burning. *Global Biogeochemical Cycles*, *15*(4), 955–966. https://doi.org/10.1029/2000GB001382

Bell, D. M., Gregory, M. J., Kane, V., Kane, J., Kennedy, R. E., Roberts, H. M., et al. (2018). Multiscale divergence between Landsat-and lidar-based biomass mapping is related to regional variation in canopy cover and composition. *Carbon Balance and Management*, *13*(1), 1–14.

Boschetti, L., & Roy, D. P. (2009). Strategies for the fusion of satellite fire radiative power with burned area data for fire radiative energy derivation. *Journal of Geophysical Research*, *114*(D20302). https://doi.org/10.1029/2008JD011645

Bourgeau-Chavez, L. L., Grelik, S. L., Billimire, M. G., Jenkins, L., Kasischke, E. S., & Turetsky, M. R. (2020). Mapping boreal peatland fire severity and assessing their potential vulnerability to early season wildland fire. *Frontiers in Forests and Global Change*, *3*(20). https://doi.org/10.3389/ffgc.2020.00020

Bowman, D. M. J. S., Balch, J. K., Artaxo, P., Bond, W. J., Carlson, J. M., Cochrane, M. A., et al. (2009). Fire in the Earth system. *Science*, *324*(5926), 481–484. https://doi.org/10.1126/science.1163886

Byun, D., & Schere, K. L. (2006). Review of the governing equations, computational algorithms, and other components of the Models-3 Community Multiscale Air Quality (CMAQ) modeling system. *Applied Mechanics Reviews*, *59*(2), 51–77. https://doi.org/10.1115/1.2128636

Carlson, J., Bradshaw, L. S., Nelson, R. M., Bensch, R. R., & Jabrzemski, R. (2007). Application of the Nelson model to four timelag fuel classes using Oklahoma field observations: Model evaluation and comparison with National Fire Danger Rating System algorithms. *International Journal of Wildland Fire*, *16*(2), 204–216. https://doi.org/10.1071/WF06073

Chowdhury, E. H., & Hassan, Q. K. (2015). Operational perspective of remote sensing-based forest fire danger forecasting systems. *ISPRS Journal of Photogrammetry and Remote Sensing*, *104*, 224–236. https://doi.org/10.1016/j.isprsjprs.2014.03.011

Cochrane, M. A., Alencar, A., Schulze, M. D., Souza, C. M., Nepstad, D. C., Lefebvre, P., et al. (1999). Positive feedbacks in the fire dynamic of closed canopy tropical forests. *Science*, *284*(5421), 1832–1835. https://doi.org/10.1126/science.284.5421.1832

de Groot, W. J., Pritchard, J., & Lynham, T. J. (2009). Forest floor fuel consumption and carbon emissions in Canadian boreal forest fires. *Canadian Journal of Forest Research*, *39*, 367–382. https://doi.org/10.1139/X08-192

Dickinson, M. B., Hudak, A. T., Zajkowski, T., Loudermilk, E. L., Schroeder, W., Ellison, L., et al. (2016). Measuring radiant emissions from entire prescribed fires with ground, airborne and satellite sensors: RxCADRE 2012. *International Journal of Wildland Fire*, *25*(1), 48. https://doi.org/10.1071/wf15090

Dozier, J. (1981). A method for satellite identification of surface temperature fields of sub-pixel resolution. *Remote Sensing of Environment*, *11*, 221–229. https://doi.org/10.1016/0034-4257(81)90021-3

European Environment Agency (2019). *Trends and drivers of EU greenhouse gas emissions*. Luxembourg: Publications Office of the European Union. https://www.eea.europa.eu/publications/trends-and-drivers-of-eu-ghg/download.

Finney, M. A. (1998). *FARSITE: Fire Area Simulator: Model development and evaluation, Report RMRS-RP-4*. (RMRS-RP-4. Fort Collins, CO: U.S. Department of Agriculture, Forest Service, Rocky Mountain Research Station.

Fortin, M. (2021). Comparison of uncertainty quantification techniques for national greenhouse gas inventories. *Mitigation and Adaptation Strategies for Global Change*, *26*(2), 7. https://doi.org/10.1007/s11027-021-09947-4

French, N. H. F., Goovaerts, P., & Kasischke, E. S. (2004). Uncertainty in estimating carbon emissions from boreal forest fires. *Journal of Geophysical Research*, *109*, D14S08. https://doi.org/10.1029/2003JD003635

French, N. H. F., Kasischke, E. S., & Williams, D. G. (2002). Variability in the emission of carbon-based trace gases from wildfire in the Alaska boreal forest. *Journal of Geophysical Research*, *107*, 8151. https://doi.org/10.1029/2001JD000480

French, N. H. F., McKenzie, D., Erickson, T., Koziol, B., Billmire, M., Endsley, K. A., et al. (2014). Modeling regional-scale fire emissions with the Wildland Fire Emissions Information System. *Earth Interactions*, *18*, 1–26. https://doi.org/10.1175/EI-D-14-0002.1

Frost, G. J., Middleton, P., Tarrasón, L., Granier, C., Guenther, A., Cardenas, B., et al. (2013). New directions: GEIA's 2020 vision for better air emissions information. *Atmospheric Environment*, *81*, 710–712. https://doi.org/10.1016/j.atmosenv.2013.08.063

Giglio, L., Kendall, J. D., & Mack, R. (2003). A multi-year active fire dataset for the tropics derived from the TRMM VIRS. *International Journal of Remote Sensing*, *24*(22), 4505–4525. https://doi.org/10.1080/0143116031000070283

Giglio, L., Randerson, J., & van der Werf, G. (2013). Analysis of daily, monthly, and annual burned area using the fourth generation Global Fire Emissions Database (GFED4). *Journal of Geophysical Research*. https://doi.org/10.1002/jgrg.20042.c

Giglio, L., van der Werf, G. R., Randerson, J. T., Collatz, G. J., & Kasibhatla, P. (2006). Global estimation of burned area using MODIS active fire observations. *Atmospheric Chemistry and Physics*, *6*, 957–974. https://doi.org/10.5194/acp-6-957-2006

Gurney, K. R., Liang, J., Patarasuk, R., Song, Y., Huang, J., & Roest, G. (2020). The Vulcan Version 3.0 high-resolution fossil fuel CO_2 emissions for the United States. *Journal of Geophysical Research: Atmospheres*, *125*(19). https://doi.org/10.1029/2020jd032974

Hall, J. V., Loboda, T. V., Giglio, L., & McCarty, G. W. (2016). A MODIS-based burned area assessment for Russian croplands: Mapping requirements and challenges. *Remote Sensing of Environment*, *184*, 506–521. https://doi.org/10.1016/j.rse.2016.07.022

Hawbaker, T. J., Vanderhoof, M. K., Schmidt, G. L., Beal, Y.-J., Picotte, J. J., Takacs, J. D., et al. (2020). The Landsat Burned Area algorithm and products for the conterminous United States. *Remote Sensing of Environment*, *244*, 111801. https://doi.org/10.1016/j.rse.2020.111801

Hoelzemann, J. J., Schultz, M. G., Brasseur, G. P., Granier, C., & Simon, M. (2004). Global Wildland Fire Emission Model (GWEM): Evaluating the use of global area burnt satellite data. *Journal of Geophysical Research-Atmospheres*, *109*(D14), D14S04. https://doi.org/10.1029/2003JD003666.

Hoffa, E. A., Ward, D., Hao, W., Susott, R., & Wakimoto, R. (1999). Seasonality of carbon emissions from biomass burning in a Zambian savanna. *Journal of Geophysical Research*, *104*(D11), 13841–13853. https://doi.org/10.1029/1999JD900091

Holder, A. (2019). An overview of the wildland fire emissions inventory at EPA. https://cfpub.epa.gov/si/si_public_record_report.cfm?Lab=CEMM&dirEntryId=348217

Huang, Y., Wu, S., Dubey, M. K., & French, N. H. F. (2013). Impact of aging mechanism on model simulated carbonaceous aerosols. *Atmospheric Chemistry and Physics*, *13*(13), 6329–6343. https://doi.org/10.5194/acp-13-6329-2013

Hudak, A. T., Dickinson, M. B., Bright, B. C., Kremens, R. L., Loudermilk, E. L., O'Brien, J. J., et al. (2015). Measurements relating fire radiative energy density and surface fuel consumption: RxCADRE 2011 and 2012. *International Journal of Wildland Fire*, *25*(1), 25–37. https://doi.org/10.1071/WF14159

Hudak, A. T., Fekety, P. A., Kane, V. R., Kennedy, R. E., Filippelli, S. K., Falkowski, M. J., et al. (2020). A carbon monitoring system for mapping regional, annual aboveground biomass across the northwestern USA. *Environmental Research Letters*, *15*(9), 095003. https://doi.org/10.1088/1748-9326/ab93f9

Hudak, A. T., Lefsky, M. A., Cohen, W. B., & Berterretche, M. (2002). Integration of lidar and Landsat ETM+ data for estimating and mapping forest canopy height. *Remote Sensing of Environment*, *82*(2–3), 397–416. https://doi.org/10.1016/S0034-4257(02)00056-1

Hyer, E. J., & Reid, J. S. (2009). Baseline uncertainties in biomass burning emission models resulting from spatial error in satellite active fire location data. *Geophysical Research Letters*, *36*. https://doi.org/10.1029/2008GL037145

Ichoku, C., & Ellison, L. (2014). Global top-down smoke-aerosol emissions estimation using satellite fire radiative power measurements. *Atmospheric Chemistry and Physics*, *14*(13), 6643–6667. https://doi.org/10.5194/acp-14-6643-2014

IPCC. (2007). *2006 IPCC guidelines for national greenhouse gas inventories*. http://www.ipcc-nggip.iges.or.jp/public/2006gl/index.html

Ito, A., & Penner, J. E. (2004). Global estiamtes of biomass burning emissions based on satellite imagery for the year 2000. *Journal of Geophysical Research*, *109*, D14S05. https://doi.org/10.1029/2003JD004423

Johnston, J. M., Johnston, L. M., Wooster, M. J., Brookes, A., McFayden, C., & Cantin, A. S. (2018). Satellite detection limitations of sub-canopy smouldering wildfires in the North American boreal forest. *Fire*, *1*(2), 28. https://doi.org/10.3390/fire1020028

Kaiser, J. W., Heil, A., Andreae, M. O., Benedetti, A., Chubarova, N., Jones, L., et al. (2012). Biomass burning emissions estimated with a global fire assimilation system based on observed fire radiative power. *Biogeosciences*, *9*(1), 527–554. https://doi.org/10.5194/bg-9-527-2012

Kasischke, E. S., & Penner, J. E. (2004). Improving global estimates of atmospheric emissions from biomass burning. *Journal of Geophysical Research-Atmospheres*, *109*(D14), D14S01. https://doi.org/10.1029/2004JD004972

Kaufman, Y. J., Justice, C. O., Flynn, L. P., Kendall, J. D., Prins, E. M., Giglio, L., et al. (1998). Potential global fire monitoring from EOS-MODIS. *Journal of Geophysical Research: Atmospheres*, *103*(D24), 32215–32238.

Keane, R. E. (2012). Describing wildland surface fuel loading for fire management: A review of approaches, methods and systems. *International Journal of Wildland Fire*, *22*(1), 51–62. https://doi.org/10.1071/WF11139

Keane, R. E. (2015). Surface and ground fuels. In *Wildland fuel fundamentals and applications* (pp. 39–56). Springer.

Keane, R. E., Gray, K., Bacciu, V., & Leirfallom, S. (2012). Spatial scaling of wildland fuels for six forest and rangeland ecosystems of the northern Rocky Mountains, USA. *Landscape Ecology*, *27*(8), 1213–1234. https://doi.org/10.1007/s10980-012-9773-9

Kennedy, R. E., Ohmann, J., Gregory, M., Roberts, H., Yang, Z., Bell, D. M., et al. (2018). An empirical, integrated forest biomass monitoring system. *Environmental Research Letters*, *13*(2), 025004. https://doi.org/10.1088/1748-9326/aa9d9e

Key, C. H., & Benson, N. C. (2006). *Landscape analysis (LA) sampling and analysis methods, RMRS-GTR-164-CD*.

Klauberg, C., Hudak, A., Bright, B., Boschetti, L., Dickinson, M., Kremens, R., et al. (2018). Use of ordinary kriging and Gaussian conditional simulation to interpolate airborne fire radiative energy density estimates. *International Journal of Wildland Fire*, *27*(4), 228–240. https://doi.org/10.1071/WF17113

Kleinman, L. I., Sedlacek, A. J., III, Adachi, K., Buseck, P. R., Collier, S., Dubey, M. K., et al. (2020). Rapid evolution of aerosol particles and their optical properties downwind of wildfires in the western U.S. *Atmospheric Chemistry and Physics*, *20*(21), 13319–13341. https://doi.org/10.5194/acp-20-13319-2020

Knorr, W., Lehsten, V., & Arneth, A. (2012). Determinants and predictability of global wildfire emissions. *Atmospheric Chemistry and Physics*, *12*(15), 6845–6861. https://doi.org/10.5194/acpd-12-4243-2012

Korontzi, S. (2005). Seasonal patterns in biomass burning emissions from southern African vegetation fires for the year 2000. *Global Change Biology*, *11*(10), 1680–1700. https://doi.org/10.1111/j.1365-2486.2005.001024.x

Larkin, N., & O'Neill, S. M. (2018). *BlueSky playground v 3.5*. https://info.airfire.org/playground

Larkin, N. K., O'Neill, S. M., Solomon, R., Raffuse, S., Strand, T., Sullivan, D. C., et al. (2009). The BlueSky smoke modeling framework. *International Journal of Wildland Fire*, *18*(8), 906–920. https://doi.org/10.1071/WF07086

Larkin, N. K., Raffuse, S. M., Huang, S., Pavlovic, N., Lahm, P., & Rao, V. (2020). The comprehensive fire information reconciled emissions (CFIRE) inventory: Wildland fire emissions developed for the 2011 and 2014 U.S. National Emissions Inventory. *Journal of the Air and Waste Management Association*, *70*(11), 1165–1185. https://doi.org/10.1080/10962247.2020.1802365

Leite, R. V., Silva, C. A., Broadbent, E. N., Do Amaral, C. H., Liesenberg, V., De Almeida, D. R. A., et al. (2022). Large scale multi-layer fuel load characterization in tropical savanna using GEDI spaceborne lidar data. *Remote Sensing of Environment*, *268*, 112764.

Levine, J. S. (Ed.) (1991). *Global biomass burning: Atmospheric, climatic, and biospheric implications*. Cambridge, MA: MIT Press.

Mathews, B. J., Strand, E. K., Smith, A. M., Hudak, A. T., Dickinson, B., & Kremens, R. L. (2016). Laboratory experiments to estimate interception of infrared radiation by tree canopies. *International Journal of Wildland Fire*, *25*(9), 1009–1014. https://doi.org/10.1071/WF16007

McCarley, T. R., Hudak, A. T., Sparks, A. M., Vaillant, N. M., Meddens, A. J. H., Trader, L., et al. (2020). Estimating wildfire fuel consumption with multitemporal airborne laser scanning data and demonstrating linkage with MODIS-derived fire radiative energy. *Remote Sensing of Environment*, *251*, 112114. https://doi.org/10.1016/j.rse.2020.112114

McCarty, J. L., Aalto, J., Paunu, V.-V., Arnold, S. R., Eckhardt, S., Klimont, Z., et al. (2021). Reviews and syntheses: Arctic fire regimes and emissions in the 21st century. *Biogeosciences*, *18*(18), 5053–5083. https://doi.org/10.5194/bg-18-5053-2021

Michalek, J. L., French, N. H. F., Kasischke, E. S., Johnson, R. D., & Colwell, J. E. (2000). Using Landsat TM data to estimate carbon release from burned biomass in an Alaskan spruce complex. *International Journal of Remote Sensing*, *21*(2), 323–338. https://doi.org/10.1080/014311600210858

Morton, D., Le Page, Y., DeFries, R., Collatz, G., & Hurtt, G. (2013). Understorey fire frequency and the fate of burned forests in southern Amazonia. *Philosophical Transactions of the Royal Society B: Biological Sciences*, *368*(1619), 20120163. https://doi.org/10.1098/rstb.2012.0163

Morton, D. C., DeFries, R. S., Nagol, J., Souza, C. M., Jr., Kasischke, E. S., Hurtt, G. C., et al. (2011). Mapping canopy damage from understory fires in Amazon forests using annual time series of Landsat and MODIS data. *Remote Sensing of Environment*, *115*(7), 1706–1720. https://doi.org/10.1016/j.rse.2011.03.002

Nelson, R. M., Jr. (2000). Prediction of diurnal change in 10-h fuel stick moisture content. *Canadian Journal of Forest Research*, *30*(7), 1071–1087. https://doi.org/10.1139/x00-032

Nelson, R. M., Jr. (2001). Water relations of forest fuels. In E. A. Johnson & K. Miyanishi (Eds.), *Forest fires: Behavior and ecological effects*, (pp. 79–149). San Diego, CA : Academic Press.

NOAA (2021). F*ire influence on regional to global environments and air quality (FIREX-AQ)*. https://csl.noaa.gov/projects/firex-aq/

O'Brien, J. J., Loudermilk, E. L., Hiers, J. K., Pokswinski, S. M., Hornsby, B., Hudak, A. T., et al. (2016). Canopy-derived fuels drive patterns of in-fire energy release and understory plant mortality in a longleaf pine (*Pinus palustris*) sandhill in northwest Florida, USA. *Canadian Journal of Remote Sensing*, *42*(5), 489–500. https://doi.org/10.1080/07038992.2016.1199271

O'Neill, S., & Raffuse, S. (2021). Advances in satellite data for wildfire smoke forecasting. *EOS*, *102*. https://doi.org/10.1029/2021EO155076

O'Neill, S. M., Diao, M., Raffuse, S., Al-Hamdan, M., Barik, M., Jia, Y., et al. (2021). A multi-analysis approach for estimating regional health impacts from the 2017 Northern California wildfires. *Journal of the Air and Waste Management Association*, *71*(7), 791–814. https://doi.org/10.1080/10962247.2021.1891994

Olson, J. S. (1963). Energy storage and the balance of producers and decomposers in ecological systems. *Ecology*, *44*(2), 322–331.

Ottmar, R. D. (2014). Wildland fire emissions, carbon, and climate: Modeling fuel consumption. *Forest Ecology and*

Management, *317*, 41–50. https://doi.org/10.1016/j.foreco.2013.06.010

Ottmar, R. D., Sandberg, D. V., Riccardi, C. L., & Prichard, S. J. (2007). An overview of the Fuel Characteristic Classification System (FCCS): Quantifying, classifying, and creating fuelbeds for resource planning. *Canadian Journal of Forest Research*, *37*, 2383–2393. https://doi.org/10.1139/X07-077

Pettinari, M. L., & Chuvieco, E. (2016). Generation of a global fuel data set using the Fuel Characteristic Classification System. *Biogeosciences*, *13*(7), 2061–2076. https://doi.org/10.5194/bg-13-2061-2016

Pettinari, M. L., Ottmar, R. D., Prichard, S. J., Andreu, A. G., & Chuvieco, E. (2013). Development and mapping of fuel characteristics and associated fire potentials for South America. *International Journal of Wildland Fire*. https://doi.org/10.1071/WF12137

Pfister, G. G., Avise, J., Wiedinmyer, C., Edwards, D. P., Emmons, L. K., Diskin, G. D., et al. (2011). CO source contribution analysis for California during ARCTAS-CARB. *Atmospheric Chemistry and Physics*, *11*(15), 7515–7532. https://doi.org/10.5194/acp-11-7515-2011

Popescu, S., Wynne, R., & Nelson, R. (2003). Measuring individual tree crown diameter with lidar and assessing its influence on estimating forest volume and biomass. *Canadian Journal of Remote Sensing*, *29*(5), 564–577. https://doi.org/10.5589/m03-027

Pouliot, G., Pace, T. G., Roy, B., Pierce, T., & Mobley, D. (2008). Development of a biomass burning emissions inventory by combining satellite and ground-based information. *Journal of Applied Remote Sensing*, *2*(16 May 2008), 021501. https://doi.org/10.1117/1.2939551

Prichard, S. J., Keane, R., Rowell, E., Hudak, A., Loudermilk, E. L., Lutes, D., et al. (2021). Fuels and consumption. In D. L. Peterson, S. M. McCaffrey, & T. Patel-Weyand (Eds.), *U.S. Forest Service national smoke assessment*. Springer.

Prichard, S. J., Kennedy, M. C., Andreu, A. G., Eagle, P., & French, N. H. F. (2019). Next-generation biomass mapping for regional emissions and carbon inventories: Incorporating uncertainty in wildland fuel characterization. *Journal of Geophysical Research*, *134*. https://doi.org/10.1029/2019JGR005083

Prichard, S. J., O'Neill, S. M., Eagle, P., Andreu, A. G., Drye, B., Dubowy, J., et al. (2020). Wildland fire emission factors in North America: Synthesis of existing data, measurement needs and management applications. *International Journal of Wildland Fire*, *29*(2), 132–147. https://doi.org/10.1071/WF19066

Prichard, S. J., Ottmar, R. D., & Anderson, G. A. (2009). *CONSUME 3.0 users guide*. Pacific Wildland Fire Sciences Laboratory, USDA Forest Service. Seattle, WA. http://www.fs.fed.us/pnw/fera/research/smoke/consume/consume30_users_guide.pdf

Prichard, S. J., Ottmar, R. D., & Anderson, G. K. (2006). *CONSUME 3.0 user's guide*. Fire and Environmental Research Applications Team, USDA Forest Service. Seattle, WA. http://www.fs.fed.us/pnw/fera/research/smoke/consume/consume30_users_guide.pdf

Randerson, J. T., Chen, Y., van der Werf, G. R., Rogers, B. M., & Morton, D. C. (2012). Global burned area and biomass burning emissions from small fires. *Journal of Geophysical Research: Biogeosciences*, *117*(G4), n/a-n/a. https://doi.org/10.1029/2012JG002128

Rappold, A. G., Stone, S. L., Cascio, W. E., Neas, L. M., Kilaru, V. J., Carraway, M. S., et al. (2011). Peat bog wildfire smoke exposure in rural North Carolina is associated with cardiopulmonary emergency department visits assessed through syndromic surveillance. *Environmental Health Perspectives*, *119*(10), 1415–1420.

Reid, J. S., Hyer, E. J., Prins, E. M., Westphal, D. M., Zhang, J., Wang, J., et al. (2009). Global monitoring and forecasting of biomass-burning smoke: Description of and lessons from the Fire Locating and Modeling of Burning Emissions (FLAMBE) program. *IEEE Journal of Selected Topics in Applied Earth Observations and Remote Sensing*, *2*(3). https://doi.org/10.1109/JSTARS.2009.2027443

Reinhardt, E. D., Keane, R. E., & Brown, J. K. (1997). *First Order Fire Effects Model: FOFEM 4.0, user's guide*. USDA Forest Service, General Technical Report INT- GTR- 344.

Roberts, G., Wooster, M., Freeborn, P., & Xu, W. (2011). Integration of geostationary FRP and polar-orbiter burned area datasets for an enhanced biomass burning inventory. *Remote Sensing of Environment*, *115*(8), 2047–2061. https://doi.org/10.1016/J.RSE.2011.04.006

Roberts, G. J., & Wooster, M. J. (2008). Fire detection and fire characterization over Africa using Meteosat SEVIRI. *IEEE Transactions on Geoscience and Remote Sensing*, *46*(4), 1200–1218. https://doi.org/10.1109/TGRS.2008.915751

Rollins, M. G. (2009). LANDFIRE: A nationally consistent vegetation, wildland fire, and fuel assessment. *International Journal of Wildland Fire*, *18*(3), 235–249. https://doi.org/10.1071/WF08088

Saatchi, S. S., Harris, N. L., Brown, S., Lefsky, M., Mitchard, E. T., Salas, W., et al. (2011). Benchmark map of forest carbon stocks in tropical regions across three continents. *Proceedings of the National Academy of Sciences*, *108*(24), 9899–9904.

Schultz, M. G., Heil, A., Hoelzemann, J. J., Spessa, A., Thonicke, K., Goldammer, J. G., et al. (2008). Global wildland fire emissions from 1960 to 2000. *Global Biogeochemical Cycles*, *22*. https://doi.org/10.1029/2007GB003031

Seiler, W., & Crutzen, P. J. (1980). Estimates of gross and net fluxes of carbon between the biosphere and atmosphere from biomass burning. *Climatic Change*, *2*, 207–247.

Seinfeld, J. H., & Pandis, S. N. (1998). *Atmospheric chemistry and physics: From air pollution to climate change*. New York. John Willey & Sons.

Selimovic, V., Yokelson, R. J., McMeeking, G. R., & Coefield, S. (2019). In situ measurements of trace gases, PM, and aerosol optical properties during the 2017 NW U.S. wildfire smoke event. *Atmospheric Chemistry and Physics*, *19*(6), 3905–3926. https://doi.org/10.5194/acp-19-3905-2019

Skowronski, N., Clark, K., Nelson, R., Hom, J., & Patterson, M. (2007). Remotely sensed measurements of forest structure and fuel loads in the Pinelands of New Jersey. *Remote Sensing of Environment*, *108*(2), 123–129. https://doi.org/10.1016/j.rse.2006.09.032

Smith, A. M. S., Tinkham, W. T., Roy, D. P., Boschetti, L., Kremens, R. L., Kumar, S. S., et al. (2013). Quantification of fuel moisture effects on biomass consumed derived from fire radiative energy retrievals. *Geophysical Research Letters*, *40*, 6298–6302. https://doi.org/10.1002/2013GL058232

Souza, J., Carlos, M., Roberts, D. A., & Cochrane, M. A. (2005). Combining spectral and spatial information to map canopy damage from selective logging and forest fires. *Remote Sensing of Environment*, *98*(2–3), 329–343. https://doi.org/10.1016/j.rse.2005.07.013

UNECE (1979). *1979 Convention on Long-range Transboundary Air Pollution*. https://unece.org/convention-and-its-achievements

UNFCCC (2021). *Greenhouse gas reporting requirements*. https://unfccc.int/process-and-meetings/transparency-and-reporting/reporting-and-review-under-the-convention/greenhouse-gas-inventories-annex-i-parties/reporting-requirements

Urbanski, S. (2014). Wildland fire emissions, carbon, and climate: Emission factors. *Forest Ecology and Management*, *317*, 51–60. https://doi.org/10.1016/j.foreco.2013.05.045

Urbanski, S. P., Hao, W. M., & Nordgren, B. (2011). The wildland fire emission inventory: Western United States emission estimates and an evaluation of uncertainty. *Atmospheric Chemistry and Physics*, *11*(24), 12973–13000. https://doi.org/10.5194/acp-11-12973-2011

Urbanski, S. P., O'Neill, S. M., Holder, A. L., Green, S. A., & Graw, R. L. (2021). Emissions. In D. L. Peterson, S. M. McCaffrey, & T. Patel-Weyand (Eds.), *U.S. Forest Service National Smoke Assessment*. Springer.

Urbazaev, M., Thiel, C., Cremer, F., Dubayah, R., Migliavacca, M., Reichstein, M., et al. (2018). Estimation of forest aboveground biomass and uncertainties by integration of field measurements, airborne LiDAR, and SAR and optical satellite data in Mexico. *Carbon Balance and Management*, *13*(1), 1–20.

USEPA (2021a). *Criteria air pollutants*. https://www.epa.gov/criteria-air-pollutants

USEPA (2021b). *U.S. greenhouse gas emissions and sinks: 1990–2019*. https://www.epa.gov/ghgemissions/inventory-us-greenhouse-gas-emissions-and-sinks

USGS (2021). *3D elevation program*. https://www.usgs.gov/core-science-systems/ngp/3dep

Vallack, H. W., & Rypdal, K. (2019). *The global atmospheric pollution forum air pollutant emission inventory manual*. Version 6.0 May 2019 revision: York.

van der Werf, G., Randerson, J. T., Giglio, L., Collatz, G. J., Kasibhatla, P. S., & Arellano, A. F. (2006). Interannual variability of global biomass burning emissions from 1997 to 2004. *Atmospheric Chemistry and Physics*, *6*, 3423–3441. https://doi.org/10.5194/acp-6-3423-2006

van der Werf, G. R., Randerson, J. T., Giglio, L., van Leeuwen, T. T., Chen, Y., Rogers, B. M., et al. (2017). Global fire emissions estimates during 1997–2016. *Earth System Science Data*, *9*(2), 697–720. https://doi.org/10.5194/essd-9-697-2017

van Leeuwen, T., Peters, W., Krol, M., & Van Der Werf, G. (2013). Dynamic biomass burning emission factors and their impact on atmospheric CO mixing ratios. *Journal of Geophysical Research: Atmospheres*, *118*(12), 6797–6815. https://doi.org/10.1002/jgrd.50478

van Leeuwen, T. T. (2014). *Modeling spatio-temporal variability in biomass burning emission factors*. PhD dissertation. Vrije Univeriteit, Amsterdam.

van Leeuwen, T. T., van der Werf, G. R., Hoffmann, A. A., Detmers, R. G., Rücker, G., French, N. H. F., et al. (2014). Biomass burning fuel consumption rates: A field measurement database. *Biogeosciences*, *11*(24), 7305–7329. https://doi.org/10.5194/bg-11-7305-2014

Viegas, D., Piñol, J., Viegas, M., & Ogaya, R. (2001). Estimating live fine fuels moisture content using meteorologically-based indices. *International Journal of Wildland Fire*, *10*(2), 223–240. https://doi.org/doi.org/10.1071/WF01022

Viegas, D., Viegas, M., & Ferreira, A. (1992). Moisture content of fine forest fuels and fire occurrence in central Portugal. *International Journal of Wildland Fire*, *2*(2), 69–86. https://doi.org/10.1071/WF9920069

Viney, N. (1991). A review of fine fuel moisture modelling. *International Journal of Wildland Fire*, *1*(4), 215–234. https://doi.org/10.1071/WF9910215

Wiedinmyer, C., Akagi, S. K., Yokelson, R. J., Emmons, L. K., Al-Saadi, J. A., Orlando, J. J., et al. (2011). The fire inventory from NCAR (FINN): A high resolution global model to estimate the emissions from open burning. *Geoscientific Model Development*, *4*(3), 625–641. https://doi.org/10.5194/gmd-4-625-2011

Wooster, M. J., Roberts, G., Smith, A. M., Johnston, J., Freeborn, P., Amici, S., et al. (2013). Thermal remote sensing of active vegetation fires and biomass burning events. In C. Kuenzer & S. Dech (Eds.), *Thermal infrared remote sensing* (pp. 347–390). Springer.

Wooster, M. J., Robertson, G., Perry, G. L. W., & Kaufman, Y. J. (2005). Retrieval of biomass combustion rates and totals from fire radiative power observations: FRP derivation and calibration relationships between biomass consumption and fire radiative energy release. *Journal of Geophysical Research*, *D24311*. https://doi.org/10.1029/2005JD006318

Wooster, M. J., Zhukov, B., & Oertel, D. (2003). Fire radiative energy for quantitative study of biomass burning: derivation from the BIRF experimental satellite and comparison to MODIS fire products. *Remote Sensing of the Environment*, *86*, 83–107. https://doi.org/10.1016/S0034-4257(03)00070-1

Wulder, M. A., Bater, C. W., Coops, N. C., Hilker, T., & White, J. C. (2008). The role of LiDAR in sustainable forest management. *The Forestry Chronicle*, *84*(6), 807–826. https://doi.org/10.5558/tfc84807-6

Yao, J., Brauer, M., & Henderson, S. B. (2013). Evaluation of a wildfire smoke forecasting system as a tool for public health protection. *Environmental Health Perspectives*, *121*(10), 1142–1147. https://doi.org/10.1289/ehp.1306768

Yebra, M., Dennison, P. E., Chuvieco, E., Riano, D., Zylstra, P., Hunt, E. R., Jr., et al. (2013). A global review of remote sensing of live fuel moisture content for fire danger assessment: Moving towards operational products. *Remote Sensing of Environment*, *136*, 455–468. https://doi.org/10.1016/j.rse.2013.05.029

Yokelson, R. J., Goode, J. G., Ward, D. E., Susott, R. A., Babbitt, R. E., Wade, D. D., et al. (1999). Emissions of formaldehyde, acetic acid, methanol, and other trace gases from biomass fires in North Carolina measured by airborne Fourier transform infrared spectroscopy. *Journal of Geophysical Research: Atmospheres*, *104*(D23), 30109–30125. https://doi.org/10.1029/1999JD900817

Zhang, G., Ganguly, S., Nemani, R. R., White, M. A., Milesi, C., Hashimoto, H., et al. (2014). Estimation of forest aboveground biomass in California using canopy height and leaf area index estimated from satellite data. *Remote Sensing of Environment*, *151*, 44–56. https://doi.org/10.1016/j.rse.2014.01.025

Zhang, X., Kondragunta, S., Da Silva, A., Lu, S., Ding, H., Li, F., & Zhu, Y. (2019). *The Blended global biomass buring emissions product from MODIS and VIIRS observations (GBBEPx), version 3.1*. https://www.ospo.noaa.gov/Products/land/gbbepx/docs/GBBEPx_ATBD.pdf

Part II
From Emissions to Concentrations

6

Surface Monitoring of Fire Pollution

Allison E. Bredder

ABSTRACT

This chapter discusses efforts to measure surface observations of air pollution at the country scale. The countries with the most comprehensive regulatory systems to monitor air pollution are the older industrial nations such as countries in the United Kingdom and the United States. Recent proliferation of low-cost air quality sensors (LCAQS) are making near-real-time air pollution monitoring more prevalent across the globe. While unique challenges exist between regulatory and LCAQS data access and usability, there are common challenges in using these data for decision support and research applications. This chapter discusses common statistical methods for estimating air pollution including spatial interpolation methods, statistical regression methods, machine learning, and chemical transport modeling.

6.1. INTRODUCTION

The impact of air quality (AQ) on health has been acknowledged by governments of individual countries and the World Health Organization (WHO) for more than half a century. The United States, the United Kingdom, and the Union of the Soviet Socialist Republics (USSR) were among the first to enact a version of a clean air act around 1955–1956 aimed at controlling air pollution and minimizing negative impacts on public health (Barker et al., 1961). In 1958, the World Health Organization (WHO) published its first technical report, *Air Pollution*, that explicitly linked exposure to high concentrations of pollutants to adverse health outcomes (WHO, 1958). Although the report neither discussed the toxicology of individual pollutants nor proposed any guidelines on concentrations, it nonetheless was a major step toward the eventual development of national and subsequently global AQ standards.

Department of Geographical Sciences, University of Maryland, College Park, Maryland, USA

Over time, many countries worldwide developed a set of rigorous science-based AQ standards, enacted laws and regulations, and established networks of monitoring stations. Reflecting the historical development of the AQ regulations, the monitoring stations are primarily focused on urban AQ with attention to populated areas (Ambient Air Quality Surveillance, 1994). In addition, considering that the primary purpose of these networks is to meet defined regulatory AQ goals from the regional to international levels, the expected accuracy of measurements and the precision of the instruments require careful cost consideration and make high-density spatial observations prohibitively expensive. While these traditional government-sponsored national air monitoring networks provide gold-standard observations for a large suite of air pollutants, they are frequently far too sparse and suboptimally located to support monitoring of air pollution associated with biomass burning (Reid et al., 2015). Globally, biomass burning is highly varied (see Chapter 2).

New advancements and global proliferation of less costly air monitors, termed *low-cost air quality sensors* or

Landscape Fire, Smoke, and Health: Linking Biomass Burning Emissions to Human Well-Being, Geophysical Monograph 280, First Edition.
Edited by Tatiana V. Loboda, Nancy H. F. French, and Robin C. Puett.

DOI: 10.1002/9781119757030.ch06

LCAQS, has dramatically increased the potential for near-real-time monitoring of smoke events by governments, researchers, and citizen scientists alike. Although the advance of LCAQS has increased the availability of stationary measurements, their spatial patterns are frequently subject to similar limitations and biases toward urban environments but to a lesser degree.

This chapter provides a brief overview of the following topics:

1. An overview of AQ monitoring networks, including established regulatory networks, global and emerging networks, and LCAQS networks
2. Common statistical methods to derive spatiotemporally resolved AQ estimates, with a focus on applications to particulate matter
3. A discussion of the challenges associated with using AQ monitoring networks for smoke pollution monitoring
4. The future directions and opportunities for monitoring smoke pollution

6.2. MONITORING NETWORKS

Air quality monitoring networks, also referred to as surveillance networks, record information about levels of air pollutants (Marć et al., 2015). Monitoring networks measure a range of ambient air pollutants. The air pollutants that are most commonly collected include particulate matter that is less than or equal to 10 and 2.5 μm in aerodynamic diameter in size (known as PM_{10} and $PM_{2.5}$, respectively), ozone (O_3), mercury (Hg), sulfur dioxide (SO_2), nitrous oxides (NO_X), nitrous dioxide (NO_2), and persistent organic pollutants (Marć et al., 2015). Monitoring networks can be classified into two categories: regulatory (or reference) monitoring networks and LCAQS networks.

6.2.1. Regulatory AQ Networks

Air pollution can contribute to a range of negative effects that impact humans, ecosystems, and human-made structures. Governments and regulatory bodies have a vested interest in monitoring AQ for economic, public health, and political reasons. Air-quality monitoring systems operated by governments have increased since the 1800s, coinciding with air pollution impacts from the Industrial Revolution, and other large-scale air pollution events that resulted in negative impacts, such as the London Smog Event of 1952 that served as a catalyst for legislative change and investment in technology to monitor goals. Countries have adopted their own systems for monitoring AQ using ground-based monitors, with the responsibility for collecting and disseminating information typically assigned to entities broadly referred to as *environmental protection agencies*.

Regulatory monitors are broadly defined here as ground-based, stationary monitors (also known as *in situ monitors*) that are deployed by or on behalf of country-level governments. This section primarily focuses on regulatory networks to monitor AQ, defined as meeting two criteria: (1) the network is mandated or sponsored by or on behalf of a country's government, and (2) the network is constructed of ground-based, stationary AQ monitors. Additionally, the focus is on networks where data are publicly available via the Internet, but other programmatic efforts are also discussed. They are typically used to meet legislative requirements such as ambient air quality standards or research purposes (Castell et al., 2017). However, with the proliferation of LCAQS, governments have also begun to invest in those to make information available in near real time to support emergency management and to provide more information to communities interested in tracking smoke events (Morawska et al., 2018).

While not discussed here, there are dedicated efforts to assessing technology and methods associated with sampling AQ (Helsen, 2005; Shaddick & Zidek., 2014) and determining optimal locations where monitors should be located for optimal spatial distribution (see Chapter 10) (Hao & Xie, 2018; Piersanti et al., 2015). Quantifying spatiotemporally resolved air pollution concentrations is critical for mapping biomass burning and understanding how biomass burning emissions are transported (see Chapter 8).

Information about country-level AQ monitoring networks was derived from peer-reviewed and grey literature that described air AQ monitoring networks in the United States by an English-speaker; therefore, a limitation for information provided in this section may be attributable to language or website accessibility from the United States.

Established National AQ networks

Overall, as can be expected, most extensive networks and the longest archives of measurements are found within wealthy countries with a long history of industrial development. The world's older industrial giants (the United States, United Kingdom, and USSR) were among the first to enact laws governing air pollution in the midtwentieth century (Barker et al., 1961). These were rapidly joined by other industrialized countries, including many European countries, Canada (Government of Canada, 2021), and Japan (Wakamatsu et al., 2013), which initialized their national monitoring networks in the late 1960s–early 1970s. Over half of the century, these networks have undergone several major improvements, including the increase in number of measured pollutants, technical advances in instrumentation, improved statistical techniques, and substantial network growth.

In the United States, the Environmental Protection Agency (EPA) is charged with collecting and disseminating

AQ information from local, state, and tribal entities using Federal Reference Methods and Federal Equivalent Methods. The EPA monitoring network consists of over 4,000 stations that are distributed across all states and territories for criteria pollutants (CO, NO_2, O_3, Pb, PM_{10}, $PM_{2.5}$, and SO_2) and 188 other toxic air pollutants (U.S. Environmental Protection Agency, 2021a). Data from the EPA monitoring cites are publicly available since 1980 for the criteria gases, 1988 for PM_{10}, and 1999 for $PM_{2.5}$. Hazardous air pollutants and toxic air pollutants are available from 1980 (U.S. Environmental Protection Agency, 2021d). While these monitors are not specifically designed for biomass burning pollution, they are often used in studies focused on assessing the health effects of pollution from biomass burning (see Chapter 10). These measurements are supplemented by over 90 Clean Air Status and Trends Network (CASTNet) deposition monitoring sites operated by EPA (U.S. Environmental Protection Agency, 2021c) and the Interagency Monitoring of Protected Visual Environments (IMPROVE) network with 160 sites as of 2019 located in national parks and in wilderness areas (Interagency Monitoring of Protected Visual Environments, 2020). In addition, the National Oceanographic and Atmospheric Administration (NOAA) Earth System Science Laboratory has measured surface ozone since 1973 at 20 sites across the world (NOAA, n.d.).

Like the EPA regulatory network across the United States, Canada operates the National Air Pollution Surveillance (NAPS) program (Environment Canada, 2020), which aims to deliver consistent high-quality observations across the nation. At present, the NAPS program boasts 286 sites in urban and rural communities across all provinces and territories. Although country-wide summaries have been published since 1972, these early reports are based on observations from a very small fraction of currently available sites. The NAPS program collects continuous and time-integrated measurements for a predetermined number of pollutants. Observations include CO, NO_2, NO, NO_X, O_3, SO_2, $PM_{2.5}$, and PM_{10}, with hourly and annual data available for CO, SO_2, NO_2, and O_3 since 1974. Particulate matter data are available since 1992 for PM_{10} and since 1995 for $PM_{2.5}$.

The European Environment Agency is responsible for establishing the policy framework for monitoring AQ across the European Union (EU) zone (Directorate-General for Environment, n.d.). Through a series of directives, the EU established standards for ambient air concentrations for several pollutants, defined the methodologies for data collection, and monitored the compliance for each of the EU Member States. The Member States are expected to monitor and report AQ data by predefined zones and agglomerations (established by the Member States following the methodology defined by the agency), as well as make the AQ information available to the public through the European Air Quality Portal. At present, the number of operational stations totals around 5,300 stations across the 41 contributing countries and territories (European Environment Agency, 2020).

In Australia, the National Clean Air Agreement establishes the framework for AQ monitoring (Commonwealth of Australia, 2015). Although Australia's urban areas are reported to have some of the best AQ in the world, biomass burning is widely acknowledged as a one of the primary sources of air pollution (Keywood et al., 2016). Similar to the EU framework, the National Environmental Protection Council administers legislation pertaining to AQ and provides scientific and policy support. Data collection, which follows predetermined standards, called National Environment Protection Measures (NEPMs), is the responsibility of provincial and state governments who are also charged with managing AQ. While there was no centralized data repository found for all Australian data across all states, each jurisdiction offers varying levels of access to AQ data.

Although the USSR was the first country in the world to define the standards for acceptable AQ (Izmerov, 1974), the data from the government-sponsored monitoring network and information about the precision of instruments, statistical methods, reporting frequency, and the number of monitoring sites do not appear to be publicly available. The Russian Federal Service for Hydrometeorology and Environmental Monitoring reports annually on the most polluted cities in Russia (Klyuev, 2019), which indicates the presence of the statewide network of monitoring stations at least across major urban areas.

Global and Emerging AQ Networks

The global awareness of health impacts from AQ in urban areas was growing from the early 1970s, when the WHO published its technical report on air quality guidelines for urban areas (WHO Expert Committee on Air Quality Criteria and Guides for Urban Air Pollutants & World Health Organization, 1972), which included contributors from Egypt, India, and Japan in addition to the European and North American experts. However, AQ monitoring networks in much of the rest of the world have been relatively slow to grow. In Central, South America, the Caribbean, and Africa, the monitoring networks are sparse (Awokola et al., 2020; Riojas-Rodríguez et al., 2016). Riojas-Rodríguez et al. (2016) found in their review that only half (17 of 33) Latin American and Caribbean countries had AQ monitoring stations. There appears to be less consistency in collected measurements across the region, for example, PM_{10} measurements are collected in 104 cities while $PM_{2.5}$ measurements are collected only in 57 cities. According to Rees et al. (2019), only 13% (7 of 54 countries) in Africa provide reliable, real-time AQ monitoring; however, it is unclear if these monitors meet the criteria of this section. Ghana, Nigeria, and Kenya each have 5 national level, manual stations (Gulia et al., 2020). South Africa is the only country in Africa with a monitoring network that was found to be available to the public. The network of 130 fully automated stations within the National Ambient Air Quality Monitoring Network (NAAQMN) of South Africa was launched in the late 2010s as a partnership between the Department of Environmental Affairs and the South African Weather Service (Gwaze & Mashele, 2018). In line with best practices from the international community, the agency monitors pollutants following established criteria and methodology and delivers the information to the public through a mobile application tool.

Air quality monitoring in Asia presents a unique set of challenges. On the one hand, expansive monitoring networks exist in some parts of Asia, with the other two largest government-run networks within Japan and South Korea. The Korean Ministry of Environment has provided real-time data at 16 locations since 2002 near the World Cup Stadium located in the capital city of Seoul and has provided public access to data in real time since 2005 on a nationwide scale for CO, NO, SO_2, and $PM_{2.5}$ and PM_{10} from 332 stations via the AirKorea website (Hwang et al., 2020). On the other hand, the two largest industrial economies of the continent, China and India, only comparatively recently launched their AQ monitoring networks. Although the China National Environmental Monitoring Center (CNEMC) was founded in 1980 by the Ministry of Ecology and Environment of China, AQ data have been collected only since 2013. The monitoring network has grown very rapidly to currently reach over 2,100 stations that monitor CO, NO_2, O_3, $PM_{2.5}$, PM_{10}, and SO_2 (China National Environmental Monitoring Centre, n.d.; Chu et al., 2021). The data are available via the CNEMC website, but the volume of observations is skewed toward eastern parts of the country. The government of India initiated the National Clean Air Program (NCAP) only in the beginning of 2019 under the oversight of the Ministry of Environment, Forests and Climate Change (International Trade Administration, 2020). The network currently includes a suite of 703 manual monitoring stations and 134 Continuous Ambient Air Quality Monitoring Stations (CAAQMS: low-cost monitoring sensors), which is expected to grow substantially in the near future to the total of 1,500 manual monitoring stations and an additional 150 CAAQMS (Sundaray, & Bhardwaj, 2019). In other parts of Asia, Vietnam has 29 fixed and mobile CAAQMS, Pakistan has 70 manual and CAAQMS, Bangladesh has 11 CAAQMS, Sri Lanka has 78 manual stations, Nepal and Bhutan have 12 and 3 CAAQMS, respectively (Gulia et al., 2020).

6.2.2. Low-Cost Air Quality Sensor (LCAQS) Networks

Technological advances of the past decades combined with the growing public awareness of health consequences of environmental pollution globally have created a favorable climate for the development of alternative approaches to the regulatory AQ monitoring stations. Fueled by investment from commercial companies, governments, nongovernmental organizations, and lay citizens, LCAQS networks have rapidly increased in number across the world. Considerably lower financial costs and expertise are required to set up and maintain these stations compared with regulatory-grade monitors, which has allowed for a manifold increase in surface measurements for a suite of pollutants deployed by government agencies and private citizens alike (Table 6.1). LCAQS networks are attractive for use in biomass burning and

Table 6.1 Selected Low-Cost Air Quality Sensor (LCAQS) networks

Network name	Pollutants measured	Sensor technology used and cost per individual sensor	Data retrieval location(s)
IQAir	$PM_{2.5}$, CO_2	AirVisual Pro Air Quality Monitor: $269	Web map: United Nations Environment Programme, n.d.
PurpleAir	$PM_{0.3}$, $PM_{0.5}$, PM_1, $PM_{2.5}$, PM_5, PM_{10}, PM_1	PurpleAir PA-I-Indoor: $199 PurpleAir PA-II: $249 PurpleAir PA-II-SD: $279	Web map: PurpleAir, n.d. API: PurpleAir, 2021
Air Quality Egg	CO, CO_2, NO_2, O_3, PM_1, $PM_{2.5}$, PM_{10}, SO_2, VOCs	Indoor: $130 Outdoor: $160	Web map: Air Quality Egg, n.d.
World Air Quality Index project	$PM_{2.5}$, PM_{10}	Aggregated from Web sources	Web map: World Air Quality Index Project, 2022

Note: Prices were retrieved in June 2021.

prescribed fire smoke exposure assessment as they offer denser and more dispersed observations and are available worldwide, often in countries that do not have robust national monitoring networks. Although mobile LCAQS are available, they offer only episodic observations frequently associated with a particular event or project. In contrast, stationary LCAQS and monitoring networks (the focus of this chapter) provide consistent observations for a given location, similar to those obtained by the regulatory networks.

The LCAQS networks contain several important components. First, the data are collected by low-cost technologies largely referenced as sensors. The investment can range roughly between tens of dollars (for a single sensor) and \$5,000 USD for more comprehensive kits (Feenstra et al., 2019; Holder et al., 2020; Rai et al., 2017). Adopting the definition from Rai et al. (2017), *low-cost sensors* refer to "anything costing less than the instrumentation cost required for demonstrating compliance with the air quality regulations" and can include single sensors or "sensing kits/nodes/platforms [that] typically include one or more sensors, microprocessor, data-logger, memory card, battery, and display" (Rai et al., 2017, pp. 692–693). Monitoring networks are constructed of sensors and typically rely upon the Internet of things, generally physical objects that are connected by the Internet (Xia et al., 2012), to disseminate access to the data collected by the sensors. For example, the PurpleAir LCAQS network collects data from Plantower PMS1003 sensors; the data collected from the sensors are made publicly available using a Web map and an application programming interface (API) for data download by end users. The number of LCAQS networks are growing rapidly: the Fire and Smoke Map, OpenAQ, and Urban Air Action Platform, and the UN's Urban Air Action Web platform help illustrate the potential capabilities of LCAQS for biomass burning AQ monitoring.

The United States has piloted a Web map called the Fire and Smoke Map (U.S. Environmental Protection Agency, 2021b) that is targeted for biomass burning exposure assessment in North America (Fig. 6.1). The Web map integrates AQ and fire information from a variety of sources. Specifically, $PM_{2.5}$ concentrations are provided from permanent monitors, which feed into the AirNow network, and temporary $PM_{2.5}$ monitors, that are deployed by governmental agencies to monitor smoke events PurpleAir data, an increasingly popular network (PurpleAir, n.d., 2021). For example, Gupta et al. (2018) used 180 PurpleAir $PM_{2.5}$ data in conjunction with satellite data to estimate $PM_{2.5}$ during California fires in 2017. The EPA led a nationwide effort of over 30 agencies at the state, local, and tribal levels to develop a nationwide correction for PurpleAir $PM_{2.5}$ measurements that are applied to the data displayed on the Fire and Smoke Map (U.S. Environmental Protection Agency, 2021b). Over 70 PurpleAir sensors were co-located with regulatory-grade monitors in the evaluation (U.S. Environmental Protection Agency, n.d.). Active fire detections from the National Oceanic and Atmospheric Administration's Hazard Mapping System (NOAA, n.d.) and large fire incidents from the U.S. National Interagency Fire Center (InciWeb, n.d.) are also available as data layers on the Web interface. The U.S. AirNow Department of State network (U.S. Department of State & U.S. Environmental Protection Agency, n.d.) provides real-time $PM_{2.5}$ data from monitors on U.S. embassies and consulates across the globe.

OpenAQ is an open-source platform that integrates reference-quality data from governments and low-cost AQ data from the Air Quality Data Commons, HabitMap, PurpleAir, and Carnegie Mellon University (OpenAQ, 2021). The platform primarily provides data regarding CO, NO_2, O_3, $PM_{2.5}$, PM_{10}, SO_2, and black carbon. The Web platform provides download capability of 2 yr of data (historic data can be retrieved from Amazon Web Services), an R wrapper, and a Python wrapper. The wrappers allow users to access the Application Programming Interface (OpenAQ, n.d.). Importantly, OpenAQ does not perform quality assessment of the data, which necessitates substantial effort in data cleaning and preprocessing when those data sets are acquired for research or management purposes.

On a global scale, the United Nations (UN) Environment Programme and UN-Habitat deployed the Urban Air Action Web platform in 2020 (United Nations Environment Programme, n.d.). The Web platform displays near-real-time $PM_{2.5}$ data in collaboration with the commercial company IQAir, wind data, world population data, and fire locations.

6.3. METHODS TO ESTIMATE AIR POLLUTION CONCENTRATIONS

Methods to develop spatially contiguous estimates of air pollution have rapidly evolved in the past nearly two decades with interest in using ground-based monitors and sensors for that exposure assessment in epidemiological studies (see Chapter 7). Four categories of methods for developing continuous measurements will be discussed below, with particular attention to particulate matter: (1) spatial interpolation methods, (2) land use regression, (3) machine learning, and (4) chemical transport models (CTMs). Biomass burning events exhibit unique characteristics in space and time, and those unique characteristics can affect which modeling approach best represents smoke concentration and is feasible given model limitations (Mirzaei et al., 2018). Most of these approaches provide some measure of uncertainty. While statistical metrics are often reported to express error and uncertainty in interpolation, machine learning, regression, and chemical

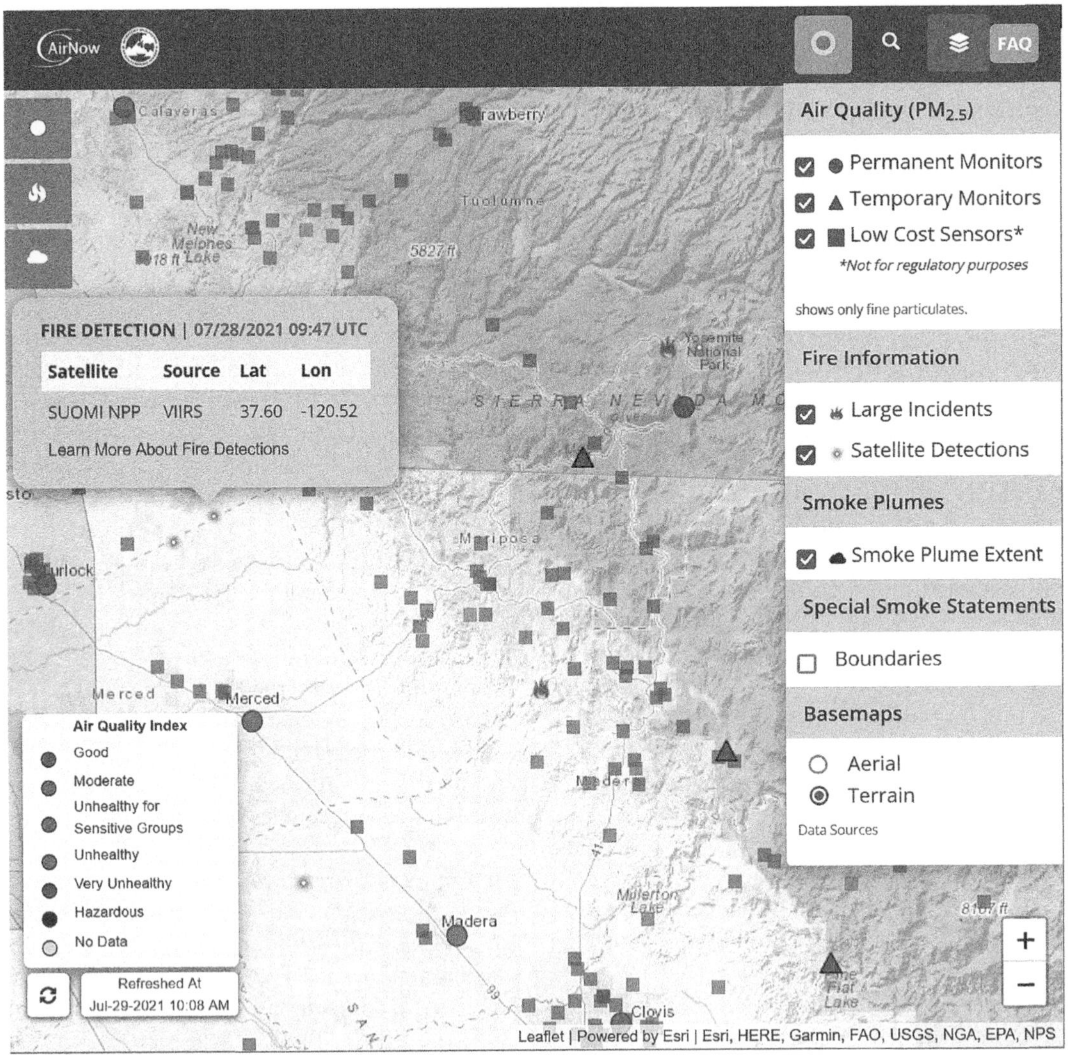

Figure 6.1 Screen capture of the Fire and Smoke Map Web portal over California, USA. Three types of air monitoring sensors are displayed with different shapes. Squares represent PurpleAir sensors. Triangle monitoring sensor is colored according to the Air Quality Index (legend shown) (U.S. Environmental Protection Agency, 2021a). Fire symbols are large fires, and smaller circles represent active fires that are detected by satellites.

transport model efforts, it is common for only a subsuite or the final chosen model to be presented and details regarding sensitivity analyses are absent (Gan et al., 2017; Hu et al., 2017; Stafoggia et al., 2019). Often, effect estimates due to model uncertainty are not reported for models that did not meet specified criteria, but this information could be useful for model selection in other applications (Arhami et al., 2013).

6.3.1. Spatial Interpolation

Spatial interpolation involves using values with known locations to predict estimates where values are not known. For AQ applications, this frequently means using AQ monitor readings at one location to predict values where AQ readings do not exist, but can also be applied to raster data, such as satellite imagery. With the most simplistic spatial interpolation methods, no other ancillary data are required (Watson et al., 2019). Spatial interpolation methods are commonly used given the primary data input is known information and popular geostatistical and mapping software such as ArcGIS, QGIS and GRASS GIS, and R readily support spatial interpolation methods through functions and packages.

Thirty-eight spatial interpolation methods and submethods exist, with progress continuing to be made in this field (Li & Heap, 2014). These methods are commonly described and categorized according to dichotomies of features (Deligiorgi & Philippopoulos, 2011; Li & Heap, 2008, 2014), including:

1. Deterministic and stochastic methods. The primary difference between the two suites of methods is that

deterministic methods do not incorporate randomness into their models while stochastic methods do. Thus, deterministic methods do not provide a measure of uncertainty, whereas stochastic methods provide error estimates.

2. Global and local methods. Global methods derive estimations using all data available in the study area whereas local methods use a sample of estimates in their calculation.

3. Exact interpolators and approximate interpolators. Exact interpolators derive values that are part of the known data whereas approximate interpolators can estimate values that are not the same as data that already exist.

To assist practitioners and researchers in determining which spatial interpolation method is best suited for the available information and desired results, Li and Heap (2014) provide a detailed decision tree that classifies spatial interpolation methods.

Two common spatial interpolation methods for wildfire AQ applications include inverse distance weighting (IDW) and kriging (Krige, 1951). Both methods realize Tobler's first law: phenomena that are closer together in space are more like each other than to things that are located farther away (Tobler, 1970). The IDW function interpolates values using existing values at a specified distance from the location without known values. Therefore, optimal application of IDW is when the known values are close in distance to unmeasured locations. Conversely, this method is less useful when predicting over areas where known values are farther away, such as remote rural areas where known values are sparse. Studies have used IDW to predict $PM_{2.5}$ using ground monitors (Wu et al., 2006; Yang et al., 2020). A large body of literature exists that is dedicated to developing new formulations for IDW (Ma et al., 2019).

Kriging also uses weights for closer values, but the weights also take into consideration the spatial patterns of known data. Currently, over 20 versions of kriging methods are in existence (Li & Heap, 2014). As a geostatistical method, kriging delivers an uncertainty metric that can be useful to assess the performance of the algorithm. Kriging has been used to estimate $PM_{2.5}$ over Washington State, USA, from reference-grade monitors (Gan et al., 2017) and over the coterminous United States and Ontario, Canada, from 1988 to 2016 from research-grade monitors.

6.3.2. Statistical Regression Methods

Common statistical models to estimate pollutant concentrations include multiple linear regression, land-use regression, mix-effects modeling, generalized additive models (GAM), and geographically weighted regression (GWR). Earlier studies that used multiple linear regression to predict PM values established the importance of improving model estimations by including meteorological covariates (Chu et al., 2016). Land use regression (LUR), an extension of multiple linear regression, refers to regression models that are used to predict AQ concentrations (as the dependent variable), using covariates of ancillary information. However, despite what the name of this technique implies, the parameters are not always associated with land use (Watson et al., 2019). In practice, LUR models commonly incorporate meteorological information, including temperature, humidity, precipitation, wind, and air related variables, topographic variables, aerosol optical depth (AOD) (see Chapter 7). For these methods, ground-level PM, ozone, or other pollutants are the dependent variable, and independent variables include AOD and other ancillary variables (Liu et al., 2005). Both multiple linear regression and land-use regression are limited in their effectiveness where covariates and ground-level PM have a nonlinear relationship. Additionally, these approaches can become difficult to handle with large amounts of data (Hu et al., 2017; Shin et al., 2020).

Another extension of the multiple linear regression, the GAM, accounts for nonlinear relationships between variables (Ma et al., 2014; Shin et al., 2020; Sorek-Hamer et al., 2013). The mix-effects modeling has largely replaced the use of MLR since 2010 (Chu et al., 2016). Fixed and random effects are incorporated into the mix-effects modeling to represent the background relationship between PM and AOD, and temporal and regional variation, respectively (Shin et al., 2020). Finally, geographically weighted regression accounts for nonstationarity and different relationships between ground-level and covariates (Luo et al., 2017; Shin et al., 2020). However, these models are highly sensitive to locations and distribution of ground stations (Shin et al., 2020) as well as the suite of ultimately selected variables. Considering that inclusion or exclusion of variables is subject to the discretion of the user, the resultant predictive capability is highly diverse as the tactics for selecting variables can vary widely among individual researchers and by discipline (Watson et al., 2019). Using a linear regression model, Yao and Henderson (2014) estimated $PM_{2.5}$ concentrations in British Columbia in areas that did not have a monitoring network. They assessed model performance on low-, moderate-, and high-smoke days.

6.3.3. Machine Learning

Machine learning refers to methods that use artificial intelligence, which fit independent variables that are spatiotemporally variant (Watson et al., 2019). Machine learning approaches to estimate smoke concentrations have quickly become a dominant method in the past few years, as they do not assume linearity between the dependent variable and covariates and are stable and efficient for processing large amounts of data, increasing the capabilities for predicting longer time series of trace gases and atmospheric pollutants (Bellinger et al., 2017). Popular machine learning techniques include kernel and tree-based approaches. Kernel-based approaches, such as support vector regression, are often used in multistage modeling (Shin et al., 2020; Song et al., 2014). Tree-based approaches rely upon decision trees to make predictions. These include classification/regression trees and random forest (RF) ensembles (Breiman, 2001), gradient boosting machines (Ferreira & Figueiredo, 2012), and extreme gradient boosting.

In 2015, Reid et al. compared 11 statistical models for predicting $PM_{2.5}$ during the 2008 biomass burning event in Northern California fires and found that the RF had among the highest cross-validated accuracy. Since this finding, machine learning algorithms, and specifically RF models, have been increasingly used to estimate the PM at regional and national scales (Chen et al., 2018a; Chen et al., 2018b; Di et al., 2019; Hu et al., 2017; Park et al., 2019; Reid et al., 2015; Stafoggia et al., 2019; Zhao et al., 2020). A more recent study showed the RF approach to predicting PM_{10} over China had better performance and improved predictive capabilities compared with traditional regression models (Chen et al., 2018b). In addition to predicting PM, machine learning has been used to predict other pollutants, including ozone exposure before and after biomass burning events (Watson et al., 2019). Cross-validation methods are common metrics to use to evaluate model performance and estimate uncertainty. A disadvantage of machine learning methods is they often rely on specialized computer coding languages that are not always publicly available (Watson et al., 2019), although a number of open-source applications, including an R-package and a Python-based implementation, are openly available and easily accessible. In addition to the steep learning curve required to implement these methods, RF models are frequently referred to as *black box* methods, which implies that the internal algorithm decisions that produce the ultimate outcome are not always transparent, and it may be difficult to interpret the results (Affenzeller et al., 2020).

6.3.4. Chemical Transport Modeling

Chemical transport models (CTMs) rely upon meteorology, emissions inventories, and chemical and physical processes to quantify spatiotemporal patterns of atmospheric gases (Engel-Cox et al., 2013). Chemical transport models have been used to estimate PM and have been shown to be effective at coarser spatial resolutions and global scales. As CTMs do not rely upon ground-based measurements, these approaches are useful in areas where ground records do not exist or are highly heterogeneous (Boys et al., 2014; Chu et al., 2016; Van Donkelaar et al., 2006). CTMs are more commonly used in multistage models for gap filling missing information, such as aerosol optical depth (Di et al., 2019; Stafoggia et al., 2019). Studies have also used CTMs to model biomass burning emissions on air pollution and to determine emission factors (Akagi et al., 2011; Garcia-Menendez et al., 2014; Hodzic et al., 2007; Konovalov et al., 2011; Wiedinmyer et al., 2006). A limitation of CTMs' utility for biomass burning smoke is limited by knowledge of fire properties such as injection height and fuel loading (Paugam et al., 2016), meteorology uncertainties, and computational limitations to integrate the information into a useful model (Lassman et al., 2017). Chapter 8 provides a full review of CTM for biomass burning smoke concentration mapping.

6.4. GAPS AND CHALLENGES IN MONITORING WILDFIRE POLLUTION

This chapter provides a selected overview of international AQ monitoring efforts based on information that is publicly available and accessible. Although these observations are undoubtedly a vital resource, comprehensive monitoring fire pollution using ground-based stations is unattainable because the task requires spatially and temporally inclusive estimates. Ultimately, the regulatory networks were never designed to monitor air pollution originating from biomass burning. Thus, they present a very limited, although valuable, source of information.

The technology that regulatory-grade monitors rely upon delivers highly accurate measurements at the point of data collection. However, the tradeoff is that the instruments are heavy, large, and expensive to construct and maintain. As a result, the spatial coverage of measurements from regulatory networks is very sparse. Fire events can be unpredictable in size, scale, and duration, making cost-effective instrumentation for effective monitoring extremely challenging. Considering the primary focus of regulatory networks on air pollution associated with industrial activity and transportation, monitors are typically found in urban centers. This positions the stations away from the majority of ongoing biomass burning events. While stationary monitoring networks are established and continue to grow (see section 6.2) and temporary monitors are deployed during smoke events (see section 6.2.1), they deliver point measurement in 3D space and time. They also require a large subsequent effort to produce spatially contiguous estimates of AQ and pollutants' concentrations.

In addition to limited spatial coverage, conventional ground-based measurements represent measurements that offer limited temporal coverage. Temporally, comprehensive AQ records rarely date back before the midtwentieth century and are extremely limited in spatial coverage. Furthermore, some regulatory measurement sites record data every few days. This frequency may not be optimal to capture fire emission concentrations that are often short, episodic events. While there are benefits for collecting more data regarding ambient AQ, especially in unmonitored areas, there has been no concerted movement to increase the spatial resolution of reference monitors (Engel-Cox et al., 2013).

Despite government investment into using LCAQS to supplement regulatory data, there are still growing concerns that they are not able to replace reference measurements for regulatory decisions. While LCAQS offer advantages to supplement regulatory-grade information and empower more people to be engaged with monitoring AQ, the novelty of these sensors for regulatory purposes presents challenges. A primary known challenge is the quality of data reported by LCAQS. Previous studies have shown that data are subject to biases, and there are important considerations for obtaining high-quality data that are comparable to reference measurements (Giordano et al., 2021). A substantial effort has been focused on developing robust statistical approaches to calibrate data collected by LCAQS to those collected by instruments at the regulatory network stations (Barkjohn et al., 2021; Delp & Singer, 2020; Liu et al., 2017; Wallace et al., 2021). However, limited consensus has been reached in the literature regarding the best calibration, and it is likely regionally dependent upon other factors such as topography, meteorology, and other contributing factors. Assessing spatially contiguous AQ from regulatory and LCAQS networks presents an additional major challenge. Even in densely populated areas where many monitors may exist, there are no agreed-upon methods for extrapolating the stationary measurements to community and regional scales (Diao et al., 2019).

A key limitation of LCAQS is the lack of access to historical data. For example, Web portals that integrate LCAQS information, such as the Fire and Smoke map, offer near-real-time information that is useful to track impact of on-going fire events. However, the tool has limited or no ability to download historic data. Therefore, these portals have very little utility in historic analysis or retrospective health studies that aim to study trends over longer time periods. Many sensors within LCAQS are owned and operated by lay citizens, which on the one hand diversifies the spatial distribution of sensors while on the other hand opens the door for potential measurement errors due to sensors that may have inaccurate location (e.g., wrong location provided to protect the owners' privacy), deployment, or maintenance of individual instruments (Barkjohn et al., 2021). Additionally, particularly for historical analysis, the global record of monitors is highly skewed toward high-income countries (World Bank Group, 2021), limiting the utility for global analysis. Even in countries such as the United States with a longer and denser network of AQ monitoring, the spatial resolution of reference monitors is generally too sparse to capture the behavior of smoke and provide decision-support information for managing decisions associated with exposure to biomass burning emissions (Reid et al., 2015; Sánchez-Balseca & Pérez-Foguet, 2020; Watson et al., 2019).

Key challenges exist for establishing and expanding AQ monitoring networks, especially at the global scale. For example, real-time AQ monitoring relies upon internet infrastructure and transportation infrastructure to support routine maintenance. In low and lower-middle-income countries (defined for fiscal year 2022 as countries that have gross national income per capita less than and between $1,046 and $4,095, respectively), both infrastructure and maintenance issues, and other reasons, present challenges to these efforts (World Bank Group, 2021).

6.5. OPPORTUNITIES AND FUTURE DIRECTIONS IN MONITORING WILDFIRE POLLUTION

The rapidly developing networks of LCAQS offer an exciting opportunity for delivering a more robust system of ground-based measurements valuable for smoke monitoring. Their potential is widely recognized by governments within developing and developed nations alike. And although outside the scope of this chapter, there is a large, growing body of literature that focuses on sensor technology, including calibration methods (Wallace et al., 2021) and performance compared to regulatory monitors during smoke events (Delp & Singer, 2020). With continuing development and improvements of cost-effectiveness among LCAQS and the improvements in the global satellites that enable Web connectivity, it is reasonable to expect that LCAQS networks will become the leading component of a global AQ monitoring system with an increased data availability in remote and sparsely populated regions where fire activity and smoke pollution are frequent and persistent. Open access to the observations from such a dense network will likely lead to substantial improvement in models delivering spatially and temporally resolved estimates of fire-related air pollution.

ACKNOWLEDGMENTS

I would like to thank Tatiana Loboda for support throughout the writing process. Her encouragement and feedback were instrumental to the completion of this chapter.

REFERENCES

Affenzeller, M., Burlacu, B., Dorfer, V., Dorl, S., Halmerbauer, G., Königswieser, T., et al. (2020). White box vs. black box modeling: On the performance of deep learning, random forests, and symbolic regression in solving regression problems. In R. Moreno-Díaz, F. Pichler, & A. Quesada-Arencibia (Eds.), *Computer aided systems theory: EUROCAST 2019* (pp. 288–295). Cham: Springer International Publishing.

Akagi, S. K., Yokelson, R. J., Wiedinmyer, C., Alvarado, M. J., Reid, J. S., Karl, T., et al. (2011). Emission factors for open and domestic biomass burning for use in atmospheric models. *Atmospheric Chemistry and Physics*, *11*(9), 4039–4072.

Ambient Air Quality Surveillance (1994). 40 C.F.R § 58.

Air Quality Egg. (n.d.). *Air quality egg: Web portal.* https://airqualityegg.com/portal/

Arhami, M., Kamali, N., & Rajabi, M. M. (2013). Predicting hourly air pollutant levels using artificial neural networks coupled with uncertainty analysis by Monte Carlo simulations. *Environmental Science and Pollution Research*, *20*(7), 4777–4789.

Awokola, B. I., Okello, G., Mortimer, K. J., Jewell, C. P., Erhart, A., & Semple, S. (2020). Measuring air quality for advocacy in Africa (MA3): Feasibility and practicality of longitudinal ambient $PM_{2.5}$ measurement using low-cost sensors. *International Journal of Environmental Research and Public Health*, *17*(19), 7243. https://doi.org/10.3390/ijerph17197243

Barker, K., Cambi, F., Catcott, E. J., Chambers, L. A., Halliday, E. C., Hasegawa, A., et al. (1961). *Air pollution / contributors,* K. Barker, *et al.* World Health Organization. https://apps.who.int/iris/handle/10665/41722

Barkjohn, K. K., Gantt, B., & Clements, A. L. (2021). Development and application of a United States-wide correction for $PM_{2.5}$; data collected with the PurpleAir sensor. *Atmospheric Measurement Techniques*, *14*(6), 4617–4637. https://doi.org/ 10.5194/amt-14-4617-2021

Bellinger, C., Mohomed Jabbar, M. S., Zaïane, O., & Osornio-Vargas, A. (2017). A systematic review of data mining and machine learning for air pollution epidemiology. *BMC Public Health*, *17*(1), 907. https://doi.org/ 10.1186/s12889-017-4914-3

Boys, B. L., Martin, R. V., van Donkelaar, A., MacDonell, R. J., Hsu, N. C., Cooper, M. J., et al. (2014). Fifteen-year global time series of satellite-derived fine particulate matter. *Environmental Science & Technology*, *48*(19), 11109–11118. https://doi.org/ 10.1021/es502113p

Breiman, L. (2001). Random forests. *Machine Learning*, *45*(1), 5–32. https://doi.org/ 10.1023/A:1010933404324

Castell, N., Dauge, F. R., Schneider, P., Vogt, M., Lerner, U., Fishbain, B., et al.(2017). Can commercial low-cost sensor platforms contribute to air quality monitoring and exposure estimates? *Environment International*, *99*, 293–302. https://doi.org/ 10.1016/j.envint.2016.12.007

Chen, G., Li, S., Knibbs, L. D., Hamm, N. A. S., Cao, W., Li, T., et al. (2018a). A machine learning method to estimate $PM_{2.5}$ concentrations across China with remote sensing, meteorological and land use information. *Science of the Total Environment*, *636*, 52–60. https://doi.org/ 10.1016/j.scitotenv.2018.04.251

Chen, G., Wang, Y., Li, S., Cao, W., Ren, H., Knibbs, L. D., et al. (2018b). Spatiotemporal patterns of PM10 concentrations over China during 2005 to 2016: A satellite-based estimation using the random forests approach. *Environmental Pollution*, *242*, 605–613. https://doi.org/ 10.1016/j.envpol.2018.07.012

China National Environmental Monitoring Centre. (n.d.). Major responsibilities (government). http://www.cnemc.cn/en/main_responsibilities/

Chu, B., Zhang, S., Liu, J., Ma, Q., & He, H. (2021). Significant concurrent decrease in PM2.5 and NO2 concentrations in China during COVID-19 epidemic. *Journal of Environmental Sciences*, *99*, 346–353. https://doi.org/ 10.1016/j.jes.2020.06.031

Chu, Y., Liu, Y., Li, X., Liu, Z., Lu, H., Lu, Y., et al. (2016). A review on predicting ground $PM_{2.5}$ concentration using satellite aerosol optical depth. *Atmosphere*, *7*(10), 129. https://doi.org/ 10.3390/atmos7100129

Commonwealth of Australia (2015). *National clean air agreement.* https://www.environment.gov.au/system/files/resources/188756ab-ed94-4a3c-9552-62763ca86a7f/files/national-clean-air-agreement.pdf

Deligiorgi, D., & Philippopoulos, K. (2011). Spatial interpolation methodologies in urban air pollution modeling: application for the greater area of metropolitan Athens, Greece. In *Advanced Air Pollution.* IntechOpen. https://doi.org/ 10.5772/17734

Delp, W. W., & Singer, B. C. (2020). Wildfire smoke adjustment factors for low-cost and professional $PM_{2.5}$ monitors with optical sensors. *Sensors*, *20*(13), 3683. https://doi.org/ 10.3390/s20133683

Di, Q., Amini, H., Shi, L., Kloog, I., Silvern, R., Kelly, J., et al. (2019). An ensemble-based model of $PM_{2.5}$ concentration across the contiguous United States with high spatiotemporal resolution. *Environment International*, *130*, 104909. https://doi.org/ 10.1016/j.envint.2019.104909

Diao, M., Holloway, T., Choi, S., O'Neill, S. M., Al-Hamdan, M. Z., Donkelaar, A. V., et al. (2019). Methods, availability, and applications of $PM_{2.5}$ exposure estimates derived from ground measurements, satellite, and atmospheric models. *Journal of the Air and Waste Management Association*, *69*(12), 1391–1414. https://doi.org/ 10.1080/10962247.2019.1668498

Directorate-General for Environment. (n. d.). *Air quality.* https://ec.europa.eu/environment/air/quality/

Engel-Cox, J., Kim Oanh, N. T., van Donkelaar, A., Martin, R. V., & Zell, E. (2013). Toward the next generation of air quality monitoring: Particulate matter. *Atmospheric Environment*, *80*, 584–590. https://doi.org/ 10.1016/j.atmosenv.2013.08.016

Environment Canada (2020). *National air pollution surveillance program.* https://www.canada.ca/en/environment-climate-change/services/air-pollution/monitoring-networks-data/national-air-pollution-program.html

European Environment Agency (EEA) (2020). Air quality assessment methods (data flow D). http://aided.apps.eea.europa.eu/?source=%7B%22query%22%3A%7B%22match_all%22%3A%7B%7D%7D%2C%22display_type%22%3A%22tabular%22%7D

Feenstra, B., Papapostolou, V., Hasheminassab, S., Zhang, H., Boghossian, B. D., Cocker, D., et al. (2019). Performance evaluation of twelve low-cost $PM_{2.5}$ sensors at an ambient air monitoring site. *Atmospheric Environment*, *216*, 116946. https://doi.org/ 10.1016/j.atmosenv.2019.116946

Ferreira, A. J., & Figueiredo, M. A. T. (2012). Boosting algorithms: A review of methods, theory, and applications. In *Ensemble machine learning*. Boston, MA: Springer. https://doi.org/10.1007/978-1-4419-9326-7_2

Gan, R. W., Ford, B., Lassman, W., Pfister, G., Vaidyanathan, A., Fischer, E., et al. (2017). Comparison of wildfire smoke estimation methods and associations with cardiopulmonary-related hospital admissions. *GeoHealth*, *1*(3), 122–136. https://doi.org/ https://doi.org/10.1002/2017GH000073

Garcia-Menendez, F., Hu, Y., & Odman, M. T. (2014). Simulating smoke transport from wildland fires with a regional-scale air quality model: Sensitivity to spatiotemporal allocation of fire emissions. *Science of the Total Environment*, *493*, 544–553. https://doi.org/10.1016/j.scitotenv.2014.05.108

Giordano, M. R., Malings, C., Pandis, S. N., Presto, A. A., McNeill, V. F., Westervelt, D. M., et al. (2021). From low-cost sensors to high-quality data: A summary of challenges and best practices for effectively calibrating low-cost particulate matter mass sensors. *Journal of Aerosol Science*, *158*, 105833. https://doi.org/10.1016/j.jaerosci.2021.105833

Government of Canada (2021). *National air pollution surveillance (NAPS) program: Open government portal*. https://open.canada.ca/data/en/dataset/1b36a356-defd-4813-acea-47bc3abd859b

Gulia, S., Khanna, I., Shukla, K., & Khare, M. (2020). Ambient air pollutant monitoring and analysis protocol for low and middle income countries: An element of comprehensive urban air quality management framework. *Atmospheric Environment*, *222*, 117120. https://doi.org/ 10.1016/j.atmosenv.2019.117120

Gupta, P., Doraiswamy, P., Levy, R., Pikelnaya, O., Maibach, J., Feenstra, B., et al. (2018). Impact of California fires on local and regional air quality: The role of a low-cost sensor network and satellite observations. *GeoHealth*, *2*(6), 172–181. https://doi.org/ https://doi.org/10.1029/2018GH000136

Gwaze, P., & Mashele, S. H. (2018). South African air quality information system (SAAQIS) mobile application tool: Bringing real time state of air quality to South Africans. *Clean Air Journal*. https://doi.org/ 10.17159/2410-972X/2018/v28n1a1

Hao, Y., & Xie, S. (2018). Optimal redistribution of an urban air quality monitoring network using atmospheric dispersion model and genetic algorithm. *Atmospheric Environment*, *177*, 222–233. https://doi.org/ 10.1016/j.atmosenv.2018.01.011

Helsen, L. (2005). Sampling technologies and air pollution control devices for gaseous and particulate arsenic: A review. *Environmental Pollution*, *137*(2), 305–315. https://doi.org/10.1016/j.envpol.2005.01.025

Hodzic, A., Madronich, S., Bohn, B., Massie, S., Menut, L., & Wiedinmyer, C. (2007). Wildfire particulate matter in Europe during summer 2003: Meso-scale modeling of smoke emissions, transport and radiative effects. *Atmospheric Chemistry and Physics*, *7*(15), 4043–4064.

Holder, A. L., Mebust, A. K., Maghran, L. A., McGown, M. R., Stewart, K. E., Vallano, D. M., et al. (2020). Field evaluation of low-cost particulate matter sensors for measuring wildfire smoke. *Sensors*, *20*(17), 4796. https://doi.org/ 10.3390/s20174796

Hu, X., Belle, J. H., Meng, X., Wildani, A., Waller, L. A., Strickland, M. J., et al. (2017). Estimating $PM_{2.5}$ concentrations in the conterminous United States using the random forest approach. *Environmental Science and Technology*, *51*(12), 6936–6944. https://doi.org/ 10.1021/acs.est.7b01210

Hwang, J., Bae, H., Choi, S., Yi, H., Ko, B., & Kim, N. (2020). Impact of air pollution on breast cancer incidence and mortality: A nationwide analysis in South Korea. *Scientific Reports*, *10*(1), 5392. https://doi.org/ 10.1038/s41598-020-62200-x

InciWeb (n.d.). *Incident information system*. https://inciWeb.nwcg.gov

Interagency Monitoring of Protected Visual Environments (2020). *Current site list spreadsheet (updated December 2019)*. http://vista.cira.colostate.edu/improve/wp-content/uploads/2020/03/IMPROVE_Sites_Active_Updated_Dec_2019.xlsx

International Trade Administration (2020). *India ambient air quality*. http://www.trade.gov/market-intelligence/india-ambient-air-quality

Izmerov, N. F. (1974). Principles underlying the establishment of air quality standards in USSR. *Science and Public Policy*, *1*(9), 230–233. https://doi.org/10.1093/spp/1.9.230

Keywood M. D., Emmerson, K. M., & Hibberd, M. F. (2016). Atmosphere: Key findings. In *Australia state of the environment 2016*. Canberra: Australian Government Department of the Environment and Energy. https://soe.environment.gov.au/theme/atmosphere/key-findings?year=96. https://doi.org/10.4226/94/58b65c70bc372

Klyuev, N. N. (2019). Air quality in Russian cities for 1991–2016. *Regional Research of Russia*, *9*(2), 204–212. https://doi.org/ 10.1134/S2079970519020072

Konovalov, I. B., Beekmann, M., Kuznetsova, I. N., Yurova, A., & Zvyagintsev, A. M. (2011). Atmospheric impacts of the 2010 Russian wildfires: Integrating modeling and measurements of an extreme air pollution episode in the Moscow region. *Atmospheric Chemistry and Physics*, *11*(19), 10031–10056.

Krige, D. G. (1951). A statistical approach to some basic mine valuation problems on the Witwatersrand. *Journal of the Southern African Institute of Mining and Metallurgy*, *52*(6), 21. https://doi.org/ https://hdl.handle.net/10520/AJA0038223X_4792

Lassman, W., Ford, B., Gan, R. W., Pfister, G., Magzamen, S., Fischer, E. V., & Pierce, J. R. (2017). Spatial and temporal estimates of population exposure to wildfire smoke during the Washington State 2012 wildfire season using blended model, satellite, and in situ data. *GeoHealth*, *1*(3), 106–121. https://doi.org/ https://doi.org/10.1002/2017GH000049

Li, J., & Heap, A. D. (2008). A review of spatial interpolation methods for environmental scientists. *Geoscience Australia*, *137*.

Li, J., & Heap, A. D. (2014). Spatial interpolation methods applied in the environmental sciences: A review. *Environmental Modeling and Software*, *53*, 173–189. https://doi.org/ 10.1016/j.envsoft.2013.12.008

Liu, D., Zhang, Q., Jiang, J., & Chen, D.-R. (2017). Performance calibration of low-cost and portable particular matter (PM) sensors. *Journal of Aerosol Science*, *112*, 1–10. https://doi.org/10.1016/j.jaerosci.2017.05.011

Liu, Y., Sarnat, J. A., Kilaru, V., Jacob, D. J., & Koutrakis, P. (2005). Estimating ground-level $PM_{2.5}$ in the eastern United States using satellite remote sensing. *Environmental Science and Technology*, *39*(9), 3269–3278. https://doi.org/ 10.1021/es049352m

Luo, J., Du, P., Samat, A., Xia, J., Che, M., & Xue, Z. (2017). Spatiotemporal pattern of $PM_{2.5}$ concentrations in mainland China and analysis of its influencing factors using geographically weighted regression. *Scientific Reports*, *7*(1), 40607. https://doi.org/ 10.1038/srep40607

Ma, J., Ding, Y., Gan, V. J. L., Lin, C., & Wan, Z. (2019). Spatiotemporal prediction of $PM_{2.5}$ concentrations at different time granularities using IDW-BLSTM. *IEEE Access*, *7*, 107897–107907. https://doi.org/ 10.1109/ACCESS.2019.2932445

Ma, Z., Hu, X., Huang, L., Bi, J., & Liu, Y. (2014). Estimating ground-level $PM_{2.5}$ in China using satellite remote sensing. *Environmental Science and Technology*, *48*(13), 7436–7444. https://doi.org/ 10.1021/es5009399

Marć, M., Tobiszewski, M., Zabiegała, B., Guardia, M. de la, & Namieśnik, J. (2015). Current air quality analytics and monitoring: A review. *Analytica Chimica Acta*, *853*, 116–126. https://doi.org/ 10.1016/j.aca.2014.10.018

Mirzaei, M., Bertazzon, S., & Couloigner, I. (2018). Modeling wildfire smoke pollution by integrating land use regression and remote sensing data: Regional multi-temporal estimates for public health and exposure models. *Atmosphere*, *9*(9), 335.

Morawska, L., Thai, P. K., Liu, X., Asumadu-Sakyi, A., Ayoko, G., Bartonova, A., et al. (2018). Applications of low-cost sensing technologies for air quality monitoring and exposure assessment: How far have they gone?. *Environment international*, *116*, 286–299.

NOAA (n.d.). *Hazard mapping system fire and smoke product*. National Oceanic and Atmospheric Administration Office of Satellite and Product Operations National Environmental Satellite, Data and Information Service. https://www.ospo.noaa.gov/Products/land/hms.html

NOAA (n.d.). *Interagency monitoring of protected visual environments. GML surface ozone*. National Oceanic and Atmospheric Administration Global Monitoring Laboratory Earth System Research Laboratories. https://gml.noaa.gov/ozwv/surfoz/

OpenAQ (n.d.). *API for OpenAQ LCS*. https://docs.openaq.org/

OpenAQ (2021). *Announcing the launch of our low-cost sensor pilot!* https://openaq.medium.com/low-cost-sensor-platform-pilot-launch-bd66ab226c6

Park, S., Shin, M., Im, J., Song, C.-K., Choi, M., Kim, J., et al. (2019). Estimation of ground-level particulate matter concentrations through the synergistic use of satellite observations and process-based models over South Korea. *Atmospheric Chemistry and Physics*, *19*(2), 1097–1113. https://doi.org/ 10.5194/acp-19-1097-2019

Paugam, R., Wooster, M., Freitas, S., & Val Martin, M. (2016). A review of approaches to estimate wildfire plume injection height within large-scale atmospheric chemical transport models. *Atmospheric Chemistry and Physics*, *16*(2), 907–925. https://doi.org/ 10.5194/acp-16-907-2016

Piersanti, A., Vitali, L., Righini, G., Cremona, G., & Ciancarella, L. (2015). Spatial representativeness of air quality monitoring stations: A grid model based approach. *Atmospheric Pollution Research*, *6*(6), 953–960. https://doi.org/ 10.1016/j.apr.2015.04.005

PurpleAir (n.d.). *Map*. https://map.purpleair.com/

PurpleAir (2021, September 11). *API–PurpleAir*. https://api.purpleair.com/

Rai, A. C., Kumar, P., Pilla, F., Skouloudis, A. N., Di Sabatino, S., Ratti, C., et al. (2017). End-user perspective of low-cost sensors for outdoor air pollution monitoring. *Science of The Total Environment*, *607–608*, 691–705. https://doi.org/ 10.1016/j.scitotenv.2017.06.266

Rees, N., Wickham, A., & Choi, Y. (2019). *Silent suffocation in Africa: Air pollution is a growing menace, affecting the poorest children the most*. New York: UNICEF Division of Data, Research and Policy. https://www.unicef.org/media/55081/file/Silent%20suffocation%20in%20africa%20air%20pollution%202019%20.pdf

Reid, C. E., Jerrett, M., Petersen, M. L., Pfister, G. G., Morefield, P. E., Tager, I. B., et al. (2015). Spatiotemporal prediction of fine particulate matter during the 2008 northern California wildfires using machine learning. *Environmental Science & Technology*, *49*(6), 3887–3896. https://doi.org/ 10.1021/es505846r

Riojas-Rodríguez, H., da Silva, A. S., Texcalac-Sangrador, J. L., & Moreno-Banda, G. L. (2016). Air pollution management and control in Latin America and the Caribbean: Implications for climate change. *Revista Panamericana de Salud Pública*, *40*, 150–159.

Sánchez-Balseca, J., & Pérez-Foguet, A. (2020). Modeling hourly spatio-temporal $PM_{2.5}$ concentration in wildfire scenarios using dynamic linear models. *Atmospheric Research*, *242*, 104999.

Shaddick, G., & Zidek, J. V. (2014). A case study in preferential sampling: Long term monitoring of air pollution in the UK. *Spatial Statistics*, *9*, 51–65. https://www.sciencedirect.com/science/article/abs/pii/S2211675314000219

Shin, M., Kang, Y., Park, S., Im, J., Yoo, C., & Quackenbush, L. J. (2020). Estimating ground-level particulate matter concentrations using satellite-based data: A review. *GIScience and Remote Sensing*, *57*(2), 174–189. https://doi.org/ 10.1080/15481603.2019.1703288

Song, L., Pang, S., Longley, I., Olivares, G., & Sarrafzadeh, A. (2014). Spatio-temporal $PM_{2.5}$ prediction by spatial data aided incremental support vector regression. *2014 International Joint Conference on Neural Networks (IJCNN)*, 623–630. https://doi.org/ 10.1109/IJCNN.2014.6889521

Sorek-Hamer, M., Strawa, A. W., Chatfield, R. B., Esswein, R., Cohen, A., & Broday, D. M. (2013). Improved retrieval of $PM_{2.5}$ from satellite data products using non-linear methods. *Environmental Pollution*, *182*, 417–423. https://doi.org/ 10.1016/j.envpol.2013.08.002

Stafoggia, M., Bellander, T., Bucci, S., Davoli, M., de Hoogh, K., de' Donato, F., et al. (2019). Estimation of daily PM10 and $PM_{2.5}$ concentrations in Italy, 2013–2015, using a spatiotemporal land-use random-forest model. *Environment International*, *124*, 170–179. https://doi.org/ 10.1016/j.envint.2019.01.016

Sundaray, S. N. K., & Bhardwaj, S. R. (2019). *National clean air programme* (p. 106). Ministry of Environment, Forest & Climate Change Government of India.

World Bank Group. (2021). *World Bank country and lending groups*. https://datahelpdesk.worldbank.org/knowledgebase/articles/906519-world-bank-country-and-lending-groups

Tobler, W. R. (1970). A computer movie simulating urban growth in the Detroit region. *Economic Geography*, *46*(sup1), 234–240. https://doi.org/ 10.2307/143141

United Nations Environment Programme (n.d.). https://www.iqair.com/unep

U.S. Department of State & U.S. Environmental Protection Agency (n.d.). *AirNow Department of State*. https://www.airnow.gov/international/us-embassies-and-consulates/

U.S. Environmental Protection Agency. (n.d.). *AirNow's fire and smoke map frequently asked questions, AirNow.gov*. https://www.airnow.gov/fire-smoke-map-faqs

U.S. Environmental Protection Agency. (2021a). *Air data basic information*. https://www.epa.gov/outdoor-air-quality-data/air-data-basic-information

U.S. Environmental Protection Agency (2021b). *Fire and smoke map*. https://fire.airnow.gov/

U.S. Environmental Protection Agency (2021c). *Clean air status and trends network (CASTNET)*. https://www.epa.gov/castnet

U.S. Environmental Protection Agency (2021d). *Pre-generated data files*. https://aqs.epa.gov/aqsWeb/airdata/download_files.html

Van Donkelaar, A., Martin, R. V., & Park, R. J. (2006). Estimating ground-level $PM_{2.5}$ using aerosol optical depth determined from satellite remote sensing. *Journal of Geophysical Research*, *111*(D21), D21201. https://doi.org/10.1029/2005JD006996

Wakamatsu, S., Morikawa, T., & Ito, A. (2013). Air pollution trends in Japan between 1970 and 2012 and impact of urban air pollution countermeasures. *Asian Journal of Atmospheric Environment*, *7*(4), 14. https://doi.org/ https://doi.org/10.5572/ajae.2013.7.4.177

Wallace, L., Bi, J., Ott, W. R., Sarnat, J., & Liu, Y. (2021). Calibration of low-cost PurpleAir outdoor monitors using an improved method of calculating $PM_{2.5}$. *Atmospheric Environment*, *256*, 118432. https://doi.org/10.1016/j.atmosenv.2021.118432

Watson, G. L., Telesca, D., Reid, C. E., Pfister, G. G., & Jerrett, M. (2019). Machine learning models accurately predict ozone exposure during wildfire events. *Environmental Pollution*, *254*, 112792. https://doi.org/ 10.1016/j.envpol.2019.06.088

WHO (1958). *Air pollution: Fifth report of the Expert Committee on Environmental Sanitation* Meeting held in Geneva from 18 to 23 November 1957. WHO IRIS. https://apps.who.int/iris/handle/10665/40416

WHO Expert Committee on Air Quality Criteria and Guides for Urban Air Pollutants & World Health Organization. (1972). Air quality criteria and guides for urban air pollutants: Report of a WHO expert committee [meeting held in Geneva from 5 to 11 April 1972]. WHO IRIS. https://apps.who.int/iris/handle/10665/40989

Wiedinmyer, C., Quayle, B., Geron, C., Belote, A., McKenzie, D., Zhang, X., et al. (2006). Estimating emissions from fires in North America for air quality modeling. *Atmospheric Environment*, *40*(19), 3419–3432.

World Air Quality Index Project (2022). Air pollution in world: Real-time air quality index visual map. https://aqicn.org/map/world/

Wu, J., Winer, A., & Delfino, R. (2006). Exposure assessment of particulate matter air pollution before, during, and after the 2003 Southern California wildfires. *Atmospheric Environment*, *40*(18), 3333–3348. https://doi.org/ 10.1016/j.atmosenv.2006.01.056

Xia, F., Yang, L. T., Wang, L., & Vinel, A. (2012). Internet of things. *International Journal of Communication Systems*, *25*(9), 1101.

Yang, C.-T., Chan, Y.-W., Liu, J.-C., & Lou, B.-S. (2020). An implementation of cloud-based platform with R packages for spatiotemporal analysis of air pollution. *The Journal of Supercomputing*, *76*(3), 1416–1437. https://doi.org/ 10.1007/s11227-017-2189-1

Yao, J., & Henderson, S. B. (2014). An empirical model to estimate daily forest fire smoke exposure over a large geographic area using air quality, meteorological, and remote sensing data. *Journal of Exposure Science and Environmental Epidemiology*, *24*(3), 328–335.

Zhao, C., Wang, Q., Ban, J., Liu, Z., Zhang, Y., Ma, R., et al. (2020). Estimating the daily $PM_{2.5}$ concentration in the Beijing-Tianjin-Hebei region using a random forest model with a 0.01° × 0.01° spatial resolution. *Environment International*, *134*, 105297. https://doi.org/ 10.1016/j.envint.2019.105297

7

Data Assimilation for Numerical Smoke Prediction

Edward J. Hyer[1], Christopher P. Camacho[1], David A. Peterson[1], Elizabeth A. Satterfield[1], and Pablo E. Saide[2]

ABSTRACT

Skillful forecasts of weather phenomena in numerical models begin with the most accurate set of initial conditions achievable from observational data sets. The process of combining observations with numerical model predictions is called *data assimilation*. This chapter describes the types of observations available for data assimilation in models that predict the transport, fate, and impacts of smoke pollution. Observation properties needed for effective data assimilation are identified based on experiences with a variety of observation types in data assimilation experiments, compiled from the published literature. The second half of the chapter surveys the data assimilation methodologies that have been applied to smoke aerosols, and describes specific problems associated with the smoke observations that require innovative techniques in data assimilation. The chapter concludes by providing an outlook for future research and development in data assimilation for smoke prediction models. Data assimilation for prediction of smoke is an emerging area of development that promises to greatly improve forecast skill as new data sets and techniques are applied.

7.1. INTRODUCTION

A cascading sequence of numerical models is used to predict downwind impacts of smoke: fire behavior and smoke emission models translate observations of fires into model-relevant quantities, plume rise models describe fine-scale vertical motion of smoke at the fire source, and atmospheric transport models simulate the evolution and transport of smoke and the effects of local, mesoscale, and synoptic meteorology on the distribution of smoke. In this chapter, we refer to an atmospheric transport model used to predict the movement and distribution of smoke plumes as a *smoke transport model*.

For use in operational forecasting, these models require very timely information about fires, smoke, and meteorological factors. Observations of smoke are used to constrain the location, thickness, and composition of atmospheric smoke plumes. Under the right circumstances, these observations can be combined with model predictions of plume extent and direction to improve predictions of downwind smoke. This process of combining numerical model predictions with observations is called *data assimilation*. Data assimilation is an essential component in skillful atmospheric models of all kinds, but it demands accurate numerical estimates of the uncertainty in both the forecast model and the relevant observations. Initializing models based on sparse observations also requires understanding how these uncertainties vary in space and time (spatial and temporal correlations).

[1] *Naval Research Laboratory, Marine Meteorology Division, Monterey, California, USA*

[2] *Department of Atmospheric and Oceanic Sciences, and Institute of the Environment and Sustainability, University of California Los Angeles, Los Angeles, California, USA*

Landscape Fire, Smoke, and Health: Linking Biomass Burning Emissions to Human Well-Being, Geophysical Monograph 280, First Edition.
Edited by Tatiana V. Loboda, Nancy H. F. French, and Robin C. Puett.

DOI: 10.1002/9781119757030.ch07

Previous chapters have discussed the observations of fires that are relevant to this problem. The condition and behavior of the fire itself are boundary conditions in the terminology of an atmospheric model. Data on fire characteristics (see Chapters 3 and 4) are used to inform submodels of fire behavior and smoke emissions (see Chapter 5) that translate events outside the model domain (fires on the ground) into changes to the smoke transport model state (the 3D, time-variable, numerical description of the model atmosphere).

As discussed in Chapter 8, models of fire behavior and plume rise may be tightly coupled to the smoke transport model, meaning that temperature, wind, and humidity predictions from the atmospheric model are fed into the submodel of fire behavior, and sensible and latent heat flux from the fire are introduced as boundary conditions to influence the smoke transport model.

The application of fire observations into models of fire behavior can be treated as a data assimilation problem. However, the formal methods of data assimilation are more fully developed for assimilating observations of the atmosphere that can be used to directly perturb the smoke transport model state to obtain a more accurate starting point for a model prediction, and thus a more accurate forecast. A different category of data assimilation problem is when observations are used to constrain emissions or other nonmeteorological parameters; this is generally referred to as *inverse modeling*. Many operational aerosol prediction models use data assimilation to improve predictions of plumes of pollution, dust, and smoke. This chapter discusses the technical considerations for improving model predictions of downwind smoke impacts through data assimilation.

The first half of this chapter discusses the types of observations available and the characteristics of those observations that make them more or less useful in the context of data assimilation for smoke prediction. Many observations that are clearly useful in managing health impacts of smoke are difficult to apply in numerical forecasts, for a variety of reasons we will discuss. We will outline some key characteristics that apply to existing and future observations affecting their potential for data assimilation.

The second half of the chapter discusses some of the mechanics of data assimilation and how they relate to the problem of smoke from fires. Many technical details in the implementation of data assimilation have very substantial consequences for the specific problem of smoke plumes.

At the end of the chapter, we briefly describe some directions of research that could lead to improved communication between observations and models, and ultimately improved forecasts of downwind smoke impacts.

7.2. MATCHING OBSERVATIONS TO FORECAST PROBLEMS

This section describes the principal types of observations relevant to the smoke forecasting problem and the domains in which each provides information to improve the forecast. The template for a smoke prediction has two main components: (1) how much smoke, and (2) for how long, or simply the magnitude and duration of smoke impacts. For most purposes, smoke impacts at the surface are the target of prediction. When smoke transport models are used to aid in these predictions, they serve two purposes. First, the model gives an estimate of conditions that is spatially continuous across the model domain. Second, the model computes integration in time including the effects of the atmosphere on movement, dilution, and removal of smoke. While there are many applications that require models in conjunction with contemporary observations (such as estimation of exposure for epidemiological studies, e.g., Hammer et al., 2020), this section specifically considers the condition where the model must make a prediction at a location and time where observations are unavailable.

If we break this problem into components, we can examine how different types of observations contribute to this smoke prediction problem. The model provides a coarse representation of a smoke plume; model resolution is often insufficient to capture much of the variation in smoke, especially in the vertical, and the model representation of smoke composition is inevitably vastly oversimplified. In spite of these simplifications, we find that all of the available observations provide limited information, leaving the model to fill in the rest. Therefore, benefit to the forecast can be increased by integration of multiple observation types, utilizing each observation's strengths and supporting weak points of observations with strong model treatments.

7.2.1. Overview of Observations Relevant to Smoke Plumes

The objective of data assimilation in a smoke transport model is to constrain key aspects of the smoke plume by observations, and thus estimate the most accurate initial condition and/or emissions for a numerical prediction of the smoke plume's evolution. The model state of the smoke prediction model is a 3D gridded representation of the atmosphere with various properties of the atmosphere, including aerosols and trace gases, represented at each grid cell. For inverse modeling, the state may include spatial and temporal patterns of emissions magnitude and composition. At a more abstract level, we can identify the properties of a smoke plume that will impact the ability of the model to correctly

Table 7.1 Observation types and some of their relevant properties for use in smoke forecasting

Observation	Representative data source	Coverage	Reduction of model state uncertainty	Key limitation for smoke prediction
Surface PM	AirNow	Sparse in CONUS; very sparse globally	Moderate-high	2 m height; no information on elevated smoke
Sun / Moon photometer	AERONET	Sparse in CONUS and globally	Moderate	Column-integrated retrieval, no vertical specificity
Satellite AOD retrieval (polar)	MODIS MxD14; MAIAC; NOAA Enterprise Aerosol Products	Global ~1x/day per satellite; daytime only	Low-moderate	Column-integrated; resolution/noise tradeoffs; retrieval failures in thick smoke
Satellite AOD (geostationary)	GOES-R ABI NOAA Enterprise Aerosol Product	Every 15 min or better within satellite domain	Moderate	Column-integrated; resolution/noise tradeoffs; retrieval failures in thick smoke
Ground-based lidar	NASA MPLNET	Very sparse	Moderate	<200 sites globally
Space-based lidar	CALIOP	Global w/16-day repeat	Moderate-high	Sparse; cannot see under thick clouds
Total column space-based trace gas retrievals	TROPOMI CO, NO_2, and HCHO	Global ~1x/day; daytime only	Moderate	Column-integrated
Space-based trace gas retrievals from sounders or profilers	MOPITT, CrIS, and IASI CO, OMI/OMPS CO, HCHO and NO2	Global coverage every ~1–3 days, day and night data	Low-moderate	Coarse resolution; limited vertical signal; limited sensitivity near surface

Note: Acronyms are defined in the front matter of this book.

predict its evolution. These properties are, roughly, position of the plume, horizontal extent, vertical extent, quantity of smoke, and smoke composition. A wide range of observations can be used to reduce uncertainties in these properties, but no single observation provides the comprehensive information needed by the model.

Table 7.1 overviews the main types of observations that have been used to constrain atmospheric models used to predict smoke. This list is not exhaustive but captures the types of data that are most likely to be practically useful in a forecast context (see section 7.2.2). Products from research satellites may provide detailed information on smoke (for instance, MISR; Noyes et al., 2020b), but are often not suited to near-real-time application. Each of these types of observations has been examined in detail in the references cited in this chapter and other chapters of this book.

The general outlook is that the greater the level of accuracy, precision, and detail provided by an observation, the more sparse those observations are. Surface PM and trace gas monitors provide the most direct observation of the smoke quantity, but are too sparsely located to constrain plume location and extent (see Chapter 6). PM measurements may also be biased relative to the intended predictand. For instance, evaporation of organic aerosol at the surface (Selimovic et al., 2019) can act as a confounding factor. Ground-based lidar observations, which observe the backward scattering of laser light, capture the vertical mixing that is a critical determinant of surface smoke, but the locations of these few instruments rarely facilitate application to prediction of a local or regional smoke event. Some lidar observations, especially low-power instruments such as ceilometers, may attenuate for thick smoke plumes, limiting their vertical coverage. Satellite instruments provide far greater numbers of observations over far greater areas, but the resolution of these observations is coarse (1 km or more), compared to the size of the active burning area. The weak signal of smoke in satellite observations of reflected sunlight often demands additional spatial aggregation to reduce noise. Aerosol optical depth (AOD) is an optical quantity, which depends not only on aerosol mass but also on other variables like relative humidity and the aerosol composition and size distribution. Conversions between aerosol mass and optical properties have been shown to vary substantially with smoke age (Kleinman et al., 2020). Additionally, multiple algorithms filter out AOD retrievals on thick smoke (see section 7.3.2), and the ones that do provide them can be uncertain due to assumptions on smoke absorption. Ongoing work seeks to advance the capability of space-based sensors to constrain aerosol composition and other properties (e.g., Rogozovsky et al., 2021).

Despite the limitations, assimilation of many of these types of observations has been shown to greatly increase the predictive skill of smoke transport models (Jianglong et al., 2008; Saide et al., 2014; Saide et al., 2015; Yumimoto et al., 2018). Much of the published literature has focused on global models, which are recognized for accurately representing long-range transport of heavy aerosol plumes. However, for many practical smoke prediction scenarios, observations that may show significant benefit in a global model may provide very little skill at the scale where smoke impacts are most significant. The next section discusses the characteristics of observations that make them impactful for real-world forecast scenarios.

7.2.2. Applicability of Observations to Real-World Smoke Forecasting

Types of Forecast Uncertainties

The goal of data assimilation is to apply observations to correct errors in a short forecast and produce the most accurate initial conditions for subsequent predictions. Therefore, a good place to start to is to consider common types of forecast errors and how the inclusion of observational information could act to mitigate such issues. This section will outline these forecast errors in detail. Subsequent sections will examine specific topics of timeliness and spatial matching.

1. Missing Fires

Without a smoke source, there is very little a numerical model can achieve for smoke prediction. Numerical smoke models use the fire observations discussed in the earlier chapters of this book to estimate smoke production (see Chapter 5). The availability of those observations to the model is a precondition for effectively utilizing observations of smoke. This is mostly a matter of timing, which we discuss in more detail below.

Errors in fire timing can arise from the temporal resolution of the fire observations. Polar-orbiting weather satellites that can detect relatively smaller fires require parameterization of fire behavior over the diurnal cycle; coarser, less-sensitive geostationary fire observations have typically been used for this purpose (e.g., Hyer et al., 2013; Mu et al., 2011; Reid et al., 2004). The errors associated with infrequent fire observations can be mitigated by assimilation of observations with better temporal resolution, but satellite observations have been shown less effective than *in situ* observations due to lack of satellite aerosol retrievals near the fire source (Saide et al., 2015a) (see section 7.3.2.3).

The spatial resolution of satellite fire observations is a significant source of error for wildfire responders on the ground and for matching fire observations to fuel maps (Hyer & Reid, 2009), but represents a small error at the scale of significant smoke plumes. Atmospheric retrievals of smoke properties such as aerosol optical depth (AOD) are almost universally less precise spatially than fire observations, so assimilation of these observations will do little to mitigate these spatial errors.

2. Too Much or Too Little Smoke

Smoke production magnitude is a recognized weakness of satellite-based fire inventories (e.g., Al-Saadi et al., 2008; French et al., 2011; Zhang et al., 2014a). Error in smoke source magnitude is the error most readily addressed by assimilation of smoke observations.

If a smoke source is present in the model at the approximately correct time and location, observations of the smoke downwind can be used to correct the smoke magnitude. As noted in section 7.4, the data assimilation system is sometimes used to correct the smoke loading in the model state, and sometimes used to apply a correction to the smoke source magnitude (i.e., an emission inversion). The higher the spatial resolution, the more precise the correction can be, but the improved description of smoke magnitude from high-resolution observations can be achieved only if errors in simulated wind speed and direction are small.

3. Vertical Placement and Transport

Ventilation of smoke is a key consideration for downwind impacts, and a key component of any prediction of smoke air quality. Complex interactions between the fire and the atmosphere are explicitly simulated in some high-resolution models, as discussed in Chapter 8 of this volume, but in most regional and global models, these interactions are represented by simple parameterizations. The energy release from the fire is relevant to this question, but studies using existing observations of fire energetics have found only low correlation with smoke vertical placement (Paugam et al., 2016; Peterson et al., 2014; Val Martin et al., 2012), due to large uncertainties in fire energy (Giglio & Schroeder, 2014), the importance of model representation of stable layers (Val Martin et al., 2012), and the complexities of 3D plume behavior (Liu et al., 2019; Nelson et al., 2012; Potter, 2012).

Observations at the surface are obviously relevant to correct errors in smoke vertical placement, but assimilation of these observations requires careful consideration of the vertical and horizontal spread of the information, represented in data assimilation as error correlations (Liu et al., 2011; Zhang et al., 2011). Profile information such as that provided by a lidar can be very impactful to improve placement of smoke (Cheng et al., 2019). Several recent papers have demonstrated the potential of passive satellite observations to retrieve some information on smoke plume height (Lyapustin et al., 2020; Choi et al., 2021; Xu et al., 2019; Xu et al., 2017; Carr et al., 2020) or smoke over cloud (Alfaro-Contrerase et al., 2014; Jethva et al., 2016; Jethva et al., 2014). This information could improve predictions of long-range transport.

Downwind of a fire, a plume will remain aloft or mix into the surface layer controlled by a variety of processes at scales from synoptic subsidence to turbulent mixing (e.g., Colarco et al., 2004). When the smoke transport model erroneously mixes or fails to mix smoke between the surface and layers aloft, only observations resolving smoke at the surface can be used to place the smoke correctly. The surface pollution may have long spatial correlation lengths in temporal averages (Kaku et al., 2018; Wang et al., 2020c), but spatial correlation lengths associated with smoke events are often short, with the implication that corrections from surface observations cannot be applied to large areas. Networks of low-cost sensors, which provide much finer spatial coverage, could be a solution to this problem (Delp & Singer, 2020; Robinson, 2020).

4. Horizontal Movement of Smoke

If errors in wind speed and direction are large, assimilation of smoke observations without vertical information will likely do little to improve the forecast. Even if the plume position, shape, and size are fully mapped with satellite data, the lack of any model prior for the vertical placement of the smoke will result in significant error in placement of aerosols in the model, even in aerosol species not corresponding to smoke.

In practical terms, wind errors will generally be small in the short model forecasts. However, in the case of small plumes and observations very near the smoke source, small errors in wind direction can be just as costly as large errors downwind. Assimilation of smoke observations onto a background with a misplaced plume will result in unrealistic depiction of the smoke in the analysis, as well as a mismatch between the smoke and other meteorological fields in the simulation (e.g., Zhang et al., 2016b). Adding or subtracting mass from the downwind tail of a plume is one of the primary means by which smoke predictions are improved by data assimilation, so correcting small errors in wind speed is a clear use for data assimilation in smoke prediction.

5. Persistence in Emission Forecasts

The current state of air quality forecast models is to continue the fire emissions estimated from the latest satellite fire detections into the next forecast cycle, which we define here as *persistence*. However, fires can often change substantially from day to day, which can generate large uncertainties in smoke forecasts (Ye et al., 2021). This results in systematic underprediction of smoke after periods of extreme fire weather and overprediction when fires extinguish rapidly (e.g., due to precipitation). Limited studies have been performed to predict changes in fire behavior and smoke emissions during the weather forecast period (e.g., Peterson, et al., 2013; Peterson et al., 2015), and this issue remains an active area of research in the smoke forecasting community.

6. Prescribed Diurnal Cycles

Many smoke emission models rely on fire observations collected only once a day. In order to distribute emissions at the hourly resolution, emission models typically assume that fires follow a prescribed diurnal pattern based on observations with better temporal resolution (Giglio, 2007; Mu et al., 2011), with a peak in fire activity in the afternoon and low emissions during evening and morning hours. Fires in practice often deviate from the climatological pattern (Saide et al., 2015a; Ye et al., 2021), especially for extreme fire behavior, which often burns throughout the night (Peterson et al., 2015). Recent studies have found strong correlation between geostationary FRP and smoke concentrations (Wiggins et al., 2020), which is motivating multiple groups to move to using these observations to distribute emissions to an hourly timescale (e.g., Reid et al., 2004; Roberts et al., 2015; Wang et al., 2006). However, there are two caveats to this. First, in order for these observations to make a difference, there needs to be a post run (see section 7.2.2.2) of the forecast or assimilation system so that corrected emissions are introduced at the time they occurred. Second, similar to the previous subsection, persistence of diurnal cycles is likely not to occur, and thus methods need to be developed to forecast emission diurnal cycles as well.

All of these different types of model errors can be corrected to some degree by assimilation of smoke observations. In each case, we see that correction for one type of model error is most effective when other errors are small. Operational prediction systems thus achieve accumulating benefits from repeated cycles of short forecasts followed by data assimilation. But for the specific case of forecasting impacts of significant smoke events, important limitation on the practical results is imposed by the logistics of timing and the placement of observations. These dependencies are discussed in the following sections.

Timeliness of Observations

The single most important characteristic for utilization of observations to support model predictions is timeliness. Models at the global scale typically run four forecast cycles each day, and at each cycle an initialization is performed using observations in a 6 hr window around the nominal forecast start time. We describe this using a notation where *T* is the valid time of the forecast initialization, so the observation window for initialization at *T* is from *T-3* to *T+3*. The delay between the time of an observation and its availability for use in the forecast is referred to as *latency*. If the forecast cycle is started 4 hr after *T* (*T+4*), and the latency is 1 hr, then all observations up to *T+3* will be used in the initialization. If the cycle is started at *T+4* and the latency is 3 hr, then each cycle will use observations only from *T-3* to *T+1*. Many global systems therefore include a final analysis at ~*T+8* (also referred to as *reanalysis* or *post run*), where all observations can be used to prepare for the following forecast cycle.

In a system configured this way, fires and smoke observed at (for example) 20 UTC may be reflected in forecasts initialized at 18 UTC, which may be available as early as 22 UTC (2 hr after observation time). Observations at 21 UTC will be included only in the post run, if latency is too high, as it often is for derived satellite products or research products. These observations will only be included in the forecast products initialized at 00 UTC, which become available no earlier than 04 UTC the following day (7 hr after observation time). This scenario leaves out several other potential sources of delay. A forecaster compiling a forecast at 12 UTC for the following day may have a numerical model output incorporating observations as late as 9 UTC, but only observations from 3 UTC or earlier are guaranteed to be incorporated in model runs available for preparing a forecast at 12 UTC.

When the observations come from a satellite in a Sun-synchronous polar orbit, such as MODIS or VIIRS, the interactions between the observation times and forecast cycles produce aliasing effects where forecast cycles have widely varying utility for specific locations. A further aliasing effect will be noted in the verification if the model is verified using observations available at specific times (such as daytime-only Sun photometer observations). This effect can be seen in (Zhang et al., 2014b) where 18 hr forecasts in many cases verified better than 12 hr forecasts; the satellite data assimilated were all collected near local noon, and the benefits of data assimilation manifested in a limited zonal range such that impacts would be missing from the daytime-only verification of the 12 hr forecast. This effect is illustrated in Figure 7.1 and Table 7.2, and will be highly noticeable in a regional prediction system.

High-resolution regional-scale models such as the NOAA RAP start new prediction cycles as often as each hour. From Lin et al. (2017, p. 1276): "RAP and HRRR use a short observation data cutoff time of about +35 min after the analysis time.. . . For RAP, this window extends back no more than 1.5 h from the model initial time." Some types of observations, such as geostationary satellite radiances, are available for assimilation within 10 min of acquisition, but observations with more than 30 min latency will frequently miss the cutoff, and observations with >2 hr latency have no path to ever be included in a system of this type.

The latency of satellite-based AOD observations is the difference between the observation time and the time that the observation is available for integration into the modeling system. Thus, downlink from the satellite, generation of data products in the satellite ground station network, and delivery of products to end users are all included in the overall latency. This question of latency, together with the common satellite infrastructure used by both NWP observations and observations of fire and smoke, explains why weather satellite data are the most commonly assimilated type in global models. Many research satellites (e.g., CALIOP, MISR) deliver data very applicable to problems of smoke forecasting, but the space and ground systems were not originally designed to deliver data within the tight latency requirements of NWP models.

At the Naval Research Laboratory Marine Meteorology Division, the MODIS Aqua and Terra AOD products (MYD04_L2 and MOD04_L2, respectively) are downloaded from NASA LANCE's near real-time servers every 30 min. The VIIRS (SNPP/NOAA-20) and ABI (GOES-16/17, full disk) AOD products (NOAA Enterprise algorithms) are downloaded from NOAA every 10 min. Maintaining these efficient download workflows helps to achieve the minimal latency for satellite observations (Fig. 7.2). The GOES AOD products exhibit the lowest latencies, with modes around 13 min and 22 min. The MODIS and VIIRS latencies are similar in their distributions, but there is an observed platform dependence for granule latency. AQUA and SNPP show a bimodal distribution, with peaks around 63–66 min and 120–135 min. TERRA and NOAA-20 latency is unimodal with means around 77 min (σ = 18) and 57 min (σ = 10), respectively. This discrepancy is related to the satellite downlink; Terra and NOAA-20 utilize the Tracking and Data Relay Satellite System to achieve more frequent downlink leading to reduced latency (https://www.nasa.gov/directorates/heo/scan/services/networks/tdrs_main).

Ground-based observation networks have made great advances in latency over the past decade due to the improvements in electronics and communication networks. The AirNow air quality observation network is a partnership between multiple U.S. federal agencies (including the Environmental Protection Agency (EPA), National Oceanic and Atmospheric Administration (NOAA), and U.S. Forest Service, among others) and local agencies. Figure 7.3 shows a snapshot of the timeliness of available hourly PM2.5 and PM10 site observations within a 3 day window. A total of 1,131 sites reported PM2.5 or PM10 during this window, with the bulk of the sites (80%) reporting within a 2 hr window. The AirNow network has far lower latencies compared with the Aerosol Robotic Network (AERONET; Holben et al., 1998). A small subset of AERONET sites (20%) is capable of reporting Level 1.5 data within 3 hr, but it may take up to 20 hr for roughly half of the data to become available. AERONET Level 2 products include additional quality control, but are not available in near real time (Giles et al., 2019).

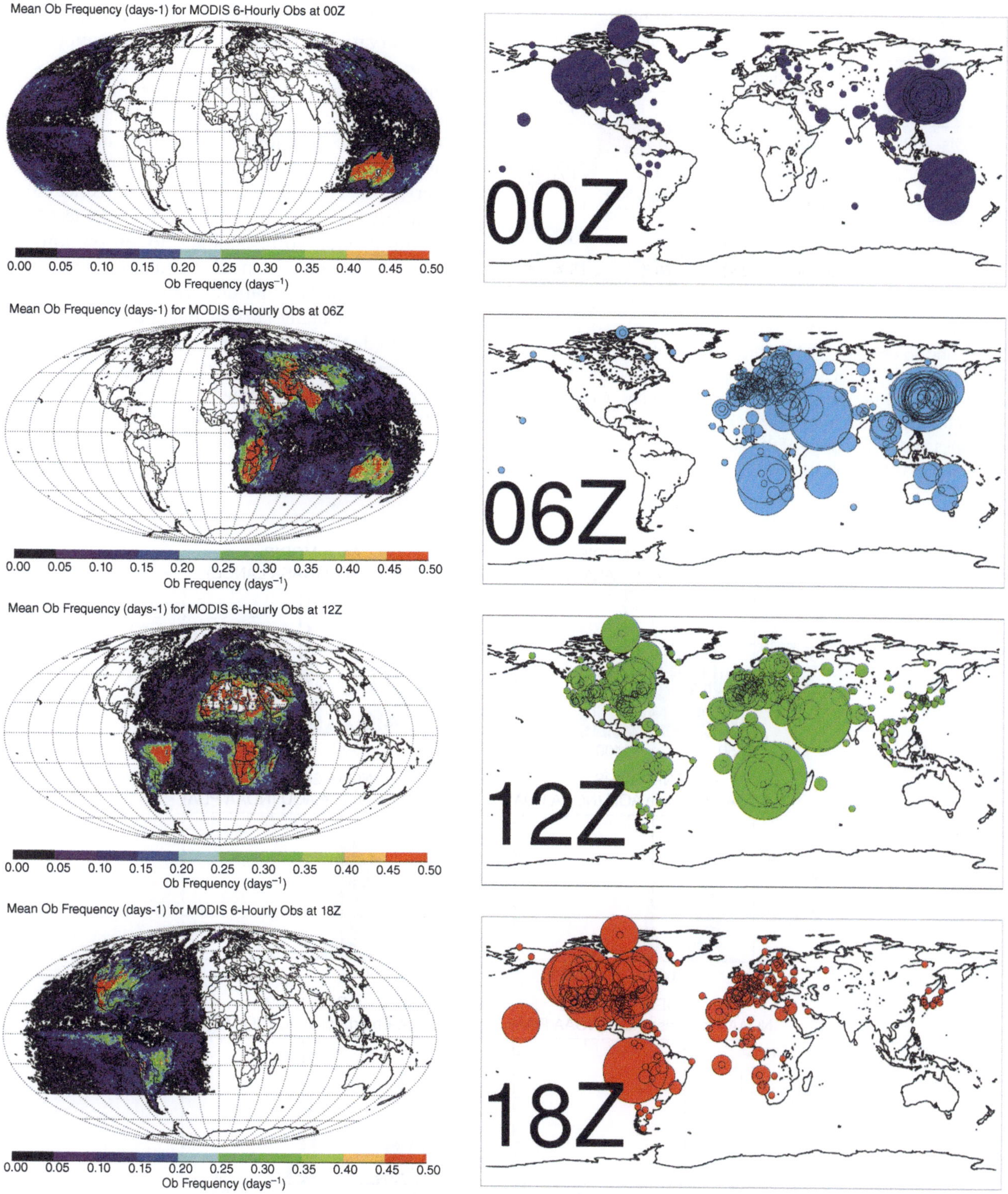

Figure 7.1 Density of AOD observations at synoptic times 00Z, 06Z, 12Z, 18Z. (Left): MODIS NRL/UND Level 3 observations assimilated into the NAAPS model; Colored grid cells indicate density of observations. (Right) AERONET AOD retrievals used for model verification. Symbols indicate locations of AERONET sites. Size of circles is proportional to data volume over the study period of 1 May 2016 to 18 June 2016.

Table 7.2 Availability of MODIS AOD for assimilation versus times/locations of AERONET forecast verification observations

		AERONET observation time			
		21Z-3Z	3Z-9Z	9Z-15Z	15Z-18Z
Analysis time	00Z	16%	10%	1%	2%
	06Z	18%	20%	9%	1%
	12Z	0%	20%	25%	14%
	18Z	12%	0%	9%	16%
	PAIRED	26%	28%	28%	25%

Notes: Blue color indicates overlap between MODIS at analysis time and AERONET observations at 24 hr, 48 hr, 72 hr, and so on, forecast lead times. Red color indicates overlap between MODIS at analysis time and AERONET at 12 hr, 36 hr, 60 hr, and so on, forecast lead times. Green colors indicate availability of MODIS AOD observations at the times and locations of AERONET observations used to verify the NAAPS analysis.

With a rapid update model and a low-latency system to deliver observations, many types of observations can contribute to smoke prediction. However, latency of observations remains the primary bounding condition for utility of numerical smoke predictions. In the best-case scenario, the first forecast incorporating observations of a fire or smoke plume will likely be 2–4 hr after the observation. The inference is that the numerical forecast is far less useful for predictions of smoke less than 2–4 hr downwind, corresponding to 30–60 km for moderate wind speeds of 4 m s-1. For a global prediction system, this radius of inutility can be much larger, exceeding 9 hr in many current operational configurations. For prediction of smoke impacts in the near field within this radius, a different type of modeling, such as trajectory modeling, may be more appropriate.

Geostationary observations mitigate this effect because observations are available throughout the day. Several methods of retrieving aerosol optical depth at nighttime

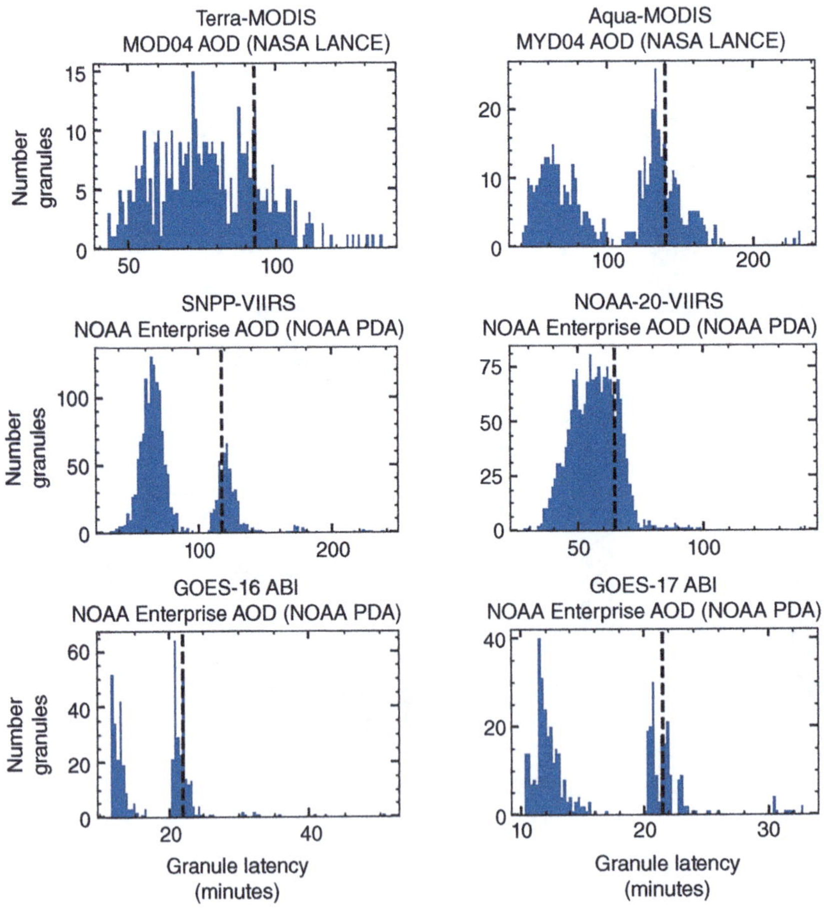

Figure 7.2 Latency of satellite aerosol products. Each plot shows the time gap between the observation time and the time the observation files were available on the NRL file system for use by the model. The dashed vertical line on each plot represents the 80th percentile of data latency. Details of the observations are in the text; plots are based on May 2021 data.

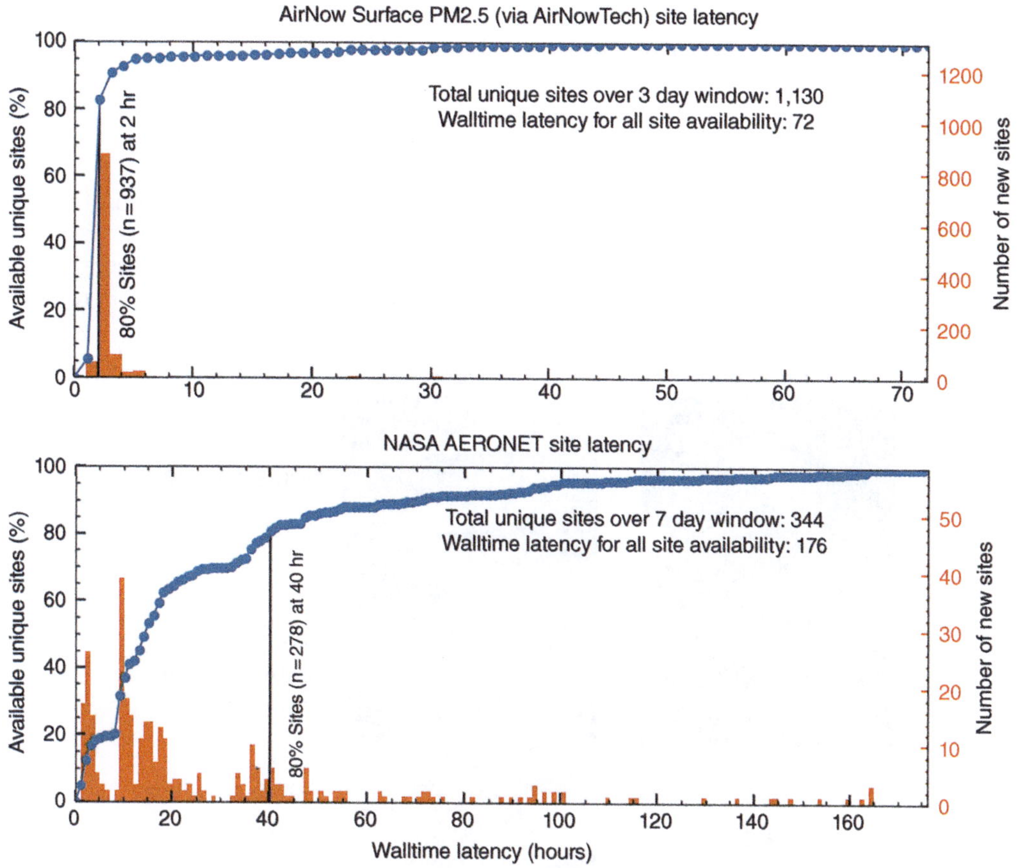

Figure 7.3 Latency of ground-based data sets. These plots show the cumulative availability of data from different stations as a function of delay. Vertical lines indicate the delay associated with 80% of the total sites reporting. Details of the observations are in the text; plots are based on May 2021 data.

have been prototyped (Fu et al., 2018; Johnson et al., 2013; McHardy et al., 2015; Wang et al., 2016; Wang et al., 2020b; Zhang et al., 2019; Zhao et al., 2016), but so far there is no operational nighttime aerosol product available. Nighttime observations also provide enhanced information on fire behavior by combining the visible signal of flame with the infrared heat signature of fires; this can provide important insight into overnight fire behavior that is very important to predict fire activity the following day (Polivka,et al., 2016; Wang et al., 2020a). From a practical standpoint, it is easy to see how nighttime observations could be very valuable, in terms of providing information about overnight progress of a fire and early insight into smoke conditions for the following day. Assimilation of CO observations from sounders shows promise for smoke prediction due to their nighttime availability. While not directly constraining aerosol, increments can be applied through covariation of smoke and CO as they are coemitted in biomass burning plumes (Arellano et al., 2010). But currently available geostationary observations of smoke are daytime only, and observations with solar elevation lower than 10° have large uncertainty, so these observations will still not be available near sunrise and sunset.

Timing of observations interacts with the practicalities of forecasting to impose important limits on the ability of observations to improve a real-time forecast. Numerical forecasts available during the morning hours may have overnight fire observations applied in the smoke source, but the smoke plume characteristics will likely reflect only smoke observations from the previous day, even with geostationary observations. Only around midday will model outputs incorporating same-day smoke observations become available.

Range From Smoke Source

The consequence of the timing considerations from the previous section is that only predictions of smoke at least 2 hr or multiple tens of kilometers from the source will see improvement from assimilation of smoke observations. The proximity of the observation to the smoke source is a key factor determining how effectively that observation can mitigate different types of errors in the forecast model.

Figure 7.4 shows an array of different satellite aerosol products for 2 days of extreme fire activity in northern California during September 2020. The top row is true color imagery giving the context of land, ocean, and smoke in these scenes. Below that, rows depict satellite AOD retrievals, ordered from fine to coarse spatial resolution, starting with nominal 750 m VIIRS data using the NOAA Enterprise aerosol retrieval (Zhang et al., 2016a), and proceeding through the nominal 1 km MAIAC (Lyapustin et al., 2011), the nominal 2 km GOES-17 ABI AOD using the NOAA Enterprise retrieval (Zhang et al., 2020), the nominal 10 km MODIS combined Dark Target/ Deep Blue retrieval (Levy et al., 2013), and the 0.5° Level 3 NRL/UND DA-quality AOD product (Hyer et al., 2011; Shi et al., 2011). For data and visualization sources, see section 7.6.

Several aspects of satellite aerosol retrievals relevant to smoke modeling are apparent in Figure 7.4. Most obviously, in all of the products, there is some part of the thickest smoke that is excluded from retrieval. At a given

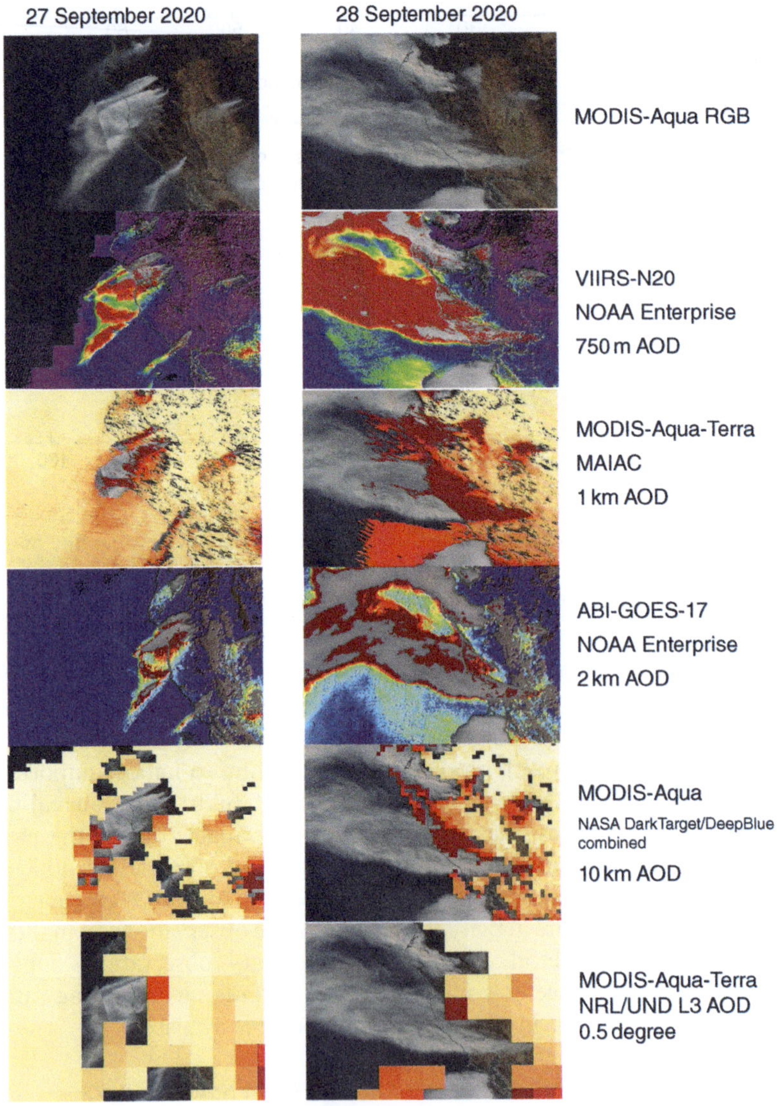

Figure 7.4 Retrieval of smoke by different AOD products for 2 days in September 2020. Data sets are described in the text. Dark red colors indicate the highest AOD; lower AODs are indicated by blue or beige colors on the maps. (Data sources as follows: MODIS RGB from NASA Worldview, http://worldview.earthdata.nasa.gov/; NOAA Enterprise VIIRS AOD from NOAA JPSS STAR Mapper, https://www.star.nesdis.noaa.gov/jpss/mapper; MODIS MAIAC, Dark Target / Deep Blue combined, and NRL/UND L3 AOD from NASA Worldview (Note: NRL/UND L3 is labeled "MODIS Combined Value-Added Aerosol Optical Depth"); NOAA Enterprise GOES ABI AOD from NOAA Aerosol Watch, https://www.star.nesdis.noaa.gov/smcd/spb/aq/AerosolWatch/. See section 7.5 for more information.)

wavelength, there are hard physical limitations on the maximum retrievable AOD because a complete retrieval requires at least some light reflected from the surface to penetrate through the plume to reach the sensor. Many sensors based on retrieval at 550 nm adopt an AOD of 5 as a retrieval limit for this reason. Longer wavelengths are more transparent to smoke, and some proposed methods exploit this to obtain retrievals even in smoke that is opaque in visible wavelengths (Eck et al., 2019).

Heavy smoke conditions can also trigger textural filters intended to ensure exclusion of clouds. The MODIS Collection 6.1 Dark Target aerosol retrievals are the basis for the bottom two panels in Figure 7.4 (the bottom row is the NRL/UND DA-quality product with additional stringent QC filtering), and the conservative approach to clouds taken by these retrievals results in the gaps in these data sets over the heaviest smoke. This is in spite of significant effort to implement a call-back method in MODIS Collection 6 aerosol retrievals to rescue retrievals in heavy smoke initially flagged as cloud (Levy et al., 2013). An experimental method based on MODIS achieved significantly improved retrieval of heavy smoke, at the cost of weakening the cloud filtering (Shi et al., 2019). The higher-resolution retrievals shown in Figure 7.4 have smaller excluded areas, and a greater ability to retrieve heavy smoke and separate smoke and cloud in the NOAA Enterprise algorithm (Laszlo & Liu, 2016) and the MAIAC (Lyapustin et al., 2012). All retrievals will exclude some part of thick smoke near the fire source. This will result in significant underestimation of heavy aerosol events (Shi et al., 2019) and is an ongoing area of improvement for aerosol retrievals.

The filtering methods applied to exclude flawed and cloud-contaminated retrievals from aerosol retrievals are necessary to assimilate retrievals on a systematic basis (Zhang et al., 2005), but these impose a low bias on the assimilated observations, which will be transferred to the model analysis if not mitigated. Some mitigation can be achieved by using an ensemble approach to obtain flow-dependent background errors that can help to preserve gradients (Rubin et al., 2017) (also see section 7.4.2); it may also be profitable to explore automated means for identifying contexts where cloud filtering can be relaxed to permit improved retrieval of thick smoke.

7.2.3. Discussion: How and When Are Aerosol Observations Helpful in Smoke Prediction?

The discussion above identifies critical properties of observations that make them useful for data assimilation in smoke prediction systems.

1. Observations must be available in a timely fashion with fully automated data flows.

2. Observations must project onto the model state (e.g., some observations may contain information on smoke composition, but this information can be assimilated only into a model that simulates composition information).

3. Observations should not exhibit systematic bias in the presence of thick smoke (although many commonly used observations do this).

4. The smoke model must adequately constrain the characteristics of the smoke plume that are not observed, such as plume height for column-integrated retrievals, plume shape for point retrievals, accurate conversions between aerosol mass and optical properties.

Based on these considerations, the trajectory of development of aerosol data assimilation systems in aerosol prediction models can be understood. The wide-area, low-latency data from meteorological satellites are the most straightforward fit into model systems. However, numerous other observation types may satisfy these criteria (e.g., aircraft observations (Saide et al., 2015a) or surface measurements (Ma et al., 2019)). Many potentially beneficial data sources have not yet been explored.

7.3. CONSIDERATIONS FOR ASSIMILATION OF SMOKE OBSERVATIONS

Equally important to what observations are included in data assimilation is the method applied to combine the observations and the model to obtain an optimal state. This section briefly surveys the published literature on aerosol data assimilation, and discusses some of the unique interactions between observations of smoke aerosols and data assimilation systems that complicate the problem and demand advanced techniques. The final section outlines some areas with future work in the near and medium term.

7.3.1. Development and Scope of Aerosol Data Assimilation Efforts

The effort to assimilate aerosol observations to improve model simulations of the atmosphere goes back 20 years (Yu et al., 2003; Wang et al. 2004). The Navy Aerosol Analysis and Prediction System (NAAPS) began to use assimilation of MODIS AOD observations for forecast initialization in 2009 (Zhang et al. 2008), making it the first operational aerosol model to include data assimilation. Since that time, many different groups have worked on aerosol data assimilation systems for operational and research models.

Many different data sets have been used in aerosol data assimilation including EPA surface monitor PM2.5 data (Schwartz et al., 2012) and surface aerosol optical measurements (Chang et al, 2021), visibility networks (Clark et al., 2008), AERONET Sun photometer data

(Rubin et al., 2017), and ground-based lidars (Liang et al., 2020; Yumimoto et al., 2008). Many kinds of satellite data also have been assimilated into aerosol models: visible/near-infrared satellite data from MODIS (Hyer et al., 2011; Liu et al., 2011), MISR (Lynch et al., 2016), UV observations from OMI (Lee et al., 2017; Zhang et al., 2021), multiangle, multipolarization observations from POLDER (Generoso et al., 2007), lidar backscatter from CALIOP (Zhang et al., 2014), and geostationary observations (Saide et al., 2014; Yumimoto et al., 2016), and even cloud retrievals (Saide et al., 2012). Assimilation methods tested include a wide variety of both variational (Benedetti et al., 2009; Yumimoto & Takemura, 2013), optimal interpolation (Tang et al., 2017), and ensemble (Dai et al., 2013; Rubin et al., 2016; Yumimoto & Takemura, 2011) methods.

While most aerosol data assimilation efforts have been aimed at forecast improvement (Xian et al., 2019), important work has also been done applying data assimilation to retrieval of aerosol sources (Dai et al., 2019a; Dubovik et al., 2008), evaluation of aerosol property assumptions used in models and remote sensing (Goto et al., 2011), long-term reanalysis of atmospheric aerosol (Gelaro et al., 2017; Lynch et al., 2016; Yumimoto et al., 2017), and study of aerosol interactions with weather (Saide et al., 2015b). A review of assimilation methods and applications for trace gases and aerosols can be found in Bocquet et al. (2015).

7.3.2. Problems Particular to Aerosol Data Assimilation

While the techniques of data assimilation are generally applicable to aerosol, obtaining the most accurate forecast results requires a data assimilation scheme that accommodates the realities of the aerosol observations as well as the realities of the atmospheric aerosol, both as it exists in nature and in the simplified form in which it is represented by the model. This section discusses a few of the specifics of aerosols and aerosol observations and how they affect data assimilation. The distinguishing characteristics of smoke plumes interact with the behavior of both aerosol measurements and models to pose specific challenges for data assimilation to improve smoke prediction.

Non-Gaussian Distribution of Aerosol-Related Quantities

Most operational data assimilation systems, whether variational, ensemble, or hybrid, employ the convenient assumptions that the errors obey Gaussian statistics and that the observation operator, which maps state variables to observation space, is linear (Amezcua & Van Leeuwen, 2014; Bannister et al., 2020, among others). The assumption of Gaussianity allows for PDFs to be defined by a mean and covariance. For variables such as AOD or atmospheric moisture, their probability distributions (PDFs) can deviate significantly from Gaussian. Nonlinear observation operators are often required for nonlocal observations. Atmospheric aerosols, measured in terms of mass concentration or optical extinction, cannot have negative values, and for this reason alone violate the assumption of Gaussian distribution. The true distribution of atmospheric aerosol quantity is generally recognized as lognormal (O'Neill et al., 2000). Detailed analysis has shown that variation in aerosol within narrow spatiotemporal bounds influenced primarily by fine-scale meteorology may appear more Gaussian but at scales where variability is driven by aerosol sources and sinks and synoptic meteorology, a lognormal distribution best matches the observations (Alexandrov et al., 2016; Sayer & Knobelspiesse, 2019). When considering specifically the errors in aerosol observations, these errors will include errors of spatial and temporal sampling, which can be modeled as draws from the lognormal regime, as well as instrumental errors and errors associated with underdetermined retrievals, which may or may not have a Gaussian distribution (Sayer et al., 2020).

The consequences of assuming Gaussianity for a non-Gaussian parameter can be significant. As discussed in Bannister et al. (2020), a Gaussian distribution with the same mean and variance will not represent the true PDF (which is especially detrimental for a multimodal case), negative values may be allowed for bounded variables, and the mode may be a physically improbably state. There are several approaches to dealing with non-Gaussianity in data assimilation. The most computationally efficient is a variable transformation. As discussed in Amezcua and Van Leeuwen (2014), using a variable transformation will only provide an approximate solution when the requirement of Gaussianity is not met in the original space; however, it is clear that some transformations are better than others. Further, Bocquet et al. (2010) discussed the potential drawback of introducing an observation operator in a new space following a typically nonlinear transform. A more difficult approach is to use a more appropriate filter that does not require a Gaussian assumption. Although such methods can be computationally expensive, this is an area of active research. Particle filters (van Leeuwen, 2009; van Leeuwen et al., 2019) generalize the Bayesian data assimilation problem by approximating PDFs through weighted samples. Bishop (2016) introduced an Ensemble Kalman Filter (EnKF) that extends the Gaussian assumption to also account for Gamma or Inverse Gamma PDFs. Fletcher and Zupanski (2006) show how a cost function can be modified to account for lognormal distributions.

Power transformation (following Yeo & Johnson, 2000) can be applied, such that a variable u is mapped to log space through the transform:

$$u \rightarrow \log(1+u). \qquad (7.1)$$

This transformation effectively pulls the tails of the distribution inward and results in a distribution that is closer to a Gaussian distribution. A number of centers use, or have previously used, log humidity as a control variable (Bannister et al., 2020; Dee & Da Silva, 2003, among others).

Examination of modeled aerosol optical depth from NAAPS and AOD observations used in data assimilation shows how the untransformed AOD data deviate very significantly from the assumed Gaussian distribution (Fig. 7.5, left column). The inset plots show the quantile-to-quantile comparison, normally distributed data will fall on the 1:1 line (red line in the figure). Thus, a variable transform could significantly improve the performance of an aerosol data assimilation system.

Several scaling-based transformations have been applied for atmospheric moisture variables, most notably Dee and DeSilva (2003) and Hólm et al. (2002). Dee and DeSilva (2003) proposed the use of pseudo relative humidity (PRH). PRH is defined by scaling the mixing ratio by the background saturation mixing. Hólm et al. (2002) defined a transformation based on normalized forecast differences, which gives a more symmetric and closer to Gaussian PDF. Both PRH and the Hólm transform have been used for atmospheric moisture analyses. It is worth noting that the approach of Hólm et al. (2002) is more generalizable and could extend to other variables of interest.

Some AOD data assimilation systems do employ a transform of the form

$$AOD' \rightarrow \log(c + AOD), \qquad (7.2)$$

with *c* taking the value of 0.01(Randles et al., 2017) or 0 (Saide et al., 2013), while others do not. Benedetti et al. (2009) explicitly tested this transform in part of their aerosol data assimilation system, and described their results this way.

> As a side note, the implementation of the logarithmic variable did not dramatically improve the analysis performance. On the contrary, the RMS is higher and the correlation lower in the analysis with the logarithmic control variable than in the analysis with total mass mixing ratio. The reason for this could lie in the fact that the logarithmic control variable is only used at the level of the minimization, whereas the rest of the model is formulated in terms of mass mixing ratio. A more effective way to handle tracers could be to formulate the whole forward model in terms of logarithmic (hence positive definite) variables. This however would involve an extensive effort in modifying and rewriting the model, and it is not a viable option at this point. The use of alternative normalized control variables with a more Gaussian error distribution can still be investigated for future developments, following existing examples (Hólm et al., 2002). (Benedetti et al., 2009, p. 8)

The right-hand column of Figure 7.5 shows that application of this transform with c = 0.01 yields a more Gaussian distribution of background, observations, and innovations. However, significant distortions remain in the distribution because of the need to truncate negative AOD observations in order to perform the log transform (see section 7.3.2.2).

Gaussian anamorphasis (GA) is another approach that has gained recent popularity in atmospheric data assimilation (e.g., Brankart et al., 2012; Simon & Bertino, 2012; Zhou et al., 2011). GA employs a transform based on the cumulative distribution (CDF) such that the moments of the transformed Gaussian variable are set to those of the original prior distribution (Bertino et al., 2003). The simplest case of GA would be to transform the state variables and observations independently. Lien et al. (2013) used GA applied to precipitation, based on its model climatology, under the assumption that a forecast variable with more Gaussian climatological distribution would result in a more Gaussian error distribution. In their study, zero precipitation was handled as a delta function located at the median of the zero-precipitation part of the normal distribution. Lien et al. (2016) further showed that this approach also showed benefit in correcting the amplitude-dependent biases. Amezcua and Van Leeuwen (2014) explored GA in joint state variable and observation space.

Satellite Retrievals Permit Negative Values of AOD

AOD values of zero or higher are physically realistic, but the retrieval of AOD from reflected radiance is constructed as a problem of solving for the aerosol contribution as a deviation from the expected clear-sky radiance (Kaufman et al., 1997). Errors in the estimation of the clear-sky radiance can be positive or negative, so this formulation of the retrieval problem will always result in some negative values of retrieved AOD. Exclusion of these observations is not appropriate, because they constrain the true AOD to a narrow range and have a high probability of being cloud free. However, truncation of these values to 0 or to a low nominal AOD value distorts the distribution of the observations.

Observation Quality Control Systematically Removes Heavy Smoke

The most important error in AOD retrievals obtained by inversion of observed radiances is contamination by cloud. Because the scattering efficiency of particles smaller than the wavelength of incident light (Rayleigh scattering) scales with the sixth power of the particle radius (r^6), the presence of even a very low number of cloud droplets (radius ~10 μm) can overwhelm the signal of scattering aerosols (radius <1 μm) (Zhang et al., 2005). Separation of the signals of cloud droplets

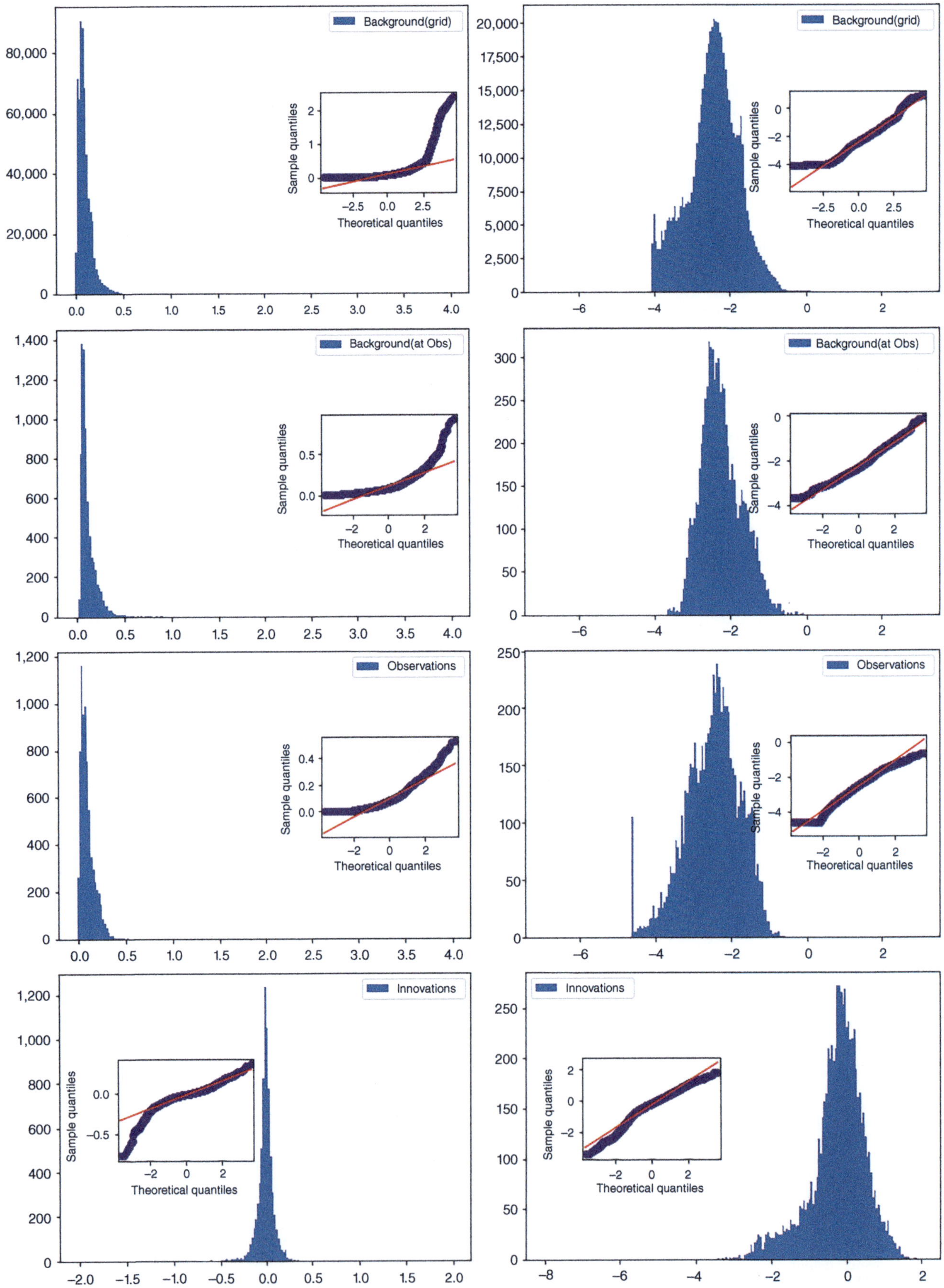

Figure 7.5 Distribution of modeled and observed AOD values, and effects of a logarithmic transformation. (Left) The distribution of AOD from (top to bottom) the NAAPS model grid, NAAPS sampled to observation locations, MODIS DA-quality observations, and lastly the innovations, computed as the differences Observation − Background. The inset plots on each histogram are the quantile-quantile plots indicating how well each distribution conforms to a Gaussian shape. (Right) The same values with the AOD′ = ln(AOD + 0.01) transform applied (see details in the text).

and aerosol particles has been attempted but the only practical results have concerned regimes where a discrete aerosol layer sits above a layer of optically opaque cloud (Jethva et al., 2016); separation of cloud and aerosol particle scattering using multispectral data is very challenging. Remotely sensed data are arbitrarily discretized in space such that condensed water particles can be present in any quantity; thus cloud clearing mechanisms have employed textural and contextual tests to exclude cloud droplets (Remer et al., 2005). These tests, to the extent that they rely on textural information in the visible scattering regime, also systematically remove heavy aerosol and smaller aerosol plumes (Levy et al., 2013), as can be seen in Figure 7.4.

This creates an asymmetric problem when assimilating strong aerosol plumes: if the model background is too high, data assimilation of observations adjacent to the core of the plume will act to reduce this error. If the model background is low, the results of data assimilation will necessarily also be low because of the exclusion of the plume core (Dai et al., 2019b). This problem can be mitigated with the use of an ensemble that includes uncertainty in the source magnitude (Rubin et al., 2016), and better results will be obtained with the use of retrievals that are able to retain more of the thick plumes (MAIAC and VIIRS NOAA Enterprise on Fig. 7.4).

Observations Do Not Constrain Aerosol Type or Vertical Structure

The atmospheric aerosol is a wildly heterogeneous mixture of particles of varying size, shape, and composition, in varying equilibria between water and chemical components in gas, liquid, and solid phases. Numerical models of aerosol reduce this complexity enormously, attempting to represent the aerosol properties that most impact targeted atmospheric outcomes such as climate forcing (Stier et al., 2013).

Optical retrieval of aerosol is underdetermined in different ways. Information can be added to the optical retrieval by incorporation of multiple wavelengths, viewing geometries, and polarization angles. Advanced methods have shown promise to obtain detailed aerosol composition information (Noyes et al., 2020a) as well information on plume height (Lyapustin et al., 2020) and combustion phase (Wang et al., 2020a) for certain fires. However, observing in all of these dimensions together still captures only a small part of the variability in the real atmosphere (Choi et al., 2021). Operational retrievals available for smoke forecasting provide a moderately uncertain estimate of column-integrated aerosol optical depth, and do not effectively constrain any other properties (Reid et al., 2013).

Some data assimilation systems simply place the speciation and vertical profile outside of the data assimilation solution itself. By transforming the model background to AOD and solving for an increment in AOD space, the critical step is applying the AOD increment to obtain a new 3D initial state for each simulated aerosol type. Simply conserving the prior was recognized early on to be inadequate because the model deficiencies corrected by assimilation of AOD project onto the speciation and vertical distribution: heavy aerosol events are unlikely to be realistically represented by linear scaling of a background aerosol. In systems that solve for a single AOD increment, a climatology is employed to more realistically distribute added aerosol in cases where the model background is strongly biased low (Zhang et al., 2008). This problem manifests differently in systems where an increment is calculated for each aerosol type, but the underlying problem remains that large increments of aerosol loading require accounting for patterns of speciation and vertical placement at high and low aerosol loadings. For models resolving aerosol size distribution, an intermediate solution involves keeping speciation fixed within each aerosol size bin. This approach reduces the degrees of freedom but still allows changes on the overall speciation (Saide et al., 2013).

7.4. FUTURE RESEARCH DIRECTIONS

The content of this chapter covered the observations available to improve smoke predictions via data assimilation, the characteristics that determine usefulness of those observations, the scenarios in which smoke predictions can benefit from data assimilation, and the interactions between the particular problems of smoke and smoke observations in data assimilation. The scientific territory of this topic is only beginning to be explored, and significant practical gains in smoke forecast skill are likely achievable over current practice. Smoke observational data sets are continuously being added and improved, with significant and rapid innovation. Advanced data assimilation methods have been developed that are relevant to the smoke prediction problem, but have not yet been applied. Techniques that maximize the effectiveness of combining multiple observation types (e.g., Saide et al., 2013; Schwartz et al., 2012; Saide et al., 2015a) need to be advanced into operational smoke prediction systems. In all of these areas, careful attention to both the scientific details of the observations and the practical realities of the forecast problem is necessary to obtain significant progress.

ACKNOWLEDGMENTS

This chapter was written with support from the Office of Naval Research code 322 as well as the NASA FIREX-AQ program.

AVAILABILITY STATEMENT

The data in Figure 7.4 were obtained from the following public website sources:

1. MODIS RGB from NASA Worldview (http://worldview.earthdata.nasa.gov/)
2. NOAA Enterprise VIIRS AOD can be viewed at NOAA JPSS STAR Mapper: https://www.star.nesdis.noaa.gov/jpss/mapper
3. MODIS MAIAC, Dark Target / Deep Blue combined, and NRL/UND L3 AOD can be viewed at NASA Worldview (NRL/UND L3 is labeled as "MODIS Combined Value-Added Aerosol Optical Depth")
4. NOAA Enterprise GOES ABI AOD can be viewed at NOAA Aerosol Watch: https://www.star.nesdis.noaa.gov/smcd/spb/aq/AerosolWatch/

Model predictions of dust, smoke, and other aerosols from the Navy Aerosol Analysis and Prediction System can be viewed at https://www.nrlmry.navy.mil/aerosol/. Data download of NASA products is available through the Earth Science Data and Information Service: http://earthdata.nasa.gov/. Download of NOAA products is available through the NOAA Comprehensive Large Array-data Stewardship System (CLASS) system: http://www.class.noaa.gov/. AirNow data are available through AirNowTech http://airnowtech.org/. AERONET data are available from NASA: http://aeronet.gsfc.nasa.gov/.

REFERENCES

Alexandrov, M. D., Geogdzhayev, I. V., Tsigaridis, K., Marshak, A., Levy, R., & Cairns, B. (2016). New statistical model for variability of aerosol optical thickness: Theory and application to MODIS data over ocean. *Journal of the Atmospheric Sciences*, *73*(2), 821–837. https://doi.gov/10.1175/jas-d-15-0130.1

Alfaro-Contreras, R., Zhang, J. L., Campbell, J. R., Holz, R. E., & Reid, J. S. (2014). Evaluating the impact of aerosol particles above cloud on cloud optical depth retrievals from MODIS. *Journal of Geophysical Research-Atmospheres*, *119*(9), 5410–5423. https://doi.gov/10.1002/2013jd021270

Al-Saadi, J., Soja, A., Pierce, R. B., Szykman, J., Wiedinmyer, C., Emmons, L., et al. (2008). Intercomparison of near-real-time biomass burning emissions estimates constrained by satellite fire data. *Journal of Applied Remote Sensing*, *2*.

Amezcua, J., & Van Leeuwen, P. J. (2014). Gaussian anamorphosis in the analysis step of the EnKF: a joint state-variable/observation approach. *Tellus Series a-Dynamic Meteorology and Oceanography*, *66*, 1–18. https://doi.gov/10.3402/tellusa.v66.23493

Arellano, A. F., Hess, P. G., Edwards, D. P., & Baumgardner, D. (2010). Constraints on black carbon aerosol distribution from Measurement of Pollution in the Troposphere (MOPITT) CO. *Geophysical Research Letters*, *37*, 5. https://doi.gov/10.1029/2010gl044416

Bannister, R. N., Chipilski, H. G., & Martinez-Alvarado, O. (2020). Techniques and challenges in the assimilation of atmospheric water observations for numerical weather prediction towards convective scales. *Quarterly Journal of the Royal Meteorological Society*, *146*(726), 1–48. https://doi.gov/10.1002/qj.3652

Benedetti, A., Morcrette, J. J., Boucher, O., Dethof, A., Engelen, R. J., Fisher, M., et al. (2009). Aerosol analysis and forecast in the European Centre for Medium-Range Weather Forecasts Integrated Forecast System: 2. Data assimilation. *Journal of Geophysical Research-Atmospheres*, *114*. https://doi.gov/D1320510.1029/2008jd011115

Bertino, L., Evensen, G., & Wackernagel, H. (2003). Sequential data assimilation techniques in oceanography. *International Statistical Review*, *71*(2), 223–241. https://doi.gov/10.1111/j.1751-5823.2003.tb00194.x

Bishop, C. H. (2016). The GIGG-EnKF: Ensemble Kalman filtering for highly skewed non-negative uncertainty distributions. *Quarterly Journal of the Royal Meteorological Society*, *142*(696), 1395–1412. https://doi.gov/10.1002/qj.2742

Bocquet, M., Elbern, H., Eskes, H., Hirtl, M., Zabkar, R., Carmichael, G. R., et al. (2015). Data assimilation in atmospheric chemistry models: Current status and future prospects for coupled chemistry meteorology models. *Atmospheric Chemistry and Physics*, *15*(10), 5325–5358. https://doi.gov/10.5194/acp-15-5325-2015

Bocquet, M., Pires, C. A., & Wu, L. (2010). Beyond Gaussian statistical modeling in geophysical data assimilation. *Monthly Weather Review*, *138*(8), 2997–3023. https://doi.gov/10.1175/2010mwr3164.1

Brankart, J. M., Testut, C. E., Beal, D., Doron, M., Fontana, C., Meinvielle, M., et al. (2012). Towards an improved description of ocean uncertainties: Effect of local anamorphic transformations on spatial correlations. *Ocean Science*, *8*(2), 121–142. https://doi.gov/10.5194/os-8-121-2012

Carr, J. L., Wu, D. L., Daniels, J., Friberg, M. D., Bresky, W., & Madani, H. (2020). GEO-GEO stereo-tracking of atmospheric motion vectors (AMVs) from the geostationary ring. *Remote Sensing*, *12*(22), 47. https://doi.gov/10.3390/rs12223779

Chang, W. Y., Zhang, Y., Li, Z. Q., Chen, J., & Li, K. T. (2021). Improving the sectional Model for Simulating Aerosol Interactions and Chemistry (MOSAIC) aerosols of the Weather Research and Forecasting-Chemistry (WRF-Chem) model with the revised Gridpoint Statistical Interpolation system and multi-wavelength aerosol optical measurements: The dust aerosol observation campaign at Kashi, near the Taklimakan Desert, northwestern China. *Atmospheric Chemistry and Physics*, *21*(6), 4403–4430. https://doi.gov/10.5194/acp-21-4403-2021

Cheng, Y. M., Dai, T., Goto, D., Schutgens, N. A. J., Shi, G. Y., & Nakajima, T. (2019). Investigating the assimilation of CALIPSO global aerosol vertical observations using a four-dimensional ensemble Kalman filter. *Atmospheric Chemistry and Physics*, *19*(21), 13445–13467. https://doi.gov/10.5194/acp-19-13445-2019

Choi, M., Sander, S. P., Spurr, R. J. D., Pongetti, T. J., van Harten, G., Drouin, B. J., et al. (2021). Aerosol profiling using radiometric and polarimetric spectral measurements in the O-2 near infrared bands: Estimation of information content and measurement uncertainties. *Remote Sensing of Environment*, *253*, 20. https://doi.gov/10.1016/j.rse.2020.112179

Clark, P. A., Harcourt, S. A., Macpherson, B., Mathison, C. T., Cusack, S., & Naylor, M. (2008). Prediction of visibility and aerosol within the operational Met Office Unified Model. I: Model formulation and variational assimilation. *Quarterly Journal of the Royal Meteorological Society*, *134*(636), 1801–1816. https://doi.gov/10.1002/qj.318

Colarco, P. R., Schoeberl, M. R., Doddridge, B. G., Marufu, L. T., Torres, O., & Welton, E. J. (2004). Transport of smoke from Canadian forest fires to the surface near Washington, D. C.: Injection height, entrainment, and optical properties. *Journal of Geophysical Research-Atmospheres*, *109*(D6), D06203. https://doi.gov/06210.01029/02003JD004248

Dai, T., Cheng, Y. M., Goto, D., Schutgens, N. A. J., Kikuchi, M., Yoshida, M., et al. (2019a). Inverting the East Asian dust emission fluxes using the Ensemble Kalman Smoother and Himawari-8 AODs: A case study with WRF-Chem v3.5.1. *Atmosphere*, *10*(9). https://doi.gov/10.3390/atmos10090543

Dai, T., Cheng, Y. M., Suzuki, K., Goto, D., Kikuchi, M., Schutgens, N. A. J., et al. (2019b). Hourly aerosol assimilation of Himawari-8 AOT using the four-dimensional local ensemble transform Kalman filter. *Journal of Advances in Modeling Earth Systems*, *11*(3), 680–711. https://doi.gov/10.1029/2018ms001475

Dai, T., Schutgens, N. A. J., & Nakajima, T. (2013). Applying a local ensemble transform Kalman filter assimilation system to the NICAM-SPRINTARS model. In R. F. Cahalan & J. Fischer (Eds.), *Radiation processes in the atmosphere and ocean* (pp. 744–747). AIP Conference Proceedings 1531.

Dee, D. P., & Da Silva, A. M. (2003). The choice of variable for atmospheric moisture analysis. *Monthly Weather Review*, *131*(1), 155–171. https://doi.gov/10.1175/1520-0493(2003)131<0155:tcovfa>2.0.co;2

Delp, W. W., & Singer, B. C. (2020). Wildfire smoke adjustment factors for low-cost and professional PM(2.5)monitors with optical sensors. *Sensors*, *20*(13), 21. https://doi.gov/10.3390/s20133683

Dubovik, O., Lapyonok, T., Kaufman, Y. J., Chin, M., Ginoux, P., Kahn, R. A., & Sinyuk, A. (2008). Retrieving global aerosol sources from satellites using inverse modeling. *Atmospheric Chemistry and Physics*, *8*(2), 209–250.

Eck, T. F., Holben, B. N., Giles, D. M., Slutsker, I., Sinyuk, A., Schafer, J. S., et al. (2019). AERONET remotely sensed measurements and retrievals of biomass burning aerosol optical properties during the 2015 Indonesian burning season. *Journal of Geophysical Research: Atmospheres*, *124*(8), 4722–4740. https://doi.gov/10.1029/2018jd030182

Fletcher, S. J., & Zupanski, M. (2006). A data assimilation method for log-normally distributed observational errors. *Quarterly Journal of the Royal Meteorological Society*, *132*(621), 2505–2519. https://doi.gov/10.1256/qj.05.222

French, N. H. F., de Groot, W. J., Jenkins, L. K., Rogers, B. M., Alvarado, E., Amiro, B., et al. (2011). Model comparisons for estimating carbon emissions from North American wildland fire. *Journal of Geophysical Research-Biogeosciences*, *116*. https://doi.gov/G00k05 10.1029/2010jg001469

Fu, D., Xia, X., Duan, M., Zhang, X., Li, X., Wang, J., et al. (2018). Mapping nighttime PM2.5 from VIIRS DNB using a linear mixed-effect model. *Atmospheric Environment*, *178*, 214–222. https://doi.gov/10.1016/j.atmosenv.2018.02.001

Gelaro, R., McCarty, W., Suarez, M. J., Todling, R., Molod, A., Takacs, L.,et al. (2017). The modern-era retrospective analysis for research and applications, version 2 (MERRA-2). *Journal of Climate*, *30*(14), 5419–5454. https://doi.gov/10.1175/jcli-d-16-0758.1

Generoso, S., Breon, F. M., Chevallier, F., Balkanski, Y., Schulz, M., & Bey, I. (2007). Assimilation of POLDER aerosol optical thickness into the LMDz-INCA model: Implications for the Arctic aerosol burden. *Journal of Geophysical Research: Atmospheres*, *112*(D2), 15. https://doi.gov/10.1029/2005jd006954

Giglio, L. (2007). Characterization of the tropical diurnal fire cycle using VIRS and MODIS observations. *Remote Sensing of Environment*, *108*(4), 407–421.

Giglio, L., & Schroeder, W. (2014). A global feasibility assessment of the bi-spectral fire temperature and area retrieval using MODIS data. *Remote Sensing of Environment*, *152*(0), 166–173. https://doi.gov/10.1016/j.rse.2014.06.010

Giles, D. M., Sinyuk, A., Sorokin, M. G., Schafer, J. S., Smirnov, A., Slutsker, I., et al. (2019). Advancements in the Aerosol Robotic Network (AERONET) Version 3 database: Automated near-real-time quality control algorithm with improved cloud screening for Sun photometer aerosol optical depth (AOD) measurements. *Atmospheric Measurement Techniques*, *12*(1), 169–209. https://doi.gov/10.5194/amt-12-169-2019

Goto, D., Schutgens, N. A. J., Nakajima, T., & Takemura, T. (2011). Sensitivity of aerosol to assumed optical properties over Asia using a global aerosol model and AERONET. *Geophysical Research Letters*, *38*. https://doi.gov/10.1029/2011gl048675

Hammer, M. S., van Donkelaar, A., Li, C., Lyapustin, A., Sayer, A. M., Hsu, N. C., et al. (2020). Global estimates and long-term trends of fine particulate matter concentrations (1998–2018). *Environmental Science and Technology*, *54*(13), 7879–7890. https://doi.gov/10.1021/acs.est.0c01764

Holben, B. N., Eck, T. F., Slutsker, I., Tanre, D., Buis, J. P., Setzer, A., et al. (1998). AERONET: A federated instrument network and data archive for aerosol characterization. *Remote Sensing of Environment*, *66*(1), 1–16.

Hólm, E., Andersson, E., Beljaars, A., Lopez, P., Mahfouf, J.-F., Simmons, A., et al. (2002). Assimilation and modeling of the hydrological cycle: ECMWF's status and plans. *Technical Memoranda*, *383*.

Hyer, E. J., & Reid, J. S. (2009). Baseline uncertainties in biomass burning emission models resulting from spatial error in satellite active fire location data. *Geophysical Research Letters*, *36*, https://doi.gov/10.1029/2008GL037145.

Hyer, E. J., Reid, J. S., & Zhang, J. (2011). An over-land aerosol optical depth data set for data assimilation by filtering, correction, and aggregation of MODIS Collection 5 optical depth retrievals. *Atmospheric Measurement Techniques*, *4*, 379–408. Retrieved from https://doi.gov/10.5194/amt-4-379-2011

Hyer, E. J., Reid, J. S., Prins, E. M., Hoffman, J. P., Schmidt, C. C., Miettinen, J. I., et al. (2013). Patterns of fire activity over Indonesia and Malaysia from polar and geostationary satellite observations. *Atmospheric Research*, *122*, 504–519. https://doi.gov/10.1016/j.atmosres.2012.06.011

Jethva, H., Torres, O., Remer, L., Redemann, J., Livingston, J., Dunagan, S., et al. (2016). Validating MODIS above-cloud aerosol optical depth retrieved from "color ratio" algorithm using direct measurements made by NASA's airborne AATS and 4STAR sensors. *Atmospheric Measurement Techniques*, *9*(10), 5053–5062. https://doi.gov/10.5194/amt-9-5053-2016

Jethva, H., Torres, O., Waquet, F., Chand, D., & Hu, Y. X. (2014). How do A- train sensors intercompare in the retrieval of above-cloud aerosol optical depth? *A case-study-based assessment. Geophysical Research Letters*, *41*(1), 186–192. https://doi.gov/10.1002/2013gl058405

Jianglong, Z., Reid, J. S., Westphal, D. L., Baker, N. L., & Hyer, E. J. (2008). A system for operational aerosol optical depth data assimilation over global oceans. *Journal of Geophysical Research, Part D: Atmospheres*, *113*(D10). https://doi.gov/10.1029/2007jd009065

Johnson, R. S., Zhang, J., Hyer, E. J., Miller, S. D., & Reid, J. S. (2013). Preliminary investigations toward nighttime aerosol optical depth retrievals from the VIIRS day/night band. *Atmospheric Measurement Techniques*, *6*, 1245–1255. https://doi.gov/https://doi.gov/10.5194/amt-6-1245-2013

Kaku, K. C., Reid, J. S., Hand, J. L., Edgerton, E. S., Holben, B. N., Zhang, J., et al. (2018). Assessing the challenges of surface-level aerosol mass estimates from remote sensing during the SEAC(4)RS and SEARCH campaigns: Baseline surface observations and remote sensing in the southeastern United States. *Journal of Geophysical Research: Atmospheres*, *123*(14), 7530–7562. https://doi.gov/10.1029/2017jd028074

Kaufman, Y. J., Tanre, D., Remer, L. A., Vermote, E. F., Chu, A., & Holben, B. N. (1997). Operational remote sensing of tropospheric aerosol over land from EOS moderate resolution imaging spectroradiometer. *Journal of Geophysical Research: Atmospheres*, *102*(D14), 17051–17067.

Kleinman, L. I., Sedlacek, A. J., Adachi, K., Buseck, P. R., Collier, S., Dubey, M. K., et al. (2020). Rapid evolution of aerosol particles and their optical properties downwind of wildfires in the western U.S. *Atmospheric Chemistry and Physics*, *20*(21), 13319–13341. https://doi.gov/10.5194/acp-20-13319-2020

Laszlo, I., & Liu, H. (2016). *NOAA EPS Aerosol Optical Depth (AOD) algorithm theoretical basis document, version 3.0.1.* https://www.star.nesdis.noaa.gov/jpss/documents/ATBD/ATBD_EPS_Aerosol_AOD_v3.0.1.pdf

Lee, E., Zupanski, M., Zupanski, D., & Park, S. K. (2017). Impact of the OMI aerosol optical depth on analysis increments through coupled meteorology-aerosol data assimilation for an Asian dust storm. *Remote Sensing of Environment*, *193*, 38–53. https://doi.gov/10.1016/j.rse.2017.02.013

Levy, R. C., Mattoo, S., Munchak, L. A., Remer, L. A., Sayer, A. M., Patadia, F., et al. (2013). The Collection 6 MODIS aerosol products over land and ocean. *Atmospheric Measurement Techniques*, *6*(11), 2989–3034. https://doi.gov/10.5194/amt-6-2989-2013

Liang, Y. F., Zang, Z. L., Liu, D., Yan, P., Hu, Y. W., Zhou, Y., et al. (2020). Development of a three-dimensional variational assimilation system for lidar profile data based on a size-resolved aerosol model in WRF-Chem model v3.9.1 and its application in PM2.5 forecasts across China. *Geoscientific Model Development*, *13*(12), 6285–6301. https://doi.gov/10.5194/gmd-13-6285-2020

Lien, G. Y., Kalnay, E., & Miyoshi, T. (2013). Effective assimilation of global precipitation: simulation experiments. *Tellus Series a-Dynamic Meteorology and Oceanography*, *65*. https://doi.gov/10.3402/tellusa.v65i0.19915

Lien, G. Y., Kalnay, E., Miyoshi, T., & Huffman, G. J. (2016). Statistical properties of global precipitation in the NCEP GFS model and TMPA observations for data assimilation. *Monthly Weather Review*, *144*(2), 663–679. https://doi.gov/10.1175/mwr-d-15-0150.1

Lin, H. D., Weygandt, S. S., Benjamin, S. G., & Hu, M. (2017). Satellite radiance data assimilation within the hourly updated rapid refresh. *Weather and Forecasting*, *32*(4), 1273–1287. https://doi.gov/10.1175/waf-d-16-0215.1

Liu, Y. Q., Kochanski, A., Baker, K. R., Mell, W., Linn, R., Paugam, R., et al. (2019). Fire behaviour and smoke modeling: Model improvement and measurement needs for next-generation smoke research and forecasting systems. *International Journal of Wildland Fire*, *28*(8), 570–588. https://doi.gov/10.1071/wf18204

Liu, Z. Q., Liu, Q. H., Lin, H. C., Schwartz, C. S., Lee, Y. H., & Wang, T. J. (2011). Three-dimensional variational assimilation of MODIS aerosol optical depth: Implementation and application to a dust storm over East Asia. *Journal of Geophysical Research-Atmospheres*, *116*. https://doi.gov/D2320610.1029/2011jd016159

Lyapustin, A., Korkin, S., Wang, Y., Quayle, B., & Laszlo, I. (2012). Discrimination of biomass burning smoke and clouds in MAIAC algorithm. *Atmospheric Chemistry and Physics*, *12*(20), 9679–9686. https://doi.gov/10.5194/acp-12-9679-2012

Lyapustin, A., Wang, Y., Laszlo, I., Kahn, R., Korkin, S., Remer, L., et al. (2011). Multiangle implementation of atmospheric correction (MAIAC): 2. *Aerosol algorithm. Journal of Geophysical Research: Atmospheres*, *116*, 15. https://doi.gov/10.1029/2010jd014986

Lyapustin, A., Wang, Y. J., Korkin, S., Kahn, R., & Winker, D. (2020). MAIAC thermal technique for smoke injection height from MODIS. *IEEE Geoscience and Remote Sensing Letters*, *17*(5), 730–734. https://doi.gov/10.1109/lgrs.2019.2936332

Lynch, P., Reid, J. S., Westphal, D. L., Zhang, J. L., Hogan, T. F., Hyer, E. J., et al. (2016). An 11-year global gridded aerosol optical thickness reanalysis (v1.0) for atmospheric and climate sciences. *Geoscientific Model Development*, *9*(4), 1489–1522. https://doi.gov/10.5194/gmd-9-1489-2016

Ma, C. Q., Wang, T. J., Mizzi, A. P., Anderson, J. L., Zhuang, B. L., Xie, M., et al. (2019). Multiconstituent data assimilation with WRF-Chem/DART: Potential for adjusting anthropogenic emissions and improving air quality forecasts over eastern China. *Journal of Geophysical Research: Atmospheres*, *124*(13), 7393–7412. https://doi.gov/10.1029/2019jd030421

McHardy, T., Zhang, J., Reid, J. S., Miller, S. D., Hyer, E. J., & Kuehn, R. E. (2015). An improved method for retrieving nighttime aerosol optical thickness from the VIIRS Day/Night Band. *Atmospheric Measurement Techniques*, *8*, 4773–4783.

Mu, M., Randerson, J., van der Werf, G., Giglio, L., Kasibhatla, P., Morton, D., et al. (2011). Daily and hourly variability in global fire emissions and consequences for atmospheric model predictions of carbon monoxide. *Journal of Geophysical Research*. https://doi.gov/10.1029/2011JD016245

Nelson, R. M., Butler, B. W., & Weise, D. R. (2012). Entrainment regimes and flame characteristics of wildland fires.

International Journal of Wildland Fire, *21*(2), 127–140. https://doi.gov/10.1071/wf10034

Noyes, K. J., Kahn, R., Sedlacek, A., Kleinman, L., Limbacher, J., & Li, Z. Q. (2020a). Wildfire smoke particle properties and evolution, from space-based multi-angle imaging. *Remote Sensing*, *12*(5), 23. https://doi.gov/10.3390/rs12050769

Noyes, K. T. J., Kahn, R. A., Limbacher, J. A., Li, Z. Q., Fenn, M. A., Giles, D. M., et al. (2020b). Wildfire smoke particle properties and evolution, from space-based multi-angle imaging II: The Williams Flats fire during the FIREX-AQ campaign. *Remote Sensing*, *12*(22), 27. https://doi.gov/10.3390/rs12223823

O'Neill, N. T., Ignatov, A., Holben, B. N., & Eck, T. F. (2000). The lognormal distribution as a reference for reporting aerosol optical depth statistics: Empirical tests using multi-year, multi-site AERONET sunphotometer data. *Geophysical Research Letters*, *27*(20), 3333–3336.

Paugam, R., Wooster, M., Freitas, S., & Martin, M. V. (2016). A review of approaches to estimate wildfire plume injection height within large-scale atmospheric chemical transport models. *Atmospheric Chemistry and Physics*, *16*(2), 907–925. https://doi.gov/10.5194/acp-16-907-2016

Peterson, D., Hyer, E., & Wang, J. (2013). A short-term predictor of satellite-observed fire activity in the North American boreal forest: Toward improving the prediction of smoke emissions. *Atmospheric Environment*, *71*, 304–310. https://doi.gov/10.1016/j.atmosenv.2013.01.052

Peterson, D., Hyer, E., & Wang, J. (2014). Quantifying the potential for high-altitude smoke injection in the North American boreal forest using the standard MODIS fire products and subpixel-based methods. *Journal of Geophysical Research: Atmospheres*, *119*(6), 3401–3419. https://doi.gov/10.1002/2013jd021067

Peterson, D. A., Hyer, E. J., Campbell, J. R., Fromm, M. D., Hair, J. W., Butler, C. F., et al. (2015). THE 2013 RIM FIRE implications for predicting extreme fire spread, pyroconvection, and smoke emissions. *Bulletin of the American Meteorological Society*, *96*(2), 229–247. https://doi.gov/10.1175/bams-d-14-00060.1

Polivka, T. N., Wang, J., Ellison, L. T., Hyer, E. J., & Ichoku, C. M. (2016). Improving nocturnal fire detection with the VIIRS Day-Night Band. *Ieee Transactions on Geoscience and Remote Sensing*, *54*(9), 5503–5519. https://doi.gov/10.1109/tgrs.2016.2566665

Potter, B. E. (2012). Atmospheric interactions with wildland fire behavior: I. Basic surface interactions, vertical profiles and synoptic structures. *International Journal of Wildland Fire*, *21*(7), 779–801. https://doi.gov/10.1071/wf11128

Randles, C. A., da Silva, A. M., Buchard, V., Colarco, P. R., Darmenov, A., Govindaraju, R., et al. (2017). The MERRA-2 aerosol reanalysis, 1980 onward. Part I: System description and data assimilation evaluation. *Journal of Climate*, *30*(17), 6823–6850. https://doi.gov/10.1175/jcli-d-16-0609.1

Reid, J. S., Hyer, E. J., Johnson, R. S., Holben, B. N., Yokelson, R. J., Zhang, J. L., et al. (2013). Observing and understanding the Southeast Asian aerosol system by remote sensing: An initial review and analysis for the Seven Southeast Asian Studies (7SEAS) program. *Atmospheric Research*, *122*, 403–468. https://doi.gov/10.1016/j.atmosres.2012.06.005

Reid, J. S., Prins, E. M., Westphal, D. L., Schmidt, C. C., Richardson, K. A., Christopher, S. A., et al. (2004). Real-time monitoring of South American smoke particle emissions and transport using a coupled remote sensing/box-model approach. *Geophysical Research Letters*, *31*(6), 5. https://doi.gov/10.1029/2003gl018845

Remer, L. A., Kaufman, Y. J., Tanre, D., Mattoo, S., Chu, D. A., Martins, J. V., et al. (2005). The MODIS aerosol algorithm, products, and validation. *Journal of the Atmospheric Sciences*, *62*(4), 947–973.

Roberts, G., Wooster, M. J., Xu, W., Freeborn, P. H., Morcrette, J. J., Jones, L., et al. (2015). LSA SAF Meteosat FRP products: Part 2: Evaluation and demonstration for use in the Copernicus Atmosphere Monitoring Service (CAMS). *Atmospheric Chemistry and Physics*, *15*(22), 13241–13267. https://doi.gov/10.5194/acp-15-13241-2015

Robinson, D. L. (2020). Accurate, low cost PM(2.5)measurements demonstrate the large spatial variation in wood smoke pollution in regional Australia and improve modeling and estimates of health costs. *Atmosphere*, *11*(8), 21. https://doi.gov/10.3390/atmos11080856

Rogozovsky, I., Ansmann, A., Althausen, D., Heese, B., Engelmann, R., Hofer, J., et al. (2021). Impact of aerosol layering, complex aerosol mixing, and cloud coverage on high-resolution MAIAC aerosol optical depth measurements: Fusion of lidar, AERONET, satellite, and ground-based measurements. *Atmospheric Environment*, *247*, 12. https://doi.gov/10.1016/j.atmosenv.2020.118163

Rubin, J. I., Reid, J. S., Hansen, J. A., Anderson, J. L., Collins, N., Hoar, T. J., et al. (2016). Development of the Ensemble Navy Aerosol Analysis Prediction System (ENAAPS) and its application of the Data Assimilation Research Testbed (DART) in support of aerosol forecasting. *Atmospheric Chemistry and Physics*, *16*(6), 3927–3951.

Rubin, J. I., Reid, J. S., Hansen, J. A., Anderson, J. L., Holben, B. N., Xian, P., et al. (2017). Assimilation of AERONET and MODIS AOT observations using variational and ensemble data assimilation methods and its impact on aerosol forecasting skill. *Journal of Geophysical Research: Atmospheres*, *122*(9), 4967–4992. https://doi.gov/10.1002/2016jd026067

Saide, P. E., Carmichael, G. R., Liu, Z., Schwartz, C. S., Lin, H. C., Da Silva, A. M., et al. (2013). Aerosol optical depth assimilation for a size-resolved sectional model: Impacts of observationally constrained, multi-wavelength and fine mode retrievals on regional scale forecasts. *Atmospheric Chemistry and Physics*, *13*, 10425–10444. https://doi.gov/https://doi.gov/10.5194/acp-13-10425-2013

Saide, P. E., Carmichael, G. R., Spak, S. N., Minnis, P., & Ayers, J. K. (2012). Improving aerosol distributions below clouds by assimilating satellite-retrieved cloud droplet number. *Proceedings of the National Academy of Sciences of the United States of America*, *109*(30), 11939–11943. https://doi.gov/10.1073/pnas.1205877109

Saide, P. E., Kim, J., Song, C. H., Choi, M., Cheng, Y., & Carmichael, G. R. (2014). Assimilation of next generation geostationary aerosol optical depth retrievals to improve air quality simulations. *Geophysical Research Letters*, *41*(24), 9188–9196. https://doi.gov/10.1002/2014gl062089

Saide, P. E., Peterson, D. A., da Silva, A., Anderson, B., Ziemba, L. D., Diskin, G., et al. (2015a). Revealing important nocturnal and day-to-day variations in fire smoke emissions through a multiplatform inversion. *Geophysical Research Letters*, *42*(9), 3609–3618. https://doi.gov/10.1002/2015gl063737

Saide, P. E., Spak, S. N., Pierce, R. B., Otkin, J. A., Schaack, T. K., Heidinger, A. K., et al. (2015b). Central American biomass burning smoke can increase tornado severity in the U.S. *Geophysical Research Letters*, *42*(3), 956–965. https://doi.gov/10.1002/2014gl062826

Sayer, A. M., & Knobelspiesse, K. D. (2019). How should we aggregate data? Methods accounting for the numerical distributions, with an assessment of aerosol optical depth. *Atmospheric Chemistry and Physics*, *19*(23), 15007–15032. https://doi.gov/10.5194/acp-19-15023-2019

Sayer, A. M., Govaerts, Y., Kolmonen, P., Lipponen, A., Luffarelli, M., Mielonen, T., et al. (2020). A review and framework for the evaluation of pixel-level uncertainty estimates in satellite aerosol remote sensing. *Atmospheric Measurement Techniques*, *13*(2), 373–404. https://doi.gov/10.5194/amt-13-373-2020

Schwartz, C. S., Liu, Z. Q., Lin, H. C., & McKeen, S. A. (2012). Simultaneous three-dimensional variational assimilation of surface fine particulate matter and MODIS aerosol optical depth. *Journal of Geophysical Research: Atmospheres*, *117*. https://doi.gov/D1320210.1029/2011jd017383

Selimovic, V., Yokelson, R. J., McMeeking, G. R., & Coefield, S. (2019). In situ measurements of trace gases, PM, and aerosol optical properties during the 2017 NW US wildfire smoke event. *Atmospheric Chemistry and Physics*, *19*(6), 3905–3926. https://doi.gov/10.5194/acp-19-3905-2019

Shi, Y., Zhang, J., Reid, J. S., Holben, B., Hyer, E. J., & Curtis, C. (2011). An analysis of the collection 5 MODIS over-ocean aerosol optical depth product for its implication in aerosol assimilation. *Atmospheric Chemistry and Physics*, *11*(2), 557–565. https://doi.gov/10.5194/acp-11-557-2011

Shi, Y. X. R., Levy, R. C., Eck, T. F., Fisher, B., Mattoo, S., Remer, L. A., et al. (2019). Characterizing the 2015 Indonesia fire event using modified MODIS aerosol retrievals. *Atmospheric Chemistry and Physics*, *19*(1), 259–274. https://doi.gov/10.5194/acp-19-259-2019

Simon, E., & Bertino, L. (2012). Gaussian anamorphosis extension of the DEnKF for combined state parameter estimation: Application to a 1D ocean ecosystem model. *Journal of Marine Systems*, *89*(1), 1–18. https://doi.gov/10.1016/j.jmarsys.2011.07.007

Stier, P., Schutgens, N. A. J., Bellouin, N., Bian, H., Boucher, O., Chin, M., et al. (2013). Host model uncertainties in aerosol radiative forcing estimates: Results from the AeroCom Prescribed intercomparison study. *Atmospheric Chemistry and Physics*, *13*(6), 3245–3270. https://doi.gov/10.5194/acp-13-3245-2013

Tang, Y. H., Pagowski, M., Chai, T. F., Pan, L., Lee, P., Baker, B., et al. (2017). A case study of aerosol data assimilation with the Community Multi-scale Air Quality Model over the contiguous United States using 3D-Var and optimal interpolation methods. *Geoscientific Model Development*, *10*(12), 4743–4758. https://doi.gov/10.5194/gmd-10-4743-2017

Val Martin, M., Kahn, R. A., Logan, J. A., Paugam, R., Wooster, M., & Ichoku, C. (2012). Space-based observational constraints for 1-D fire smoke plume-rise models. *Journal of Geophysical Research*, *117*(D22), 27. https://doi.gov/10.1029/2012JD018370

Van Leeuwen, P. J. (2009). Particle filtering in geophysical systems. *Monthly Weather Review*, *137*(12), 4089–4114. https://doi.gov/10.1175/2009mwr2835.1

Van Leeuwen, P. J., Kunsch, H. R., Nerger, L., Potthast, R., & Reich, S. (2019). Particle filters for high-dimensional geoscience applications: A review. *Quarterly Journal of the Royal Meteorological Society*, *145*(723), 2335–2365. https://doi.gov/10.1002/qj.3551

Wang, J., Aegerter, C., Xu, X. G., & Szykman, J. J. (2016). Potential application of VIIRS Day/Night Band for monitoring nighttime surface PM2.5 air quality from space. *Atmospheric Environment*, *124*, 55–63. https://doi.gov/10.1016/j.atmosenv.2015.11.013

Wang, J., Christopher, S. A., Nair, U. S., Reid, J. S., Prins, E. M., Szykman, J., et al. (2006). Mesoscale modeling of Central American smoke transport to the United States: 1. "Top-down" assessment of emission strength and diurnal variation impacts. *Journal of Geophysical Research: Atmospheres*, *111*(D5).

Wang, J., Nair, U. S., & Christopher, S. A. (2004). GOES 8 aerosol optical thickness assimilation in a mesoscale model: Online integration of aerosol radiative effects. *Journal of Geophysical Research: Atmospheres*, *109*(D23), D23203. https://doi.gov/23210.21029/22004JD004827

Wang, J., Roudini, S., Hyer, E. J., Xu, X. G., Zhou, M., Garcia, L. C., et al. (2020a). Detecting nighttime fire combustion phase by hybrid application of visible and infrared radiation from Suomi NPP VIIRS. *Remote Sensing of Environment*, *237*, 14. https://doi.gov/10.1016/j.rse.2019.111466

Wang, J., Zhou, M., Xu, X. G., Roudini, S., Sander, S. P., Pongetti, T. J., et al. (2020b). Development of a nighttime shortwave radiative transfer model for remote sensing of nocturnal aerosols and fires from VIIRS. *Remote Sensing of Environment*, *241*, 16. https://doi.gov/10.1016/j.rse.2020.111727

Wang, Y. Z., Bechle, M. J., Kim, S. Y., Adams, P. J., Pandis, S. N., Pope, C. A., et al. (2020c). Spatial decomposition analysis of NO2 and PM2.5 air pollution in the United States. *Atmospheric Environment*, *241*, 8. https://doi.gov/10.1016/j.atmosenv.2020.117470

Wiggins, E. B., Soja, A. J., Gargulinski, E., Halliday, H. S., Pierce, R. B., Schmidt, C. C., et al. (2020). High temporal resolution satellite observations of fire radiative power reveal link between fire behavior and aerosol and gas emissions. *Geophysical Research Letters*, *47*(23), 12. https://doi.gov/10.1029/2020gl090707

Xian, P., Reid, J. S., Hyer, E. J., Sampson, C. R., Rubin, J. I., Ades, M., et al. (2019). Current state of the global operational aerosol multi-model ensemble: An update from the International Cooperative for Aerosol Prediction (ICAP). *Quarterly Journal of the Royal Meteorological Society*, *145*, 176–209. https://doi.gov/10.1002/qj.3497

Xu, X. G., Wang, J., Wang, Y., Zeng, J., Torres, O., Reid, J. S., et al. (2019). Detecting layer height of smoke aerosols over vegetated land and water surfaces via oxygen absorption bands: Hourly results from EPIC/DSCOVR in deep space. *Atmospheric Measurement Techniques*, *12*(6), 3269–3288. https://doi.gov/10.5194/amt-12-3269-2019

Xu, X. G., Wang, J., Wang, Y., Zeng, J., Torres, O., Yang, Y. K., et al. (2017). Passive remote sensing of altitude and optical depth of dust plumes using the oxygen A and B bands: First results from EPIC/DSCOVR at Lagrange-1 point. *Geophysical Research Letters*, *44*(14), 7544–7554. https://doi.gov/10.1002/2017gl073939

Ye, X. X., Arab, P., Ahmadov, R., James, E., Grell, G. A., Pierce, B., et al. (2021). Evaluation and intercomparison of wildfire smoke forecasts from multiple modeling systems for the 2019 Williams Flats fire. *Atmospheric Chemistry and Physics*, *21*(18), 14427–14469. https://doi.gov/10.5194/acp-21-14427-2021

Yeo, I. K., & Johnson, R. A. (2000). A new family of power transformations to improve normality or symmetry. *Biometrika*, *87*(4), 954–959. https://doi.gov/10.1093/biomet/87.4.954

Yu, H. B., Dickinson, R. E., Chin, M., Kaufman, Y. J., Holben, B. N., Geogdzhayev, I. V., et al. (2003). Annual cycle of global distributions of aerosol optical depth from integration of MODIS retrievals and GOCART model simulations. *Journal of Geophysical Research-Atmospheres*, *108*(D3).

Yumimoto, K., & Takemura, T. (2011). Direct radiative effect of aerosols estimated using ensemble-based data assimilation in a global aerosol climate model. *Geophysical Research Letters*, *38*, 6. https://doi.gov/10.1029/2011gl049258

Yumimoto, K., & Takemura, T. (2013). The SPRINTARS version 3.80/4D-Var data assimilation system: Development and inversion experiments based on the observing system simulation experiment framework. *Geoscientific Model Development*, *6*(6), 2005–2022. https://doi.gov/10.5194/gmd-6-2005-2013

Yumimoto, K., Nagao, T. M., Kikuchi, M., Sekiyama, T. T., Murakami, H., Tanaka, T. Y., et al. (2016). Aerosol data assimilation using data from Himawari-8, a next-generation geostationary meteorological satellite. *Geophysical Research Letters*, *43*(11), 5886–5894. https://doi.gov/10.1002/2016gl069298

Yumimoto, K., Tanaka, T. Y., Oshima, N., & Maki, T. (2017). JRAero: The Japanese Reanalysis for Aerosol v1.0. *Geoscientific Model Development*, *10*(9), 3225–3253. https://doi.gov/10.5194/gmd-10-3225-2017

Yumimoto, K., Tanaka, T. Y., Yoshida, M., Kikuchi, M., Nagao, T. M., Murakami, H., et al. (2018). Assimilation and forecasting experiment for heavy Siberian wildfire smoke in May 2016 with Himawari-8 aerosol optical thickness. *Journal of the Meteorological Society of Japan*, *96B*, 133–149. https://doi.gov/10.2151/jmsj.2018-035

Yumimoto, K., Uno, I., Sugimoto, N., Shimizu, A., Liu, Z., & Winker, D. M. (2008). Adjoint inversion modeling of Asian dust emission using lidar observations. *Atmospheric Chemistry and Physics*, *8*(11), 2869–2884. https://doi.gov/10.5194/acp-8-2869-2008

Zhang, F., Wang, J., Ichoku, C., Hyer, E. J., Yang, Z., Ge, C., et al. (2014a). Sensitivity of mesoscale modeling of smoke direct radiative effect to the emission inventory: A case study in northern sub-Saharan African region. *Environmental Research Letters*, *9*(7), 075002 (075014 pp.)-075002 (075014 pp.). https://doi.gov/10.1088/1748-9326/9/7/075002

Zhang, H., Kondragunta, S., Laszlo, I., & Zhou, M. (2020). Improving GOES Advanced Baseline Imager (ABI) aerosol optical depth (AOD) retrievals using an empirical bias correction algorithm. *Atmospheric Measurement Techniques*, *13*(11), 5955–5975. https://doi.gov/10.5194/amt-13-5955-2020

Zhang, H., Kondragunta, S., Laszlo, I., Liu, H. Q., Remer, L. A., Huang, J. F., et al. (2016a). An enhanced VIIRS aerosol optical thickness (AOT) retrieval algorithm over land using a global surface reflectance ratio database. *Journal of Geophysical Research: Atmospheres*, *121*(18), 10717–10738. https://doi.gov/10.1002/2016jd024859

Zhang, J., Campbell, J. R., Hyer, E. J., Reid, J. S., Westphal, D. L., & Johnson, R. S. (2014b). Evaluating the impact of multisensor data assimilation on a global aerosol particle transport model. *Journal of Geophysical Research: Atmospheres*, *119*(8), 4674–4689. https://doi.gov/10.1002/2013jd020975

Zhang, J., Reid, J. S., Westphal, D. L., Baker, N. L., & Hyer, E. J. (2008). A system for operational aerosol optical depth data assimilation over global oceans. *Journal of Geophysical Research: Atmospheres*, *113*(D10). https://doi.gov/D10208 10.1029/2007jd009065

Zhang, J. L., Campbell, J. R., Hyer, E. J., Reid, J. S., Westphal, D. L., & Johnson, R. S. (2014b). Evaluating the impact of multisensor data assimilation on a global aerosol particle transport model. *Journal of Geophysical Research: Atmospheres*, *119*(8), 4674–4689. https://doi.gov/10.1002/2013jd020975

Zhang, J. L., Campbell, J. R., Reid, J. S., Westphal, D. L., Baker, N. L., Campbell, W. F., et al. (2011). Evaluating the impact of assimilating CALIOP-derived aerosol extinction profiles on a global mass transport model. *Geophysical Research Letters*, *38*. https://doi.gov/L148011 0.1029/2011gl047737

Zhang, J. L., Jaker, S. L., Reid, J. S., Miller, S. D., Solbrig, J., & Toth, T. D. (2019). Characterization and application of artificial light sources for nighttime aerosol optical depth retrievals using the Visible Infrared Imager Radiometer Suite Day/Night Band. *Atmospheric Measurement Techniques*, *12*(6), 3209–3222. https://doi.gov/10.5194/amt-12-3209-2019

Zhang, J. L., Reid, J. S., & Holben, B. N. (2005). An analysis of potential cloud artifacts in MODIS over ocean aerosol optical thickness products. *Geophysical Research Letters*, *32*(15), 4. https://doi.gov/10.1029/2005gl023254

Zhang, J. L., Reid, J. S., Christensen, M., & Benedetti, A. (2016b). An evaluation of the impact of aerosol particles on weather forecasts from a biomass burning aerosol event over the midwestern United States: Observational-based analysis of surface temperature. *Atmospheric Chemistry and Physics*, *16*(10), 6475–6494. https://doi.gov/10.5194/acp-16-6475-2016

Zhang, J. L., Spurr, R. J. D., Reid, J. S., Xian, P., Colarco, P. R., Campbell, J. R., et al. (2021). Development of an ozone monitoring instrument (OMI) aerosol index (AI) data assimilation scheme for aerosol modeling over bright surfaces: A step toward direct radiance assimilation in the UV spectrum. *Geoscientific Model Development*, *14*(1), 27–42. https://doi.gov/10.5194/gmd-14-27-2021

Zhao, X. R., Shi, H. Q., Yu, H., & Yang, P. L. (2016). Inversion of nighttime PM2.5 mass concentration in Beijing based on the VIIRS Day-Night Band. *Atmosphere*, *7*(10). https://doi.gov/10.3390/atmos7100136

Zhou, H. Y., Gomez-Hernandez, J. J., Franssen, H. J. H., & Li, L. P. (2011). An approach to handling non-Gaussianity of parameters and state variables in ensemble Kalman filtering. *Advances in Water Resources*, *34*(7), 844–864. https://doi.gov/10.1016/j.advwatres.2011.04.014

8

A Review of Modeling Approaches Used to Simulate Smoke Transport and Dispersion

Derek V. Mallia[1] and Adam K. Kochanski[2]

ABSTRACT

A variety of smoke model frameworks is used to simulate smoke for research and forecast applications. Here, a comprehensive summary is provided, which covers the many different smoke models that are available, while simultaneously highlighting some of the strengths and weaknesses of each model, along with the uncertainties surrounding each of these frameworks. This review also provides an in-depth discussion on coupled wildfire-atmosphere models, which is a relatively newer smoke modeling tool not previously discussed in other review papers. Key processes related to smoke transport and dispersion, such as the wildfire plume rise, are also discussed at length. This review wraps up with a discussion of future smoke modeling needs and potential new research directions for smoke transport and dispersion models.

8.1. INTRODUCTION

Smoke is a product of the combustion process, which contains various chemical species and particulates, which can degrade air quality across a broad range of spatio-temporal scales (Goodrick et al., 2012). Increased fire activity due to climate change (Westerling et al., 2006; Spracklen et al., 2009) and robust population growth across the western United States is expected to expose 82 million Americans to smoke in the coming decades (Liu et al., 2016a). As wildfires increase in frequency and intensity, it is imperative that tools be developed and improved upon for studying and forecasting smoke and combustion products detrimental to health. This includes particulate matter with a diameter less than 2.5 μm ($PM_{2.5}$) and precursors for ozone (O_3) formation downwind of the fire (Jaffe & Widger, 2012). Some objectives of smoke modeling include limiting the public's exposure to unhealthy concentrations of smoke and determining how smoke could impact fire management operations during active wildfires and prescribed burns (Kochanski et al., 2018; Peterson et al., 2020). Prescribed burns, for example, are used to manage forests, combat wildfires, and mitigate public exposure to smoke (Rappold et al., 2014). Igniting fires in a controlled setting limits the intensity and fuel consumption of wildfires, and therefore reduces smoke production relative to uncontrolled wildfires with no fuel thinning (Haikerwal et al., 2015). Smoke models can also be used to forecast meteorology that can favorably disperse smoke from prescribed burns and to inform burn decisions so that the public's exposure to unhealthy concentrations of smoke is limited (Lahm, 2015). Finally, smoke models are also needed to elucidate processes that govern the chemical makeup and transport of smoke. Such processes range from

[1] *Department of Atmospheric Sciences, University of Utah, Salt Lake City, Utah, USA*

[2] *Department of Meteorology and Climate Science, San Jose State University, San Jose, California, USA*

Landscape Fire, Smoke, and Health: Linking Biomass Burning Emissions to Human Well-Being, Geophysical Monograph 280, First Edition.
Edited by Tatiana V. Loboda, Nancy H. F. French, and Robin C. Puett.

DOI: 10.1002/9781119757030.ch08

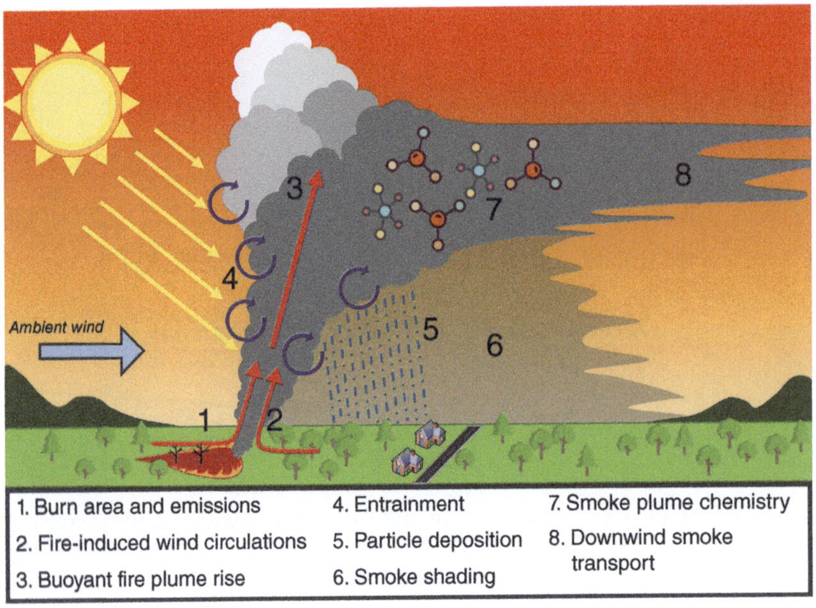

Figure 8.1 A schematic of important smoke transport processes. Definitions and details are in the text.

small-scale mechanisms that drive the wildfire plume rise (Mallia et al., 2020a) to the global impacts of smoke aerosols on climate (Peterson et al., 2018; Christian et al., 2019).

Forecasting and simulating smoke is an inherently difficult task as wildfires and smoke transport are multiscale phenomena with many interconnected processes (Fig. 8.1). For example, the wildfire plume rise, which is responsible for injecting smoke in the atmosphere, is controlled by many factors such as atmospheric stability, wind shear, heat fluxes, and fire geometry (Fig. 8.1; Freitas et al., 2007; 2010). Pyroconvective plumes can also affect local meteorology by increasing near surface winds (Clements et al., 2018; Kochanski et al., 2019), shading areas underneath the plume (Fig. 8.1) from direct insolation (Robock, 1988, 1991; Lareau & Clements, 2015; Walter et al., 2016; Kochanski et al., 2019); and in rarer cases, plumes can initiate fire-generated thunderstorms (i.e., pyrocumulonimbus; Fromm et al., 2010). These processes can often feed back to the local meteorology at the fire line and impact wildfire behavior.

The amount of smoke and heat emitted by the fire serves as essential input for smoke models (see #1 in Fig, 8.1). Smoke emissions are used to determine the mass flux of chemical species into the atmosphere while the heat flux and fire area are important variables for determining how far up smoke might be lofted into the atmosphere. However, accurately quantifying fire emissions and heat fluxes from wildfires remains challenging. Estimating smoke emissions and heat fluxes requires information on the exact location and geospatial context of active burning, a description of fuels that are being consumed by the fire, and how intensively that fuel burns (see Chapter 5). Previous work has estimated that uncertainties associated with $PM_{2.5}$ emissions from fires could be as high as 64% (Urbanski et al., 2011). Errors in emission estimates often stem from errors in the estimated burned area, which can be difficult to quantify due to ambiguities associated with differentiating the burned from unburned areas within and around the fire perimeter (Battye & Battye, 2002). Emission factors for different chemical species and aerosols emitted by the fire are yet another major source of uncertainty and can exhibit significant variability due to heterogeneous fuel type and conditions (Urbanski, 2014), as well as combustion characteristics (flaming versus smoldering) (Lobert, 1991; Yokelson et al., 1996; Chen et al., 2007; McKeeking et al., 2009; Burling et al., 2010). Lastly, burn severity, which is related to fuel consumption, can also affect emission estimates and add uncertainties to smoke emission inventories (Urbanski et al., 2011).

Forecasting smoke emissions and heat fluxes adds additional challenges, as it requires a model to make future projections for fuel consumption, on top of the underlying assumptions needed to convert burned biomass into emissions for different chemical species and aerosol particles (see Chapter 7). The uncertainties surrounding fire-emitted fluxes can also influence the vertical plume extent and plume dynamics (Freitas et al., 2007), which, in turn, can impact how smoke is transported and dispersed from the fire.

Many atmospheric and chemical modeling frameworks can be used to simulate the transport of smoke from wildfires and prescribed burns. These smoke modeling frameworks range from simple box and Gaussian plume models (Lavdas, 1996) to more sophisticated modeling systems that can simulate smoke on an atmospheric grid with full physics and photochemistry (Liu et. al., 2009; Hodzic et al., 2007; Grell et al., 2011, Larkin et al., 2009; Kochanski et al., 2016). The primary difference between smoke models is how they account for physical processes that govern smoke transport and dispersion (Fig. 8.1), along with other underlying processes such as fire emissions and burn area, fire-atmosphere interactions, plume entrainment, atmospheric chemistry, aerosol physics and particle deposition, and plume entrainment (Fig. 8.1; Table 8.1). These models can also differ in terms of the reference frame that they use to simulation smoke, that is, the Eulerian versus the Lagrangian perspective.

The type of model used to simulate smoke often depends on the application and/or the scientific question that the researcher or fire manager is addressing. For example, a Gaussian plume model could be more practical for a case where the wind field is unidirectional and constant between the smoke source and receptor. However, if the user attempts to model a case where there is a large wildfire in complex terrain, with erratic wind fields generating intense pyroconvection, a Gaussian plume model may not be sufficient; thus, necessitating the need for a more complex and computationally demanding modeling framework such as a coupled fire-atmosphere model.

In the following sections, we will provide a brief overview of important smoke-related processes (section 8.2) while highlighting the different modeling frameworks used to simulate smoke transport (section 8.3). Section 8.3 will be divided by smoke transport model type. This section will then be followed up with a list of plume-rise models, which are often integrated within various smoke transport models to vertically distribute smoke emissions (section 8.4). Finally, section 8.5 will summarize some of the major discussion points of sections 8.2, 8.3, and 8.4, while section 8.6 will discuss future smoke modeling needs and directions.

8.2. SMOKE-RELATED PROCESSES

As discussed in the previous section, smoke models need to represent many critical fire and atmospheric processes, such as (1) fire growth or burned area; (2) smoke emissions; (3) the buoyant rise plume rise driven by the fire; (4) mixing between the smoke plume and the ambient air outside of it, often referred to as entrainment; (5) deposition processes; (6) downwind smoke dispersion; and (7) plume chemistry (Fig. 8.1). It should be emphasized that these processes do not operate independently, and are sometimes dynamically linked together (Fromm et al., 2010; Lareau & Clements, 2016, 2017; Kochanski et al., 2019; Mallia et al., 2020a). For example, heat fluxes generated by the fire can sometimes result in intense pyroconvection. If the smoke plume reaches a high enough altitude, water vapor will condense into liquid cloud water that can aid in the formation of pyrocumulus (pyroCu) or pyrocumulonimbus (pyroCb) clouds. There have been several documented cases of pyroCbs reaching altitudes of 15 km or more (Fromm et al., 2010; Peterson et al., 2018). The range of scales involved in the dynamics of fire-generated plumes is immense as it encompasses small-scale processes driving combustion and heat release at a fire front, up through large-scale global weather patterns, which are responsible for driving long-range smoke transport. The processes discussed above are conceptualized in Figure 8.1.

Currently, most smoke modeling frameworks are developed to deal with smoke transport targeted at specific spatiotemporal scales. It should be emphasized that assumptions made within one model may not necessarily be valid for another model that deals with smoke transport at a different scale. Thus, there is no single model that encompasses the full range of scales needed to explicitly resolve smoke generation, plume rise, and dispersion. This concept is conceptualized in Table 8.1, where individual smoke models have niches in the continuum of spatial and temporal scales. Combustion-resolving models such as Wildland-Urban Interface Fire Dynamics Simulator (WFDS; Mell et al., 2007) and Firetech (Linn & Cunningham, 2005) operate at smaller scales while, models such as Daysmoke (Achtemeier et al., 2011) or WRF-SFIRE/WRF-FIRE (Mandel et al., 2011; Coen et al. 2013) and others focus on simulating smoke at larger spatiotemporal scales, but at the expense of small-scale processes that need to be simplified as parameterizations. Chemical transport models, such as GEOS-CHEM, resolve the coarsest processes, but simulate smoke at the largest scale possible (global). Aside from spatiotemporal scales, smoke models can be classified based on how they represent critical smoke-related processes and the frame of reference used to simulate smoke.

8.2.1. Fire Burn Area and Emissions

The fire burned area and emissions, which are related to fire activity, are critical inputs for most smoke models (see Chapters 3 and 5). Fire burned area emissions can be represented in several different ways within smoke models. In many cases, models simply rely on external fire emission inventories such as GFED (Van der Werf et al., 2010), FINN (Wiedinmyer et al., 2010), or MFLEI

Table 8.1 Commonly used smoke modeling frameworks for research and operational-based applications

Model	Model type	Scale	Fire activity	Plume	Meteorology	Aerosol physics	Chemistry
VSMOKE[1]	Gaussian	100–10,000 m	Prescribed	Parameterized	Parameterized	–	Combustion only, smoke treated as passive tracer
SASEM[2]	Gaussian	100–10,000 m	Prescribed	Parameterized	Parameterized	–	Combustion only, smoke treated as passive tracer
CALPUFF[3]	Puff model	1–1,000 km	Prescribed	Parameterized	Parameterized	Parameterized	Parameterized
HYSPLIT[4]	Puff model	1–1,000 km	Prescribed	Parameterized	Uncoupled	Parameterized	Parameterized
HYSPLIT[4]	LPDM	1–1,000 km	Prescribed	Parameterized	Uncoupled	Parameterized	Parameterized
FLEXPART[5]	LPDM	1–1,000 km	Prescribed	Parameterized	Uncoupled	Parameterized	Parameterized
STILT[6]	LPDM	1–1,000 km	Prescribed	Parameterized	Uncoupled	Parameterized	Parameterized
Daysmoke[7]	Hybrid	l–10 km	Parameterized	Parameterized	Parameterized or uncoupled	–	Combustion only, smoke treated as passive tracer
CMAQ-BlueSky[8]	Eulerian	1–1,000 km	Prescribed	Parameterized	Uncoupled	Parameterized	Parameterized
WRF-CHEM[9]	Eulerian	1–1,000 km	Prescribed	Parameterized	Resolved	Parameterized & coupled	Parameterized & coupled
GEOS-CHEM[10]	Eulerian	Global	Prescribed	Parameterized	Uncoupled	Parameterized	Parameterized
WRF-FIRE[11]	Coupled	100 m–1,000 km	Parameterized	Resolved	Resolved	–	Combustion only, smoke treated as passive tracer
WRF-SFIRE[12]	Coupled	100 m–1,000 km	Parameterized	Resolved	Resolved	Parameterized & coupled	Parameterized & coupled
WFDS[13]	Coupled	1 m–1 km	Resolved	Resolved	Resolved	–	Combustion only, smoke treated as passive tracer
Firetech[14]	Coupled	1 m–1 km	Resolved	Resolved	Resolved	–	Combustion only, smoke treated as passive tracer

Sources: [1]Lavdas (1996), [2]Sestak & Riebau (1988), [3]Scire et al. (2000), [4]Draxler & Hess (1997), [5]Stohl & Thomson (1999), [6]Lin et al. (2003), [7]Achtemeier et al. (2011), [8]Larkin et al. (2009), [9]Grell et al. (2005), [10]http://acmg.seas.harvard.edu/geos/, [11]Coen et al. (2013), [12]Mandel et al. (2011), [13]Mell et al. (2007), [14]Linn & Cunningham (2005)

Note: The scale column contains loose approximations of scales most appropriate for the listed model framework. Dashes within a column denote processes that are not accounted for. Gray shading intensity refers to the degree that specific processes are resolved where red is prescribed, yellow refers to processes that are parameterized, and green refers to explicitly resolve and/or parameterized and coupled.

(Urbanski et al., 2017) to provide historical estimates of smoke emissions and fire area. Some fire emission inventories, such as Missoula Fire Laboratory Emission Inventory (MFLEI), include emission uncertainty estimates using a Monte Carlo analysis (Urbanski et al., 2011). A more comprehensive list and description of fire emission inventories can be found in Chapter 4. Satellites are also playing an increasing large role to estimate fire emissions and heat fluxes. Operational smoke forecast models, such as HRRR-Smoke (Ahmodav et al., 2017), use satellite fire radiative power (FRP) to estimate smoke emissions and heat fluxes, and then scale fire activity by an average fire diurnal cycle. A subset of smoke models, mainly, coupled fire-atmosphere models, can project future fire activity based on a fire spread parameterization that accounts for local meteorology, fuel types and characteristics, and terrain.

Firetech and WFDS employ a physics-based approach for estimating fire growth and the burned area. The physics-based approach utilizes models that explicitly represent combustion, heat transfer, aerodynamic drag, and turbulence. These models can predict fire growth, which can be used to estimate the burned area at any time, along with the amount of fuel consumed, and subsequently, smoke emissions. While physics-based models represent the most realistic way to simulate combustion processes and fire progression (where and when a fire moves), they simplify smoke transport processes such that smoke is assumed to be a passive tracer, and, thus, ignore smoke chemical transformations and radiative impacts. Finally, explicitly resolving combustion requires very detailed information about fuels at high spatial resolutions (order of meters) and therefore are very computationally demanding. Ultimately, this limits the size of simulated fires to less than 100 acres for physics-based approaches (Liu et al., 2019).

An alternative method for estimating fire progression can be accomplished through empirical-based parameterizations. The most widely used fire progression parameterization is the Rothermel surface fire spread model (Rothermel, 1972), which was developed within the United States Forest Service (USFS) during the 1960s and 1970s. Unlike the physics-based approach, fire-spread models, like the Rothermel model, estimate fire growth rates through a quasi-empirical equation that relates fire spread to variables such as fuel type and characteristics, terrain slope, and wind. Since fire-spread parameterizations rely on simple algebraic formulas, they estimate fire growth rates at a more modest computational cost (Liu et al., 2019). Coupled fire-atmosphere models such as WRF-SFIRE and WRF-FIRE employ an empirical-based parameterization to estimate fire growth, fuel consumption, fire heat fluxes, and smoke emissions.

8.2.2. Plume Rise

The fire plume rise, that is, the vertical transport of smoke, is yet another important phenomenon that must be accounted for when simulating smoke transport from prescribed burns or wildfires. The plume rise is primarily driven by heat released from the fire along with the atmosphere's response to this heating (Fig. 8.1). Essentially, the fire plume rise acts as a chimney, which can loft smoke high in the atmosphere, with plume rise altitudes sometimes reaching upwards of 15 km in exceptional cases (Fromm et al., 2010; Peterson et al., 2018). The height over which the plume extends is referred to as the plume injection height and is a function of the intensity and geometry of surface fire, ambient atmospheric conditions such as stability, wind shear, moisture profile, and plume microphysics.

The smoke injection height can control the fate of smoke, among other factors, such a large-scale weather patterns and convection. For example, when smoke is lofted at a lower altitude, weaker winds near Earth's surface can limit how far the smoke is transported, while particle removal processes such as dry deposition are more dominant near the ground (Zhang et al., 2001; Emerson et al., 2020). For cases of limited smoke transport, smoke can accumulate in areas local to the smoke source region, which can further degrade the air quality, locally (Kochanski et al., 2019). Conversely, smoke that is injected higher in the atmosphere will often travel farther from the fire and can degrade air quality over a much larger geographical region. At the same time, the fire plume rise can also cause the smoke to overshoot areas near the fire, therefore limiting local impacts of smoke on air quality.

The injection height can also play a vital role on aerosol feedbacks within the climate system. Smoke that is lofted into the upper troposphere and lower stratosphere can have a much longer residence time relative to smoke aerosols injected into the lower troposphere and planetary boundary layer (PBL). Previous research has demonstrated that smoke lofted farther up in the atmosphere can have greater impact on climate forcing (Barnes & Hofmann, 1997; Robock, 2000). The few examples provided above exemplify the need to accurately resolve the fire plume rise for smoke modeling applications.

A variety of different modeling approaches currently exist for quantifying the vertical transport of smoke by the fire plume rise (Liu et al., 2010; Paugam et al., 2016). These models range from simple approximations that release smoke at altitudes that correspond climatological averages to full-physics models that can explicitly resolve the wildfire plume rise and plume rise dynamics (Trentmann et al., 2006; Kochanski et al., 2016). Sometimes plume rise models are integrated directly within smoke transport models (Larkin et al., 2009; Ahmodav et al., 2017), while other frameworks run the plume rise model in an offline setting (Mallia et al., 2018). Section 8.4 has been dedicated to describing the various plume rise modeling approaches used within smoke transport models.

Work carried out by Mallia et al. (2018) found that simulations of local-scale smoke transport were highly sensitive to the altitude in which emissions were injected. A regional-based study by Walter et al. (2016) also found that smoke transport within WRF-CHEM was sensitive to the plume injection height, with simulated aerosol optical depth values varying by as much as ±50% depending on the plume height injection scheme that was used. In both studies, the simulations that attempted to estimate vertical plume extent performed better than model configurations that injected smoke emissions at single level or at the surface.

While the work outlined above has indicated that plume rise models have improved smoke simulations, several studies have noted inconsistencies between simulated and observed plume top heights (Val Martin et al., 2012; Raffuse et al., 2012). Val Martin et al. (2012) concluded that implementing plume rise models within smoke transport models "remains a difficult proposition" given the uncertainties surrounding the formulations of plume rise parameterizations and model inputs such as fire heat fluxes and area.

8.2.3. Meteorology

Meteorological models are often needed to simulate the downwind transport of smoke. Numerical weather prediction models (NWP) are the most widely used method for characterizing the three-dimensional structure of meteorological variables such as winds, temperature, humidity, and pressure. Approximated forms of partial differential equations that describe the atmosphere are used to predict the state of the atmosphere for any given time and location. These equations are solved numerically on an atmospheric mesh that covers the simulated domain (Kalnay, 2003). Certain meteorological processes, such as cloud microphysics, land-atmosphere interactions, and solar radiation, are usually too small or too complex to be explicitly accounted for by the governing equations. Thus, most NWP models parameterize these processes using a variety of different methods (Kalnay, 2003). Depending on the grid spacing of the meteorological model, certain processes can be either parameterized (if the model resolution is too coarse to resolve them) or explicitly resolved if the model resolution is sufficient (Weisman et al., 2008; Shin & Hong, 2015). NWP models such as the Weather Research and Forecast model (WRF; Powers et al., 2017) operate across a large range of spatiotemporal scales and therefore parameterize processes such as convection in coarser domains but can explicitly resolve convective processes when run at a fine spatial resolution.

In essence, NWP models provide the inputs needed to simulate the transport of smoke from the fire source to the area(s) of interest. Some smoke modeling frameworks, such as WRF-SFIRE (Mandel et al., 2011; Kochanski et al., 2016), WRF-CHEM (Grell et al., 2005), and HRRR-Smoke, directly account for the transport of smoke within the dynamical core of the WRF. Other models compute the transport of smoke in an offline setting, where output from a NWP model, such as WRF, North American Mesoscale Forecast System (NAM), or the Global Forecast System (GFS), is used to trace the transport of smoke. Smoke modeling frameworks, such as HYSPLIT and CMAQ, use the offline method. One benefit of the offline method is that the smoke modelers do not always need to run their own meteorological models, which can be timely and computationally expensive. However, this method does not allow two-way coupling between the smoke and atmosphere, which can sometimes be important when simulating the interactions between the smoke and meteorology (Kochanski et al., 2019). Smoke models, such as VSMOKE, preclude the use of NWP models, and simply assume that the wind fields are steady state, therefore using wind data from a nearby weather station.

8.2.4. Aerosol Physics

Smoke particles directly interact with energy from the sun by scattering and absorbing incoming solar radiation due to the presence of black and organic carbon (Fig. 8.1). Interactions between smoke particles and incoming solar radiation can result in differential heating of the atmosphere that can impact atmospheric stability and/or near-surface temperatures, that is, aerosol direct effects (Bauer & Menon, 2012). For example, smoke shading effects occur when incoming energy from the Sun is blocked by the opaque smoke plume, which results in cooling at the surface (Robock, 1988, 1991; Trentmann et al., 2006). Smoke shading can impact temperature forecasts, or in more extreme circumstances, it can affect smoke transport (Segal & Arritt,1992; Kochanski et al., 2019). An observation-based field campaign in northern California found evidence of smoke-induced density currents where differential solar heating between areas with and without smoke resulting in a self-propagating, surfaced-based smoke plume that opposed the ambient wind (Lareau & Clements, 2015). A modeling-based study carried out by Kochanski et al. (2019) found that localized reductions in incoming solar radiation within smoke-filled mountain valleys reduced surface temperatures while increasing temperatures near the top of the smoke layer. In this scenario, there was evidence that smoke was responsible for cooling the surface, suppressing convective boundary layer growth, which effectively limited ventilation between the smoke-filled layer and the atmosphere. In turn, this resulted in an accumulation of smoke and, subsequently, more cooling via a nonlinear feedback mechanism.

Smoke particles can also interact with atmosphere via indirect effects where smoke particles alter cloud microphysics (Lindsey & Fromm, 2008; Lee et al., 2018). To summarize, smoke particles can promote the formation of additional cloud water droplets at the expense of larger cloud water droplets since cloud droplets have to compete for a finite amount of water vapor (Andreae et al., 2004). A simulation of a PyroCb in the Texas Panhandle found that the smoke particles played an important role toward enhancing the strength of the convective updraft (Zhang et al., 2019). These results were consistent with Grell et al. (2011) who also concluded that simulated convection over Alaska was stronger in the presence of smoke, albeit the convection produced less precipitation.

Several existing smoke modeling frameworks, such as WRF-CHEM, WRF-SFIRE, and HRRR-Smoke, are equipped to deal with some of the interactions noted above. Other processes not previously discussed, such as wet and dry deposition, represent important loss processes for atmospheric particles like smoke (Zhang et al., 2001; Saylor et al., 2019) and are parameterized within most smoke transport models that simulate particulate matter. WRF-CHEM contains a full suite of aerosol parameterizations that can account for effects ranging from direct aerosol effects to indirect effects that can impact cloud microphysics and PyroCb development (Grell et al., 2011; Zhang et al., 2019). GEOS-CHEM is another popular aerosol transport model for simulating smoke transport and for quantifying the impacts of smoke on radiative forcing at the global scale (Christian et al., 2019).

A variety of methods are used within smoke models to simulate aerosol physics. For example, some aerosol schemes use the bulk method where only the total mass of the aerosol compound is known, therefore this is no information about the particle number and aerosol size distribution (Chin et al., 2000). While this method is simple, it is numerically efficient and computationally cheaper to run. Modal aerosol schemes are slightly more complex in that they include aerosol size distributions using three or more modes that includes the Aitken, accumulation, and coarse modes (Liu et al., 2016b). The most sophisticated method for simulating aerosol physics is through a bin method where aerosols are distributed into many discrete size bins, which are simulated separately (Zaveri et al., 2008). Bin methods are typically computationally expensive to run.

8.2.5. Chemistry

Smoke plumes are made of a mixture of many chemically active species and aerosols, such as nitrogen oxides ($NO_x = NO + NO_2$), nitrous acid (HONO), volatile organic compounds (VOCs), which can impact air quality through the formation of ozone (O_3) and secondary organic aerosols (SOA) (Andreae & Merlet, 2001; Akagi et al., 2011; Jaffe & Wigder, 2011; Kochanski et al., 2016; Brey & Fischer, 2016; Peng et al., 2020). O_3 is formed through the chemical reaction between molecular oxygen (O_2) and atomic oxygen ($O(^3P)$). The supply of atomic oxygen is driven by chemical reactions involving NO_x and nonmethane organic compounds that are simultaneously exposed to sunlight photodissociate creating $O(^3P)$. Since O_2 is abundant in the atmosphere, O_3 production is typically limited by the availability of NO_x. The most common sources of NO_x are anthropogenic emission sources and wildfires (Finlayson-Pitts & Pitts, 1986).

Smoke plume chemistry is sensitive to several factors including time of day, meteorology, altitude, chemical composition of the plume, combustion efficiency, transport time, and nearby emission sources (Giglio, 2007; Jaffe et al., 2004; Lim et al., 2019; Peng et al., 2020). Smoke shading effects within the plume can also reduce O_3 production by limiting photochemical reactions (Jaffe & Wigder, 2011) producing molecular oxygen. The sequestration of NO_x as the smoke plume ages can also limit O_3 production downwind of the fire (Tanimoto et al., 2008). Due to the complex and nonlinear interactions between O_3 and other chemical processes, accurately simulating O_3 chemistry within smoke can be difficult (Jaffe & Wigder, 2011; Kochanski et al., 2016).

There are several existing modeling frameworks that have been used both to better understand smoke plume chemistry and to make air quality forecasts for chemical species such as O_3. The Community Multiscale Air Quality (CMAQ) model is a state-of-the-art air quality model that can simulate many atmospheric chemical processes related to gas, aqueous, and aerosol phase chemistry. Therefore, models such as CMAQ can simulate complex chemistry associated with O_3 and SOA. Operational air quality modeling frameworks such as AIRPACT (http://lar.wsu.edu/airpact/) are based on the CMAQ model (see Chapter 9). CMAQ generally estimates anthropogenic emissions using the Sparse Matrix Operator Kernal Emissions (SMOKE; https://www.cmascenter.org/smoke/) combined with fire emissions defined by the Satellite Mapping Automated Reanalysis Tool for Fire Incident Reconciliation (SmartFire2)–BlueSky framework (Larkin et al., 2009). WRF-CHEM and WRF-SFIRE-CHEM have also been used to simulate chemical reactions within wildfire plumes (Pfister et al., 2011; Kochanski et al., 2016). For example, work by

Kochanski et al. (2016) integrated WRF-SFIRE with WRF-CHEM's Model of Ozone and Related Chemical Tracers' (MOZART; Emmons et al., 2010) chemical mechanism to forecast O_3 for the 2007 Witch-Guejito Santa Ana fires. Chemical models such as WRF-CHEM and CMAQ also need chemical boundary conditions from either a larger-scale chemical transport model or a data assimilation product that utilize satellite observations. This is covered more in depth in Chapter 7.

8.3. SMOKE TRANSPORT MODELS

8.3.1. Box Model

Box models are among the simplest approaches used to simulate smoke exposure (Lettau, 1970). As suggested by the name, a box model assumes that air for a specified domain can be represented by a box, which is often bounded by the surface and the top of the PBL. Smoke within the box model is often assumed to be instantaneously diluted throughout the entire column, thus eliminating the need to simulate smoke dispersion and the fire plume rise. Because of the relative simplicity of box models, these models are easy to run, and require limited computational resources (Goodrick et al., 2012).

Box models have been previously used for smoke management applications in mountain valleys, where the lateral boundaries of the box are bounded by valley walls. Research presented by Brown and Bradshaw (1994) indicated that while box models struggle with predicting near-surface smoke concentrations from local fires, these models can be useful for assessing smoke loading within remote mountain valleys for prolonged smoke episodes. Another study by Pharo et al. (1976) found that box models tended to overestimate smoke concentrations near fires. It was hypothesized that overestimated smoke concentrations stemmed from the instantaneous dilution assumption made by box models.

Zero-dimensional box models are also a popular choice for simulating complex atmosphere chemistry within smoke plumes for research-based applications (Wolfe et al., 2016; Decker et al., 2021). Zero-dimensional box models are deployed by atmospheric chemists to investigate different chemical mechanisms (Archibald et al., 2010), analyze field observations (Decker et al., 2021), and for laboratory chamber experiments (Paulot et al., 2009). These models are particularly useful for understanding specific chemical processes, developing conceptual models, and testing hypotheses through sensitivity experiments (Wolfe et al., 2016).

8.3.2. Gaussian Plume Model

The Gaussian plume model represents the simplest way for simulating the downwind transport of smoke. Instead of letting smoke dilute within a targeted domain like what is done for box models, the Gaussian plume model attempts to account for atmospheric transport and dispersion (Taylor, 1922). Crosswind transport (i.e., dispersion) is parameterized as a Gaussian distribution that takes the form of a steady-state solution of the advection-diffusion equation. The direction of the smoke transport is determined by the wind speed and direction. Since winds are assumed to be constant in time and space, smoke is assumed to travel in a straight line from where it is emitted until it reaches the end point of the smoke plume or model domain. As a result, areas that frequently experience highly variable weather phenomena, such as sea breezes, frontal passages, and mountain-valley circulations, may not be appropriate for a Gaussian plume model. However, for cases where meteorological conditions are homogenous, Gaussian plume models can be an ideal tool for simulating downwind horizontal smoke transport given the limited computational demands and model inputs for these models.

As of today, there are two smoke models that utilize Gaussian plume theory to simulate smoke transport and exposure. VSMOKE (Lavdas, 1996) is often used by land managers in the southeastern United States to provide a quick and simple estimate of smoke impacts for prescribed burns based on planned fire activity and weather forecasts (Jackson et al., 2007). No wildfire plume rise is used within VSMOKE, thus the user must specify a fraction of smoke that is released near the ground and above the PBL. For smaller prescribed burns, it is generally safe to partition most of the smoke emissions within the PBL. However, for larger prescribed burns and wildfires, this assumption could be inadequate, as the fire plume rise can sometimes inject smoke into the free troposphere (Banta et al., 1992). The Simple Approach Smoke Estimation model (SASEM; Sestak & Riebau, 1988) is another Gaussian plume model that was designed to estimate smoke transport across relatively flat terrain. Like VSMOKE, SASEM can estimate ground-level smoke concentrations. SASEM can also estimate visibility impairment and the height of the fire plume rise as predicted by the Briggs (1975) plume rise model (see section 8.4b). SASEM is mostly used for prescribed burns in the State of Arizona (Goodrick et al., 2012).

8.3.3. Puff Models

Puff models represent another class of dispersion models, which reduces the number of assumptions made by Gaussian plume models (Lin, 2012). The Puff model represents the smoke plume as a collection of independent smoke puffs that are assigned an average smoke concentration that is representative of the puff's volume. Puffs are constantly released throughout the duration of a burn, with each puff having a total mass of smoke that is related to the smoke emissions at the time when the puff was emitted from the fire (Goodrick et al., 2012). Once the puffs are released into the atmosphere, they are transported by winds that can vary in time and space, unlike Gaussian plume models. Since Puff models follow a fluid parcel as it travels through time and space (e.g., moving reference frame), these models are classified as being Lagrangian.

Puff models are well suited for areas with lots of variability in winds, such as mountainous areas and coastlines. The effects of diffusion and entrainment are also accounted by Puff models. For cases where the Puff's volume increases, the smoke concentration within the Puff would decrease, while a decrease in the Puff's volume would correspond to an increase in smoke concentration. While Puff models represent a significant step forward relative to Gaussian plume modeling approaches, areas with strong wind shear and turbulence can distort puffs into non-Gaussian shapes (Lin, 2012). In these situations, ad hoc parameterizations such as puff splitting or merging are often necessary (Walcek, 2002).

CALPUFF (Scire et al., 2000) and Hybrid Single-Particle Lagrangian Integrated Trajectory (HYSPLIT; Draxler & Hess, 1997) are the most commonly used model frameworks that utilize Puff models. The CALPUFF model is driven by a diagnostic meteorological model (CALMET) that grids variables such as winds, temperature, PBL heights, friction velocity, and the Monin Obukhov length on a 3D micrometeorological domain. The 3D data are either obtained by interpolating meteorological data from nearby near-surface and upper-air observations and/or from a Eulerian NWP model. CALPUFF is commonly used by the Environmental Protection Agency to assess the impact of atmospheric pollutants on air quality for an area of interest (Scire et al., 2000). Even though CALPUFF does not explicitly resolve the plume rise, it utilizes the Brigg plume rise parameterization to estimate the injection height of atmospheric pollutants. Several studies have used CALPUFF for assessing the impacts of fires on different airsheds across North America. In one study, CALPUFF was used to quantify the impacts of agriculture burning for areas along the U.S.-Mexico border (Choi & Fernando, 2007). Converting fire activity, fuel conditions, and burn time into smoke emissions was listed as one of the major limitations of simulating smoke with CALPUFF. Jain et al. (2007) found that smoke plumes from agriculture burns in the Pacific Northwest exhibit large variability and were sensitive to fire input parameters when using CALPUFF. Despite the uncertainties associated with meteorology and fire input parameters, CALPUFF was mostly able to reproduce surface $PM_{2.5}$ concentrations when evaluated with nearby air quality stations.

HYSPLIT is another modeling framework that can be used to simulate the transport of pollutants as puffs, single trajectories, or as an ensemble air parcel trajectories, with the latter being discussed more in section 8.3. Similar to CALPUFF, an external three-dimensional NWP model needs to provide meteorological inputs, such as temperature and wind, to determine transport pathways for puffs within HYSPLIT. A joint project between the U.S. National Oceanic and Atmospheric Administration (NOAA) and Australia's Bureau of Meteorology led to the implementation of several modules, which allow HYSPLIT to account for chemical reactions in the atmosphere. HYSPLIT's puff model assumes that puffs continuously grow until they reach a size threshold that is larger than the meteorological grid cell. Once puffs reach the size threshold, they are split up into smaller puffs with identical properties in terms of pollutant concentrations.

8.3.4. Lagrangian Particle Dispersion Models

While Puff models can account for changing flow fields, these models make many assumptions regarding the expansion and contraction of the puff, along with interactions between different puffs. Lagrangian particle dispersion models (LPDMs) attempt to rectify some of these issues by simulating atmospheric transport as an ensemble of particles, with each particle representing a parcel of air with equal mass. These particles possess several important properties, such as (1) being small enough that they can follow the wind field without becoming deformed, but (2) being much larger than the average distance between air molecules, and (3) having fluid properties that are nearly identical to the surrounding air; thus, they are unaffected by gravitational settling and/or buoyancy (Lin, 2012). These particles are transported by the mean wind ($\bar{u}$) and a stochastic turbulent component (u'), which can be parameterized as a Markov process (Lin, 2012). As a result, these models are well equipped to handle cases with strong turbulence and/or wind shear. Simulated particles can also be referred to as trajectories. Since LPDM models must use thousands of particles to accurately depict turbulent dispersion (Mallia et al., 2015), these models are more computationally expensive than puff models. However, the downside of the added computational cost of simulating thousands of particles through 3D space is generally outweighed by LPDMs' ability to naturally simulate the effects of turbulence and wind shear. Several LPDM models are currently used to simulate smoke from prescribed burns to reduce human exposure to smoke or used for research-based applications. For example, LPDMs have been deployed in inverse-based studies to better understand spatiotemporal variability of fire emissions. LPDM models have also been used to identify major source regions of wildfire smoke and to quantify the role of the wildfire plume rise on smoke transport (Mallia et al., 2015; 2018).

FLEXPART (Stohl & Thomson, 1999) is a LPDM model that simulates long-range atmospheric transport and dispersion for a many atmospheric pollutants, tracers, and greenhouse gases. FLEXPART parameterizes the effects of wet and dry deposition. FLEXPART was first applied to wildfire smoke by Wotawa and Trainer (2000), who used FLEXPART to examine the impacts of Canadian wildfires on air quality in the southeastern United Sttates. Based on simulated results from FLEXPART, Wotawa and Trainer (2000) found that wildfire smoke was largely responsible for elevated concentrations of carbon monoxide (CO) during the summer of 1995. FLEXPART was integrated with the National Observatory of Athens FireHub platform (http://ocean.space.noa.gr/fires) to simulate smoke plumes over Greece. An analysis carried out Solomos et al. (2015) found that FLEXPART, driven by winds from WRF, was able to capture long-range smoke transport over Greece, along with transport near complex terrain features such as mountains and coastlines. A column-based plume rise model was integrated within FLEXPART to handle the vertical transport of smoke due to the fire plume rise.

Daysmoke (Achtemeier et al., 2011) is another model that uses Lagrangian-based framework to simulate downwind smoke transport. Daysmoke was originally built off the ASHFALL model, which was used to simulate deposition of ash particles from agriculture fires. Today, Daysmoke is used to simulate smoke dispersion to limit smoke exposure of communities downwind of prescribed burns. Daysmoke consists of four components for simulating smoke, including an entraining torrent model, a detraining particle model, a large eddy parameterization used to simulate the PBL, and a smoke emissions model, which describes the emission history of prescribed burns. The entraining torrent model handles the effects of convective uplift from the fire plume rise. In addition, the convective updraft within Daysmoke can be separated into multicore updrafts, which have weaker updrafts, smaller diameters, and more sensitivity to the entrainment. Ultimately, the separation of the convective updraft into multiple cores can limit the altitude at which smoke is injected. Thus, correctly specifying the number of updraft cores is critical when simulating the fire plume rise (Liu et al., 2010). Once the smoke particles are discharged from the smoke plume, they are traced through the atmosphere by a mean and turbulent wind component (Achtemeier et al., 2011). Like other LPDM models, the turbulent or convective mixing component is considered stochastic. Since Daysmoke employs relatively simple physics and no chemistry, the model requires less computational resources relative to other smoke modeling frameworks.

HYSPLIT is a popular tool for simulating smoke transport at larger scales (10–1,000 km), and can be run as a LPDM or, as previously mentioned, as a puff model, depending on the options selected at runtime (Draxler & Hess, 1997). HYSPLIT has been integrated with the BlueSky modeling framework (Larkin et al., 2009; O'Neill et al., 2008), which utilizes fuel maps and fire consumption rates to estimate smoke emissions (https://www.arl.noaa.gov/hysplit/smoke-prescribed-burns/). The meteorology used to drive HYSPLIT trajectories generally comes from an external NWP model, such as the many model outputs provided by the National Centers for Environmental Prediction (NCEP).

The Stochastic Time-Inverted Lagrangian Transport (STILT) model (Lin et al., 2003), which is based off HYSPLIT and has since been merged back with HYSPLIT (Loughner et al., 2021), is another LPDM that has been used to simulate the impacts of smoke on air quality across the western United States (Mallia et al., 2015). Smoke emissions used by STILT can be vertically distributed using the Freitas plume rise model (Freitas et al., 2007; Mallia et al., 2018). STILT typically uses backward trajectories to determine the origin of air that is arriving at a receptor location. Backward trajectories can be used to derive the footprint for a receptor, which can then be mapped with smoke emissions to determine contributions of smoke from upwind fires (Fig. 8.2a). The receptor-orientated approach used by STILT makes this modeling framework particularly useful for identifying fires responsible for deteriorating air quality as seen in Figure 8.2b. Since the NWP models used to drive backward trajectories are often imperfect, STILT has the unique ability to translate wind errors into modeled smoke uncertainties (Fig. 8.2b; Mallia et al., 2015).

8.3.5. Eulerian Grid Models

Smoke transport can also be simulated from the Eulerian perspective where instead of following a puff or particle in a moving coordinate system, a Eulerian grid model simulates smoke transport on a fixed reference plane. A Eulerian model can be visualized as collection of individual cubes that are stacked within a large box, with the box being representative of the lateral boundaries of the model. Equations used to describe the transport of smoke are then solved for each individual cube, which is often referred to as a model grid cell. While tracking individual smoke plumes with a Eulerian-based model

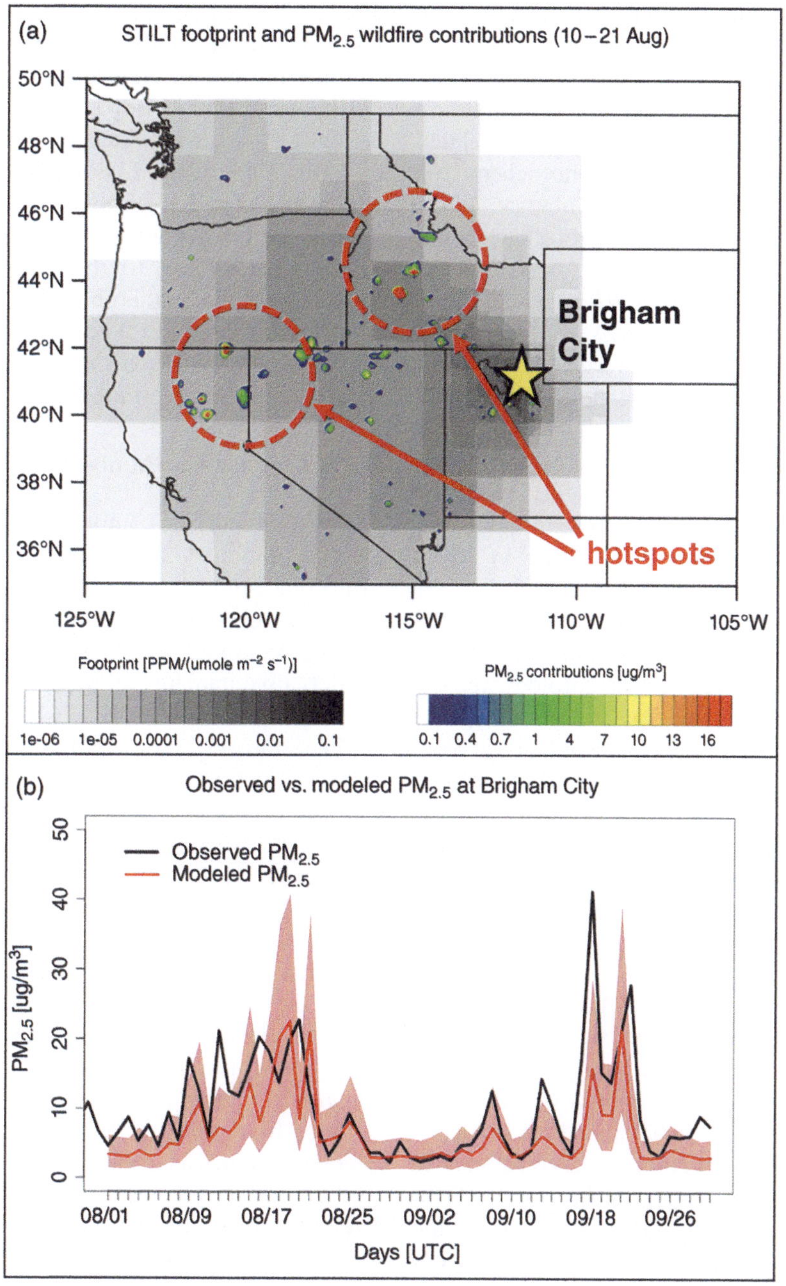

Figure 8.2 (a) STILT footprint (gray), which highlights backward trajectory transport pathways averaged between 10 and 21 August 2012. $PM_{2.5}$ contributions from wildfires toward Brigham City are shown by the color-filled contours. (b) Observed versus modeled $PM_{2.5}$ concentrations at Brigham City, Utah, for an episodic smoke event (fall 2012) with model uncertainties (+/−1σ) related to transport errors, are shaded as pink.

can be more difficult, grid models are more suited for simulating interactions between different plumes and for determining how anthropogenic emission sources might interact with these plumes to form secondary pollutants like O_3 (Goodrick et al., 2012). Eulerian grid models heavily rely on NWP models to determine how smoke is transported throughout the model domain. Meteorological data can be provided as an input for Eulerian-based smoke models or smoke transport and chemistry can be solved inline with the meteorology. One potential limitation of Eulerian-based frameworks is that emissions are assumed to be instantaneously diluted through model grid cells, which can be unrealistic, especially in coarser-scale model simulations (Goodrick et al., 2012).

CMAQ is a state-of-the-art air quality model, which is one of the most widely used tools for air quality applications. Such applications include regulatory and policy analysis, research, and operational forecasting (Byun & Schere, 2006; Baker et al., 2018). CMAQ contains a suite of atmospheric chemistry and emission routines that enables the model to simulate smoke-related chemical and aerosol processes such as photochemistry, SOA formation, and advanced aerosol physics. While CMAQ does not simulate its own meteorology, NWP model data can be provided as an input, or the model can be coupled directly with the WRF model (Zou et al., 2019). Routines exist within CMAQ, where smoke can be injected between two specified vertical levels either by the user or by an offline plume rise model. AIRPACT (http://lar.wsu.edu/airpact/), which is an operational model used to make air quality forecasts across the Pacific Northwest, is an example of an air quality modeling system that uses CMAQ driven by an external WRF model and fire emissions generated from BlueSky. More details on AIRPACT can be found in Chapter 9.

Another popular choice for simulating smoke is WRF-CHEM, which is a chemical transport modeling framework that can simultaneously model meteorology, aerosol physics, and chemical transformations in the atmosphere (Grell et al., 2005). Since the chemical and aerosol modules within WRF-CHEM are directly coupled with the dynamical core and physical parameterizations, smoke emissions can modify weather conditions through smoke shading and/or cloud microphysical processes. This type of coupling is unique to modeling frameworks like WRF-CHEM, where smoke simulations are computed in line with meteorology.

Smoke emissions within WRF-CHEM are typically provided by an external emission inventory such as the Fire Inventory from NCAR (FINN; Wiedinmyer et al., 2010), while smoke can be vertically distributed within WRF-CHEM using the Freitas et al. (2007) plume rise model. A study by Grell et al. (2011) found that smoke emissions had the potential to affect mesoscale (10–100 km) weather patterns across Alaska by changing vertical temperature and moisture profiles in areas absent of cloud cover. Sensitivity tests also revealed that high concentrations of $PM_{2.5}$ were responsible for altering cloud microphysical processes, which ultimately impacted the modeled spatiotemporal distribution of precipitation across Alaska in 2004. The National Oceanic and Atmospheric Administration's operational smoke forecast system, HRRR-Smoke, is based on WRF-CHEM v3.9, with several in-house modifications related to smoke aerosol physics (Ahmodav et al., 2017). More details on HRRR-Smoke can be found in Chapter 9.

Global-scale simulations of smoke transport can be achieved with modeling frameworks such as GEOS-CHEM (http://acmg.seas.harvard.edu/geos/), which has been used in previous work to isolate the impacts of wildfire smoke on global climate and air quality (Christian et al., 2019; Li et al., 2020). Like CMAQ and WRF-CHEM, GEOS-CHEM includes chemical and aerosol routines to simulate changes in the chemical composition of the atmosphere (Bey et al., 2001). Meteorology for GEOS-CHEM is provided as an input from an external global NWP model, while smoke emissions are estimated using GFED or FINN. Since GEOS-CHEM is a global model, the horizontal grid spacing within the model is very coarse relative to the grid spacing of the other models presented in this chapter. Despite having a coarser model resolution, GEOS-CHEM is one of the few models specifically designed to simulate the large-scale impacts of smoke on global air quality, weather, and climate.

8.3.6. Coupled Fire-Atmosphere Models

Advancements in computational facilities have led to the development and deployment of coupled fire-atmosphere models. Like Eulerian grid models, these coupled fire-atmosphere models simulate smoke transport on a three-dimensional grid. Coupled-fire atmosphere models also simulate their own meteorology using a computational fluid dynamics weather prediction model. Unlike some of the Eulerian grid models discussed in the previous section, coupled fire-atmosphere models simulate fire progression using a formula that parameterizes fire growth based on local meteorology, terrain, and fuel characteristics (Mandel et al., 2011), or through the explicit representation of combustion processes (Mell et al., 2007; Linn et al., 2002; Linn & Cunningham, 2005).

Coupled fire-atmosphere models can utilize information about the predicted burned area and fuel loading to forecast fuel consumption, heat fluxes, and smoke emissions. The heat fluxes forecasted by these models can also dynamically interact with the atmosphere, which allows coupled-fire atmosphere models to explicitly simulate phenomena such as the wildfire plume rise (see section 8.4) and fire-induced winds near the fire front. Some coupled fire-atmosphere models such as WRF-SFIRE can simulate the impacts of smoke on meteorology through aerosol physics and chemistry coupling (Kochanski et al., 2016; Kochanski et al., 2019). While coupled fire-atmosphere models represent the most sophisticated way to simulate smoke, these models can be computationally demanding compared with other models due to the computations needed to resolve near-fire circulations and plume dynamics. However, multiscale coupled fire-atmosphere models such as WRF-FIRE and WRF-SFIRE use a nested domain setup that allows these models to embed small-scale, high-resolution domains within a larger and coarser domain. Ultimately, this allows modeling frameworks like WRF-FIRE and WRF-SFIRE to simulate smoke dispersion across large distances at a relatively lower computational cost. Outside of forecasting applications, coupled fire-atmosphere models are ideal tools for studying how fire and fire behavior dynamically interact with the atmosphere.

Firetech and WFDS use a finite-volume, large eddy simulation to model fine-scale meteorological flows near the fire of interest. Here, large eddies within turbulent flow are explicitly resolved within the numerical grids of Firetech and WFDS, while smaller eddies are parameterized with subgrid scale models (Mell et al., 2007; Linn & Cunningham, 2005). Typically, the grids used by Firetech and WFDS are on the order of meters. Since Firetech and WFDS use a physics-based approach for simulating fire growth and combustion, detailed information about fuels and fuel density on a scale ~1 m are needed. This attention to detail comes at a cost as Firetech and WFDS simulations are computationally expensive to run. Furthermore, the model grid spacing used within Firetech and WFDS also limits these models to individual fire-scale problems that are typically less than 100 acres in size. Due to the domain size limitations associated with Firetech and WFDS, these models are better suited for describing fire behavior and hyper-local smoke transport (Liu et al., 2019). While models such as Firetech and WFDS can represent detailed combustion processes and the fire-atmosphere interactions, they do not account for the microphysical and radiative impacts of smoke on the atmosphere or chemical transformations of smoke downwind from the fire.

Models such as WRF-FIRE and WRF-SFIRE operate on a slightly different scale than Firetech and WFDS, so that they can be used in both research and forecast applications. To reduce computational costs, fire growth within WRF-FIRE and WRF-SFIRE is parameterized using an empirical formula instead of taking a physics-based approach (Mandel et al., 2011; Liu et al., 2019). While models such as WRF-FIRE and WRF-SFIRE parameterize fire growth rates, heat fluxes generated from the modeled fire are dynamically coupled to the atmosphere. Typically, these models resolve fire progression on scales on the order of tens of meters, while the meteorology from WRF, which is used to drive the fire and simulate smoke transport, is solved on grid with a horizontal grid spacing between 400 and 1,300 m (Kochanski et al., 2019). This allows models such as WRF-FIRE and WRF-SFIRE to simulate smoke across a larger domain compared with Firetech and WFDS. In addition, WRF-based modeling frameworks use a nested domain configuration where meteorology and smoke simulated in the innermost domain centered on the fire is fed into subsequently coarser but larger domains. Despite using a coarser atmospheric grid relative to Firetech and WFDS, both WRF-FIRE and WRF-SFIRE can explicitly resolve the wildfire plume rise and first order fire-atmosphere interactions (Liu et al., 2019). Both models treat smoke as a passive tracer, with smoke emissions being estimated based on the fuel consumed by the simulated fire.

WRF-SFIRE was recently coupled to WRF-CHEM to allow smoke generated by the fire to undergo chemical transformations while smoke aerosols can be scavenged from the atmosphere. Coupling with the WRF-CHEM's aerosol model (GOCART) also allows smoke to interact with atmospheric radiation, therefore allowing WRF-SFIRE to account for smoke shading effects. This type of coupling is unique to only WRF-SFIRE. Preliminary work carried out by Kochanski et al. (2016) found that WRF-SFIRE, coupled with WRF-Chem was able to reproduce elevated concentrations of NO_x and $PM_{2.5}$ for two large fires in southern California during the 2007 fire season. A follow-up study by Kochanski et al. (2019) found that WRF-SFIRE simulations were able to capture smoke shading effects within mountain valleys across northern California. A positive feedback mechanism was also identified where smoke aerosols resulted in cooler temperature at the surface, which allowed additional smoke aerosols to accumulate at the bottom of mountain valleys. WRF-SFIRE simulations coupled with WRF-CHEM were also used to forecast a wildfire smoke event in Salt Lake City, Utah, during the summer of 2018. WRF-SFIRE smoke simulations during this event were able to skillfully capture the orientation and shape of the plume, along with local-scale nocturnal mountain valley circulations (Fig. 8.3; Mallia et al., 2020b).

8.4. PLUME-RISE MODELS

8.4.1. Simplified Approaches

Earlier smoke modeling frameworks often assumed that smoke from biomass burning was either injected at a fixed altitude, evenly distributed throughout the PBL (Pfister et al., 2008; Hyer & Chew, 2010), assumes some type of ratio for partitioning emissions between the PBL and free troposphere (FT) (Turquety et al., 2007; Leung et al., 2007; Elguindi et al., 2020), or is prescribed based on local measurements such as satellite (Chen et al., 2009). For continental-scale smoke simulations across North America, the vertical distribution of smoke was found to be insensitive to the modeled plume injection height. It was hypothesized by Chen et al. (2009) that strong summertime convection tends to mix smoke throughout the troposphere, which limited the influence of the plume injection height on vertical smoke distributions.

8.4.2. Empirical

Another way to estimate the plume injection height is through empirical based models, which require inputs

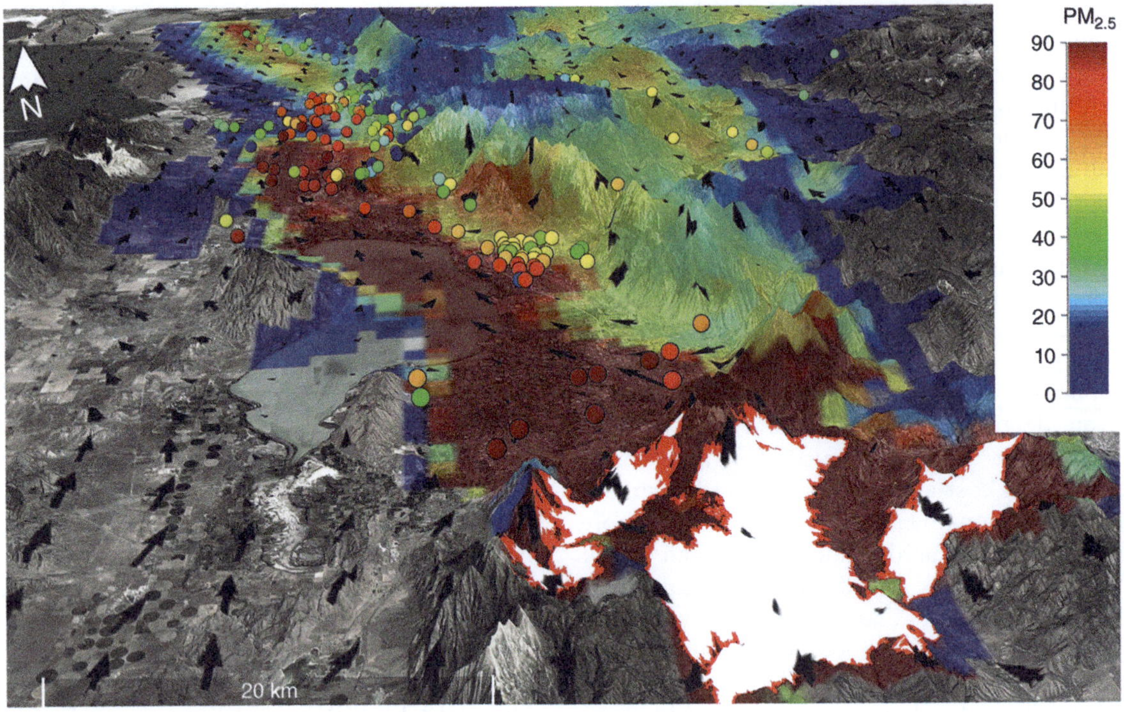

Figure 8.3 WRF-SFIRE-simulated and observed $PM_{2.5}$ concentrations for the Pole Creek and Bald Mountain fire on 15 September 2018. Simulated smoke concentrations are represented by the color-filled contours, while observed $PM_{2.5}$ concentrations are denoted by the color-filled circles. All $PM_{2.5}$ concentrations displayed here are in units of µg m^{-3}. The white polygons in the lower right represent model-estimated burned areas, and the black arrows represent simulated near-surface winds.

such as buoyancy, fire power, fire area, and/or generalized characteristics of the atmosphere such as atmospheric stability or the PBL height.

The Briggs equations (Briggs, 1975), which represent one of the first plume injection models, were originally developed to estimate the height of plumes released from smokestacks. Today, the Briggs equations are commonly used by smoke modeling frameworks such as CMAQ, BlueSky, and HYSPLIT. The Briggs model consists of a series of equations used for different stability conditions and whether the plume is momentum or buoyancy dominated. The plume injection height estimated by the Briggs model is a function of buoyancy, ambient wind speeds, and stability. Since the Briggs formulas contain no direct input for the fire heat release, the fire heat release needs to be converted into a buoyancy flux (Raffuse et al., 2012). Plume rise results with the Briggs model have been mixed, which is reasonable considering that it was originally developed to model plumes from smokestacks. Work by Raffuse et al. (2012) and Gordon et al. (2018) found that the Briggs model typically underestimated plume rises, especially for larger wildfires. Achtemeier et al. (2011) hypothesized that models like Briggs are unable to account for microphysical impacts, such as latent heat releases. As a result, the Briggs model is unable to account for extreme pyroconvection like pyrocumulus or pyrocumulonimbus clouds. However, Achtemeier et al. (2011) suggest that the Briggs model may perform better for smaller wildfires and prescribed burns.

A newer methodology for estimating wildfire smoke plume heights was presented in Sofiev et al. (2012). Like the Briggs models (Briggs, 1975), this methodology uses a semiempirical formula to estimate fire plume tops. This parameterization uses an energy-balance-based approach to estimate plume tops, similar to convective cloud parameterizations used within larger-scale atmospheric models. The plume height within the Sofiev et al. (2012) scheme is estimated from the following:

$$H_p = \alpha zi + \beta \left(\frac{FRP}{P_{f0}} \right)^{\gamma} \exp\left(\frac{-\delta N_{FT}^2}{N_0^2} \right), \tag{8.1}$$

where α is part of the PBL passed freely, β weights the contribution of fire intensity, γ determines the power-law dependence on the fire radiative power (FRP), δ weights the dependence of the stability of the FT on the plume rise height (H_p), P_{f0} is the reference fire power ($P_{fo} = 10^6$ W), N_0^2 is the Brunt-Vaisala frequency reference number ($N_0^2 = 2.5 \times 10^{-4}$ s^{-2}), and N_{FT}^2 is the Brunt-Vaisala frequency of the FT. A learning subset of satellite fire smoke plume observations from the Multiangle Imaging SpectroRadiometer (MISR) were then used to determine the value of the empirical calibration constants (α, β, γ, δ) where α represents the part of the plume that passes freely through the PBL, β accounts for the weighted contribution from the fire intensity, γ quantifies the power-law dependence of the fire's FRP on the plume height, and δ defines the plume top dependence on the atmospheric stability within the free troposphere.

Results in Sofiev at el. (2012) found that their methodology outperformed both the Briggs and 1-D column models when applied to 2,000 fire plumes from an independent MISR database across North America and Siberia. One potential limitation of this model is that the parameters defined in Sofiev at el. (2012) are primarily tuned for shallower smoke plumes, since the training data set to develop the parameters included only plume rises with heights less than 4 km (Paugnam et al., 2016).

8.4.3. Column Models

An alternative way to estimate plume top heights is through one-dimensional column models such as the Freitas et al. (2007, 2010) model. The Freitas plume rise model is based off a plume model develop by Latham (1994), which simulates the wildfire plume rise using the equations for vertical momentum, first law of thermodynamics, mass continuity, and cloud microphysics. In addition, the effects of entrainment near the edges of the plume are parameterized as two entrainment coefficients, with one accounting for the effects of turbulence at the plume edge, and the other describing for ambient wind shear effects. The final plume injection height is often used within chemical and smoke transport models such as WRF-Chem (Grell et al., 2005; Pfister et al., 2011; Sessions et al., 2011), STILT (Mallia et al., 2018), and HRRR-Smoke. One added benefit of a column-based approach is that this method can provide the vertical plume characteristics, in addition to the final injection height (Fig. 8.4). The final injection height within the Freitas model is defined as when the upward vertical velocity w reaches 0 m s^{-1}. Since the Freitas plume rise model is cloud resolving, it can simulate moist pyroconvection. The Freitas model's ability to simulate moist pyroconvection can be seen in Figure 8.4, which shows a secondary increase in vertical velocity between 2.5 and 5 km, which is collocated with an increase in liquid cloud water and latent heat releases.

Inputs for the Freitas plume rise model include a one-dimensional profile of ambient atmospheric conditions, such as temperature, relative humidity, and wind speed, along with surface boundary conditions provided by the fire (i.e., heat flux and fire area). Natively, the Freitas model assumes fire heat fluxes and area based on the vegetation type that is being burned. The heat flux and fire area are then used to compute a buoyancy flux (F) following the expression derived by Viegas (1998):

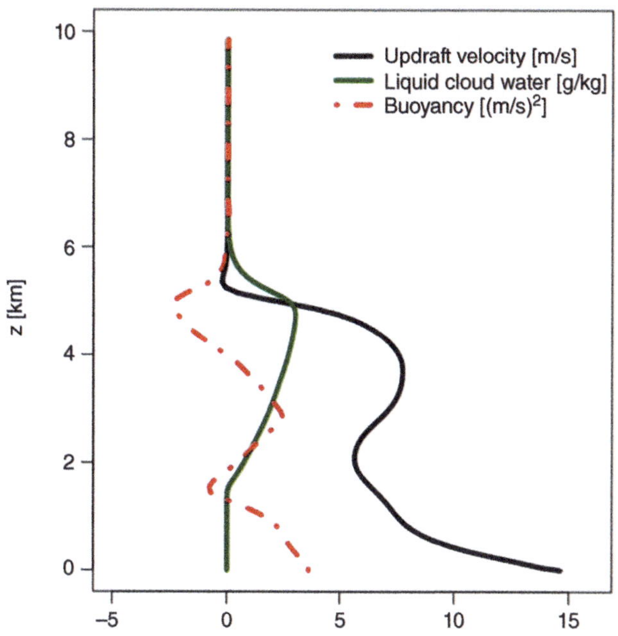

Figure 8.4 Plume rise simulation generated from the Freitas plume model for the 2012 Dry Creek fire in Alaska.

$$F = \frac{gR}{c_p p_e} r^2, \tag{8.2}$$

where g is equal to the gravity constant ($g = 9.81$ m s^{-1}), R represents the ideal gas constant (R = 287 J K^{-1} kg^{-1}), c_p denotes specific heat at a constant pressure ($c_p = 1004$ J kg^{-1}), p_e is the ambient surface pressure, and r defines the radius of the fire. The buoyancy flux can be used to compute the near-surface vertical wind velocity and temperature.

More recent work by Val Martin et al. (2012) tried alternative methods for prescribing fire input parameters, such as using satellite FRP to estimate sensible heat fluxes and aggregating satellite fire pixels to construct burned areas. This method resulted in slightly improved simulated plume rises when compared with satellite observations, however, the authors note the Freitas model was unreliable for identifying plumes that were injected into the FT. It was hypothesized that model errors in plume rises likely stemmed from uncertainties surrounding fire input parameters rather than the plume rise model formulation. A study conducted by Mallia et al. (2018) showed the Freitas model was able to realistically capture the plume rise for an extensively instrumented prescribed burn in Eglin Airforce Base, Florida, when driven by observed fire heat fluxes and burned area. It should be noted that this analysis was carried out for a single case study, for a relatively small burn (area = 1.51 km^2), where no pyroCu or Cb activity was observed.

8.4.4. Fully Physical Three-Dimensional Representation of Plume Dynamics

Continued advancements in computational resources have resulted in a newer generation of smoke models that can explicitly resolve wildfire plume rises. For the full physics method, heat fluxes from the fire are injected directly into the near-surface grid cells of a high-resolution three-dimensional atmospheric model (Goodrick et al., 2012). Like the column-based approach, the atmosphere will respond to the fire heating by generating a buoyant convective plume. One way that this approach is different from the column-based approach is that here the plume is resolved in three-dimensional space instead of one-dimensional. In addition, these models typically run at a fine enough resolution that key processes such as entrainment, multiple plume cores, pyroconvection, and upward smoke transport are directly simulated by the model instead of being parameterized. The full physics approach typically requires the model to have a sufficiently fine grid-spacing so that the model can explicitly resolve plume rise dynamics while simultaneously having the volume needed to encompass the convective plume (Goodrick et al., 2012). As such, short model time steps, combined with fine grid cells, often covering a large volume, can drastically increase computational resources needed to explicitly resolve the wildfire plume rise.

Directly simulating the wildfire plume rise using a fully physical approach was first pioneered by Trentmann et al. (2006) and Luderer et al. (2006), who used a high-resolution atmospheric model to simulate extreme pyroconvection over the Chisholm wildfire located in Alberta, Canada. The plume rise associated with the Chisholm wildfire reached an altitude of 13 km according to radar observations. The plume rise simulations carried out by Trentmann et al. (2006) and Luderer et al. (2006) were able to replicate the intense pyroconvection observed during this event while also illustrating how meteorological dynamics are coupled with large wildfires. Coupled fire-atmosphere models such as WRF-SFIRE and WRF-FIRE also explicitly resolve the wildfire plume rise. WRF-SFIRE simulations conducted by Kochanski et al. (2016, 2018) and Mallia et al. (2020a) found that modeled plume top heights compared reasonably well with satellite-estimated plume heights (average error = ± 500 m). An example of a full physics simulation of a wildfire plume rise by WRF-SFIRE can be seen in Figure 8.5.

8.5. SUMMARY

The number of large and devastating fires is expected to increase in the coming decades, which will expose communities to poor air quality. Therefore, smoke models will be an important tool for limiting the public's exposure to degraded air quality through smoke forecasts and for determining the optimal time for igniting prescribed burns. Wildfires are also projected to emit more aerosols into the atmosphere, which can affect weather and climate (Peterson et al., 2018).

Within this chapter, we've provided a brief introduction to the different types of smoke models that are available for researchers and air quality and land/fire managers alike. This chapter reviews models that range from simple box and Gaussian plume models to more sophisticated modeling systems that can simulate smoke on an atmospheric grid with full physics and photochemistry. Also provided in this chapter is an in-depth discussion on coupled fire-atmosphere models, which has not been included in other review articles. This chapter also reviews processes that are important in the context of smoke transport and how these fundamental processes are resolved within smoke modeling frameworks.

While we attempt to cover all smoke models that are available to researchers and managers, covering every smoke model would be an exhaustive effort that could probably be a book in itself! Nonetheless, we attempt to present a description of a diverse range of smoke transport and dispersion models to the reader. Ultimately, there is no smoke modeling tool that can be treated as a silver bullet as each of the models presented here has strengths and weaknesses that are dependent on the application that the model is being used for. Thus, we emphasize that determining the best smoke model for any given application will be dependent on the needs of the users and what they need the smoke model to do.

8.6. FUTURE DIRECTIONS

Fundamentally, the processes that govern smoke transport and dispersion are well understood, especially in the absence of significant pyroconvection (Goodrick et al., 2012). However, processes related to the fire plume rise (Val Martin et al., 2012; Paugam et al., 2016), aerosol microphysics (Forrister et al., 2015; Xie et al., 2017), and plume chemistry (Jaffe & Widger, 2012) are less understood. Recent field campaigns, such as the NASA-NOAA FIREX-AQ campaign and Joint Fire Science's Fire Smoke and Model Evaluation Experiment (Prichard et al., 2019; Liu et al., 2019) have started unraveling some of the unknowns associated with smoke plume chemistry

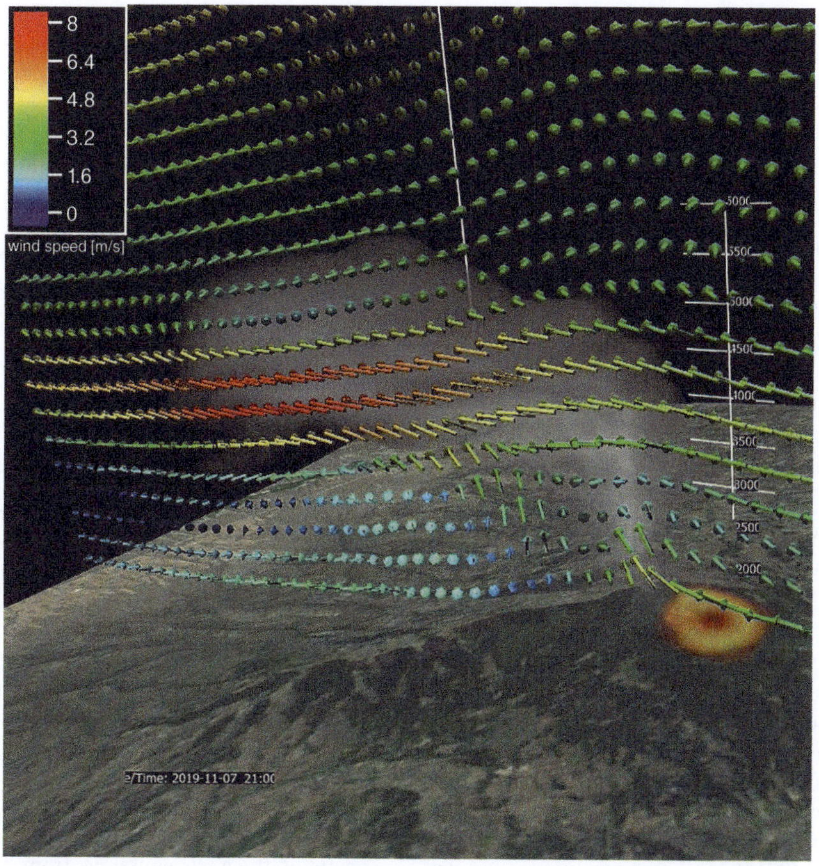

Figure 8.5 WRF-SFIRE simulated wildfire plume rise for the Anabella Reservoir prescribed burn. Warm-colored surface contours represent the modeled burn area; the vectors represent simulated cross-sectional winds. Smoke is denoted by the transparent gray isosurface.

and aerosol physics, however, work is still needed that integrates observations with smoke models. For example, shading from smoke aerosols can limit O_3 production in smoke plumes despite wildfires emitting chemical precursors that are conducive for O_3 production (Verma et al., 2009). There are also outstanding questions surrounding the timescale it takes for NO_x to be converted into peroxyacetyl nitrate and then back to NO_x, which can be used to form O_3 (Alvarado et al., 2010). These are just a few of the many questions that need to be addressed regarding smoke plume chemistry.

Properly evaluating fire plume rise models also continues to be a challenging proposition. There are a limited number of observational data sets that measure the plume height and properties, while simultaneously constraining surface fire characteristics, such as the fire heat flux, burned area, and fuel consumption. Incomplete observational data sets makes it difficult to disentangle whether a simulated plume rise result is erroneous due to assumptions made within the model, or if the errors stem from the model inputs (Val Martin et al., 2012).

Finally, work is also needed to better project future fire behavior and emissions. Outside of coupled fire-atmosphere models, most smoke modeling frameworks either use a persistence assumption (smoke today will be the same as smoke yesterday) or scale current fire heat fluxes and emissions using a diurnal curve. Therefore, most smoke models are unable to account for weather-driven fire effects. Recent studies have indicated that climate change is now impacting how some fires behave during the nighttime (Chiodi et al., 2021), which could further limit the usefulness of diurnal scaling techniques. While running a coupled fire-atmosphere model for every wildfire may not be practical with today's computing resources, new approaches could be developed that scale short-term fire activity based on changes in meteorology.

As we head farther into the future, our ability to monitor fires will continue to improve as remote sensing products become more advanced. New and exciting tools are also emerging that synthesize remote sensing products with machine learning techniques. These sorts of tools can provide detailed fire information at a high spatiotemporal resolution, therefore reducing some of the uncertainties described earlier in this chapter (Farguell et al., 2021). Such tools will be critical for providing accurate inputs into smoke models. It is expected that these emerging technologies, combined with data assimilation and improved computational resources, will play an increasingly important role toward improving the representation of smoke by smoke transport models.

ACKNOWLEDGMENTS

The authors acknowledge financial support from the National Science Foundation (NSF ICER-1664175) and the Utah Division of Air Quality. The authors also thank the WRF-SFIRE modeling team for feedback related to the coupled fire-atmosphere modeling components of this chapter. Finally, the authors are grateful for the reviewer's comments and suggestions.

REFERENCES

Achtemeier, G. L., GoodrickS. A., Liu, Y., Garcia-Menendez, F., Hu, Y., & Odman, M. T. (2011). Modeling smoke plume rise and dispersion from southern United States prescribed burns with daysmoke. *Atmosphere*, *2*, 358–388.

Ahmadov, R., Grell, G., James, E., Freitas, S., Pereira, G., Csiszar, I., et al. (2017). A high-resolution coupled meteorology-smoke modeling system HRRR-Smoke to simulate air quality over the CONUS domain in real time. *Geophysical Research Abstracts*, *19*, 10841.

Akagi, S. K., Yokelson, R. J., Wiedinmyer, C., Alvarado, M. J., Reid, J. S., Karl, T., et al. (2011). Emission factors for open and domestic biomass burning for use in atmospheric models. *Atmospheric Chemistry and Physics*, *11*, 4039–4072.

Alvarado, M. J., et al. (2010). Nitrogen oxides and PAN in plumes from boreal fires during ARCTAS-B and their impact on ozone: an integrated analysis of aircraft and satellite observations. *Atmospheric Chemistry and Physics*, *10*, 9739–9760.

Andreae, M. O., & Merlet, P. (2001). Emission of trace gases and aerosols from biomass burning. *Global Biogeochemical Cycles*, *15*, 955–966.

Andreae, M. O., Rosenfeld, D., Artaxo, P., Costa, A. A., Frank, G. P., Longo, K. M., et al. (2004). Smoking rain clouds over the Amazon. *Science*, *303*(5662), 1337–1342.

Archibald, A. T., Jenkin, M. E., & Shallcross, D. E. (2010). An isoprene mechanism intercomparison. *Atmospheric Environment*, *44*, 5356–5364.

Baker, K., et al. (2018). Photochemical model evaluation of 2013 California wild fire air quality impacts using surface, aircraft, and satellite data. *Science of the Total Environment*, *637*, 1137–1149.

Banta R. M., Olivier, L. D., Holloway, E. T., Kropfli, R. A., Bartram, B. W., Cupp, R. E., et al. (1992). Smoke-column observations from two forest fires using Doppler lidar and Doppler radar. *Journal of Applied Meteorology*, *31*, 1328–1349.

Barnes, J. E., & Hofmann, D. J., (1997). Lidar measurements of stratospheric aerosol over Mauna Loa observatory. *Geophysical Research Letters*, *24*, 1923–1926.

Battye, W., & Battye, R. (2002). Final report: Development of emissions inventory methods for wildland fire. *EPA*. https://www.epa.gov/ttn/chief/ap42/ch13/related/firerept.pdf

Bauer, S. E., & Menon, S. (2012). Aerosol direct, indirect, semidirect, and surface albedo effects from sector contributions based on the IPCC AR5 emissions for preindustrial and present-day conditions. *Journal of Geophysical Research*, *117*, 1–15.

Bey, I., Jacob, D. J., Yantosca, R. M., Logan, J. A., Field, B. D., Fiore, A. M., et al. (2001). Global modeling of tropospheric chemistry with assimilated meteorology: Model description and evaluation. *Journal of Geophysical Research*, *106*(D19), 23073–23095.

Brey, S. J., & Fischer, E. V. (2016). Smoke in the city: How often and where does smoke impact summertime ozone in the United States? *Environmental Science & Technology*, *50*, 1288–94.

Briggs, G. A. (1975). *Plume rise predictions: Lectures on air pollution and environmental impact analyses*, 72–73. *American Meteorological Society*, Boston, MA, USA.

Brown J. K., & Bradshaw, L. S. (1994), Comparisons of particulate emissions and smoke impacts from presettlement, full suppression, and prescribed natural fire periods in the Selway-Bitterroot Wilderness. *International Journal of Wildland Fire*, *4*(3), 143–155.

Burling, I. R., Yokelson, R. J., Griffith, D. W. T., Johnson, T. J., Veres, P., Roberts, J. M., et al. (2010). Laboratory measurements of trace gas emissions from biomass burning of fuel types from the southeastern and southwestern United States. *Atmospheric Chemistry and Physics*, *10*, 1115–11130.

Byun, D., & Schere, K. L. (2006). Review of the governing equations, computational algorithms, and other components of the Models-3 Community Multiscale Air Quality (CMAQ) modeling system. *Applied Mechanics Reviews*, *59*, 51–77.

Chen, L.-W. A., Moosmüller, H., Arnott, W. P., Chow, J. C., Watson, J. G., Susott, R. A., et al. (2007). Emissions from laboratory combustion of wildland fuels: emission factors and source profiles. *Environmental Science & Technology*, *2*, 4317–4325.

Chen, Y., Li, Q, Randerson, J. T., Lyons, E. A., Kahn, R. A., Nelson, D. L., et al. (2009). The sensitivity of CO and aerosol transport to the temporal and vertical distribution of North American boreal fire emissions. *Atmospheric Chemistry and Physics*, *9*, 6559–6580.

Chin, M., Rood, R. B., Lin, S.-J., Muller, J. F., & Thomspon, A. M. (2000). Atmospheric sulfur cycle in the global model GOCART: Model description and global properties. *Journal of Geophysical Research*, *105*, 24671–24687.

Chiodi, A. M., Potter, B. E., & Larkin, N. K. (2021). Multidecadal change in Western U.S. nighttime vapor pressure deficit. *Geophysical Research Letters*, *48*, e2021GL092830.

Choi, Y-J, & Fernando, H. J. S. (2007). Simulation of smoke plumes from agricultural burns: application to the San Luis/Rio Colorado airshed along the U.S/.Mexico border. *Science of the Total Environment*, *388*(1–3), 270–289.

Christian, K., Wang, J., Ge, C., Peterson, D., Hyer, E., Yorks, J., et al. (2019). Radiative forcing and stratospheric warming of pyrocumulonimbus smoke aerosols: First modeling results with multisensor (EPIC, CALIPSO, and CATS) views from space. *Geophysical Research Letters*, *46*, 10061–10071.

Clements, C. B., et al. (2018). The rapid deployments to wildfires experiment (RaDFIRE): observations from the fire zone. *Bulletin of the American Meteorological Society*, *99*(12), 2539–2559. https://journals.ametsoc.org/view/journals/bams/99/12/bams-d-17-0230.1.xml

Coen, J. L., Cameron, M., Michalakes, J., Patton, E. G., Riggan, P. J., & Yedinak, K. M. (2013). WRF-Fire: Coupled weather-wildland fire modeling with the Weather Research and Forecasting model. *Journal of Applied Meteorology and Climatology*, *52*, 1, 16–38.

Decker, Z. C., et al. (2021). Nighttime and daytime dark oxidation chemistry in wildfire plumes: An observation and model analysis of FIREX-AQ aircraft data. *Atmospheric Chemistry and Physics Discussion.* https://doi.org/10.5194/acp-2021-267

Draxler, R. R., & Hess, G. D. (1997). *Description of the HYSPLIT_4 modeling system.* NOAA Technical Memorandum ERL, ARL-224.

Elguindi, N., Clark, H., Ordóñez, C., Thouret, V., Flemming, J., Stein, O., et al. (2020). Current status of the ability of the GEMS/MACC models to reproduce the tropospheric CO vertical distribution as measured by MOZAIC. *Geoscentific Model Development*, *3*, 501–518.

Emerson, E. W., Hodshire, A. L., DeBolt, H. M., Bilback, K. R., Pierce, J. R., McMeeking, G. R., et al. (2020). Revisiting particle dry deposition and its role in radiative effect estimates. *PNAS*, *117*(42), 26076–26082.

Emmons, L. K., et al. (2010). Description and evaluation of the model for ozone and related chemical tracers, version 4 (MOZART-4). *Geoscientific Model Development*, *3*(1), 43–67.

Farguell, A., Mandel, J., Haley, J., Mallia, D. V., Kochanski, A., & Hillburn, K. (2021). Machine learning estimation of fire arrival time from level-2 active fire satellite data. *Remote Sensing*, *13*(11), 2203.

Finlayson-Pitts, B. J., & Pitts, J. N. (1986). *Atmospheric chemistry: Fundamentals and experimental techniques.* New York: John Wiley and Sons.

Forrister, H., Liu, J., Scheuer, E., Dibb, J., Ziemba, L., Thornhill, K. L., et al. (2015). Evolution of brown carbon in wildfire plumes. *Geophysical Research Letters*, *42*, 4623–4630.

Freitas, S. R., Longo, K. M., Chatfield, R., Latham, D., Dias, M. A. F. S., Andreae, M. O., et al. (2007). Including the sub-grid scale plume rise of vegetation fires in low resolution atmospheric transport models. *Atmospheric Chemistry and Physics*, *7*, 3385–3398.

Freitas, S. R., Longo, K. M., Trentmann, J., & Latham, D. (2010). Technical note: Sensitivity of 1-D smoke plume rise models to the inclusion of environmental wind drag. *Atmospheric Chemistry and Physics*, *10*, 585–594.

Fromm, M., Lindsey, D. T., Servranckx, R., Yue, G., Trickl, T., Sica, R., et al. (2010). The untold story of the pyrocumulonimbus. *BAMS*, *91*(9), 1193–1210.

Giglio, L. (2007). Characterization of the tropical diurnal fire cycle using VIRS and MODIS observations. *Remote Sensing of Environment*, *108*(4), 407–421.

Goodrick, S. L., Achtemeier, G. L., Larkin, N. K., Liu, Y., & Strand, T. (2012). Modeling smoke transport from wildland fires: A review. *International Journal of Wildland Fire.* http://dx.doi.org/10.1071/WF11116

Grell, G., Freitas, S. R., Stuefer, M., & Fast, J. (2011). Inclusion of biomass burning in WRF-Chem: Impact of wildfires on weather forecasts. *Atmospheric Chemistry and Physics Discussion*, *10*, 30 613–30 650.

Grell, G. A., Peckham, S. E., McKeen, S., Schmitz, R., Frost, G., Skamarock, W. C., et al. (2005). Fully coupled "online" chemistry within the WRF model. *Atmospheric Environment*, *39*, 6957–6975.

Haikerwal, A., Reisen, F., Sim, M. R., Abramson, M. J., Meyer, C. P., Johnston, F. H., et al. (2015). Impact of smoke from prescribed burning: Is it a public health concern? *Journal of the Air & Waste Management Association*, *65*, 592–598.

Hodzic, A., Madronich, S., Bohn, B., Massie, S., Menut, L., & Wiedinmyer, C. (2007). Wildfire particulate matter in Europe during summer 2003: Meso-scale modeling of smoke emissions, transport and radiative effects. *Atmospheric Chemistry and Physics Discussions*, *7*(2), 4705–4760.

Hyer, E. J., & Chew, B. N. (2010). Aerosol transport model evaluation of an extreme smoke episode in Southeast Asia. *Atmospheric Environment*, *44*, 1422–1427.

Jackson, W. A., Achtemeier, G. L., & Goodrick, S. L. (2007). A technical evaluation of smoke dispersion from the Brush Creek prescribed fire and the impacts on Asheville, North Carolina. http://www.nifc.gov/smoke/documents/Smoke_Incident_Impacts_Asheville_NC.pdf

Jaffe, D., Bertschi, I., Jaegle, L., Novelli, P.,Reid, J. S., Tanimoto, H., et al. (2004). Long-range transport of Siberian biomass burning emissions and impact on surface ozone in western North America. *Geophysical Research Letters*, *31*, L16106.

Jaffe, D. A., & Wigder, N. L. (2012). Ozone production from wildfires: A critical review. *Atmospheric Environment*, *51*, 1–10.

Jain R, Vaughan, J., Kyle, H., Ramosa, C., Clalborn, C., Maarten, S., et al. (2007). Development of the ClearSky smoke dispersion forecast system for agricultural field burning in the Pacific northwest. *Atmospheric Environment*, *41*, 6745–6761.

Kalnay, E. (2003). *Atmospheric modeling, data assimilation and predictability*. Cambridge University Press.

Kochanski, A. K., Fournier, A., & Mandel, J. (2018). Experimental design of a prescribed burn instrumentation. *Atmosphere*, *9*, 296.

Kochanski, A. K., Jenkins, M. A., Yedinak, K., Mandel, J., Beezley, J., & Lamb, B. (2016). Toward an integrated system for fire, smoke, and air quality simulations. *International Journal of Wildland Fire*, *25*, 558–568.

Kochanski, A. K., Mallia, D. V., Fearon, M., Brown, T., & Mandel, J. (2019). Modeling wildfire smoke feedback mechanisms using a coupled fire-atmosphere model with a radiatively active aerosol scheme. *Journal of Geophysical Research*, *124*(16), 9099–9116.

Lahm, P. (2015). *Wildfire, prescribed fire and the Clean Air Act: The latest challenges and opportunities.* 6th International Fire Ecology and Management Congress: Advancing ecology in fire management, November 16–20, San Antonio, TX.

Lareau, N., & Clements, C. (2017). The mean and turbulent properties of a wildfire convective plume. *Journal of Applied Meteorology and Climatology*, *56*(8), 2289–2299.

Lareau, N. P., & Clements, C. B. (2015). Cold smoke: Smoke-induced density currents cause unexpected smoke transport near large wildfires. *Atmospheric Chemistry and Physics*, *15*, 11513–11520.

Lareau, N. P., & Clements, C. B. (2016). Environmental controls on pyrocumulus and pyrocumulonimbus initiation and development. *Atmospheric Chemistry and Physics*, *16*, 4005–4022.

Larkin, N. K., O'Neill, S. M., Solomon, R., Raffuse, S., Strand, T., Sullivan, D. C., et al. (2009). The BlueSky Smoke Modeling Framework. *International Journal of Wildland Fire*, *18*, 906–920.

Latham, D. (1994). *PLUMP: A one-dimensional plume predictor and cloud model for fire and smoke managers*. General Technical Report INT-GTR-314. Fort Collins, CO: U.S. Department of Agriculture.

Lavdas, L. G. (1996). *Program VSMOKE: Users manual*. USDA Forest Service, Southeastern Forest Experiment Station, General Technical Report SRS-6. (Macon GA). https://doi.org/10.2737/SRS-GTR-6

Lee, H., Jeong, S.-J., Kalashnikova, O., Tosca, M., Kim, S.-W., & Kug, J.-S. (2018). Characterization of wildfire-induced aerosol emissions from the Maritime Continent peatland and Central African dry savannah with MISR and CALIPSO aerosol products. *Journal of Geophysical Research*, *123*, 3116–3125.

Lettau, H. H. (1970) Physical and meteorological basis for mathematical models of urban diffusion processes. In A. C. Stern (Ed.), *Proceedings, symposium on multiple source urban diffusion models* (pp. 2.1–2.26). U.S. Environmental Protection Agency, Number AP-86, Research Triangle Park, NC.

Leung, F.-Y. T., Logan, J. A., Park, R., Hyer, E., Kasischke, E., Streets, D., et al. (2007). Impacts of enhanced biomass burning in the boreal forests in 1998 on tropospheric chemistry and the sensitivity of model results to the injection height of emissions. *Journal of Geophysical Research: Atmospheres*, *112*, D10313.

Li, Y., Loretta, M. J., Liu, P., & Kaplan, J. O. (2020). Trends and spatial shifts in lightning fires and smoke concentrations in response to 21st century climate over the national forests and parks of the western United States. *Atmospheric Chemistry and Physics*, *20*, 8827–8838.

Lim, C. Y., et al. (2019). Secondary organic aerosol formation from the laboratory oxidation of biomass burning emissions. *Atmospheric Chemistry and Physics*, *19*, 12797–12809.

Lin, J. C., & Brunner, D.(Eds.) (2012). *Lagrangian modeling of the atmosphere*. AGU Geophysical Monograph (Vol. 200). Hoboken, NJ: John Wiley.

Lin, J. C., et al. (2003). A near-field tool for simulating the upstream influence of atmospheric observations: The Stochastic Time-Inverted Lagrangian Transport (STILT) model. *Journal of Geophysical Research: Atmospheres*, *108*(D16). https://doi.org/10.1029/2002JD003161

Lindsey, D. T., & Fromm, M. (2008). Evidence of the cloud lifetime effect from wildfire-induced thunderstorms. *Geophysical Research Letters*, *35*, L22809.

Linn, R. R., & Cunningham, P. (2005). Numerical simulations of grass fires using a coupled atmosphere-fire model: Basic fire behavior and dependence on wind speed. *Journal of Geophysical Reearch.*, *110*, D13107.

Linn R. R., Reisner, J., Colmann, J. J., & Winterkamp, J. (2002). Studying wildfire behavior using FIRETEC. *International Journal of Wildland Fire*, *11*, 233–246.

Liu, J. C., et al. (2016a). Particulate air pollution from wildfires in the Western U.S. under climate change. *Climate Change*, *138*, 655–666.

Liu, X., Ma, P.-L., Wang, H., Tilmes, S., Singh, B., Easter, R. C. et al. (2016b). Description and evaluation of a new 4-mode version of Modal Aerosol Module (MAM4) within version 5.3 of the Community Atmosphere Model. *Geoscientific Model Development*, *9*, 505–522.

Liu, Y., Goodrick, S., Achtemeier, G., Jackson, W., Qu, J., & Wang, W. (2009). Smoke incursions into an urban area: Simulation of a Georgia prescribed burn. *International Journal of Wildland Fire*, *18*(3), 336–348.

Liu, Y., Kochanski, A., Bakerv, K., Mell, W., Linn, R., Paugam, R., et al. (2019). Fire behavior and smoke modeling: Model improvement and measurement needs for next-generation smoke research and forecasting systems. *International Journal of Wildland Fire*, *28*, 570–588.

Liu, Y. Q., Achtemeier, G. L., Goodrick, S. L., & Jackson, W. A. (2010). Important parameters for smoke plume rise simulation with Daysmoke. *Atmospheric Pollution Research*, *1*, 250–259.

Lobert, J. (1991). Experimental evaluation of biomass burning emissions: Nitrogen and carbon containing compounds. In *Global biomass burning: atmospheric, climatic and biospheric implications* (pp. 289–304). Cambridge: MIT Press.

Loughner, C., Fasoli, B., Stein, A. F., & Lin, J. C. (2021). Incorporating features from the Stochastic Time-Inverted Lagrangian Transport (STILT) model into the Hybrid Single-Particle Lagrangian Integrated Trajectory (HYSPLIT) model: A unified dispersion model for time-forward and time-reversed applications. *Journal of Applied Meteorology and Climatology*, *60*, 799–810.

Luderer, G., Trentmann, J., Winterrath, T., Textor, C., Herzog, M., Graf, H.-F., et al. (2006). Modeling of biomass smoke injection into the lower stratosphere by a large forest fire (Part II): Sensitivity studies. *Atmospheric Chemistry and Physics*, *6*, 5261–5277.

Mallia, D. V., Kochanski, A., Urbanski, S., & Lin, J. C. (2018). Optimizing smoke and plume rise modeling approaches at local scales. *Atmosphere*, *9*, 116.

Mallia, D. V., Kochanski, A., Urbanski, S., Mandel, J., Farguell, A., & Krueger, S. (2020a). Incorporating a canopy parameterization within a coupled fire-atmosphere model to improve a smoke simulation for a prescribed burn. *Atmosphere*, *11*(8), 832.

Mallia, D. V., Kochanski, A. K., Kelly, K. E., Whitaker, R., Xing, W., Mitchell, L., et al. (2020b). Evaluating wildfire smoke transport within a coupled fire-atmosphere model using a high-density observation network for an episodic smoke event along Utah's Wasatch Front. *Journal of Geophysical Research*, *125*, e2020JD032712.

Mallia, D. V., Lin, J. C., Urbanski, S., Ehleringer, J., & Nehrkorn, T. (2015). Impacts of upwind wildfire emissions on CO_2, CO, and $PM_{2.5}$ concentrations in Salt Lake City, *Utah. Journal of Geophysical Research*, *120*(1), 147–166.

Mandel, J., Beezley, J. D., & Kochanski, A. K. (2011). Coupled atmosphere-wildland fire modeling with WRF 3.3 and SFIRE 2011. *Geoscientific Model Development*, *4*, 591–610.

Mell, W., Jenkins, M. A., Gould, J., & Cheney, P. (2007). A physics-based approach to modeling grassland fires. *International Journal of Wildland Fire*, *16*, 1–22.

O'Neill, S. M., Larkin, N. K, Hoadley, J., Mills, G., Vaughan, J., Draxler, R., et al. (2008). Regional real-time smoke prediction systems. In A. Bytnerowicz et al. (Eds.), *Wildland fires and air pollution* (pp. 499–534). Developments in Environmental Science Series, Vol. *8*. Elsevier.

Paugam, R., Wooster, M., Freitas, S., & Val Martin, M. (2016). A review of approaches to estimate wildfire plume injection height within large-scale atmospheric chemical transport model. *Atmospheric Chemistry and Physics*, *16*, 907–925.

Paulot, F., Crounse, J. D., Kjaergaard, H. G., Kroll, J. H., Seinfeld, J. H., & Wennberg, P. O. (2009). Isoprene photooxidation: new insights into the production of acids and organic nitrates. *Atmospheric Chemistry and Physics*, *9*, 1479–1501.

Peng, Q., et al. (2020). HONO emissions from western U.S. wildfires provide dominant Radical source in fresh wildfire smoke. *Environmental Science & Technology*, *54*(10), 5954–5963.

Peterson, D. A., Campbell, J. R., Hyer, E. J., Fromm, M. D., Kablick, G. P., Cossuth, J. H., et al. (2018). Wildfire-driven thunderstorms cause a volcano-like stratospheric injection of smoke. *Climate and Atmospheric Science*, *1*(1), 30.

Peterson, J., et al. (2020). NWCG smoke management guide for prescribed fire. *National Wildfire Coordinating Group*. PMS 420-3.

Pfister, G. G., Avise, J., Wiedinmyer, C., Edwards, D. P., Emmons, L. K., Diskin, G. D., et al. (2011). CO source contribution analysis for California during ARCTAS-CARB. *Atmospheric Chemistry and Physics*, *11*(15), 7515–7532.

Pfister, G. G., Wiedinmyer, C., & Emmons, L. K. (2008). Impacts of the fall 2007 California wildfires on surface ozone: Integrating local observations with global model simulations. *Geophysical Research Letters*, *35*.

Pharo J. A., Lavdas, L. G., & Bailey, P. M. (1976). Smoke transport and dispersion. In H. E. Mobley (Ed.), *Southern forestry and smoke management guidebook* (pp. 45–55). USDA Forest Service, Southeastern Research Station, General Technical Report SE-10.

Powers, J., et al. (2017). The Weather Research and Forecasting Model: Overview, system efforts, and future direction. *Bulletin of the American Meteorological Society*, *98*(8), 1717–1737.

Prichard, S., et al. (2019). The fire and smoke model evaluation experiment: A plan for integrated, large fire–atmosphere field campaigns. *Atmosphere*, *10*(2), 66.

Raffuse, S. M., Craig, K. J., Larkin, N. K., Strand, T. T., Sullivan, D. C., Wheeler, N. J. M., et al. (2012). An evaluation of modeled plume injection height with satellite-derived observed plume height. *Atmosphere*, *3*, 103–123.

Rappold, A. G., Fann, N. L., Crooks, J., Huang, J., Cascio, W. E., Devlin, R. B., et al. (2014). Forecast-based interventions can reduce the health and economic burden of wildfires. *Environmental Science & Technology*, *48*(8), 10571–10579.

Robock, A. (1988). Enhancement of surface cooling due to forest fire smoke. *Science*, *242*, 911–913.

Robock, A. (1991). Surface cooling due to forest fire smoke. *Journal of Geophysical Research*, *96*, 2156–2202.

Robock, A. (2000). Volcanic eruptions and climate. *Reviews of Geophysics*, *38*, 191–219.

Rothermel, R (1972). *A mathematical model for predicting fire spread in wildland fuels*. Technical Report Research Paper INT 115, USDA Forest Service, Intermountain Forest and Range Experiment Station, Ogden, UT.

Saylor, R. D., Barry, D., Lee, P., Tong, D., Pan, L. & Hicks, B. B. (2019). The particle dry deposition component of total deposition from air quality models: Right, wrong or uncertain? *Tellus B: Chemical and Physical Meteorology*, *71*(1), 1550324.

Scire, J. S., Strimaitis, D. G., & Yamartino, R. J. (2000). *A user's guide for the CALPUFF Dispersion Model (version 5.0)*. Concord, MA: Earth Tech, Inc.

Segal, M., & Arritt, R. W. (1992). Nonclassical mesoscale circulations caused by surface sensible heat-flux gradients. *Bulletin of the American Meteorological Society*, *73*, 1593–1604.

Sessions, W. R., Fuelberg, H. E., Kahn, R. A., & Winker, D. M. (2011). An investigation of methods for injecting emissions from boreal wildfires using WRF-Chem during ARCTAS. *Atmospheric Chemistry and Physics*, *11*(12), 5719–5744.

Sestak, M. L., & Riebau, A. R. (1988). *SASEM, Simple Approach Smoke Estimation Model*. US Bureau of Land Management, Technical Note 382.

Shin, H. H., & Hong, S.-Y. (2015). Representation of the subgrid-scale turbulent transport in convective boundary layers at gray-zone resolutions. *Monthly Weather Review*, *143*, 250–271.

Sofiev, M., Ermakova, T., & Vankevich, R. (2012). Evaluation of the smoke-injection height from wild-land fires using remote-sensing data. *Atmospheric Chemistry and Physics*, *12*, 1995–2006.

Solomos, S., Amiridis, V., Zanis, P., Gerasopoulos, E., Sofiou, F. I., Herekakis, T., et al. (2015). Smoke dispersion modeling over complex terrain using high resolution meteorological data and satellite observations: The Fire Hub platform. *Atmospheric Environment*, *119*, 348–361.

Spracklen, D. V., Mickley, L. J., Logan, J. A., Hudman, R. C., Yevich, R., Flannigan, M. D., et al. (2009). Impacts of climate change from 2000 to 2050 on wildfire activity and carbonaceous aerosol concentrations in the western United States. *Journal of Geophysical Research*, *114*, D20301.

Stohl, A., & Thomson, D. J. A (1999). Density correction for Lagrangian particle dispersion models. *Boundary-Layer Meteorology*, *90*, 155–167. https://doi.org/10.1023/A:1001741110696https://link.springer.com/article/10.1023/A:1001741110696

Tanimoto, H., Matsumoto, K., & Uematsu, M. (2008). Ozone-CO correlations in Siberian wildfire plumes observed at Rishiri Island. *Sola*, *4*, 65–68.

Taylor, G. I. (1922). Diffusion by continuous movements. *Proceedings of the London Mathematical Society*, *20*, 196–212.

Trentmann, J., Luderer, G., Winterrath, T., Fromm, M. D., Servranckx, R., Textor, C., et al. (2006). Modeling of biomass smoke injection into the lower stratosphere by a large forest fire (Part I): Reference simulation. *Atmospheric Chemistry and Physics*, *6*, 5247–5260.

Turquety, S., Logan, J. A., Jacob, D. J., Hudman, R. C., Leung, F. Y., Heald, C. L., et al. (2007). Inventory of boreal fire emissions for North America in 2004: Importance of peat burning and pyroconvective injection. *Journal of Geophysical Research: Atmospheres*, *112*, D12S03.

Urbanski, S. (2014). Wildland fire emissions, carbon, and climate: Emission factors. *Forest Ecology and Management*, *317*, 51–60.

Urbanski, S. P., Hao, W. M., & Nordgren, B. (2011). The wildland fire emission inventory: Western United States emission estimates and an evaluation of uncertainty. *Atmospheric Chemistry and Physics*, *11*, 12973–1300.

Urbanski, S. P., Reeves, M. C., Corley, R. E., Hao, W. M., & Silverstein, R. P. (2017). Missoula Fire Lab Emission Inventory (MFLEI) for CONUS. Fort Collins, CO: Forest Service Research Data Archive. https://doi.org/10.2737/RDS-2017-0039

Val Martin, N., Kahn, R., Logan, J. A., Paugnam, R., Wooster, M., & Ichoku, C. (2012). Space-based observational constraints for 1-D fire smoke plume-rise models. *Journal of Geophysical Research*, *117*, D22204.

Van der Werf, G. R., Randerson, J. T., Giglio, L., Collatz, G. J., Mu, M., Kasibhatla, P. S. et al. (2010). Global fire emissions and the contribution of deforestation, savanna, forest, agricultural, and peat fires (1997–2009). *Atmospheric Chemistry and Physics*, *10*(23), 11707–11735.

Verma, S., et al. (2009). Ozone production in boreal fire smoke plumes using observations from the Tropospheric Emission Spectrometer and the Ozone Monitoring Instrument. *Journal of Geophysical Research*, *114*, D02303.

Viegas, D. X. (1998). Convective processes in forest fires. In E. J. Plate et al. (Eds.), *Buoyant convection in geophysical flows* (pp. 401–420). Dordrecht, The Netherlands: Kluwer Academic Publishers.

Walcek, C. J. (2002). Effects of wind shear on pollution dispersion. *Atmospheric Environment*, *36*, 511–517.

Walter, C., Freitas, S. R., Kottmeier, C., Kraut, I., Rieger, D., Vogel, H., et al. (2016). The importance of plume rise on the concentrations and atmospheric impacts of biomass burning aerosol. *Atmospheric Chemistry and Physics*, *16*(14), 9201–9219.

Wiedinmyer, C., Akagi, S. K., Yokelson, R. J., Emmons, L. K., Al-Saadi, J. A., Orlando, J. J., et al. (2010). The Fire INventory from NCAR (FINN): A high resolution global model to estimate the emissions from open burning. *Geoscentific Model Development Discussions*, *3*, 2439–2476.

Weisman, M. L., Davis, C., Wang, W., Manning, K. W., & Klemp, J. B. (2008). Experiences with 0–36-h explicit convective forecasts with the WRF-ARW Model. *Weather Forecasting*, *23*, 407–437.

Westerling, A. L., Hidalgo, H. G., Cayan, D. R., & Swetnam, T. W. (2006). Warming and earlier spring increases western U.S. forest fire activity. *Science*, *313*, 940–943.

Wolfe, G. M., Marvin, M. R., Roberts, S. J., Travis, K. R. & Liao, J. (2016). The framework for 0-D atmospheric modeling (F0AM) v 3.1, *Geoscientific Model Development*, *9*(9), 3309–3319.

Wotawa, G., & Trainer, M. (2000). The influence of Canadian forest fires on pollutant concentrations in the United States. *Science*, *288*, 324–328.

Xie, M., Hays, M. D., & Holder, A. L. (2017). Light-absorbing organic carbon from prescribed and laboratory biomass burning and gasoline vehicle emissions. *Scientific Reports*, *7*, 7318.

Yokelson, R., Griffith, D., & Ward, D. (1996). Open-path Fourier transform infrared studies of large-scale laboratory biomass fires. *Journal of Geophysical Research: Atmospheres*, *101*, 21067–21080.

Zaveri, R. A., Easter, R. C., Fast, J. D., & Peters, L. K. (2008). Model for simulating aerosol interactions and chemistry (MOSAIC). *Journal of Geophysical Research*, *113*, D13204.

Zhang, L. M., Gong, S. L., Padro, J., & Barrie, L. (2001). A size-segregated particle dry deposition scheme for an atmospheric aerosol module. *Atmospheric Environment*, *35*, 549–560.

Zhang, Y., Fan, J., Logan, T., Li, Z., & Homeyer, C. R. (2019). Wildfire impact on environmental thermodynamics and severe convective storms. *Geophysical Research Letters*, *46*, 10082–10093.

Zou, Y., et al. (2019). Machine learning-based integration of high-resolution wildfire smoke simulations and observations for regional health impact assessment. *International Journal of Environmental Research and Public Health*, *16*, 2137.

9

Profiles of Operational and Research Forecasting of Smoke and Air Quality Around the World

Susan M. O'Neill[1], Peng Xian[2], Johannes Flemming[3], Martin Cope[4], Alexander Baklanov[5], Narasimhan K. Larkin[1], Joseph K. Vaughan[6], Daniel Tong[7], Rosie Howard[8], Roland Stull[8], Didier Davignon[9], Ravan Ahmadov[10,11], M. Talat Odman[12], John Innis[13], Merched Azzi[14], Christopher Gan[15], Radenko Pavlovic[9], Boon Ning Chew[15], Jeffrey S. Reid[2], Edward J. Hyer[2], Zak Kipling[3], Angela Benedetti[3], Peter R. Colarco[16], Arlindo Da Silva[16], Taichu Tanaka[17], Jeffrey McQueen[18], Partha Bhattacharjee[19], Jonathan Guth[20], Nicole Asencio[20], Oriol Jorba[21], Carlos Pérez García-Pando[21,22], Rostislav Kouznetsov[23], Mikhail Sofiev[23], Melissa E. Brooks[24], Jack Chen[25], Eric James[10,11], Fabienne Reisen[4], Alan Wain[26], Kerryn McTaggart[27], and Angus MacNeil[28]

ABSTRACT

Biomass burning has shaped many of the ecosystems of the planet and for millennia humans have used it as a tool to manage the environment. When widespread fires occur, the health and daily lives of millions of people can be affected by the smoke leading to a range of health consequences such as respiratory issues, cardiovascular issues, and mortality. It is critical to include smoke and its consequences in atmospheric modeling systems to meet needs such as informing and protecting the public during smoke episodes. This chapter profiles many of the global and regional smoke prediction systems available. It is not an exhaustive list, but rather a profile of many of the systems to give examples of the creativity and complexity needed to simulate the phenomenon of smoke. The global smoke prediction systems are advanced, and many are self-organizing into a powerful ensemble. Regional and national systems are being developed independently for example in Europe (11 systems), North America (7 systems), and Australia (3 systems). Finally, the World Meteorological Organization is bringing together global and regional systems to form an ensemble to support countries with smoke issues and who lack resources. For each system we discuss how fire activity information is obtained, how fire emissions are calculated, and how atmospheric transport and chemical transformation of the smoke plume is treated.

[1] *Pacific Northwest Research Station, United States Forest Service, Seattle, Washington, USA*

[2] *United States Naval Research Laboratory, Marine Meteorology Division, Monterey, California, USA*

[3] *European Centre for Medium-Range Weather Forecasts, Reading, United Kingdom*

[4] *CSIRO Climate Science Centre, Aspendale, Victoria, Australia*

[5] *World Meteorological Organization, Geneva, Switzerland*

[6] *Department of Civil and Environmental Engineering, Washington State University, Pullman, Washington, USA*

[7] *Department of Atmospheric, Oceanic and Earth Sciences, George Mason University, Fairfax, Virginia, USA*

[8] *Earth, Ocean and Atmospheric Sciences Department, The University of British Columbia, Vancouver, British Columbia, Canada*

[9] *Environment and Climate Change Canada, Dorval, Quebec City, Canada*

[10] *CIRES, University of Colorado Boulder, Boulder, Colorado, USA*

[11] *NOAA Global Systems Laboratory, Boulder, Colorado, USA*

[12] *School of Civil and Environmental Engineering, Georgia Institute of Technology, Atlanta, Georgia, USA*

[13] *EPA Tasmania, Hobart, Tasmania, Australia*

(Continued)

Landscape Fire, Smoke, and Health: Linking Biomass Burning Emissions to Human Well-Being, Geophysical Monograph 280, First Edition.
Edited by Tatiana V. Loboda, Nancy H. F. French, and Robin C. Puett.

DOI: 10.1002/9781119757030.ch09

9.1. INTRODUCTION

Biomass burning has shaped many of the ecosystems of the planet (Bowman et al., 2009), and for millennia humans have used it as a tool to manage the environment (Pyne, 2001). When widespread fires occur, the health and daily lives of millions of people can be affected by the smoke, often at unhealthy to hazardous levels (Jaffe et al., 2020) leading to a range of short-term and long-term health consequences such as respiratory issues (Alman et al., 2016; Henderson & Johnston, 2012; Lipner et al., 2019; Rappold et al., 2012), cardiovascular issues (Chen et al., 2021; Gan et al., 2017; Wettstein et al., 2018), and mortality (Doubleday et al., 2020; Johnston et al., 2012; Reid et al., 2016; Xi et al., 2020). It is critical to adequately represent and include smoke and its consequences in atmospheric modeling systems to meet needs such as addressing the global climate carbon budget and informing and protecting the public during smoke episodes.

Smoke impacts from biomass burning are highly episodic, with great variability from day to day and year to year, making it extremely difficult to simulate and predict. Smoke plumes can linger close to the ground where people breathe or be lofted high into the atmosphere, and transported long distances. Further, fire interacts with the atmosphere, creating its own fire weather, such as smoke-induced density currents (Clements et al., 2018) and pyrocumulonimbus clouds (Peterson et al., 2017). On regional to global scales, smoke interacts with the atmosphere, changing the radiative budget of the atmosphere, modifying winds and temperature, and interacting with cloud processes (Grell et al., 2011). Within the smoke plume, the rich mixture of trace gases and aerosols transform continuously through a variety of chemical and physical processes (Akagi et al., 2011; Crutzen & Andreae, 1990; Hatch et al., 2017; Hodshire et al., 2019).

Previous chapters of this book outline the scientific and technical challenges associated with modeling the complex phenomenon of smoke. Variability in fire emissions estimates has an order of magnitude level of uncertainty, depending upon vegetation type, natural fuel heterogeneity across the landscape, and fuel combustion processes (French, Chapter 5 this volume; Prichard, Chapter 4 this volume). Methods for quantifying fire emissions also vary from ground/vegetation-based methods to those based on remotely sensed fire radiative power data. These emission estimates are input into dispersion and air quality modeling systems, where their vertical allocation associated with plume rise, and temporal release parameterizations influence transport patterns, and in turn affect chemical transformation and interaction with other sources. These processes lead to another order of magnitude of variability on the downwind estimates of trace gases and aerosol concentrations (Mallia & Kochanski, Chapter 8 this volume). Unlike retrospective modeling studies, forecast modeling systems have a particularly challenging task because they cannot use observations to set fire locations, fire behavior, emissions composition, timing of emissions, or how high emissions are lofted and distributed vertically in the atmosphere. They need to make assumptions, such as persistence, where the fire information from yesterday is assumed to apply to every future day in the forecast time period. Coupled atmosphere-fire behavior systems (Coen et al., 2013; Linn et al., 2002; Mandel et al., 2014; Mell et al., 2007) are promising as they can track the evolution of ongoing fires but as yet are too computationally intensive to implement on regional and global scales. Data assimilation techniques combine numerical model predictions with observational data sets to provide a powerful means of initializing model runs to address some of these forecasting challenges (Hyer, Chapter 7 this volume).

Here we merge the science and data of the previous chapters into smoke prediction systems that operate at global, regional, and local scales to simulate smoke impacts. This chapter profiles many of the global and regional smoke prediction systems currently operational or quasi-operational in real time or near real time. It is not an exhaustive list of systems, but rather is a profile of many of the systems in use to give examples of the creativity and complexity needed to simulate the phenomenon of smoke.

[14] *Department of Planning and Environment, Government of New South Wales, Sydney, Australia*

[15] *Centre for Climate Research Singapore, Meteorological Service Singapore, Singapore*

[16] *NASA Goddard Space Flight Center, Greenbelt, Maryland, USA*

[17] *Meteorological Research Institute, Japan Meteorological Agency, Tsukuba, Japan*

[18] *NOAA National Centers for Environmental Prediction, College Park, Maryland, USA*

[19] *I. M. Systems Group, NWS/NCEP/EMC, College Park, Maryland, USA*

[20] *Météo-France, Toulouse, France*

[21] *Barcelona Supercomputing Center, Barcelona, Spain*

[22] *Catalan Institution for Research and Advanced Studies, Barcelona, Spain*

[23] *Atmospheric Composition Unit, Finnish Meteorological Institute, Helsinki, Finland*

[24] *Met Office, Exeter, United Kingdom*

[25] *Environment and Climate Change Canada, Ottawa, Ontario, Canada*

[26] *Australian Bureau of Meteorology, Melbourne, Victoria, Australia*

[27] *Department of Environment, Land, Water and Planning, Government of Victoria, Melbourne, Victoria, Australia*

[28] *Forest Practices Authority, Hobart, Tasmania, Australia*

This chapter, and the systems described, reflect the needs of different agencies and regions, where the various systems are tailored to the best available science to address challenges of a specific region. Smoke forecasting requirements range from warning and informing the public about potential smoke impacts to planning burn activities for hazard reduction or resource benefit. Different agencies also have different mandates, and the lines blur between the missions of quasi-operational organizations (e.g., research institutions) and agencies with operational mandates.

The global smoke prediction systems are advanced and many are self organizing into a powerful ensemble, as discussed in section 9.2. Regional and national systems are being developed independently and are discussed in sections 9.3–9.5 for Europe (11 systems), North America (7 systems), and Australia (3 systems). Finally, the World Meteorological Organization (WMO) effort (section 9.6) is bringing together global and regional systems to form an ensemble and building Vegetation Fire and Smoke Pollution Warning Advisory and Assessment Systems (VFSP-WAS) to support countries with smoke issues who lack resources. For each system, we discuss how fire activity information is obtained, how fire emissions are calculated, and how atmospheric transport and chemical transformation (if included) of the smoke plume is treated.

9.2. GLOBAL SYSTEMS AND THE INTERNATIONAL COOPERATIVE FOR AEROSOL PREDICTION (ICAP)

The International Cooperative for Aerosol Prediction (ICAP) community is a grassroots organization founded in 2010 by developers from major operational and research centers. The mission of ICAP is to promote community development of global aerosol observations, data assimilation, and prediction technologies that can support operational aerosol forecasting (Benedetti et al., 2011; Colarco et al., 2014; J. S. Reid et al., 2011). The core participants in the ICAP model have grown from the original five to the current nine operational/research centers. One of ICAP's most significant contributions to the community is the development of the ICAP-Multi Model Ensemble (ICAP-MME; Sessions et al., 2015; Xian et al., 2019), a global multi-model aerosol forecasting ensemble consensus for basic research and the baselining of operational products. The ICAP-MME provides a testbed of probabilistic aerosol forecasts, helps to identify challenging areas for aerosol modeling, and forges valuable collaborations among forecast centers.

9.2.1. Participating Centers and Core Models

As of June 2021, the ICAP modeling system includes nine operational or quasi-operational global aerosol forecast models maintained by major numerical weather prediction (NWP) or research centers from North America, Europe, and Asia. These models include the following eight comprehensive global aerosol models and one dust-only model (Table 9.1):

1. European Center for Medium-Range Weather Forecasts (ECMWF)-Copernicus Atmosphere Monitoring Service (CAMS; Rémy et al., 2019)
2. Fleet Numerical Meteorology and Oceanography Center (FNMOC)/Naval Research Laboratory (NRL)-Navy Aerosol Analysis and Prediction System (NAAPS; Lynch et al., 2016)
3. Japan Meteorological Agency (JMA)-Model of Aerosol Species in the Global Atmosphere (MASINGAR; Tanaka & Ogi, 2018)
4. NASA Global Modeling and Assimilation Office (GMAO)-Goddard Earth Observing System (GEOS; Colarco et al., 2010)
5. NOAA Global Ensemble Forecast System (GEFS), Aerosols (Zhang et al., 2022)
6. Barcelona Supercomputing Center (BSC)- Multiscale Online Nonhydrostatic AtmospheRe CHemistry model (MONARCH; Badia et al., 2017; Klose et al., 2021)
7. Météo France Modèle de Chimie Atmospherique à Grande Echelle (MOCAGE; Guth et al., 2016)
8. Finnish Meteorological Institute (FMI)-System for Integrated modeling of Atmospheric coMposition (SILAM; Sofiev et al., 2015)
9. U.K. Met Office (UKMO) dust model (Johnson et al., 2011)

9.2.2. Variations in Model Configuration

The ICAP models are driven mostly by independent operational/quasi-operational meteorological models developed at each NWP/research center. Aerosol variables are either computed concurrently with the meteorological fields (inline) or are run in separate calculations forced by stored NWP fields (offline). Depending on the resolution of the underlying meteorology, the aerosol models have different horizontal and vertical resolutions, ranging from 0.1° × 0.1° latitude/longitude and 137 vertical layers to 1.0° × 1.0° and 40 layers; in several cases, the aerosol forecast is run at a coarser resolution than the operational NWP model. The comprehensive models carry a full set of aerosol species normally considered in operational models, including dust, sea salt, biomass burning (BB) smoke (combined black carbon and particulate organic matter (POM) from some models) and varying forms of pollution aerosols (sulfate, possible nitrates, and anthropogenic organic aerosols). All of the ICAP models simulate the full aerosol life cycle, including sources, sinks, microphysics, chemistry, and transport, which make them different from tracer models. The treatment of these aerosol processes differs considerably between models, with only a few shared components (see below and Table 9.1).

Table 9.1 Global systems included in the International Cooperative for Aerosol Prediction (ICAP)

Forecastsystem	Operational/ research center	Status*	Smoke products	Meteorology model	Grid spacing (lat × lon)	Forecast frequency and length	BB emission system	Fire info	Plume rise	Aerosol data assimilation	Notes and websites
CAMS	ECMWF	O	Smoke included in OC, BC tracers. Concentration all levels, PM1/2.5/10 near surface	Inline IFS	0.4 × 0.4	12 hrly; 0Z/12Z: 120 hr	GFAS 1.4	MODIS FRP	GFAS Plume Rise model	Yes. Assimilate total AOD from MODIS & PMAp	Coupled to full tropospheric chemistry based on CB05 https://atmosphere.copernicus.eu/ Rémy et al., 2019
GEOS	NASA	QO	Smoke included in OC, BC, SU, NI tracers. AOD, PM2.5, PM10	Inline GEOS	C720, output at 0.25 × 0.33	6 hrly; 0Z: 240 hr 6Z: 18 hr 12Z: 120 hr 18Z: 18 hr	QFED	MODIS FRP	PBL	Yes. Assimilate total AOD from Neural Net MODIS	https://fluid.nccs.nasa.gov/weather/ Colarco et al., 2010
NAAPS	FNMOC/ NRL	O	Smoke AOD, Concentration all levels, visibility	Offline NAVGEM	0.33 × 0.33	6 hrly; 120 hr	FLAMBE	MODIS FRP	PBL	Yes. Assimilate DAQ total AOD from MODIS	https://www.nrlmry.navy.mil/aerosol/ Lynch et al., 2016
MASINGAR	JMA	QO/O	Smoke included in OC, BC tracers; AOD, concentration all levels	Inline MRI- AGCM	0.375 × 0.375	Daily; 120 hr	GFAS	MODIS FRP	GFAS	Yes. Assimilate DAQ total AOD from MODIS	Tanaka & Ogi, 2018
GEFS-Aerosols	NOAA	O	Smoke included in OC, BC tracers; AOD, Concentration all levels	Inline GEFS	C384 output at 0.25 × 0.25	6 hrly; 120 hr	GBBEPx	VIIRS and MODIS FRP	Dynamic plume rise model	No.	https://www.emc.ncep.noaa.gov/emc/pages/numerical_forecast_systems/gefs_aero-test.php#
MOCAGE	Météo-France	O	Smoke included in OC, BC tracers; AOD, Concentration all levels	Offline ARPEGE	1.0 × 1.0	Daily; 96 hr	GFAS	MODIS FRP		No.	Guth et al., 2016
MONARCH	BSC	QO	Smoke included in OC, BC tracers; AOD, Concentration all levels	Inline NMMB	0.7 × 0.5	Daily; 120 hr	GFAS	MODIS FRP	Plume rise model	No.	Badia et al., 2017 Klose et al., 2021
SILAM	FMI	O	Smoke AOD, near-surface PM concentration	Offline IFS	0.1 × 0.1	Daily; 120 hr	IS4FIRES	MODIS FRP	Dynamic hourly for each fire.	No.	http://silam.fmi.fi Sofiev et al., 2015
UKMO	UK Met Office	QO	AOD, near-surface PM concentration	Inline UK Unified Model	0.1 × 0.1 to 0.2 × 0.2	TBD.	GFAS	MODIS FRP	Plume rise model	For dust only.	Experimental-run for specific campaigns.
ICAP-MME	NRL	QO	Smoke AOD	-	1.0 × 1.0	Daily; 00Z 120 hr	Website https://www.nrlmry.navy.mil/aerosol/icap.0001.php Data https://nrlgodae1.nrlmry.navy.mil/cgi-bin/datalist.pl?dset=nrl_icap_mme&summary=Go				

* Status here represents operational (O) or quasi-operational (QO).

9.2.3. Biomass Burning Aerosol Species and Emission Sources

BB smoke particles can dramatically affect regional total aerosol distributions, significantly affecting energy balance and air quality. To account for their ever-changing impact, dynamic smoke sources are included in nearly all operational global aerosol models. Black carbon (BC) and organic carbon (OC) are the two major particle components of BB smoke. Because of their drastically different optical properties, many models carry BC and OC as separate aerosol species, and smoke resulting from BC and OC are often combined with those emitted from anthropogenic pollution sources because of the computational burden of extra aerosol types. Only two current ICAP models, NAAPS and SILAM, carry BB smoke as a stand-alone aerosol species, although some modeling centers (e.g., NASA GMAO) have plans to separate biomass-burning BC from other anthropogenic sources to facilitate smoke-specific research and applications. Nonetheless, for the models without specific smoke species, BC and OC/POM are combined to approximate BB smoke in the ICAP-MME. Although not ideal, this approximation serves the purpose of BB smoke analysis and forecast on regional to global scales, because of BB smoke's dominant contribution to total aerosols both locally and along its transport pathways with the occurrence of fires. This is illustrated by the World Meteorological Office (WMO) application of the ICAP BB smoke product in their Vegetation Fire and Smoke Pollution Warning and Advisory System (section 9.6).

Five global BB emission systems are used across the current ICAP models. They are developed and maintained by the same operational or research centers for atmospheric constituent simulations.

1. The Global Fire Assimilation System (GFAS; Kaiser et al., 2012) developed at ECMWF and used by CAMS, MASINGAR, MOCAGE, MONARCH, and UKMO
2. The Fire Locating and Modeling of Burning Emissions (FLAMBE; Reid et al., 2009) system developed at NRL, used by NAAPS
3. The Quick Fire Emissions Dataset (QFED; Koster et al., 2015) from NASA, used by GEOS
4. The Integrated monitoring and modeling System for wildland FIRES (IS4FIRES; Sofiev et al., 2009) from FMI, used by SILAM
5. The Blended Global Biomass Burning Emissions Product (GBBEPx) from NOAA, used by GEFS-Aerosols

These fire emission systems all use fire pixel, or gridded spatial resolutions and hourly or daily temporal resolutions that meet operational production requirements (normally within 3 hr latency). All are based on satellite fire radiative power (FRP) information because of its rapid availability for operational use. All use MODIS FRP because of its near-daily global coverage and its stability and reliability over time (French, Chapter 5 this volume).

MODIS instruments are on board two polar-orbiting satellites, Terra and Aqua, which have local equator-crossing times of 10:30 a.m./p.m. and 13:30 a.m./p.m. local time. Despite the use of similar fundamental data sources, overall source functions are underdetermined, and processing algorithms can yield quite large differences in smoke emissions (e.g., Hyer et al., 2013; Pan et al., 2020). To mitigate the lack of observation of the full diurnal cycle, prescribed diurnal cycles are often applied on MODIS fire information to yield hourly BB emission. Sometimes thick cloud cover or plume prevents detections of fire hotspots from space. Additional artifacts occur from day-to-day shifts in the orbital pattern of the polar-orbiting satellites and reduce data coverage in the tropics. The potential of geostationary fire information to improve BB emission systems is widely recognized, and several groups are testing methods of integrating geostationary observations into operational global aerosol models (FLAMBE used geostationary satellite data for North and South America from 2007 to 2017). Because these models make predictions several days into the future, it is necessary to estimate the future behavior of fires. Although there is significant research on dynamic prediction of fires (Mallia & Kochanski, Chapter 8 this volume.), the global models of ICAP continue to use an assumption of persistence, leading to misrepresentation of emissions for days with dramatic fire evolution, such as illustrated in section 9.2.7.

9.2.4. Smoke Injection Height

Another challenge of modeling BB emission is emitting smoke at the correct altitudes (i.e., plume injection height), which influences how fire emissions are ultimately transported and the life cycle of smoke. Some ICAP models simply assume emissions within the planetary boundary layer (PBL) based on observational statistics, while others use plume rise models that consider the environmental status. Despite available studies on development of plume rise models (e.g., review by Paugam et al., 2016), there are unknowns in fire science (e.g., whether the fire activity and the plume dynamics are linked) and uncertainties in characterizing the overall environment. Also, none of the current ICAP models have explicit treatment of fire-induced convection or pyroconvection, which can result in smoke lofting high in the troposphere or even well into the stratosphere in extreme cases (Fromm et al., 2010; Peterson et al., 2018). Smoke injection height is strongly influenced by fire diurnal cycles (Roberts et al., 2009). Lower nocturnal injection heights are often observed because of weakened fire intensity, stable atmosphere, and lower PBL at nighttime (Sofiev et al., 2013). However intense and fast-evolving fires introduce large variabilities in diurnal cycle of plume rise, which is very challenging to represent in models.

9.2.5. Aerosol Data Assimilation

To increase the accuracy of aerosol forecasts, several centers have used data assimilation (DA) of satellite and/or ground-based observations of aerosol optical depth (AOD), which is the most widely available and evaluated aerosol parameter. Examples include 2D variational assimilation (2D var; Zhang et al., 2008), 3D var (Randles et al., 2017), 4D var (Benedetti et al., 2009), and Ensemble Kalman filter (e.g. Khade et al., 2013; Pagowski & Grell, 2012; Rubin et al., 2017; Schutgens et al., 2010; Schwartz et al., 2014). These models often show differences in assimilation methods and assimilated AOD observations, including treatments of observations before assimilation (e.g., quality control, bias correction, aggregation, sampling). These differences occur despite consistent use of data from the MODIS (Remer et al., 2005) across all models. Work is in process to include Visible Infrared Imaging Radiometer Suite (VIIRS; Sayer et al., 2018) in some models' daily global spatial coverage. Since satellite-retrieved AOD is a column-integrated observation, aerosol speciation and vertical distribution are not constrained by assimilation of AOD. The conversion from AOD to 3D speciated aerosol concentration/mixing ratio, and vice versa, are represented differently in the different ICAP models, adding another layer of diversity across models that include data assimilation.

Assimilation of observations of plume extent and AOD can partly correct for emission deficiencies in operational models, but extreme wildfires pose unique challenges. Often satellite retrieval algorithms fail because thick plume is misclassified as cloud. Even with retrievals, the large difference between the model state before DA and the observation creates mathematical challenges for the data assimilation system. For some fires occurring in complex terrains, the relatively coarse spatial and vertical resolution of the global models may not be able to resolve fire evolution, and this can result in transport error. For extreme fire events, even the data assimilation models have large diversity in AOD distribution and magnitude (Xian et al., 2019). AOD assimilation also does not provide vertical constraints for aerosol plume. With the uncertainties in emission height in models, assimilation of lidar-observed vertical extinction profile (e.g., from CALIOP) seemed promising, and many centers put efforts into assimilating CALIOP backscatter or extinction profiles (e.g., Cheng et al., 2019; Sekiyama et al., 2010; Zhang et al., 2011). However, no center is currently assimilating it in its operational mode because of limited data coverage, data latency for operational use, and requirement of high data quality for DA purposes.

9.2.6. ICAP Multimodel Ensemble (MME) and Its Application

The motivation for developing the ICAP-MME is based on NWP studies that have shown the usefulness of ensemble-based predictions in understanding systematic errors that arise from the imperfect nature of models and the sensitivity of models to initial conditions. ICAP-MME is a consensus-style ensemble, in which all models are weighted equally. Current ICAP-MME products include analysis and 120 hr forecast of daily global speciated AOD consensus mean and spread. The smoke product is applied in the WMO Vegetation Fire and Smoke Pollution Warning and Advisory System. ICAP-MME data have demonstrated their value as a reference data set for research and are used by many centers for their internal model evaluations. Evaluations of ICAP-MME performance in terms of fine, coarse, and total AODs show that ICAP-MME forecasts are statistically better than any individual component model overall (Sessions et al., 2015; Xian et al., 2019). The performance of ICAP-MME is relatively stable and reliable over time, even as the component models undergo significant changes. For example, AOD RMSE of the ICAP-MME is not always the lowest for a site or year or scenario, but it is relatively low and stable, unlike any of its members. Consensus MME wins in the long run because of its averaging of independent models. For probabilistic forecasting, the leading predictors for AOD forecast error are the consensus mean and spread (Xian et al., 2019). Preliminary verification of ICAP-MME using surface $PM_{2.5}$ and PM_{10} measurements shows that ICAP-MME is the top performer among all models as well, despite more challenges in surface PM and larger divergence among the models compared with AOD simulations. Extreme smoke events remain a challenge to all models for the reasons discussed.

9.2.7. 2020 Western United States Fires as an Example of Challenges

The 2020 fire season in the western United States was an excellent example of extreme events that pose a variety of challenges to operational forecast systems. The first regional event occurred in association with a mostly dry lightning subtropical disturbance on the early morning of 16 August 2020. Due in part to the high number of initiated fires, suppression crews were overwhelmed and the fires grew rapidly. With meteorology conducive for burning, the smoke spanned a total of 2,000 km a day later. In this scenario, the operational models are constantly playing catch up with emissions. This situation demonstrates the pressing need to incorporate high-fidelity prognostic emission modeling into large-scale systems. Fire prevalence and strength further intensified into 12 September 2020, leading to thick smoke cover over the West Coast and eastern Pacific Ocean (Fig. 9.1a), with AOD at AERONET sites in California at unprecedentedly high levels (>10). The extreme AODs resulted in retrieval failures of the densest smoke (Fig. 9.1b). It is not surprising that smoke simulations varied highly between centers, even after AOD assimilation (Fig. 9.1c). Owing to evolving fire characteristics and PyroCB development, plume injection heights were highly variable from near surface to 12 km, with mid- to upper-level smoke misclassified as cirrus (e.g., CALIOP, Fig. 9.1d), with model misrepresentation leading to advection errors. The UV-based aerosol index, a mainstay of significant upper troposphere/lower stratosphere (UTLS) biomass burning event monitoring, is nevertheless semiquantitative in regard to assimilation, because of interdependencies on underlying clouds, single scattering albedo, and height (e.g., OMI Aerosol index, Fig. 9.1e, Zhang et al., 2021). Despite all these challenges, the ICAP-MME was useful in forecasting smoke distribution (Fig. 9.1f, g) and issuing warnings (Fig. 9.1h) with forecast uncertainty estimates (Fig. 9.1i) in such an extreme event.

9.2.8. Future Plans

Individual ICAP contributing centers have their own plans for future aerosol model developments, with each focus depending on customer needs and current model status. These plans may include separating BB smoke components to permit improved data assimilation and analysis, update of BB emission systems, improved treatments of plume rise, addition/update of aerosol data assimilation, increased model resolution, and improved parameterization of physical, chemical processes, and/or optical properties. The next steps for ICAP-MME are to develop quasi-operational surface concentration and PM ensemble forecast and increase horizontal resolution. Future advances in fire science, observation, and representation in global atmospheric constituent models are expected to improve the accuracy of smoke and air quality forecasts from the component global aerosol models as well as the ICAP-MME.

9.3. THE COPERNICUS ATMOSPHERE MONITORING SERVICE (CAMS): GLOBAL AND REGIONAL SYSTEMS OF EUROPE

The Copernicus Atmosphere Monitoring Service (CAMS) is a service of the European Union's Copernicus program that uses satellite Earth observations, *in situ* (nonsatellite) data, and modeling to provide information about the composition of the atmosphere at both the global and the European scale. CAMS is implemented by the European Centre for Medium-Range Weather Forecasts (ECMWF) on behalf of the European Commission. The CAMS forecasts are implemented and carried out with the same operational commitment as the ECMWF NWP forecasts to ensure a reliable provision of the atmospheric composition forecasts.

9.3.1. The Global CAMS Modeling and Data Assimilation System

CAMS provides forecasts of atmospheric composition at the global scale using the ECMWF integrated forecasting system (IFS) and at the European scale based on an ensemble of 11 regional air quality models. Satellite retrievals of ozone, nitrogen dioxide, aerosol optical depth, and carbon monoxide, as well as volcanic SO_2, are assimilated by the global system to increase the realism of the initial conditions of the forecast. The global CAMS forecast system uses the ECWMF's numerical weather prediction system at a resolution of 40 km. The atmospheric composition processes are simulated online to allow a tight coupling between weather and atmospheric composition as well as using the same 4D variational data assimilation (4D-VAR) algorithm for the assimilation of satellite retrievals of weather and atmospheric composition. The IFS in CAMS configuration uses the CB05 chemistry scheme and the AER aerosol scheme (Flemming et al., 2015; Rémy et al., 2019).

Primary organic matter (OM) and black carbon (BC) originating from both biomass burning and anthropogenic sources are represented as hydrophilic and hydrophobic fractions emitted at the same rate. The ageing of OM and BC is simulated by the conversion of the hydrophobic to the hydrophilic fraction using a lifetime of 1.2 days (Rémy et al., 2019). Data assimilation of aerosol optical depth retrievals from MODIS instruments on Terra and Aqua and the Polar Multi-Sensor Aerosol product (PMAp) product is applied to improve the realism of the initial conditions of the aerosol forecast. Because AOD observations do not provide information about the aerosol speciation and vertical profiles, all aerosol components (sea salt, desert dust, OM, BC, sulphates, nitrates, and ammonia) are modified by the same fraction to match the assimilated AOD observations (Benedetti et al., 2009). This means that the AOD assimilation cannot create fire plumes, but can only increase or decrease plumes already simulated by the model.

9.3.2. The Regional CAMS System

The regional air quality production of the CAMS is based on an ensemble of 11 state-of-the-art numerical

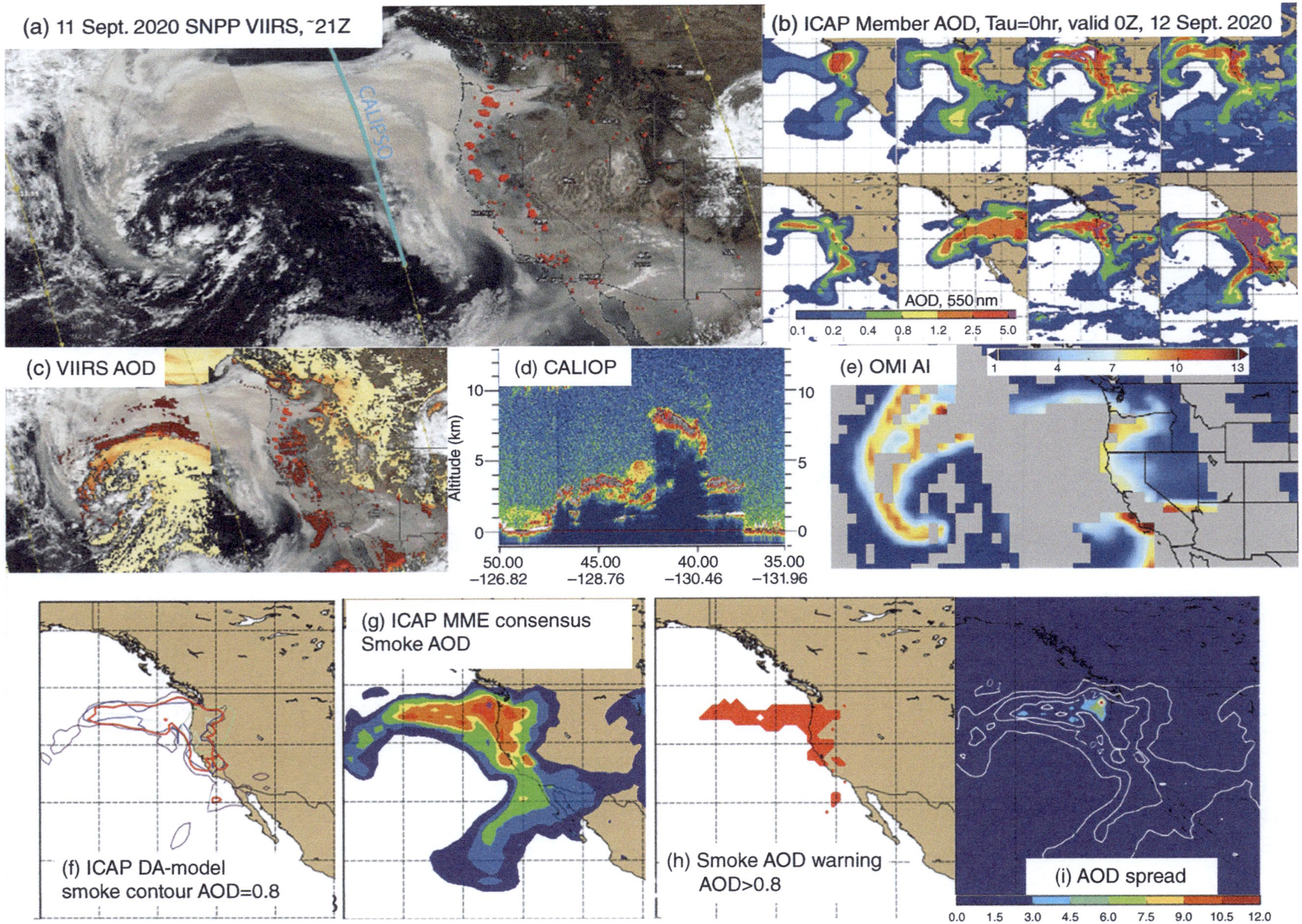

Figure 9.1 Mosaic of images and data products associated with 11 September 2020 smoke event from the western United States: (a) SNPP VIIRS RGB (NASA); (b) VIIRS AOD showing observation extent (NASA Worldview); (c) 12 September 2020 0Z ICAP member model analyses of smoke AOD; (d) CALIOP expedited browse backscatter cross-section (mapped to blue track on (a)); (e) OMI Aerosol Index highest quality assurance; (f) ICAP DA-model smoke AOD contours at 0.8; (g) ICAP DA model consensus; (h) high smoke AOD (> 0.8) warning; and (i) ICAP smoke AOD spread.

chemical transport models developed in Europe: (1) CHIMERE from INERIS (France) , (2) EMEP from MET Norway (Norway), (3) EURAD-IM from Jülich IEK (Germany), (4) LOTOS-EUROS from KNMI and TNO (Netherlands), (5) MATCH from SMHI (Sweden), (6) MOCAGE from METEO-FRANCE (France), (7) SILAM from FMI (Finland), (8) DEHM from AARHUS UNIVERSITY (Denmark), (9) GEM-AQ from IEP-NRI (Poland), (10) MONARCH from BSC (Spain), and (11) MINNI from ENEA (Italy).

The models of the regional ensemble use the same anthropogenic and fire emissions and meteorological driver data. The meteorological driver data are obtained from the global CAMS system, and the same fire emissions are used in the global system and the regional ensemble. The regional models differ substantially in their chemical and aerosol schemes, deposition schemes, and the simulation of transport, diffusion, and planetary boundary layer (PBL) processes (Marécal et al., 2015). The regional models apply data assimilation of surface air quality (AQ) observation of the European AQ network for the initialization of their forecasts. The boundary conditions of the regional models are taken from the global CAMS system to take into account long-range transport of desert dust, fire smoke plumes, and anthropogenic sources outside Europe. The forecasts of the individual regional models as well as an ensemble-mean product are presented to users. The spread of the ensemble provides information about the variable uncertainty of the regional forecast due to the aspects that vary between the regional models.

9.3.3. Global Fire Assimilation System (GFAS) Fire Emissions

The emissions from biomass burning are derived from satellite observation by the Global Fire Assimilation System (GFAS) in a timely manner and are used both in the global and regional systems. GFAS assimilates fire radiative power (FRP) retrievals from satellite-based sensors to produce daily estimates of biomass burning emissions (Kaiser et al., 2012). These daily fire emission estimates are persisted forward in the modeling systems. FRP observations currently assimilated in GFAS are the NASA Terra MODIS and Aqua MODIS active fire products (http://modis-fire.umd.edu/). GFAS data products are FRP, dry matter consumed, and biomass burning emissions for a large range of chemical and aerosol species. GFAS simulates fire injection heights, based on fire intensity and meteorological conditions using the plume rise model of Freitas et al. (2010). The GFAS data are available globally on a regular latitude-longitude grid at a horizontal resolution of 0.1 degrees from 2003 to the present. Figure 9.2 shows GFAS FRP from fires in Portugal in September 2020 (top left panel) and the impacts of those fires on AOD and near-surface $PM_{2.5}$ concentrations across Europe from the CAMS system.

9.3.4. Data Access and Presentation

All CAMS forecast and reanalysis data, GFAS fire emission, and anthropogenic emissions are available from the CAMS atmosphere data store (https://ads.atmosphere.copernicus.eu). User support is provided for queries about scientific and technical aspects of the CAMS data. The CAMS global and regional forecast systems do not consider smoke pollution from fires in isolation. Rather, they provide that information to users integrated in atmospheric composition and air quality products. The fire signal is especially strong in the organic matter and black carbon aerosol, as well as in the carbon monoxide data. Further, the regional forecast systems distribute a specific "PM_{10}, wildfires only" product to support the attribution of the forecast PM_{10} to source processes. The CAMS global and regional model and data assimilation systems and the input data are updated at irregular intervals about every 6 to 18 months. The changes introduced by the updates are documented on the CAMS website. The CAMS global forecasts contribute to community efforts such as ICAP (section 9.2), and the World Meteorological Organization (WMO) efforts described in section 9.6. The CAMS forecast is presented at www.windy.com for a wide range of users.

9.3.5. Challenges and Future Plans

The challenges to derive accurate fire emission data, such as the correct retrievals of FRP, and the wide uncertainty of both vegetation distribution and the biome-specific and fire-type specific emission factors, affect the performance of the global and regional CAMS air quality forecast. Of concern for smoke forecasting is the underestimation of the GFAS organic matter and black carbon fire emissions, which is compensated by applying a global correction factor of 3.4 (Kaiser et al., 2012). Although other global systems have to apply similar corrections to the OM and BC fire emissions, more research and optimization is needed to reduce the amount of the correction, for example to better understand the role of secondary aerosol formation from fire emissions.

A specific challenge of the forecast application is the prediction of the fires over the forecast period. The CAMS global and regional systems assume persistency of the GFAS fire emissions, which are derived from satellite observations from the day before the forecast start time. Progress has been made to link the weather-fire index forecasts to fire emission, especially to fill gaps based on previous FRP retrievals (e.g., Di Giuseppe et al., 2018). GFAS's next version upgrade (1.4)

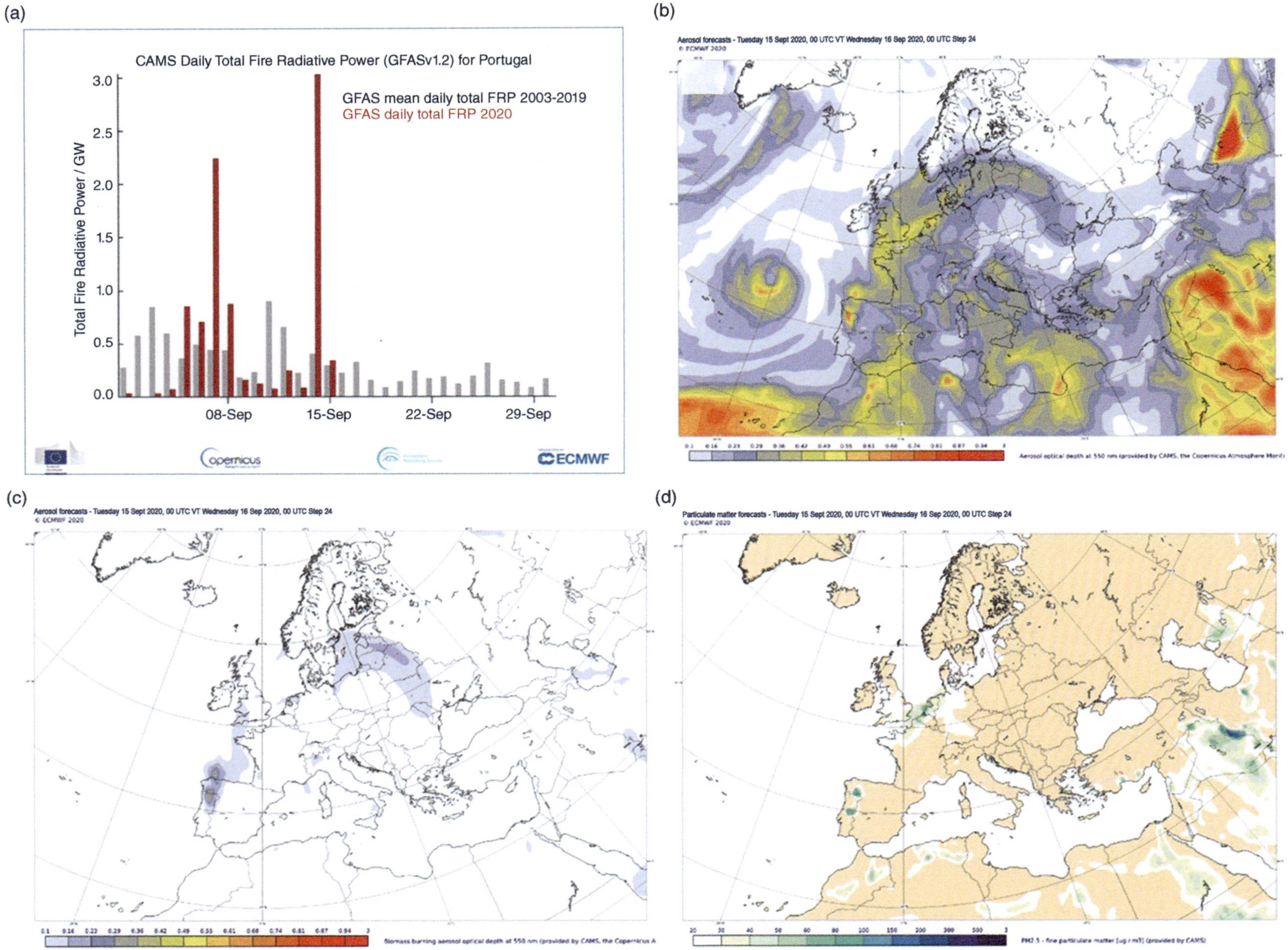

Figure 9.2 (a) Intensive wildfires in Portugal in September 2020 led to increased AOD and $PM_{2.5}$ values. Red: GFAS total radiative fire (MODIS) values in Portugal in September 2020; grey: daily mean values for 2003–2019. The next three panels show the 24 hr forecast over Europe for 16 September 2020: (b) global CAMS system for AOD, (c) biomass burning AOD, and (d) total $PM_{2.5}$.

comprises two improvements: an increase in the temporal resolution to hourly data and an updated approach of the injection height calculation (Sofiev et al., 2012). However, the correct prediction of the duration of observed fires and the ignition of new fires remains an active area of study.

Wildfires, especially in boreal regions, are often smaller than the current horizontal resolution for the global (40 km) or regional (10 km) grid box scales as well as the GFAS resolution of 0.1°. Better parameters are needed to assess the effect of subgrid scale processes on plume transport, convection, and chemical conversion.

The effect of smoke and other aerosols on weather has been successfully demonstrated with the global CAMS system. For example, smoke plumes from the intensive wildfires in the western United States in 2020 led to local temperature reduction by 1–3 K because of the reduced radiation at the surface and reduced the bias of the 2 m temperature forecasts. However, more research and evaluation are needed before this forecasted aerosol data could improve high-resolution NWP applications in a computationally affordable way.

9.4. NORTH AMERICAN SYSTEMS

Profiled in this section are seven smoke prediction systems developed in the United States and Canada by operational agencies and by universities and research groups involved in smoke and fire (Fig. 9.3). These systems operate at scales ranging from 1 to 12 km, offering forecasts that extend 24 to 84 hours into the future (Table 9.2). The following sections are organized in approximately the system development timeline that also reflects system interconnections and collaborations. First is the BlueSky system, conceptualized in 2001 in the northwestern United States when land managers and air quality agency personnel had the genesis of an idea: to build a system that relied on the many years of land manager research into forest and rangeland fuels and combustion to calculate fire emissions and input them into readily available air quality models. These fire emissions calculations and codes were then shared with the AIRPACT and NOAA NAQFC systems operating full chemical transport models. BlueSky was then ported to Canada and its development and operations were customized by the University of British Columbia (UBC). Continued years of large wildfire and smoke impacts prompted the Environment and Climate Change Canada (ECCC) to implement its operational FireWork system, building in part on the experience of NOAA NAQFC and Canadian fuel consumption research. Advancements in chemical transport modeling and computing power led to the NOAA HRRR-Smoke system, which relies on satellite fire emission estimation techniques and a coupled WRF-Chem system. Finally, the HiRes system in the southeastern United States is tackling the challenge of prescribed burning smoke impacts using the CMAQ air quality modeling system. This section gives an overview of each of these systems, and Table 9.2 summarizes many of their parameters.

9.4.1. U.S. BlueSky Smoke Modeling Framework and Smoke Prediction System

The BlueSky Smoke Modeling Framework (Larkin et al., 2009), first released in 2002, was developed by the U.S. Forest Service Research, in conjunction with partners, to compute fire emissions and smoke impacts. BlueSky's original motivation was the need to better predict potential prescribed fire smoke impacts in order to mitigate or avoid them prior to ignition. However, BlueSky has evolved and is now used in both research and in operational contexts for both prescribed burns and wildfires, and it is incorporated into multiple operational daily forecasts and tools.

One unique aspect of BlueSky is that it is open source and modularly combines databases and models across seven modeling steps: fire information, available fuels, consumption, emission factors, time release, plumes, and atmospheric dispersion and chemistry, providing specific application programming interfaces (APIs) for incorporation of new modules. This allows for the addition of newer models and databases as they become available. Because BlueSky is a framework that incorporates multiple model options at each step, caveats and assumptions are pathway dependent. The largest assumptions for fire concern the ability to correctly identify burning areas, the amount of fuel available; for smoke impacts, key assumptions concern the timing of the emission throughout the day and their plume injection heights (e.g., Drury et al., 2014; Larkin et al., 2014; Larkin et al., 2012; Prichard et al., 2019; Raffuse et al., 2012). For retrospective analyses when assumptions can be better specified, the system performs with Pearson correlation of about 0.65 and a positive bias of 7–9 μg/m^3. Data assimilation improves these results (O'Neill et al., 2021; Zou et al., 2019) but with a negative bias tendency (O'Neill et al., 2021). Analyses such as these helped justify the need for new coherent multifaceted observational campaigns to help advance the state of science in these areas, such as the Fire and Smoke Model Evaluation Experiment (Brown et al., 2014; Prichard et al., 2019).

The identification of burning areas has been a large source of development, with various approaches that combine satellite data and ground reports in both real-time and retrospective ways, including SmartFire v1 and v2 (Raffuse et al., 2009), and the newer Fire Information

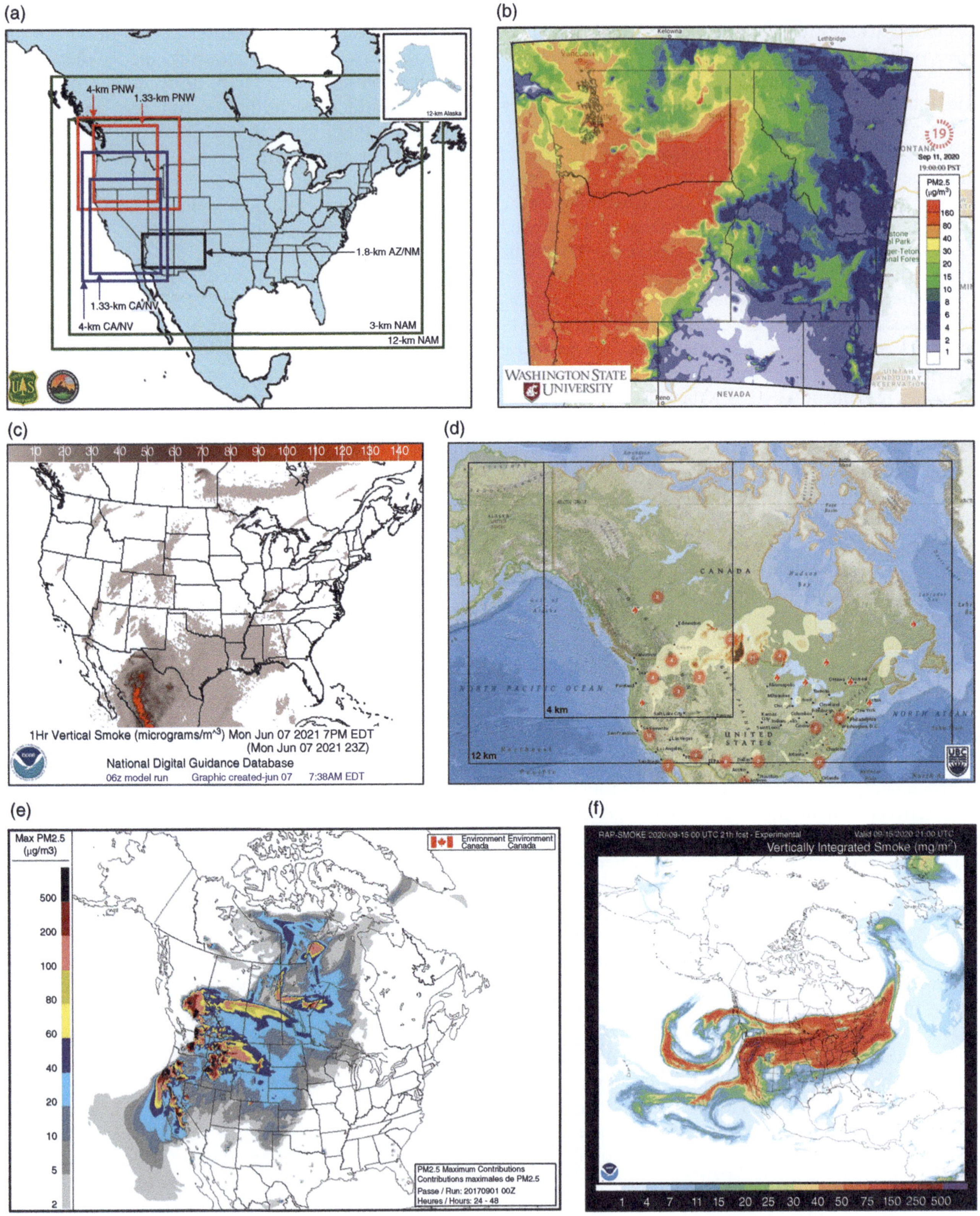

Figure 9.3 Domains of the north American smoke and air quality prediction systems summarized in this section. (a) U.S. BlueSky smoke modeling framework and smoke prediction system, (b) Air Indicator Report for Public Access and Community Tracking (AIRPACT) system, (c) U.S. National Air Quality Forecast Capability (NAQFC) system, (d) BlueSky Canada smoke forecasting system and Canadian wildland fire information system, (e) FireWork Canada, (f) U.S. NOAA operational Rapid Refresh-Smoke (RAP-Smoke) system, (g) High-Resolution Rapid Refresh-Smoke (HRRR-Smoke) System, (h) southeastern U.S. RiRes system.

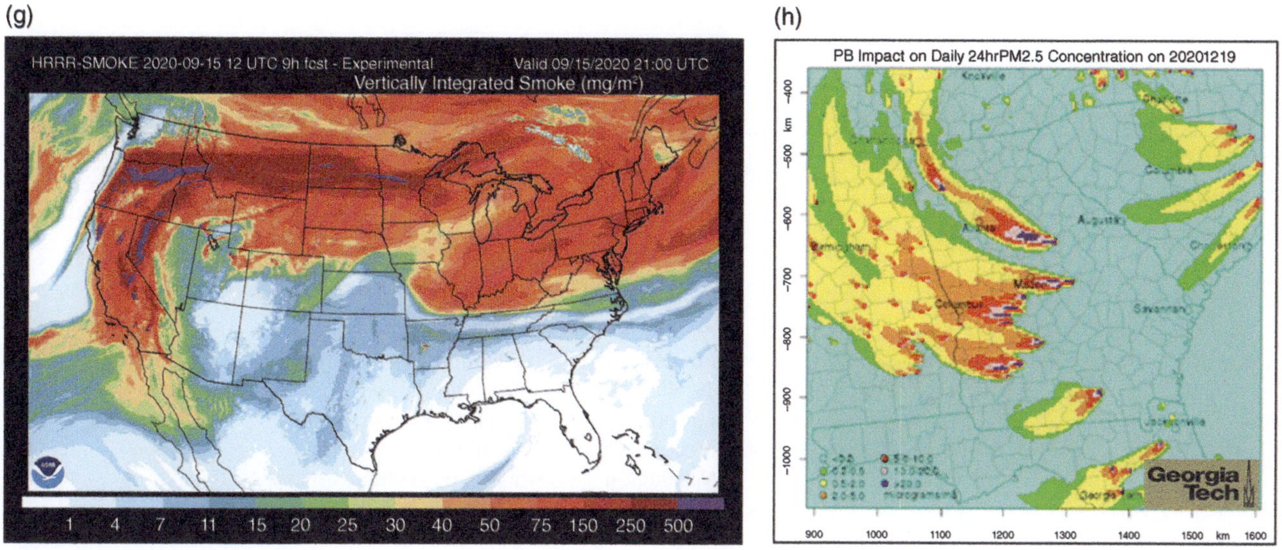

Figure 9.3 (Continued)

System (FIS; Marsha & Larkin, 2022). FIS aggregates fire detection information from the MODIS, VIIRS, and GOES-16 ABI instruments to create air quality model-ready fire activity data sets; these account for redundant fire identifications from the multiple sources, false positive detections from anthropogenic sources, and dropouts when smoke obscures the satellite ability to see the fire. FIS includes a statistical analysis of multiple years of satellite fire detections from the individual satellite instrument data streams to estimate a fire size per detection. Fire radiative power (FRP) is carried through the data stream and also aggregated to an independent grid, both of which can be used in downstream fire emission and plume rise calculations.

BlueSky has multiple operational applications. It is used as an emission processor for the biomass burning component of the U.S. National Emissions Inventory (Larkin et al., 2020). It is also used in Web tools such as BlueSky Playground that allow fire managers to customize and interactively calculate fire emissions and smoke impacts for specific fires of concern (both prescribed burns and wildfires). This functionality is also automated in some prescribed fire reporting systems. BlueSky is also incorporated into multiple operational smoke forecasting systems with different types of fire information and purposes.

The U.S. Forest Service uses BlueSky to perform a number of daily runs that focus on the needs of incident support. Over 30 model runs are performed daily using a variety of meteorological model forecasting domains across the country. Some of these domains are large-scale (e.g., from the U.S. National Weather Service's Global Forecast System), but the emphasis is on high-resolution forecasting at typical scales of around 1.33 km grid resolution. Meteorological forecasts for these high-resolution runs come from a variety of partners, including regional modeling consortia such as the Northwest Regional Modeling Consortium and the California and Nevada Smoke and Air Committee (CANSAC). These forecasts are supplemented during heavy fire and smoke periods by movable high-resolution (1.27 km) fire weather domains managed by the U.S. National Weather Service to cover affected areas.

All of these runs, as well as those from other systems, are used by deployed smoke specialists from the U.S. Interagency Wildland Fire Air Quality Response Program (IWFAQRP; Lahm & Larkin, 2020) to develop smoke outlooks that provide consistent messaging and forecasting in affected areas. These smoke specialists use the model runs to develop these tailored forecasts, which are then used by the fire, health, and air quality agencies, and the public. A unique aspect of this system is that the smoke specialists can affect the BlueSky Daily Runs by obtaining ground information (e.g., fuel types, loadings, and fuel moistures) observed by fire personnel, as well as fire growth projections done in support of the incident; this information can then be used within the context of the daily runs, creating customized smoke model predictions. These customized runs, the standard daily runs, and other BlueSky-related tools, such as the U.S. BlueSky Playground, are available through the IWFAQRP tools page (https://wildlandfiresmoke.net/).

9.4.2. Air Indicator Report for Public Access and Community Tracking (AIRPACT)

The Air Indicator Report for Public Access and Community Tracking (AIRPACT) project (http://lar.wsu.edu/airpact), begun in 2000, built upon air pollution

Table 9.2 North American smoke prediction systems

Forecast system	Operational/ research center	Smoke products	Domain	Meteorology model	Grid spacing	Forecast frequency and length
BlueSky US	USDA-FS	Ground-level $PM_{2.5}$, visibility	Various: North America, US, regional	WRF (offline) source: NWS, UW, DRI, UofA	Various: 1km to 12 km	Various: 00Z, 06Z, 12Z, 18Z Various: 36 hr to 84 hr
AIRPACT5	WSU, NW-AIRQUEST	Ground-level: $PM_{2.5}$, O_3, NOx, CO, visibility, SO_2, NH_3, other VOCs	Northwestern US, including states of Idaho, Oregon, and Washington	UW WRF runs on 4 km grid (nested within 12 km and 36 km runs) initialized with GFS at 00Z. (offline)	4 km	Daily; 48 hr
NAQFC	NOAA	$PM_{2.5}$, OC, BC, O_3, NOx, CO, NH_3, AOD	Continental US	Various. Always driven by NWS operational weather models (offline)	12 km	Daily; 48 hr
BlueSky Canada	UBC	Ground-level $PM_{2.5}$	North America, west coast of North America	WRF	12 km; 4 km	12 km: 00Z, 06Z, 12Z, 18Z 4 km: 00Z 51 hr
FireWork / RAQDPS-FW v021	ECCC	Surface and vertically integrated $PM_{2.5}$ (public); O_3, NO_2, PM_{10}, NO, CO, surface visibility etc. (by request). Offline analysis for health effects.	North America, Canada	Global Environmental Multiscale model (GEM v5.0) (inline)	10 km	00Z, 12Z 72 hr
HRRR-Smoke	NOAA	Weather, $PM_{2.5}$ (ground-level and total column), surface visibility	Continental US and Alaska	High-Resolution Rapid Refresh (Inline)	3 km	Every hour a new forecast is produced. Forecast lengths are 48 hr every 6 hr, otherwise 18 hr.
HiRes2	Georgia Tech	O_3, $PM_{2.5}$, health effects	South-eastern US	WRF 3.8 with GFS 0.25° forecasts (offline)	4 km	Daily; 24 hr

Air quality/ dispersion model	Fire activity linformation	Fire emissions	Other sources	Plume rise	Website
HYSPLIT, CMAQ (no chemistry, smoke $PM_{2.5}$ as tracer)	Fire Information System (FIS): VIIRS (S-NPP and NOAA-20) and MODIS (Terra and Aqua), GOES-16 ABI. Persistence	BlueSky pipeline	None	FEPS-Briggs	https://tools.airfire.org
CMAQ v 5.0.2, CB05 chem, aero5 aerosol physics	USDA-FS Fire Information System (FIS). Persistence	BlueSky pipeline	NEI 2017, MOVES, Biogenic emissions from MEGAN 2.10, WACCM boundary conditions	DEASCO3 from WRAP with IDEQ parameter-ization for heat.	http://lar.wsu.edu/airpact/
CMAQ v 5.0.2/v5.3	HMS; VIIRS, MODIS and GOES-16. Persistence	BlueSky, GBBEPx	NEI2016v2; BEIS for biogenic, and FENGSHA for dust	Briggs (1969) and Sofiev et al. (2012)	https://airquality.weather.gov
HYSPLIT	AVHRR, MODIS, VIIRS (from CWFIS/SF2). Persistence	CWFIS/ BlueSky	None	FEPS-Briggs	https://firesmoke.ca
GEM-MACH v3.0	VIIRS (S-NPP and NOAA-20) and MODIS (Terra and Aqua), from the Canadian Wildland Fire Information System (CWFIS). Persistence (past 24 hr hotspots)	CWFIS/ CFFEPS	Emission set 3.1.2: APEI 2013 v1, 2017 projection of NEI 2011 v3, 2008 Mexico Biogenic BEIS v3.09	Canadian Forest Fire Emission Predic-tion System CFFEPS v2.06	Public: https://weather.gc.ca/firework Open data: https://eccc-msc.github.io/open-data/msc-data/nwp_raqdps-fw/readme_raqdps-fw_en
WRF-Chem, Smoke $PM_{2.5}$ tracer. No chemistry. Includes smoke feedback effects on radiation and surface visibility.	VIIRS (S-NPP and NOAA-20) and MODIS (Terra and Aqua). Persistence	FRP	None	Freitas (2007)	https://rapidrefresh.noaa.gov/hrrr/HRRRsmoke/ https://rapidrefresh.noaa.gov/hrrr/HRRR-AKsmoke/,
CMAQv5.0.2 with DDM	Statistical analysis: GA burn permit and GBBEP fire information in the US southeast for 2010–2016	Mapped fuels (FCCS), measured fuel moisture, emission factors	NEI-2011	Empiri-cal model of Liu (2014)	https://forecast.ce.gatech.edu & https://sipc.ce.gatech.edu/SIPFIS/map/

modeling experience at the Laboratory for Atmospheric Research (LAR) (Barna et al., 2000; O'Neill & Lamb, 2005; O'Neill et al., 2006; Richter et al., 2004; Snow et al., 2003) to create a forecasting system for the U.S. Pacific Northwest (Vaughan et al., 2004). In 2003, the Northwest International Air Quality Environmental Science and Technology (NW-AIRQUEST) Consortium was created to obtain scientific input for air quality management decisions in the Pacific Northwest, providing funding and agency-oriented guidance for the AIRPACT project (https://lar.wsu.edu/nw-airquest/engineers and air-quality scientists.

AIRPACT evolved with improvements to emissions inventories, modeling systems, and computing power, which allowed greater complexity, resolution, and domain extent as well as longer forecasts (Chen et al., 2008; Al Mahmud, 2005; Vaughan et al., 2004). AIRPACT version 5 (AIRPACT5) has a 4 km grid that covers Washington, Oregon, Idaho, and surrounding areas. It uses the Community Multiscale Air Quality (CMAQ; Byun & Schere, 2006) modeling system with the CB-05 chemical mechanism, a full emission inventory of natural and anthropogenic sources, and meteorology from the Weather Research and Forecast model (WRF v4.1.3; Skamarock, 2004) from the University of Washington (Mass et al., 2003; Ovens & Mass, 2020). Anthropogenic emissions are from the EPA 2017 National Emissions Inventory (NEI; Larkin et al., 2020). Biogenic emissions are from MEGAN 2.10 (Guenther et al., 2006), as parallelized by LAR for timely execution to meet forecasting time constraints. Dynamic boundary conditions are obtained from the Whole Atmosphere Community Climate Model (WACCM) daily forecast results (Gettelman et al., 2019). Initial conditions for each AIRPACT run are obtained from the previous run. Forecast length is 48 hr; model runs begin at ~22:00 local time and finish the first of the two forecast days by 07:00 local time the next morning.

AIRPACT5 gets wildfire emissions from the U.S. Forest Service Research fire information system (FIS) linked with the BlueSky system with three options: (1) *dropouts*, to include fires detected recently but not in the latest satellite imagery; (2) *persisted*, to project current fires forward in time; and (3) *mean area*, to specify fire area per fire location. Heat released is estimated from fire area and Fuel Characteristic Classification (FCCS; McKenzie et al., 2007) fuel type, then applied in a plume rise scheme based on the Deterministic and Empirical Assessment of Smoke's Contribution to Ozone Project (DEASCO3) algorithm (Moore et al., 2013) as parameterized by the Idaho Department of Environmental Quality (DEQ, 2017).

Wildfire smoke has recently become intense and common enough that WRF forecasts for AIRPACT5 showed surface temperature biases due to unaccounted-for aerosol feedbacks. To better capture the temperature effects of smoke, since August 2020 AIRPACT5 WRF runs use aerosol optical depth and the aerosol Angstrom exponent fields from NASA's GEOS system in the radiation scheme (Ovens & Mass, 2020).

For each monitoring site, the AIRPACT5 website includes monthly performance evaluation statistics and flexible charting for ozone and $PM_{2.5}$. Munson et al. (2021) documented the evolution of components and model versions for AIRPACT3, 4, and 5, and forecast skill for 2009–2018. For AIRPACT3, 4, and 5, respectively, $PM_{2.5}$ normalized mean bias was −25%, −31%, and −35%, and $PM_{2.5}$ normalized mean error was 31%, 35%, and 44%. The statistics trend downward (worsen) over time, likely reflecting progressively greater effects of wildfires over that decade.

All components of a deterministic smoke forecasting system contribute to forecast error, so LAR recently augmented AIRPACT's forecasting with two other approaches, bias correction and machine learning. AIRPACT5 24 hr average $PM_{2.5}$ biases are computed using a Kalman filter for all $PM_{2.5}$ monitoring sites, then interpolated using Kriging to create a continuous bias correction field for correction of the AIRPACT5 daily 24 hr $PM_{2.5}$ forecast (June et al., 2021). The Kalman filter bias correction shows improvements for both winter and wildfire month biases (about −50% ± 6% annualized), and the corrected results also had much smaller mean absolute errors (typically <20%). Machine learning applies a random forest classifier and multiple linear regression models, trained on recent history, to forecast O_3 and $PM_{2.5}$ at monitoring sites (Fan et al., 2023).

9.4.3. U.S. National Air Quality Forecast Capability (NAQFC)

The U.S. National Air Quality Forecast Capability (NAQFC) forecasts, operated by the National Weather Service (NWS), provide real-time prediction of ozone, fine particles ($PM_{2.5}$), smoke, and dust over the Continental United States (CONUS), Alaska (AK), and Hawaii (HI) (Huang et al., 2017; Lee et al., 2017; Pan et al., 2014; Pan et al., 2020; Tong et al., 2015). NAQFC is one of the key tools widely used by state and local agencies to protect the public from exposure to elevated levels of air pollutants. During wildfire events, accurate prediction of fire emissions, chemical transformation, and transport are critical for air quality forecasters to issue early warning to protect human health (Tong & Tang, 2018). The NAQFC system uses various versions of the Community Multiscale Air Quality (CMAQ) model (Byun & Schere, 2006) to provide next-two-day prediction of surface O_3 and $PM_{2.5}$ concentrations over all 50 U.S. states, using three regional model domains (CONUS, AK, HI). Inputs to the CMAQ model include meteorological data from NOAA's operational weather forecasting model, anthropogenic emissions from national emission inventories from the United States, Canada and Mexico, natural source emissions derived using satellite retrievals or emission model estimation, and lateral boundary conditions from global chemical transport models (Lee et al., 2017; Pan et al., 2020; Tang et al., 2021; Tong et al., 2015).

Several products of biomass burning have been used or are being tested under NAQFC to predict wildfire air quality (Li et al., 2020; Pan et al., 2020). Previously, the current operational forecasting was based on the BlueSky emission algorithm (Larkin et al., 2009) driven by fire detection from the NOAA Hazard Mapping System (HMS) (Lee et al., 2017; Pan et al., 2020). The HMS fire detection products used to be produced manually by a team of human operators. The operator-based fire screening has since been replaced by an automatic fire detection algorithm. This switch has sharply increased the false alarm rate (false fire detection), causing the NAQFC to temporarily remove biomass burning emissions to avoid widespread degradation of model performance for surface O_3 and $PM_{2.5}$. The team has assessed alternative fire emission products for large wildfire events during the 2018 Camp Fires in California (Li et al., 2020). In July 2021, the wildfire emissions in NAQFC were updated based on the Global Biomass Burning Emissions Product (GBBEPx; Zhang et al., 2012; Zhang et al., 2014), and plume rise algorithm based on Sofiev et al. (2012).

9.4.4. The BlueSky Canada Smoke Forecasting System and Canadian Wildland Fire Information System

The Weather Forecast Research Team (WFRT) at UBC in Vancouver, Canada, adapted the BlueSky smoke forecasting system (Larkin et al., 2009) for Canada (BlueSky Canada, BSC) and has operated it since 2010. BSC uses inputs of fire information and meteorology to estimate fire emissions, initial smoke plume-rise, and subsequent smoke dispersion. Meteorology is from the Weather Research and Forecast model (WRF v4.2.1; Powers et al., 2017; Skamarock et al., 2019), also operated by the WFRT, with three nested domains: 36- m grid spacing covering North America, 12 km over most of Canada and the United States, and 4 km over the complex terrain of western Canada. The hourly evolution of emissions for each fire, initial plume rise, and meteorology is input to the HYSPLIT dispersion model (Draxler, 1999; Draxler & Hess, 1997, 1998; Stein et al., 2015). HYSPLIT is run with two domains: at a 0.1° (~11 km) grid spacing covering North America using the 12 km meteorology, and at a 0.05° (~5 km) grid spacing covering western Canada using the 4 km meteorology. The smoke forecasts have a 51 hr forecast horizon. BSC runs four times daily for the larger domain (initialized with meteorology spaced 6 hr apart) and once a day for the smaller domain (Table 9.2).

The Canadian Wildland Fire Information System (CWFIS; NRCan, 2020) provides fire location and fuel consumption to BSC. It relies on satellite data from the Advanced Very High-Resolution Radiometer (AVHRR), the Moderate Resolution Imaging Spectroradiometer (MODIS), and the Visible Infrared Imaging Radiometer Suite (VIIRS) instruments. Hotspots are sometimes obscured from the view of satellites by clouds or even smoke from fires, or they may be too small to be detected. Thus, the CWFIS also includes ground reports of fire locations. The CWFIS produces fire weather and fire behavior products and comprises a number of subsystems, two of which BSC depends upon:

1. Fire Monitoring, Mapping, and Modeling (Fire M3) system. Actively burning fires are detected by infrared satellite imagery and used to estimate daily and annual area burned, as well as to model fire behavior and forecast smoke.
2. Fire Behavior Prediction (FBP) system. Fire behavior parameters such as rate of spread, fuel consumption, and fire intensity are predicted from inputs of fire weather indices, along with fuel types, topography, and other factors.

During the fire season, Fire M3 is run four times daily to generate updated fire location data. Consumption information is obtained from the FBP system. For ease of modeling and to prevent ingesting duplicate fires, the raw fire information runs through the Satellite Mapping Automated Reanalysis Tool for Fire Incident Reconciliation (SMARTFIRE), version 2 (SF2). SF2 associates fire locations into clusters to minimize duplicates from multiple satellite data sources, links them with known fire events, converts the cluster to point data, and assigns an area burned to the points (Raffuse et al., 2009). The output from SF2 is a CSV file that includes SF2 fire perimeters, locations, and fire fuel consumption, all of which is input to BSC. Each time BSC is run, it retrieves the most recent and up-to-date SF2 data.

The fire growth module within BSC uses a simple persistence model, in which the area growth on the first day also occurs on the next two days. Each day of the forecast, BSC assumes that all fires grow linearly between 9:00 and 21:00 local time. Knowing the area growth for that period, and the total fuel consumption for that area, fire emissions are estimated using the Fire Emission Production Simulator (FEPS; Anderson et al., 2004), which gives hourly measures of $PM_{2.5}$ emissions and heat release. FEPS also computes the initial smoke plume rise, based on modeled heat release using the Brigg's plume-rise model (Briggs, 1972, 1982).

The BSC output of smoke $PM_{2.5}$ concentrations at 10 m above ground level is available at a public website (https://firesmoke.ca/) and in several formats. The most popular is the animated Leaflet display, widely used by provincial government health and environmental agencies to issue health warnings and advisories, emergency responders, Canadian residents and tourists, and the media. The outputs are also available as three kinds of files: (1) KMZ files, which can be loaded into Google Earth and compared with satellite images; (2) geoserver files for input by provincial agencies to use in their fire-data management systems; and (3) netCDF files for scientific analysis. A near-real-time quantitative verification system helps identify major issues, such as missing fires. Future plans include a new fire weather index (FWI) and FBP system, outputting column-integrated smoke, including the process of wet deposition of smoke during dispersion, and implementing a new plume-rise model (Moisseeva & Stull, 2021).

9.4.5. FireWork Canada

Boreal forests, which compose approximately 75% of Canada's landmass, frequently experience biomass burning as part of the natural ecosystem life cycle. However, with the signature of a changing climate, these fires are becoming larger and more intense, and are occurring closer to the wildland urban interface (Tymstra et al., 2020; Wotton et al., 2017). ECCC meteorologists responsible for issuing the air quality forecast under the Air Quality Health Index program (AQHI; Stieb et al., 2008) needed better numerical guidance to quantify and predict the contribution of biomass burning to $PM_{2.5}$ concentrations. There was also a growing expectation for the Meteorological Service of Canada (MSC) to deliver smoke dispersion products similar to those successfully produced by the BlueSky Canada initiative (section 4.4). Since 2016, the MSC, part of ECCC, has operated a regional air quality deterministic prediction system with fire emissions, known as FireWork (Chen et al., 2019; Pavlovic et al., 2016).

The development of FireWork took full advantage of the MSC's operational NWP supercomputing environment and leveraged existing air quality forecast model infrastructure. FireWork uses the Canadian GEM-MACH chemical transport model (Moran et al., 2018; Moran et al., 2019; Moran et al., 2014), coupled online with the GEM numerical weather prediction model (Côté et al., 1998; Girard et al., 2014). Its codebase and configuration are synced with those used for the GEM-driven Regional Deterministic Prediction System, so FireWork inherits regular innovations to the dynamics and physics components, as well as weather data assimilation. Here we describe version 2.0 of FireWork, implemented in operations in 2019 as the Regional Air Quality Deterministic Prediction System with Near-Real-Time Wildfire Emissions (RAQDPS-FW) version 021 (Chen et al., 2019).

Similar to WFRT's BSC system described previously (section 9.4.4), FireWork relies on Canadian Forest Service's (CFS) CWFIS for near-real-time input of fire location, fuel parameters, and estimated daily burn areas. However, unlike BSC, FireWork implements the Canadian Forest Fire Emissions Prediction System (CFFEPS) for near-real-time calculations of fire emissions and plume injection heights for input to the GEM-MACH chemical transport model. CFFEPS was developed jointly by ECCC and the CFS, NRCan, and uses a bottom-up, process-based approach to compute hourly wildfire emissions. Biomass fuel consumption is estimated with hourly input forecast meteorology, and surface fuel parameters. Crown and surface fuel consumptions are calculated by the Canadian Forest Fire Behavior prediction system (FBP; Wotton et al., 2009) for flaming, smoldering, and residual combustion fractions. These quantities are converted to hourly emissions with emission factors by combustion processes. Additionally, smoke injection height and emission vertical distribution are parameterized based on fire-energy thermodynamics and forecast temperature profiles at individual fire hotspot locations. Emissions enter the GEM-MACH air quality model as major point sources, with treatment for chemical speciation differentiated for smoldering and flaming combustion phases. Wildfire emissions are added to other anthropogenic and biogenic emissions to provide a complete picture of air pollutant sources across the North American domain.

To capture all significant emission sources that affect Canada, FireWork is implemented on a continental domain that includes all of the contiguous United States and most of Canada and Alaska. A twin system is run without fire emissions, which allows a first-order approximation of the attribution of forecast pollution to wildfire sources through the calculation of differences in forecast concentrations. FireWork inherits the multiphase, multi-pollutant chemistry from GEM-MACH, which includes gas-phase, aqueous-phase, and heterogeneous chemistry; aerosol processes; secondary organic aerosol chemistry; and deposition processes. This enables FireWork to capture the evolution and impact of wildfire pollution over both regional and continental scales.

Twice a day (00 and 12 UTC), the system produces a North America-wide, 72 hr model forecast of air pollution, with $PM_{2.5}$ surface and total column concentrations, as well as surface NO_2 and O_3. The simulation results provide numerical guidance on regional air quality conditions to MSC forecasters in regional offices. The forecasters issue public air quality forecasts in the form of the Air Quality Health Index (Stieb et al., 2008) as well as smoke-related special weather bulletins. Products such as maps of extinction coefficient, and time series of visibility degradation in statute miles at airports are also becoming available (So et al., 2018).

FireWork is routinely evaluated with surface observations as part of system performance tracking, and fire emissions and plume injection heights are evaluated with available remote sensing measurements (Adams et al., 2019; Griffin et al., 2020). Additional applications also include using it in a long-term surface analysis of fire-$PM_{2.5}$ concentrations (Munoz-Alpizar et al., 2020). These hourly analyses provide a valuable option as a source of long-term exposure data to the health community. FireWork forecast products for the attribution of surface $PM_{2.5}$ concentrations to fires have also been used to assess the cost of fire smoke impacts on health in Canada (Matz et al., 2020).

Active research is underway toward coupling CFFEPS with GEM-MACH for closer linkages between fire behavior and meteorology, and accounting for the direct and indirect feedbacks between air quality and meteorology (Makar et al., 2020; Ye et al., 2021). These research studies are part of the planned improvements to FireWork operational implementation, which also include the introduction of dynamical interaction of plume aerosols with meteorology, hotspot aggregation methods, a data assimilation cycle for atmospheric constituents, and finer resolution domains. For the latest information on the operational FireWork system, please refer to the Meteorological Service of Canada open-data documentation at https://eccc-msc.github.io/open-data/.

9.4.6. U.S. NOAA Operational Rapid Refresh-Smoke (RAP-Smoke) and High-Resolution Rapid Refresh-Smoke (HRRR-Smoke) Forecast Systems

The U.S. NOAA Global Systems Laboratory (GSL) is one of the world leaders in development of numerical weather and air quality forecast models. NOAA/GSL developed the Rapid Refresh (RAP) and High-Resolution Rapid Refresh (HRRR) weather forecast systems (Benjamin et al., 2016). The RAP domain covers all of North and Central America (13.5 km domain resolution), and the HRRR grids cover the CONUS and Alaska domains (3 km domain resolution; Fig. 9.3f,g). An important feature of these models is their rapid update cycle. Every hour a new RAP and HRRR CONUS forecast starts by assimilating the latest meteorological observations, including the radar data to provide storm-scale weather forecasts to stakeholders. For HRRR-Alaska, the refresh cycle is 3 hr.

In 2016, NOAA/GSL began experimental smoke forecasting by implementing a smoke tracer (fine particulate matter from fires) to the HRRR model (HRRR-Smoke) over CONUS, then expanded this capability to the RAP and HRRR-Alaska models. In both RAP-Smoke and HRRR-Smoke models, the biomass burning (BB) emissions are estimated in real time by using the fire radiative power (FRP) data from four satellite platforms: two from VIIRS (S-NPP and NOAA-20) and two from MODIS (Aqua and Terra) satellite platforms (Ahmadov et al., 2017). The FRP data are also used to estimate the fire heat flux and fire size to simulate the plume rise of the fires in the models (Freitas et al., 2007). The advection, boundary layer and convective mixing, and dry and wet deposition processes for smoke are included. The RAP-Smoke model provides lateral boundary conditions of smoke to the HRRR domains in real time. Thus, HRRR-Smoke can capture smoke transported from the wildfires in Canada and Central America (RAP domain) into the CONUS (HRRR) domain. The simulated smoke tracer is cycled between the subsequent model forecasts.

Because the HRRR has a high spatial resolution (3 km), advanced dynamics, and storm-scale forecasting capabilities, it simulates the smoke transport in a convective environment and over the complex terrain reasonably well. The hourly refresh cycle of RAP/HRRR-Smoke helps to initialize newly started wildfires into the model after they are detected by the polar orbiting satellites with short delays.

The coupled framework of the RAP/HRRR-Smoke modeling systems enables simulation of the effect of smoke on solar radiation, which, for example, can cool near-surface temperature. This ability helps improve the weather forecasting skill of the RAP/HRRR models in regions affected by dense smoke. NOAA/GSL has shown significant improvement in air temperature and surface visibility forecasting during wildfire and smoke events that have recently occurred in the United States, particularly in the western regions.

After extensive testing and improvements, in December 2020 NOAA/GSL transitioned the smoke forecasting capabilities to NOAA's operational RAP/HRRR models. The hourly operational forecast cycles now produce multiple meteorological, smoke and visibility forecast products. Respectively, the RAP-Smoke and HRRR-Smoke forecasts go out to 21 and 18 hours, with forecast lead times of 51 and 48 hours four times a day. The operational RAP/HRRR-Smoke forecast models provide hourly 3D smoke forecast and diagnostic products (e.g., vertically integrated smoke) to users within the standard model output files in GRIB2 format. The near-surface smoke and forecast products are displayed on NOAA/GSL's public website (https://rapidrefresh.noaa.gov/hrrr/HRRRsmoke/).

Weather and air quality forecasters, air resource advisors, private sector forecasters, and the public all use the model forecasts, especially the fine-scale HRRR-Smoke forecasts over the CONUS domain. During the intense fire seasons in 2018 and 2020 (Fig. 9.3f,g), the HRRR-Smoke forecasts were a critical tool for numerous U.S. stakeholders. In addition to forecasts of smoke impacts on air quality and visibility, HRRR-Smoke forecasts are used to determine the impact of dense smoke on weather.

NOAA/GSL, in collaboration with other modeling and satellite teams, is working to transition the smoke forecasting capability to the next-generation regional NWP model based on the Finite-Volume Cubed-Sphere dynamical core. Besides the new model dynamics and physics schemes, other major improvements are planned to implement in the future FV3- based smoke forecasting model, such as ingesting the GOES-16/17 FRP data and improving the diurnal cycle of the BB emissions. A new hourly fire weather index is also being developed for use with the next-generation NWP system, both to predict wildfire danger and to modulate future smoke emissions from existing fires.

9.4.7. Southeastern United States

Prescribed fire is the leading source of $PM_{2.5}$ emissions in the southeast United States (U.S.EPA, 2017), and its share is growing as the use of agricultural and silvicultural burning increases while other sources are subject to stringent controls. Especially during the burning season (January–April), the region's air quality is adversely affected by prescribed fire smoke. HiRes2 (Odman et al., 2018) was designed to forecast prescribed fire smoke and air quality in a portion of the southeastern United States centered around Georgia. Supported initially by the U.S. Environmental Protection Agency and later by the NASA Applied Sciences Program and the Joint Fire Science Program, its goal is to provide air quality and forest managers with the air quality information they need for day-to-day management of burning operations.

Prescribed burning decisions are typically made on short notice based on weather forecasts. Compared with wildfires, prescribed burns are of shorter duration and produce relatively low heat. This makes both their detection and estimation of their emissions by satellite very difficult (Huang et al., 2018). Because of these factors, persistence of fire activity information into the future is a particularly poor assumption for prescribed fires. To overcome these difficulties, HiRes2 starts with a forecast of prescribed fire activity based on (1) the weather forecast, generated by running WRF (and later used to drive CMAQ), and (2) historic burning patterns, derived from the 2010–2016 burn permit databases for Georgia and 2015–2016 GBBEP (Zhang et al., 2012, 2014) data for other states. This unique aspect of HiRes2 is enabled by a decision tree model that yields potential burned areas for likely burn locations. The prescribed fire emissions are calculated from the FCCS fuel bed maps at those locations, fuel moistures reported by the Fire Weather Network, and emission factors for southeastern U.S. fuels published by the U.S. Forest Service (Urbanski et al., 2009). An empirical plume rise model (Liu, 2014) is used to distribute those emissions to the vertical layers of CMAQ. Emissions from other anthropogenic sources are derived from the EPA National Emissions Inventory (NEI), and biogenic emissions are obtained by running the Biogenic Emission Inventory System (BEIS; Fehsenfeld et al., 1992) with the WRF-generated weather forecast. The contributions of prescribed fire emissions are tracked using the Direct Decoupled Method (DDM; Napelenok et al., 2008) in CMAQ.

The fire activity forecast is compared daily with the NOAA Hazard Mapping System (HMS) Fire and Smoke Analyses (Hu et al., 2016) and evaluated qualitatively, based on the agreement of the location and density of the fires. The agreement is generally good on days with heavy fire activity. Quantitative evaluations are also performed periodically through comparisons with data in prescribed fire records. The R^2 of linear regression between the forecast and permitted daily statewide total burn areas in Georgia during the 2015 burn season was 0.34 (Hu et al., 2019). Forecast contributions to $PM_{2.5}$ air quality are evaluated through comparisons to $PM_{2.5}$ observations at regional monitors; this is done on days of reporting exceedances of the national ambient air quality standards (daily average $PM_{2.5} > 35\ \mu g\ m^{-3}$). The precision, recall, and F1 score for correctly predicting the prescribed fire impacts in exceedances during the 2016 burn season were 19%, 36%, and 25%, respectively (Odman et al., 2018).

Through the permitting systems already in place in several southeastern U.S. states, prescribed burns can be restricted on expected poor air quality days and encouraged when meteorological conditions are more favorable. The use of HiRes2 forecasts in this manner may lead to increased burn capacity while reducing the impacts on air quality and human health from exposure to high levels of smoke (Huang et al., 2019).

HiRes2 is undergoing a major update and expansion; the new system will be called HiRes-X. The domain of the forecast will be extended and all modeling components will be updated to their latest versions. The decision tree model for the burn activity forecast will be replaced with a random forest machine learning algorithm (Zou et al., 2019), trained by updated fire information. Current health impact forecasts are based on studies of exposure to wildfire smoke (Alman et al., 2016). To document relationships between smoke from prescribed fire and health, the association between prescribed fire smoke and rates of ED visits across the southeast United States will be analyzed for 2015–2021, and exposure-response profiles specific to prescribed fire smoke will be derived for cardiopulmonary and other diseases.

9.5. SMOKE FORECASTING IN AUSTRALIA

Wildfires (e.g., bushfires) have been a significant natural feature of the Australian landscape for millennia, as illustrated by the extent of savanna burning in northern Australia (Fig, 9.4). Annual fire frequencies are linked to regional weather patterns and drought as well as more than 45,000 yr of land management by First Nation communities (Bowman et al., 2020). More recently, the number of extreme fire events have increased (BOM, 2020b), culminating in an unprecedented bushfire catastrophe (over 10 Mha burned) in 2019–2020 (Bowman et al., 2021; Fig. 9.4: area burned along the Australian eastern seaboard and southwest of Adelaide).

Prescribed burning (also called *hazard reduction burning* or *planned burning*) is a commonly used tool to manage bushfire risk, because it is a cost-effective method of reducing under-canopy fuel loads in savanna and dry Eucalypt forests. However, prescribed fires can also be a significant source of smoke. In southern Australia, population exposure to smoke can be compounded by the overlap of the prescribed burning season with the Southern Hemisphere autumn, when atmospheric ventilation rates are reduced, keeping smoke closer to the surface; wood combustion heaters add an additional source of smoke into the lower atmosphere.

The smoke hazards associated with bushfires and prescribed burning are well recognized in Australia (Horsley et al., 2018), and air pollution forecasting systems

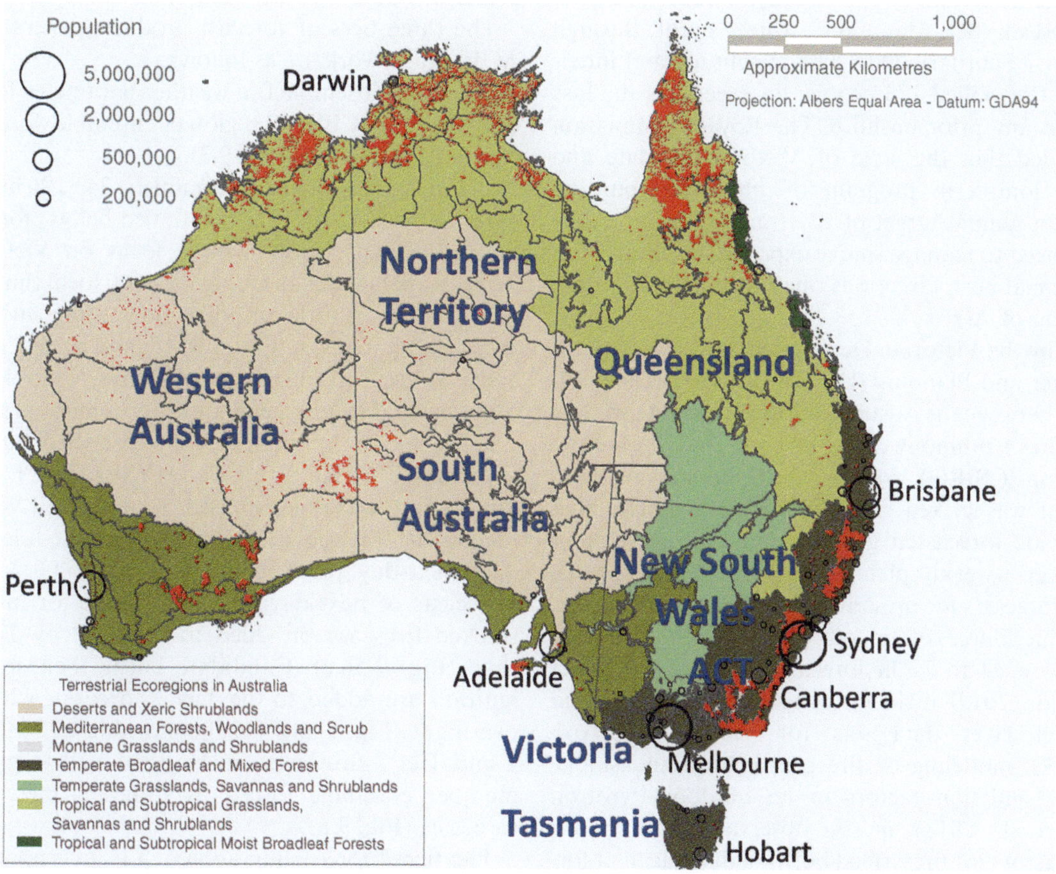

Figure 9.4 Map of Australia showing states and territories (ACT: Australian Capital Territory), capital cities, significant population centers, the major terrestrial ecoregions, and burned areas (red) for July 2019–March 2020. Data sources: The National Reserve System/Australia's ecoregions (https://www.environment.gov.au/land/nrs/science/ibra/australias-ecoregions); National Indicative Aggregated Fire Extent Data Sets (https://www.environment.gov.au/fed/catalog/search/resource/details.page?uuid=%7B9ACDCB09-0364-4FE8-9459-2A56C792C743%7D); Australian Bureau of Statistics: regional population (https://www.abs.gov.au/statistics/people/population/regional-population/2019-20).

coupled with effective now-casting tools and population alert systems have been adopted by many environmental agencies in Australia's states and territories. In the following sections, we summarize three of the Australian smoke forecasting systems and frameworks: (1) the Air Quality Forecasting (AQFx) system, managed by the Bureau of Meteorology with a focus on forecasting smoke from prescribed burns in the state of Victoria (Fig. 9.4); (2) the Coordinated Smoke Management System (CSMS), managed by the Tasmanian Forest Practices Authority for planned burn management in that state; and (3) the air quality forecasting system for the state of New South Wales (NSW) operated by the NSW Department of Planning, Industry and the Environment (DPIE).

9.5.1. The Air Quality Forecasting System (AQFx)

AQFx was developed following a Victorian Royal Commission report (http://royalcommission.vic.gov.au/finaldocuments/summary/PF/VBRC_Summary_PF.pdf) into the Black Saturday fires, which swept through Victoria on 7 February 2009. The magnitude and intensity of the fires killed 173 people, far exceeding the loss of life from any prior bushfire. The Royal Commission recommended that the state of Victoria mandate and support a long-term program of prescribed burning based on an annual target of 5% (minimum) of public land. The need to manage smoke exposure resulting from such an annual burn target was one of the drivers for the development of AQFx.

Funded by the Victorian Department of Environment, Land, Water and Planning (DELWP), AQFx was a collaboration between the Australian Bureau of Meteorology (BOM), the Commonwealth Science and Industrial Organization (CSIRO), and four Australian universities. The project was tasked with developing a system that could provide forecasted guidance on the fire weather outlook over a weekly planning cycle, as well as smoke transport forecasts for prescribed burns in the context of other existing sources of air pollution within the Victorian region over a 24 to 72 hr forecast period. The project (Cope et al., 2019) included construction of fine and coarse fuel layer data sets for Victoria (Volkova et al., 2018); modeling of fire propagation and smoke constituent emission factors in an outdoor Pyrotron (Sullivan et al., 2018); *in situ* observations of smoke emission factors for prescribed burns in dry Eucalypt forests (Reisen et al., 2018); adaptation, deployment, and verification of numerical forecasting tools; and training of Emergency Management Victoria personnel in the use of the system (e.g., https://delwp.publish.viostream.com/delwp-video-embed/media?v=ny1ykcsnp8oks6).

The AQFx framework generates three cascading time levels of weather and smoke forecasting output. Table 9.3 summarizes the AQFx technology and data sets, which are built around a suite of daily numerical meteorological forecast products generated by BOM using the Australian Community Climate and Earth-System Simulator (ACCESS). In Victoria and New South Wales, bushfire behavior is modeled using Phoenix Rapid Fire (Table 9.3), which is linked to fire agency fire locality data. Prescribed burn behavior (Victoria) is modeled using Phoenix Fire Flux (Table 9.3) and is linked to the DELWP daily schedule of prescribed burns. Satellite hotspot data are used to estimate the daily area burned by active fires in the other states and territories (Meyer et al., 2008). Smoke emissions from the flaming and smoldering components of an active fire are modeled using the CSIRO Smoke Emissions Model (C-SEM; Table 9.3). Smoke-plume rise, transport, and chemical transformation are modeled using the CSIRO chemical transport model (C-CTM; Table 9.3) operating in a forecast configuration (e.g., Table 12.7 of WMO, 2020); or with HYSPLIT for prescribed burn forecast outlooks of over 36 hr.

The three tiers of forecast products generated by the AQFx framework are as follows.

1. Tier 1: Ensemble fire weather parameter forecasting using the ACCESS-G3 global ensemble forecasts (18 ensemble members; Table 9.3).
2. Tier 2: Daily single deterministic 24 to 96 hr (depending upon model grid size mentioned below) forecasts for the Australian region, with a focus on Victoria (Fig, 9.5a), available each morning by 09:00 local time. Tier 2 is used to forecast total air pollution loading within an airshed, with a focus on $PM_{2.5}$, PM_{10}, and ozone; it includes active prescribed burns and bushfires, windblown dust, other natural sources, and anthropogenic sources.
3. Tier 3: 24 hr forecasts of prescribed burn smoke (extendable to 96 hr using HYSPLIT). The Tier 3 forecasting cycle is coupled to the DELWP smoke management process. AQFx polls the schedule of candidate next-day prescribed burns by 15:00 local time. Forecasts of next-day smoke exposure for individually tracked fires are provided to DELWP by 17:00 local time (Fig. 9.5b,c). Candidate burns then selected for ignition are added to the Tier 2 forecast, which is run overnight (Fig. 9.5d). On the morning of ignition, Tier 2 and Tier 3 forecasts can be combined to give a two-member ensemble forced by different meteorological forecasts (Fig. 9.5).

The tiered forecasting approach is a compromise between the treatment of forecast uncertainty for extended outlooks (5 days, and thus the need for an ensemble approach), and the computational demands of numerical air quality forecasting over large, nested domains over a 24 hr (1–3 km model grid size) to 72 hr (9–27 km model grid size) outlook. BOM began experimental AQFx forecasts in January 2016 and upgraded AQFx to operational

Table 9.3 Summary of the technologies and data used by the Australia AQFx Air Quality Forecasting Framework

Technology/data	Description	Australia AQFx application
Meteorological forecasts	Bureau of Meteorology ACCESS forecasting products. current generation ASP3 http://www.bom.gov.au/australia/charts/bulletins/opsbull_G3GE3_external_v3.pdf Global ensemble. 33 km model grid size; 70 vert levels; 18 members 10-day forecasts Global deterministic, 12 km model grid size; 70 vert levels 10-day forecast City deterministic; 1.5 km model grid size; 80 vert levels; hourly data assimilation Graphical Forecast Editor (GFE)- bias corrected surface weather	Tier 1: Global ensemble for the calculation of fire weather and fire danger parameters Tier 2: Global deterministic used for C-CTM (see below) 27 km and 9 km domains. City deterministic used for C-CTM 3 km domains Tier 3: City deterministic for C-CTM 1.6 km domains Global and City deterministic for HYSPLIT (see below) GFE is used to drive Phoenix Rapid Fire and Phoenix Fire Flux (see below).
Anthropogenic air pollution emissions	State EPA air emission inventories. For example, https://www.epa.nsw.gov.au/your-environment/air/air-emissions-inventory	Tier 2. Australia-wide anthropogenic emissions are estimated using emission factors derived from the NSW inventory and population-based spatial weighting. Bottom-up inventories from the state EPAs take precedence where available.
Natural and geogenic emissions	Emissions of dust, sea salt, biogenic volatile organic compounds, and soil nitrogen compounds	Tier 2. Calculated inline during the CTM integration (e.g., Emmerson et al., 2018).
Bushfire fire propagation	Phoenix Rapid Fire (https://firepredictions.atlassian.net/wiki/spaces/PH/pages/405176343/PHOENIX%2BTechnical%2BReference%2BGuide	Tier 2 forecasting for Victoria and New South Wales.
Phoenix Fire Flux	Phoenix Fire Flux (Cope et al., 2019, pp. 36–38, Appendix C)	Tier 2, Tier 3 for Victoria.
Smoke emissions	C-SEM Smoke Emissions Model (Cope et al., 2019, pp. 40–42)	Tier 2, Tier 3 for Victoria.
Chemical Transport model	CSIRO chemical transport model (C-CTM; Guérette et al., 2020; Lawson et al., 2017)	Tier 2, Tier 3. See Table 12.1 of WMO ETR-26 (https://library.wmo.int/index.php?lvl=serie_see&id=240) for a summary of a specific a C-CTM forecasting configuration.
HYSPLIT	Stein et al. (2015)	Tier 3. HYSPLIT is used, on request, for prescribed burn smoke transport forecasts with an outlook extending beyond 48 hours. Tracer transport with unit emission rates.
AQVx	AQFx visualization system (Williamson & Lucani, 2020)	Cloud-based app used to visualise AQFx near real time forecast, observations, satellite data, and contextual ground-based data.

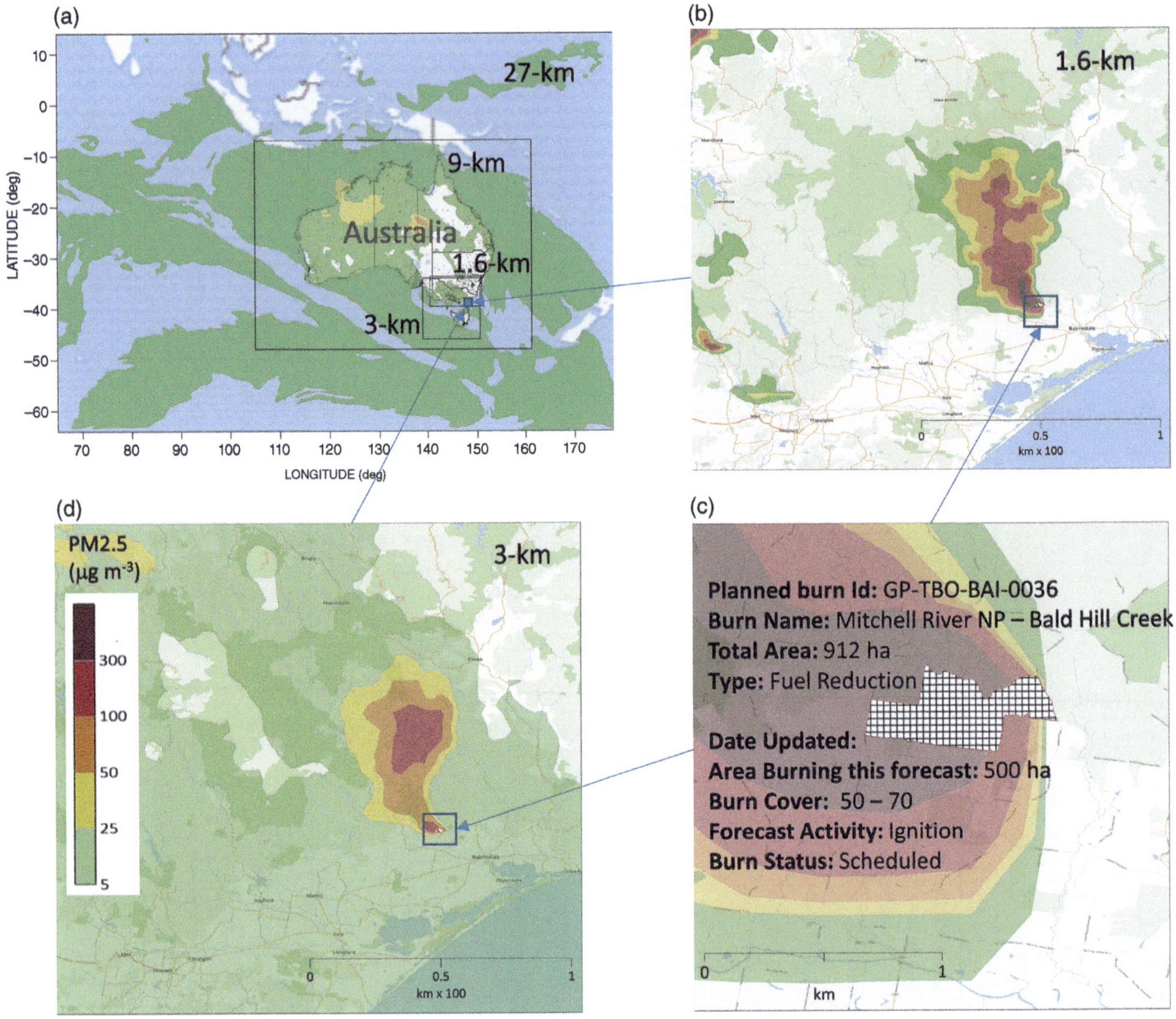

Figure 9.5 (a) Forecast of near-surface $PM_{2.5}$ for 1000 UTC 31 March 2021. The three domains (27 km, 9 km, and 3 km model grid size; 19 vertical levels) used for Tier 2 forecasting (all air pollution sources) for a 24 to 72 hr outlook. The 3 km domain here is configured to provide high-resolution forecasts for the Victoria and Tasmania in southeastern Australia. The fourth domain (1.6 km) is used for prescribed burn forecasting (tracer mode configuration), shown here for Victoria. (b) Prescribed burn forecast (1.6 km domain) for a small region in southeast Victoria for the same date and time. (c) The characteristics of the prescribed burn in (b), including a shapefile of the proposed burn area. (d) The full chemical forecast for the same time. The $PM_{2.5}$ forecasts in (b) and (d) differ because of the domain resolution differences, the inclusion of other particle sources (including secondary aerosols) in the chemical forecast (d), and because the 3 km forecast is undertaken with more recent meteorological forecast (12 UTC) than the tracer forecast (00 UTC).

status for Victoria in 2017 and NSW in 2018. For examples of AQFx performance, see Cope (2018) and Cope et al. (2019).

Future Plans

The Southern Hemisphere Australian 2019–2020 bushfire season was characterized by the greatest warming of Indian Ocean sea surface temperatures on record, coupled with the northern movement of the belt of strong westerly winds surrounding Antarctica, which decreased rainfall in southeast Australia. These seasonal factors combined with a significant hydrological drought to generate extreme fire weather conditions across the continent. The national average Forest Fire Danger Index (FFDI) for spring 2019 was the highest on record, and high to extreme levels of FFDI extended through

January 2020 in parts of South Australia, Victoria, and New South Wales (BOM, 2020a). These conditions led to an unprecedented bushfire season in eastern Australia (Bowman et al., 2021) from August 2019 to March 2020. Bushfires burned over 10 million Ha, including 18% of all Australian dry and wet Eucalypt forest (Fig. 9.4), resulting in the direct loss of life for 33 people and the estimated deaths of nearly one billion mammals, birds, and reptiles (Dickman & McDonald, 2020).

The most densely populated regions of Australia were exposed to significant concentrations of smoke for extended periods during the 2019–2020 bushfire season. Johnston et al. (2021) estimated that population exposure to $PM_{2.5}$ from bushfire smoke caused more than 1,500 asthma emergency attendances, hospital admissions for over 2,000 respiratory and 1,100 cardiovascular conditions, and over 420 premature deaths. Total health costs were AU$1.95 billion, nearly an order of magnitude more than the median of the previous 19 bushfire seasons. Tier 2 AQFx forecasting capability was extensively used throughout the season and was provided on request to fire agencies and environment departments for all impacted states and territories.

A Royal Commission into National Natural Disaster Arrangements (https://naturaldisaster.royalcommission.gov.au/) undertook an extensive investigation into the 2019–2020 bushfires (and other natural disasters) and was tasked to provide recommendations to optimize Australia's national natural disaster coordination. Of note is the commission's Recommendation 14.2: "Australian, state and territory governments should develop national air quality forecasting capabilities, which include broad coverage of population centers and apply to smoke and other airborne pollutants, such as dust and pollen, to predict plume behavior."

An outcome of Recommendation 14.2 is that AQFx is being expanded to generate national bushfire smoke forecasts. An initial prototype system was deployed for the 2021–2022 Australian bushfire season and included developments such as the following.

1. Review AQFx forecasting capability for the 2019–2020 bushfire season.
2. Engage closely with the national air quality community, encouraging state and territory government practitioners to provide guidance and feedback during the prototype system development.
3. Deploy a nationally consistent fire behavior forecasting methodology based on the Australian Fire Danger Rating System (AFDRS) national fuel layer (https://www.afac.com.au/initiative/afdrs) and the SPARK fire propagation platform (https://research.csiro.au/spark/).
4. Deploy an assimilation package that will leverage existing (assimilated) global chemical atmosphere forecasts at the national scale, and a local/regional scheme that constrains wind-blown dust and smoke emissions (e.g., Dai et al., 2019).
5. Deploy a forecast-data blending scheme (e.g., Majumder et al., 2021) for the near-real-time analysis and dissemination of spatial surface-layer $PM_{2.5}$ concentration fields through government-facing and public cloud-based apps (Campbell et al., 2020; Williamson & Lucani, 2020).
6. In support of the above, deploy a low-cost fine particle sensor (https://ecos.csiro.au/smog/) network, with a focus on improving coverage in existing smoke-prone but sparsely monitored regions.

A review of the performance of the AQFx prototype forecasting system is under way, and system components will be further refined as required. Modules from the AQFx prototype will then become available for incorporation into the BOM operational system.

9.5.2. Tasmania's Coordinated Smoke Management System (CSMS)

Tasmania, the island state of Australia, is located in temperate latitudes and has a relatively small and distributed human population of 500,000. Planned burning is conducted by organizations such as forestry companies (postharvest regeneration burning), the state Parks and Wildlife Service (land-management burns), the Tasmania Fire Service (hazard-reduction burns), councils, and private land holders (agricultural and hazard-reduction burning). Burning is mostly conducted in autumn and spring. Tasmania also has a large tourism base and a well-developed wine industry. Smoke from planned burning can be a problem for tourist operators and vintners as well as the general community. To address this concern, the Tasmanian Forest Practices Authority (FPA) and Environmental Protection Authority (EPA) Tasmania created the Coordinated Smoke Management System (CSMS) and the Base-Line Air Network of EPA Tasmania (BLANkET) real-time ambient air monitoring network (Fig. 9.6).

The CSMS strategy limits the amount of smoke produced in a given area of Tasmania on a given day. Under the CSMS, Tasmania is divided into airsheds (e.g., regions; Fig. 9.6). Each airshed has a maximum daily burn-unit allowance. Burners wishing to conduct a burn on a given day place a bid to burn a certain number of burn-units by 07:30 local time via a Web server. The server then calculates a burn-unit allocation (or daily quota) in the relevant airshed. The daily allowance in each airshed varies with the forecast ventilation index and (if present) the inversion layer height for that day and region from Australia's Bureau of Meteorology forecast system. The number of applicable burn-units for a given burn is calculated based on the burn area and fuel type, as specified in a guidance document. Software on the Web

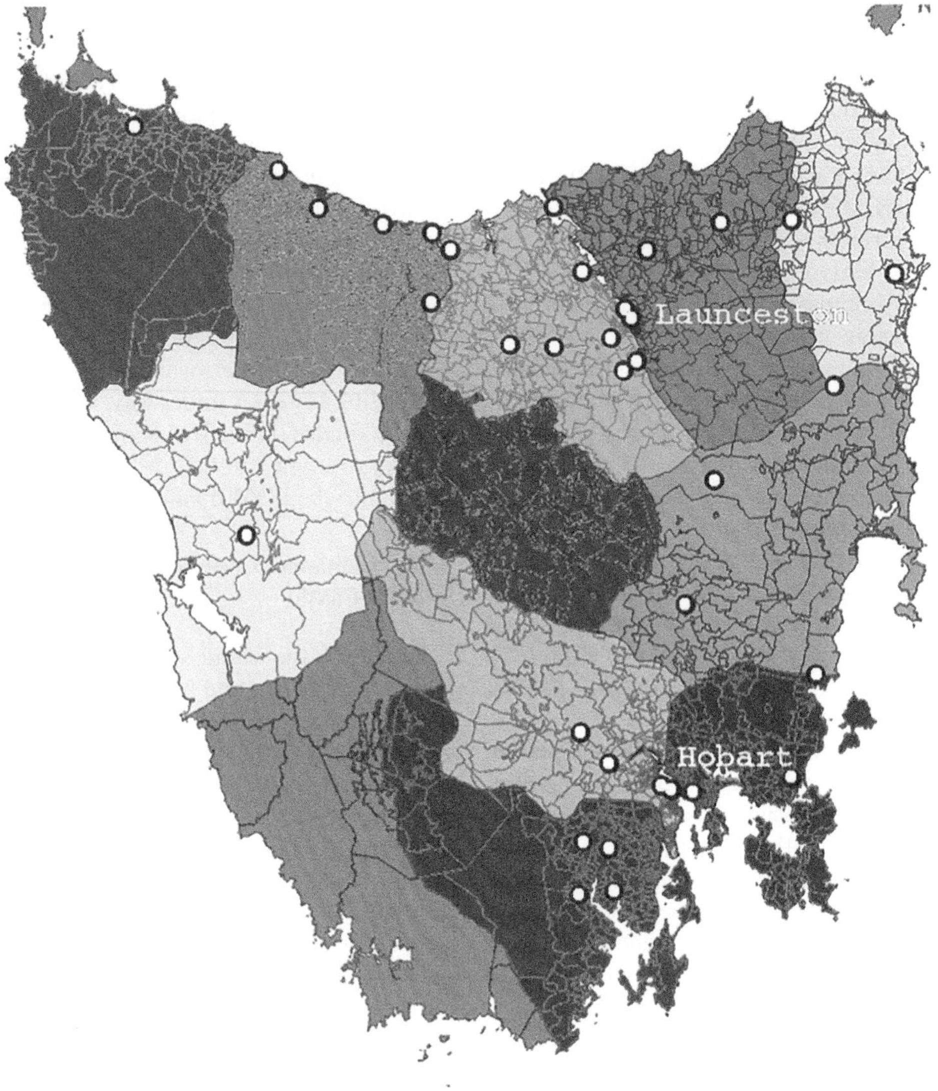

Figure 9.6 Map of Tasmania. Colored regions: Coordinated Smoke Management System (CSMS) airsheds. Black circles: locations on EPA Tasmania's real-time air monitoring stations. Monitors are located primarily in towns and larger population centers.

server allocates units to the burners daily. If the total number of units requested for an airshed exceeds the daily quota, all bids for that airshed are reduced proportionally to the allowed total. On days when very poor dispersion is forecast (low ventilation index), the quota may be zero.

Analysis of the air quality data from the expanded network has identified valuable information on smoke movement and dispersal in Tasmania and circumstances where smoke may affect communities. For example, katabatic winds (cold air drainage) may transport smoke from fires smoldering overnight to population centers, which can be a significant mechanism for smoke exposure in many areas of Tasmania. The katabatic winds tend to be more prominent during regionally calm conditions. Information of this nature is provided to burners and forms part of the documented guidance principles. Another feature of the system is the designation of "'no burn days": if the measured $PM_{2.5}$ concentration in an airshed at 09:00 local time exceeds 25 $\mu g/m^3$, the default is that no further burning takes place in that airshed on that day. This rule has significantly reduced the severity and duration of smoke impacts on populated areas, particularly in the Huon Valley south of Hobart.

Although the CSMS is a voluntary framework, it has been adopted by the major burn agencies and organizations in Tasmania as best-practice smoke management. A steering committee chaired by the FPA with membership from the burners, EPA Tasmania, and the Department of Health meets to review burning seasons and to consider whether revisions are needed. Adoption of the CSMS, and the establishment of the expanded air monitoring network, have provided a sound basis for improving planned-burn smoke management in Tasmania. Since the CSMS began, the number of public complaints about planned-burn smoke received by EPA Tasmania have significantly decreased and for some areas of the state, measured peak smoke levels in autumn have also decreased. The CSMS is likely to have played a significant role in these changes.

9.5.3. New South Wales (NSW) Air Quality Modeling System

The New South Wales Air Quality Forecast Framework was established in 2015 (Jiang et al., 2015) with the goal to produce accurate early warnings about when and where ambient air pollutant concentrations may exceed recommended levels. The NSW Department of Planning, Industry and Environment (NSW DPIE) operates an air quality modeling system that provides 72 hr forecasts of air pollutant concentrations for the Greater Metropolitan Region of New South Wales. Air quality alerts are issued on days when pollution levels are forecast to be unhealthy or worse. The system consists of two modeling systems. The first is a trajectory and dispersion system using the Hybrid Single-Particle Lagrangian Integrated Trajectory model (HYSPLIT; Stein et al., 2015) developed by the NOAA and tailored for NSW, known as *HYSPLIT in NSW* (Chang et al., 2021). The second is an air quality modeling system, which includes the coupled Conformal Cubic Atmospheric Model and Chemical Transport Model system (Chang et al., 2018), with meteorology from the Australian Community Climate and Earth-System Simulator (ACCESS) Numerical Weather Prediction (NWP) system (http://www.bom.gov.au/nwp/doc/access/NWPData.shtml).

HYSPLIT in NSW is driven by high spatial resolution (4–12 km model grid cell size) ACCESS regional weather forecast data and streamlined to provide automated atmospheric trajectories and dispersion forecasts. It has been operational since July 2019 and provides forecasts twice daily. The system driven by the higher-resolution meteorological data (e.g., 4 km grids) improves certainty when identifying areas that may be affected by smoke from bushfires (Watt et al., 2017), regional dust storms, and industrial incidents. Figure 9.7a shows the 24 hr back-trajectories starting at 10:00 local time on 29 December 2019 from locations in Sydney and Wollongong. The back-trajectories indicate that these air parcels travelled through major smoke plumes over eastern New South Wales (Fig. 9.7b) in the 24 hr before they arrived at Sydney. During the 2019–2020 bushfire season, routine evaluations of the HYSPLIT in NSW trajectories was

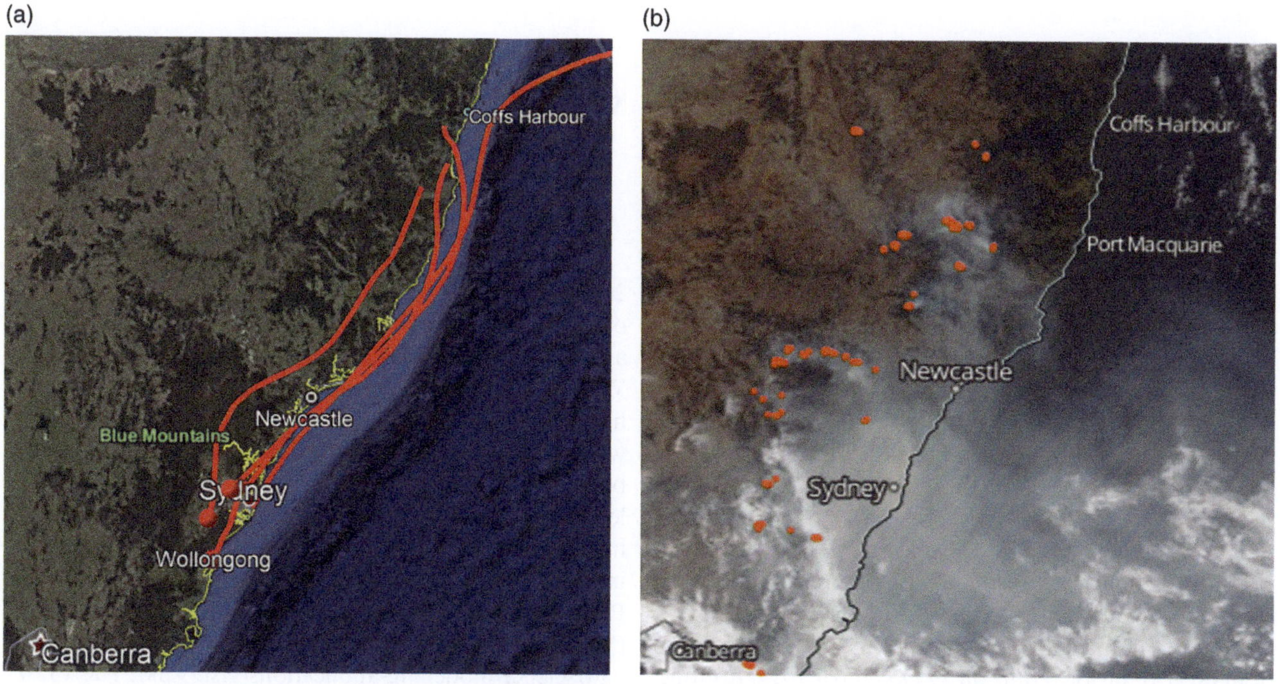

Figure 9.7 (a) HYSPLIT in NSW 24 hr back-trajectories starting at 1000 29 December 2019 from locations in Sydney and Wollongong; (b) Terra/MODIS satellite imagery on the same day showing hot spots (fires) and smoke plumes on the northern coast and in eastern New South Wales.

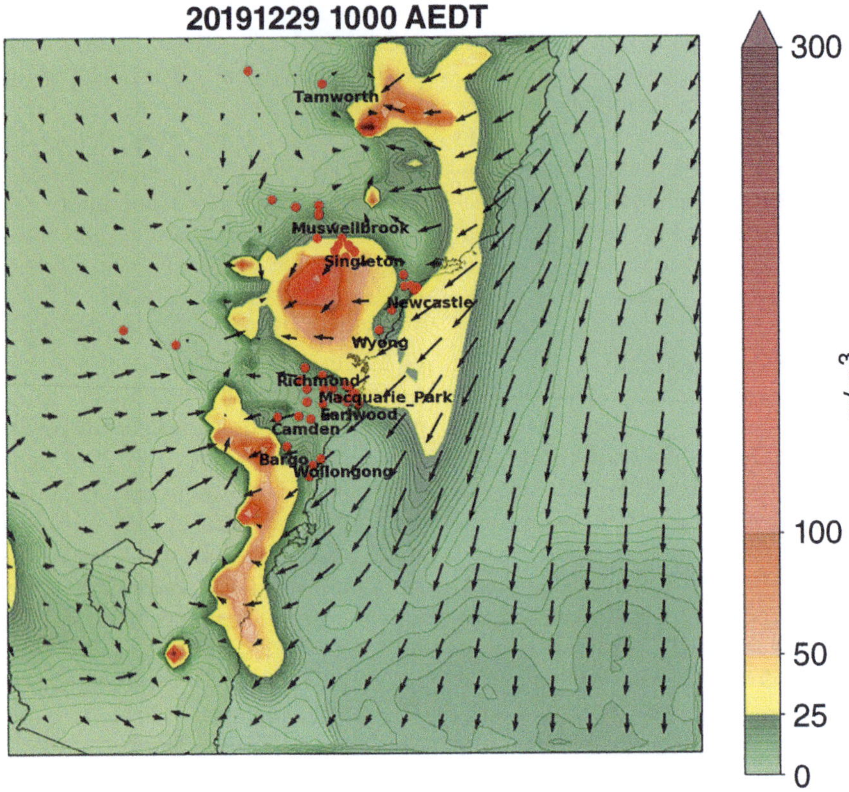

Figure 9.8 Examples of the numerical modeling forecasting products from the New South Wales Department of Planning, Industry and Environment (NSW-DPIE), showing the hourly $PM_{2.5}$ concentrations (μg/m^3) and surface winds valid at 1000 on 29 December 2019 for the 1 km spatial resolution domain.

conducted by visually comparing smoke plumes tracked by Moderate Resolution Imaging Spectroradiometer (MODIS) Corrected Reflectance imagery.

The numerical regional airshed model consists of a meteorological module, emission modules, and a chemical transport module. It was configured by the NSW DPIE to run twice a day since 2018 to support the routine daily air quality forecasts. The NSW Greater Metropolitan Region Air Emissions Inventory documents anthropogenic emissions used in the regional airshed modeling; natural sources, including wind-blown dust, sea salt, and biogenic emissions, are estimated in-line within the model. Emissions of pollutant species from wildfires and hazard-reduction burns (HRBs) are estimated by the Smoke Emission Module (Monk et al., 2019) running outside of the airshed modeling. Figure 9.8 gives an example of daily numerical modeling forecasts, showing hourly $PM_{2.5}$ concentrations on a 1 km spatial resolution domain at 10:00 local time on 29 December 2019. Routine evaluation is conducted with monitoring data from the comprehensive Air Quality Monitoring Network operated by the NSW DPIE.

9.6. WORLD METEOROLOGICAL ORGANIZATION (WMO) VEGETATION FIRE AND SMOKE POLLUTION WARNING ADVISORY AND ASSESSMENT SYSTEM (VFSP-WAS)

The Eighteenth World Meteorological Congress in June 2019 endorsed an ambitious plan to advance the integration of weather, climate, water, and environmental applications and services for health, as well as to work closely with the World Health Organization (WHO) to reduce risks to human health. Populations both near and downwind of wildfires need better warnings about serious threats posed by both the fires themselves and related air quality risk levels. The World Meteorological Organization (WMO) has responded to urgent requests for assistance in several impacted regions by initiating a Vegetation Fire and Smoke Pollution Warning Advisory and Assessment System (VFSP-WAS) (https://community.wmo.int/activity-areas/gaw/science/modelling-applications/vfsp-was). The VFSP-WAS provides guidance to address both smoke and fire danger and proposes to support the potential foundation of regional centers (WMO, 2018).

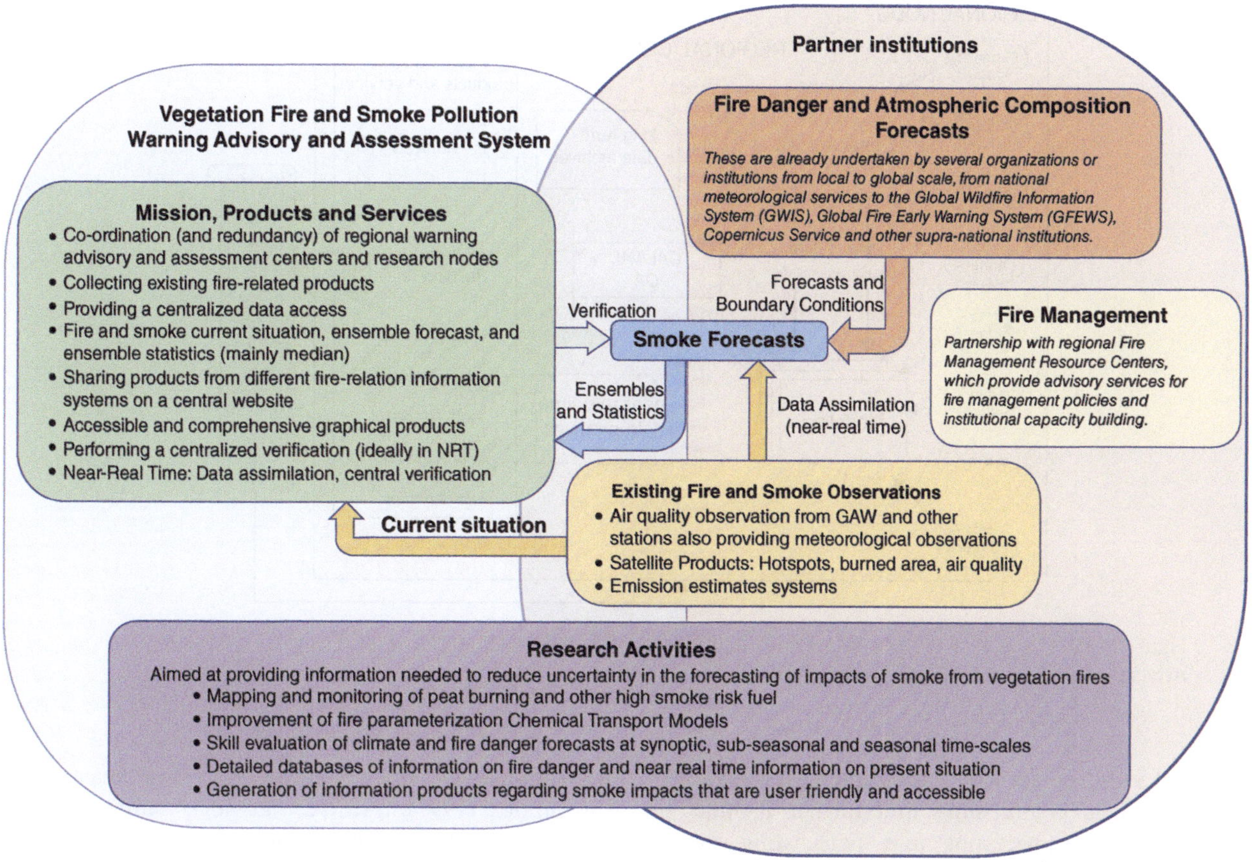

Figure 9.9 Overview of a Vegetation Fire and Smoke Pollution Warning Advisory and Assessment System (VFSP-WAS), including its mission, products, and services; research activities; smoke observations and forecasts; and connections with fire management centers and other potential end users.

Figure 9.9 shows the functional organization of a VFSP-WAS, including the interactions between the system's components and activities. The proposed warning and advisory system builds upon a number of comparable initiatives, such as CAMS, the WMO Regional Specialized Meteorological Centers (RSMCs), the Association of Southeast Asian Nations (ASEAN) Specialised Meteorological Centre (ASMC) in Singapore, the Sand and Dust Storm Warning Advisory and Assessment System (SDS-WAS) with regional centers in Barcelona and Beijing, the Global Air Quality Forecasting and Information System (GAFIS, https://community.wmo.int/activity-areas/gaw/science-for-services/gafis), and the International Cooperation for Aerosol Prediction (ICAP).

At the regional level, VFSP-WAS is organized as a federation of regional partners contributing to Nodes and realized through Regional VFSP-WAS Centers (VFSP-WAC) (Baklanov et al., 2021; WMO, 2018; Fig. 9.10). Each potential Node is an open federation of partners from interested countries in the region, with equal votes of each partner. A federated approach allows flexibility, growth, and evolution, while preserving the autonomy of individual institutions. Regional Nodes can involve the cooperation of, and provide benefits to, a variety of participants (e.g., universities, research organizations, meteorological services, emergency management bodies, health organizations). The organization of research and development and forecasting activities at a Node is defined and led by a Regional Steering Group and practically realized by the Regional VFSP-WAC. The VFSP-WAC can be hosted by one or several countries/organizations of the region, based on agreement of the Node members, and it focuses on technical realization of the Regional VFSP-WAS. A scheme of the governance structure of Regional VFSP-WAC is presented in Figure 9.10. Regional VFSP-WAC provides information, including warnings and advice, to support fire and emergency management to help partners react appropriately to harmful fire episodes. Because fire and smoke predictions still need considerable development, these centers also aim to bridge gaps between research and operational work.

The first research and development phase of the VFSP-WAS was launched in 2018. Before an operational Regional Specialized Meteorological Centre (RSMC) for

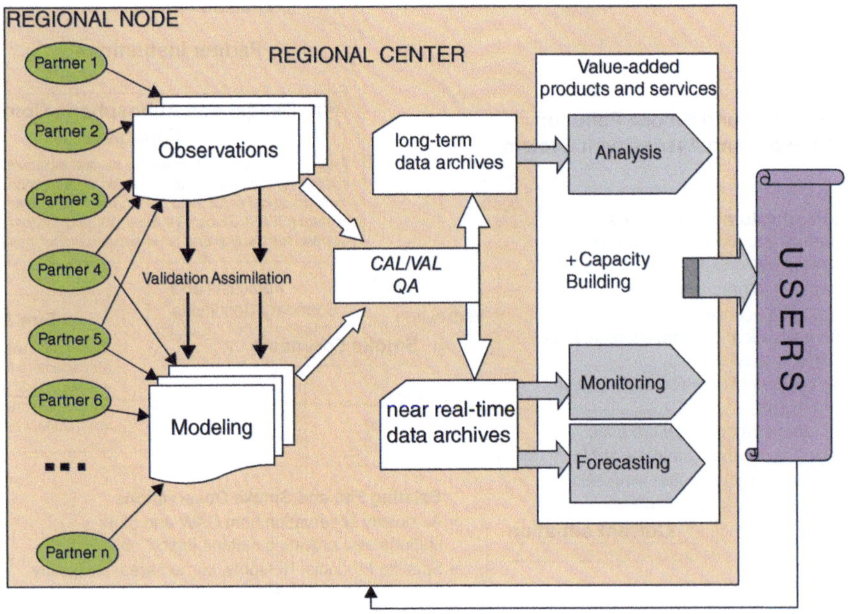

Figure 9.10 Scheme of the governance structure of a Regional Node and Fire and Smoke Pollution Warning Advisory and Assessment Center (VFSP-WAC).

VFSP-WAS can be approved, the center is evaluated by WMO technical commissions and further included in the Global Data-Processing and Forecasting System (GDPFS; WMO, 2017). Several regions (North America, Southeast Asia, Latin America, and northern Europe) are interested or have demonstrated infrastructure necessary to host such regional centers for the realization of VFSP-WAS, and they also have good experience and research.

This section describes the first and most established two VFSP-WACs on the research and advanced development stage, which are recommended as prototypes for Regional VFSP-WACs. (1) In Singapore, regional models are run and forecasts from various centers are collated to produce multimodel ensemble smoke forecasts for Southeast Asia. (2) In North America, Environmental and Climate Change Canada (ECCC) in Canada has established a demonstration VFSP-WAS center. Both centers produce ensemble particulate matter and AOD forecast products, with real-time evaluation either functioning or in development. They also include fire weather or fire risk forecast products as well as a section on current conditions with applicable observational data.

9.6.1. Singapore VFSP-WAS Regional Center for Southeast Asia

The first regional VFSP-WAC is hosted by ASMC for Southeast Asia. ASMC operates under the Meteorological Service Singapore (MSS). It was designated in 1997 as the official ASEAN center to monitor fires and transboundary smoke haze in Southeast Asia (Singapore, 1997). ASMC provides information on regional weather and haze, and issues alerts when biomass burning smoke is expected to affect any ASEAN country. ASMC also serves as technical advisor for ASEAN committees on transboundary haze and conducts capability development programs for regional users in environment, meteorological, and related agencies.

The Southeast Asia VFSP-WAC was established based on the experience of the WMO SDS-WAS (https://sds-was.aemet.es/) and ICAP, with collaboration from regional and global partners providing relevant products and services. In 2019, a pilot website for the Southeast Asia VFSP-WAS was launched in research and development phase to establish a multimodel ensemble with near-real-time forecast evaluation. In 2021, an upgraded version of this website (https://www.mss-int.sg/vfsp-was/) was deployed with a suite of enhancements, including longer-term evaluation metrics, observations, and fire-risk products.

The Southeast Asia VFSP-WAC website also includes a review of the current weather and smoke haze situation, as well as a short-term outlook issued by ASMC. Three categories of products are on the website: forecasts (including performance evaluation), observations, and fire-risk products.

Ensemble Particulate Matter and AOD Forecast Products

The VFSP-WAS multimodel ensemble includes five global models: ECMWF-CAMS (Europe), JMA

MASINGAR (Japan), NASA GEOS (USA), NOAA GEFS-Aerosols (USA), and FMI SILAM (Finland). It also includes one regional model, the MSS-United Kingdom Meteorological Office Numerical Atmospheric-dispersion Modelling Environment (MSS-UKMO NAME) model. The forecast products include the ensemble and member predictions of smoke aerosol optical depth (AOD) and surface concentrations of PM_{10} and $PM_{2.5}$. Regional weather forecast information includes subseasonal outlooks of rainfall and temperature issued by ASMC for Southeast Asia and NWP forecasts of winds at various height levels.

The VFSP-WAC retrieves available forecasts at 00 and 12 UTC from its partners and represents them over a regional domain covering major burning areas and smoke transport pathways at a common grid resolution of 0.5° × 0.5°. The ensemble and member forecasts are provided at 3 hr intervals up to 48 hr ahead. Figure 9.11 shows example products from the multimodel ensemble median forecast of smoke AOD and $PM_{2.5}$ surface concentration. Two products describe product centrality (multimodel median, mean) and two describe product spread (standard deviation, range of variation). The ensemble mean and standard deviation forecast products are also compared with ICAP products derived for the region. The ICAP multimodel ensemble (Sessions et al., 2015; Xian et al., 2019) functions as a benchmark for the VFSP-WAC ensemble air quality forecast.

The MSS-UKMO NAME (Hansen et al., 2019; Hertwig et al., 2015) is the only regional and Lagrangian-based model in the ensemble. Fire radiative power (FRP) and smoke injection height products are obtained from CAMS-GFAS. The FRP products are subsequently applied over a high-resolution land cover and peatland map (Miettinen et al., 2012), which provides detailed information of fuel loading in Southeast Asia (Hertwig et al., 2015), thus allowing the derivation of smoke emissions calibrated for the region. Peatland burning is an important source of smoke emissions: about 250,000 km^2 of peatlands (56% of total tropical peatland) are in Southeast Asia, containing 70 Gt of carbon (i.e., 77% of tropical peat carbon and 11% of the world's total peat carbon pool; Page et al. (2011)). Other sources include three types of regional emissions: anthropogenic (MIX; Li et al., 2017), biogenic (CAMS-GLOB-BIO; Granier et al., 2019; Sindelarova et al., 2014), and shipping (EDGAR Version 4.3; Janssens-Maenhout et al., 2012). The NAME chemistry scheme includes the formation of secondary particulate matter; it is now used to generate realistic background pollutant levels, compared with earlier system versions described in Hertwig et al. (2015). In contrast to the other members in the ensemble, which employ more comprehensive aerosol optical models, NAME AOD forecasts are computed based on $PM_{2.5}$ aerosol optical properties derived from an empirical study of biomass burning smoke over Singapore (Lee et al., 2016).

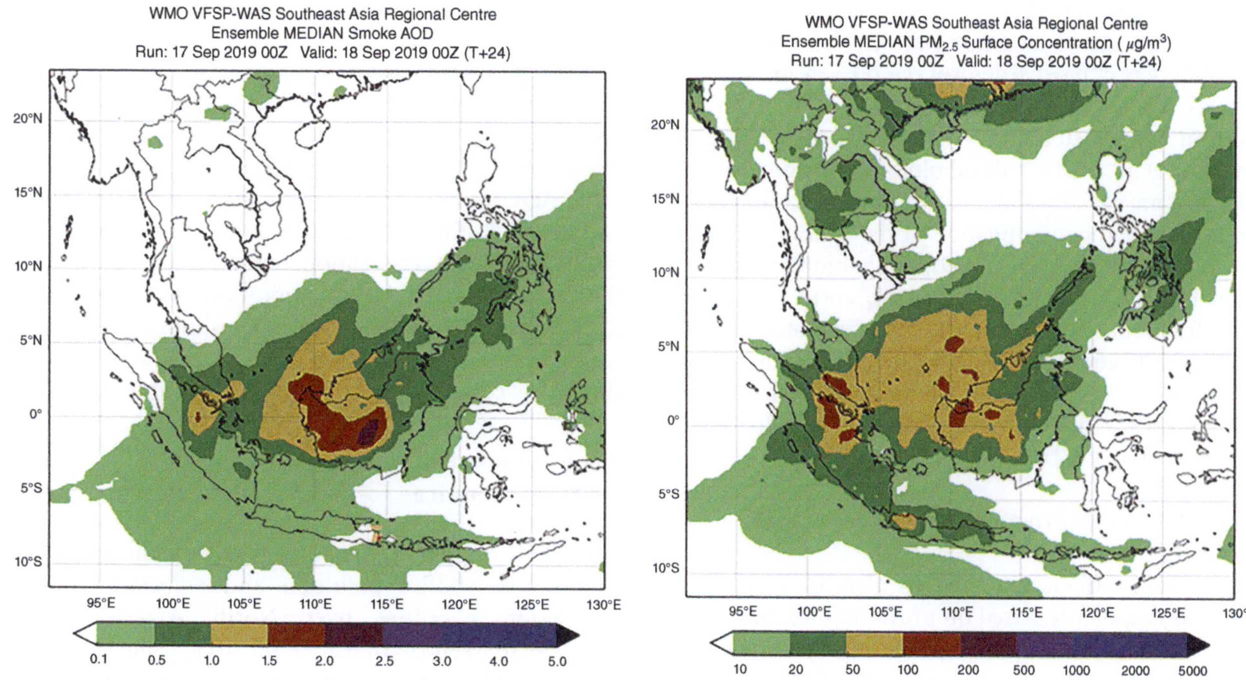

Figure 9.11 Example of Southeast Asia VFSP-WAC multimodel ensemble median 1 day forecast of smoke AOD (left) and $PM_{2.5}$ surface concentration (right) for Southeast Asia in September 2019.

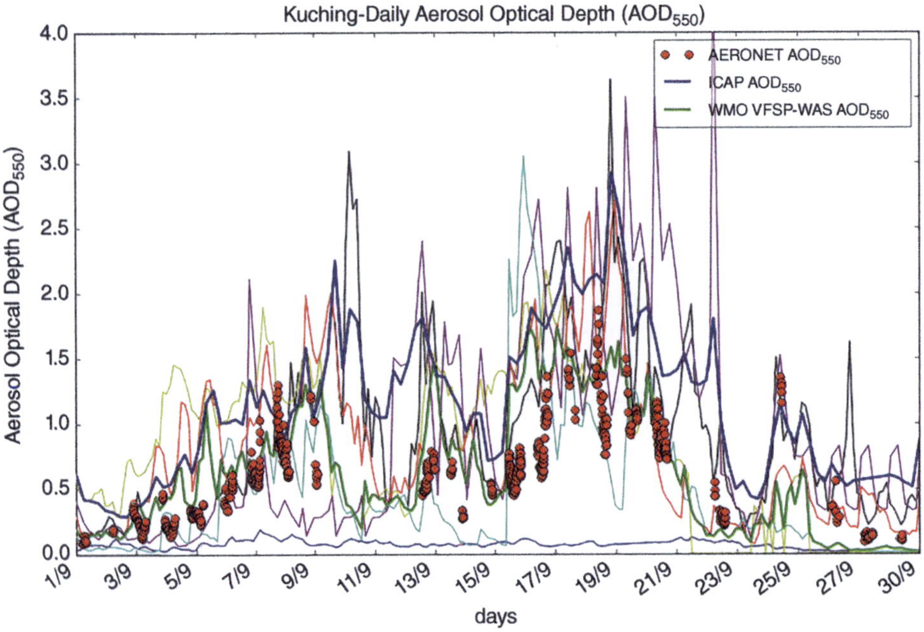

Figure 9.12 Near-real-time Southeast Asia VFSP-WAC forecast evaluation over Kuching, East Malaysia, during the peak of fire activity in September 2019. Unlabeled colored lines represent six individual model members of the Southeast Asia VFSP-WAC ensemble.

Performance Evaluation

Near-real-time evaluations compare ensemble and member modeled AOD forecasts to NASA AERONET (Holben et al., 1998) Level 1.5 (Version 3) observations at 22 sites in the region; these are updated daily on the VFSP-WAC website (Fig. 9.12). Clouds, especially cirrus clouds, are endemic in Southeast Asia and a source of potential bias in passive aerosol remote sensing data sets, such as AERONET (e.g., Chew et al., 2011). Thus, in addition to the standard AERONET cloud screening (Smirnov et al., 2000), a threshold of Angstrom exponent (440–870 nm) >0.75 is applied to screen AOD measurements for cloud contamination in the region (e.g., Chew et al., 2013; Salinas et al., 2009). Evaluation metrics (e.g., mean bias, root mean square error, correlation coefficient, fractional gross error) are also generated monthly, seasonally (3-month periods associated with regional burning processes and dominant monsoonal circulations) and annually; preliminary evaluation results are in Baklanov et al. (2021).

Fire-Risk Forecast Products and Observational Data Sets

Fire-risk products are used to identify areas in the region at risk of fire based on prevailing weather conditions. The ASEAN Fire Danger Rating System (FDRS) and Global Fire Weather Database (GFWED) products are on the VFSP-WAC website. The ASEAN FDRS has been calibrated for Southeast Asia (De Groot et al., 2007) and is based on the Fire Weather Index (FWI) System developed by Canada. This product is produced daily by the Malaysian Meteorological Department. GFWED integrates different weather factors influencing the likelihood of a vegetation fire starting and spreading, and is also based on the FWI System (Field, 2020; Field et al., 2016).

Satellite images, weather information, and air quality station measurements are on the VFSP-WAC website. Satellite images include products from geostationary (Himawari-8) and polar-orbiting (NOAA20, SUOMI-NPP, AQUA, TERRA) satellites, such as true/false color and natural, fire temperature, and night images. The ground weather observation reports collected through the WMO Global Telecommunication System (GTS) are displayed on a map showing wind, visibility, and weather conditions. The website also displays available site measurements of surface particulate matter concentrations.

9.6.2. WMO VFSP-WAS Regional Center for North America

The government of Canada recently conducted several studies on wildfire pollution and associated human population exposure over North America (Matz et al., 2020; Munoz-Alpizar et al., 2017) using the operational ECCC air quality system, FireWork (Chen et al., 2019; Pavlovic et al., 2016; see section 9.4). The importance of these consequences of wildfire pollution is a key driver for improving international collaboration in data sharing in support of wildfire preparedness and response.

In 2019, ECCC responded to the WMO's initiative and volunteered to create a North American (NA) Regional VFSP-WAC. This center began disseminating products in October 2020 (https://hpfx.collab.science.gc.ca/~svfs000/na-vfsp-was/public/dist/) and is the world's second regional VFSP-WAC. The creation of the North American regional center involves close and effective collaboration with multiple national and international organizations (e.g., National Oceanic and Atmospheric Administration (NOAA), National Aeronautics and Space Administration (NASA), U.S. Forest Service (USFS); National Resources Canada (NRCan); European Centre for Medium-Range Weather Forecasts (ECMWF); Japan Meteorological Agency (JMA); Finnish Meteorological Institute (FMI)). This partnership is essential for data collection and product tailoring for different national and international user needs.

The NA VFSP-WAC disseminates products related to smoke and fire, relying on contributions of participating agencies and ensembling of those products. The product suite includes information for smoke forecasts ($PM_{2.5}$, PM_{10}, AOD) and fire weather (the Canadian Fire Weather Index and monthly forecast estimates of precipitation and temperature anomalies). Current-condition products include maps of satellite-detected fire hot spots and fire danger estimates.

Ensemble Particulate Matter and AOD Forecast Products

The multimodel ensemble particulate matter forecast is based on six members. Two are regional: ECCC FireWork (Canada) and NOAA NCEP NAQFC (USA). Four are global: ECMWF CAMS-IFS (Europe), FMI SILAM (Finland), JMA MASINGAR (Japan), and NASA GEOS-FP (USA). In addition to the ensemble, forecasts from each member are presented separately. The common lead time for these ensemble members is 48 hr, and the related forecast products will be updated once daily. These ensemble products are available over three North American windows: CONUS, Mexico, and Canada and Alaska (Fig. 9.13a,b).

The ICAP Multimodel Ensemble smoke AOD products are constructed from the following four aerosol forecast systems: ECMWF CAMS-IFS, NASA GEOS-FP, NRL NAAPS, and JMA MASINGAR. The ICAP AOD products are tailored over North America (Fig. 9.13c) and, similar to the Singapore regional center, function as a benchmark for the VFSP-WAC North America ensemble air quality forecast.

Performance evaluation for particulate matter concentration forecasts will be done with ground-level measurements from North American monitoring networks, using ECCC's Verification of Air QUality Models (VAQUM) System. VAQUM was developed in 2017 by ECCC in collaboration with ECMWF-Copernicus and NOAA and is used for operational AQ multimodel performance analysis over North America (Pavlovic et al., 2018).

Fire Weather Forecast Products

The Canadian FWI equation, provided by NRCan, calculates FWI over North America using ECCC's Operational Regional Deterministic Weather Forecasts, which is at 10 km horizontal grid spacing and is launched twice daily, producing 72 hr lead-time forecasts. This FWI product has maps estimating the FWI 24, 48, and 72 hr into the future. These plots are still in development while partners discuss options for the provision of risk forecast products.

ECCC also produces subseasonal precipitation and temperature anomalies maps using Global Ensemble Prediction System (GEPS). These anomalies serve as an indicator for potential wildfires, especially when conditions are both drier than usual and warmer than usual. Forecasted anomalies are calculated from a 20 yr climatology (1998–2017) of this prediction system obtained from a reanalysis. The monthly forecast is updated every Thursday and covers the subsequent 28 days (Fig. 9.13d), starting on Monday.

Hotspots and Fire Danger Maps Over North America (Current Conditions)

Hotspot maps are available for the period covering the last 24 hr and previous 7 day period. This product provides information about current and most recent wildfire activity as detected by multiple satellites and processed throughout the CWFIS and GBBEPx systems. The fire danger maps are still in development and are now produced daily by combining the fire danger maps produced for Canada and Mexico through the Canadian Wildfire Information System (CWFIS) and from the U.S. Wildland Fire Assessment System (WFAS). This map displays fire danger as classified by Canadian provincial, territorial, and U.S. state fire management agencies.

9.7. SUMMARY

Here we profiled many of the global and regional smoke prediction systems currently operational or quasi-operational in real time or near real time. It is not an exhaustive list of systems, but rather is a profile of many of the systems in use to give examples of the creativity and complexity needed to model the phenomenon of smoke. These systems rely on science and data discussed in the previous chapters of this book and reflect the different needs of agencies and regions, where the various systems are tailored to the best available science for a

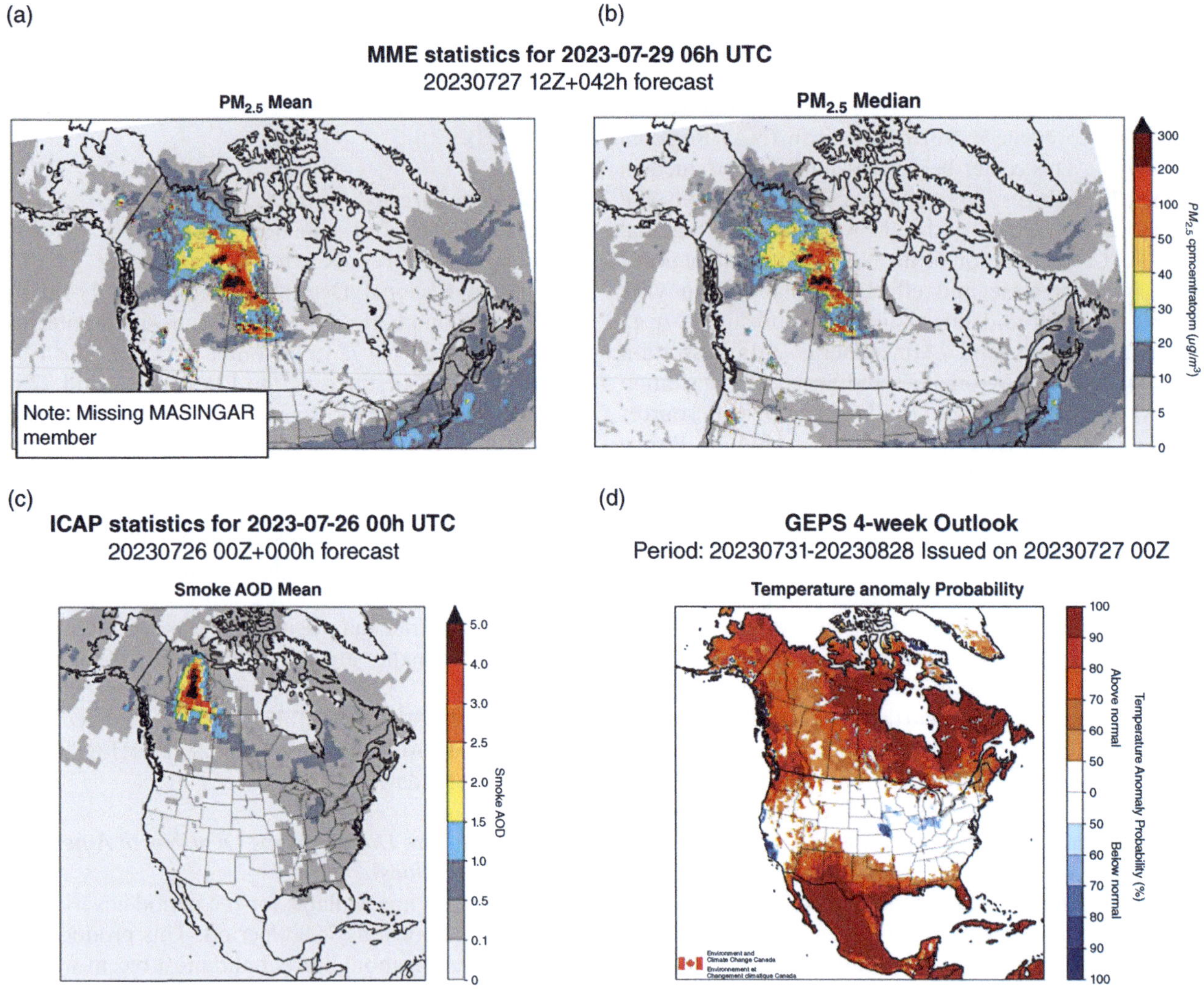

Figure 9.13 The multimodel ensemble (MME): (a) median and (b) mean PM2.5 forecast over Canada and Alaska, valid at 2023-07-29 06 UTC, from the 2023-07-27 12 UTC initialized run. (c) The ICAP average Smoke AOD valid at 2023-07-26 00 UTC, from the 2023-07-26 00 UTC initialized run. (d) The forecasted temperature monthly anomaly issued on 2023-07-27 00 UTC. Data sources: NRCan-CWFIS (Canada/Mexico); USFS-WFAS (contiguous USA), and MesoWest (Alaska). Some of these maps are already disseminated by USFS-WFAS (US Forest Service, Wildland Fire Assessment System).

region and the specific challenges of the region. Smoke forecasting needs range from warning and informing the public about potential smoke impacts to planning burn activities for hazard reduction or resource benefit. Different agencies also have different mandates, and the lines blur between the missions of quasi-operational organizations (e.g., research institutions) and agencies with operational mandates.

The global smoke prediction systems are advanced, and many are self-organizing into the ICAP community with nine operational/research centers. One of ICAP's most significant contributions to the community is the development of the ICAP-Multimodel Ensemble, which provides a testbed of probabilistic aerosol forecasts, helps to identify challenging areas for aerosol modeling, and forges valuable collaborations among forecast centers. The 11 regional air quality modeling systems in Europe utilize the global CAMS system, which is an ICAP participant, for initial and boundary conditions. Together, these 11 regional systems are another powerful ensemble. Biomass burning emissions are derived from satellite observation of fire radiative power by the GFAS and are used both in the global and regional systems. The regional models differ substantially in their chemical and aerosol schemes, deposition schemes, and the simulation of transport, diffusion, and planetary boundary layer processes. The 7 North American air quality and/or smoke predictions systems profiled are all independently developed by federal agencies or consortiums supporting university-based systems. Some simulate all pollutant sources in a chemical transport modeling system, while others only simulate smoke using a dispersion model or a chemical transport model (with coupled and offline cases). Biomass burning emissions are derived from ground-based fuel maps and consumption models or based on fire radiative power. Most of the systems are designed for wildfire conditions and are used by air quality and health agencies, but two systems are also designed specifically to aid prescribed fire operations, helping land managers to determine burning opportunities based on air quality concerns. In Australia, there are three smoke and air quality modeling systems addressing both wildfires (e.g., bushfires) and prescribed fire. The systems offer multiple modeling options, from chemical transport modeling of all sources including fires, to dispersion modeling of select prescribed burns to aid decision making. Finally, the World Meteorological Organization (WMO) effort brings together global and regional systems to form an ensemble to support countries with smoke issues and who lack resources.

ACKNOWLEDGMENTS

The authors thank Kirk Baker, U.S. EPA, for his assistance and contributions to the chapter. The authors also acknowledge the editing assistance of Patti Loesche and the helpful comments of the book editors and two anonymous reviewers. AIRPACT had been supported by NW-AIRQUEST, by NASA grants NNA06CN04A and NNX11AE57G and by Joint Fire Science Program Cooperative Agreement L15AC00170.

DISCLAIMER

The systems profiled in this chapter are continually under development as the science and technology of air quality and smoke forecasting improves. Therefore, the information provided here is a snapshot of the current state of these systems at the time of this writing. Further, this is not an exhaustive compilation of all systems globally; rather it represents many of the systems in use to give examples of the creativity and complexity needed to model the phenomenon of smoke.

REFERENCES

Adams, C., McLinden, C. A., Shephard, M. W., Dickson, N., Dammers, E., Chen, J., et al. (2019). Satellite-derived emissions of carbon monoxide, ammonia, and nitrogen dioxide from the 2016 Horse River wildfire in the Fort McMurray area. *Atmospheric Chemistry and Physics, 19*(4), 2577–2599.

Ahmadov, R., Grell, G., James, E., Csiszar, I., Tsidulko, M., Pierce, B., et al. (2017). *Using VIIRS fire radiative power data to simulate biomass burning emissions, plume rise and smoke transport in a real-time air quality modeling system.* Paper presented at the 2017 IEEE International Geoscience and Remote Sensing Symposium (IGARSS).

Akagi, S. K., Yokelson, R. J., Wiedinmyer, C., Alvarado, M. J., Reid, J. S., Karl, T., et al. (2011). Emission factors for open and domestic biomass burning for use in atmospheric models. *Atmospheric Chemistry and Physics, 11*(9), 4039–4072.

Al Mahmud, A. A. (2005). *Evaluation of the AIRPACT2 air quality forecast system for the Pacific Northwest.* Doctoral dissertation, Washington State University.

Alman, B. L., Pfister, G., Hao, H., Stowell, J., Hu, X., Liu, Y., & Strickland, M. J. (2016). The association of wildfire smoke with respiratory and cardiovascular emergency department visits in Colorado in 2012: A case crossover study. *Environmental Health, 15*(1), 64. https://www.ncbi.nlm.nih.gov/pubmed/27259511

Anderson, G. K., Sandberg, D. V., & Norheim, R. A. (2004). *Fire Emission Production Simulator (FEPS) user's guide.* Seattle, WA, USA.

Badia, A., Jorba, O., Voulgarakis, A., Dabdub, D., Pérez García-Pando, C., et al. (2017). Description and evaluation of the Multiscale Online Nonhydrostatic AtmospheRe Chemistry model (NMMB-MONARCH) version 1.0: Gas-phase chemistry at global scale. *Geoscientific Model Development, 10*(2), 609–638.

Baklanov, A., Chew, B. N., Frassoni, A., Gan, C., Goldammer, J., Keywood, M., et al. (2021). The WMO Vegetation Fire and Smoke Pollution Warning Advisory and Assessment System (VFSP-WAS): Concept, current capabilities, research and development challenges and the way ahead. *Biodiversidade Brasileira: BioBrasil(2)*, 179–201. https://doi.org/10.37002/biobrasil.v11i2.1738

Barna, M., Lamb, B., O'Neill, S., Westberg, H., Figueroa-Kaminsky, C., Otterson, S., et al. (2000). Modeling ozone formation and transport in the Cascadia region of the Pacific northwest. *Journal of Applied Meteorology, 39*(3), 349–366. https://journals.ametsoc.org/view/journals/apme/39/3/1520-0450_2000_039_0349_mofati_2.0.co_2.xml

Benedetti, A., Morcrette, J. J., Boucher, O., Dethof, A., Engelen, R. J., Fisher, M., et al. (2009). Aerosol analysis and forecast in the European centre for Medium-Range Weather Forecasts Integrated Forecast System: 2. Data assimilation. *Journal of Geophysical Research, 114*(D13).

Benedetti, A., Reid, J. S., & Colarco, P. R. (2011). International cooperative for aerosol prediction workshop on aerosol forecast verification. *Bulletin of the American Meteorological Society, 92*(11), ES48–ES53.

Benjamin, S. G., Weygandt, S. S., Brown, J. M., Hu, M., Alexander, C. R., Smirnova, T. G., et al. (2016). A North American hourly assimilation and model forecast cycle: The rapid refresh. *Monthly Weather Review, 144*(4), 1669–1694.

BOM (2020a). *Special climate statement 73: Extreme heat and fire weather in December 2019 and January 2020.* http://www.bom.gov.au/climate/current/statements/scs73.pdf

BOM (2020b). *State of the climate* 2020. http://www.bom.gov.au/state-of-the-climate/documents/State-of-the-Climate-2020.pdf

Bowman, D. M. J. S., Balch, J. K., Artaxo, P., Bond, W. J., Carlson, J. M., Cochrane, M. A., et al. (2009). Fire in the Earth system. *Science*, *324*(5926), 481–484. https://www.ncbi.nlm.nih.gov/pubmed/19390038

Bowman, D. M. J. S., Kolden, C. A., Abatzoglou, J. T., Johnston, F. H., van der Werf, G. R., & Flannigan, M. (2020). Vegetation fires in the Anthropocene. *Nature Reviews Earth and Environment*, *1*(10), 500–515.

Bowman, D. M. J. S., Williamson, G. J., Gibson, R. K., Bradstock, R. A., & Keenan, R. J. (2021). The severity and extent of the Australia 2019–20 eucalyptus forest fires are not the legacy of forest management. *Nature Ecology and Evolution*. https://www.ncbi.nlm.nih.gov/pubmed/33972737

Briggs, G. A. (1969). *Plume rise: A critical survey* (No. TID-25075). Air Resources Atmospheric Turbulence and Diffusion Lab., Oak Ridge, TN. https://www.osti.gov/servlets/purl/4743102

Briggs, G. A. (1972). Chimney plumes in neutral and stable surroundings. *Atmospheric Environment (1967)*, *6*(7), 507–510.

Briggs, G. A. (1982). Plume rise predictions. In *Lectures on air pollution and environmental impact analyses*. Boston: American Meteorological Society.

Brown, T., Clements, C., Larkin, N. K., Anderson, K., Butler, B., Goodrick, S., et al. (2014). *Validating the next generation of wildland fire and smoke models for operational and research use: A national plan*. Final report to the Joint Fire Science Program, Project #13-S-1-1. http://www.firescience.gov

Byun, D., & Schere, K. L. (2006). Review of the governing equations, computational algorithms, and other components of the Models-3 Community Multiscale Air Quality (CMAQ) modeling system. *Applied Mechanics Reviews*, *59*(2).

Campbell, S. L., Jones, P. J., Williamson, G. J., Wheeler, A. J., Lucani, C., Bowman, D. M. J. S., et al. (2020). Using digital technology to protect health in prolonged poor air quality episodes: A case study of the AirRater app during the Australian 2019–20 fires. *Fire*, *3*(3).

Chang, L. T.-C., Barthelemy, X., Watt, S., Jiang, N., Riley, M., & Azzi, M. (2021). *The use of HYSPLIT in NSW in air quality Management and forecasting*. Paper presented at the Clean Air Society of Australia and New Zealand (CASANZ) Conference, Online.

Chang, L. T.-C., Duc, H., Scorgie, Y., Trieu, T., Monk, K., & Jiang, N. (2018). Performance evaluation of CCAM-CTM regional airshed modelling for the New South Wales Greater Metropolitan Region. *Atmosphere*, *9*(12), 486. http://www.mdpi.com/2073-4433/9/12/486

Chen, H., Samet, J. M., Bromberg, P. A., & Tong, H. (2021). Cardiovascular health impacts of wildfire smoke exposure. *Particle and Fibre Toxicology*, *18*(1), 2. https://www.ncbi.nlm.nih.gov/pubmed/33413506

Chen, J., Anderson, K., Pavlovic, R., Moran, M. D., Englefield, P., Thompson, D. K., et al. (2019). The FireWork v2.0 air quality forecast system with biomass burning emissions from the Canadian Forest Fire Emissions Prediction System v2.03. *Geoscientific Model Development*, *12*(7), 3283–3310.

Chen, J., Vaughan, J., Avise, J., O'Neill, S., & Lamb, B. (2008). Enhancement and evaluation of the AIRPACT ozone and PM2.5 forecast system for the Pacific Northwest. *Journal of Geophysical Research: Atmospheres*, *113*(D14). https://agupubs.onlinelibrary.wiley.com/doi/abs/10.1029/2007JD009554

Cheng, Y., Dai, T., Goto, D., Schutgens, N. A. J., Shi, G., & Nakajima, T. (2019). Investigating the assimilation of CALIPSO global aerosol vertical observations using a four-dimensional ensemble Kalman filter. *Atmospheric Chemistry and Physics*, *19*(21), 13445–13467.

Chew, B. N., Campbell, J. R., Reid, J. S., Giles, D. M., Welton, E. J., Salinas, S. V., et al. (2011). Tropical cirrus cloud contamination in sun photometer data. *Atmospheric Environment*, *45*(37), 6724–6731.

Chew, B. N., Campbell, J. R., Salinas, S. V., Chang, C. W., Reid, J. S., Welton, E. J., et al. (2013). Aerosol particle vertical distributions and optical properties over Singapore. *Atmospheric Environment*, *79*, 599–613.

Clements, C. B., Lareau, N. P., Kingsmill, D. E., Bowers, C. L., Camacho, C. P., Bagley, R., et al. (2018). The Rapid Deployments to Wildfires Experiment (RaDFIRE): Observations from the fire zone. *Bulletin of the American Meteorological Society*, *99*(12), 2539–2559.

Coen, J. L., Cameron, M., Michalakes, J., Patton, E. G., Riggan, P. J., & Yedinak, K. M. (2013). WRF-Fire: Coupled weather–wildland fire modeling with the Weather Research and Forecasting Model. *Journal of Applied Meteorology and Climatology*, *52*(1), 16–38.

Colarco, P., Benedetti, A., Reid, J., & Tanaka, T. (2014). Using EOS data to improve aerosol forecasting: the International Cooperative for Aerosol Research (ICAP). *The Earth Observer*, *26*, 14–19.

Colarco, P., da Silva, A., Chin, M., & Diehl, T. (2010). Online simulations of global aerosol distributions in the NASA GEOS-4 model and comparisons to satellite and ground-based aerosol optical depth. *Journal of Geophysical Research*, *115*(D14).

Cope, M. E. (2018). *Smoke forecasting. Sydney Hazard Reduction Burning*. Research Note, Climate Science Centre. https://www.researchgate.net/publication/326427008_Smoke_Forecasting_Sydney_Hazard_Reduction_Burning

Cope, M. E., Lee, S., Meyer, M., Reisen, F., Trindade, C., Sullivan, A., et al. (2019). *Smoke emission and transport modelling*. June 2019, Research Report 102. https://www.ffm.vic.gov.au/__data/assets/pdf_file/0027/420759/Final-Report_May_2019_v1.1-1.pdf

Côté, J., Gravel, S., Méthot, A., Patoine, A., Roch, M., & Staniforth, A. (1998). The Operational CMC-MRB Global Environmental Multiscale (GEM) model. Part I: Design considerations and formulation. *Monthly Weather Review*, *126*(6), 1373–1395.

Crutzen, P. J., & Andreae, M. O. (1990). Biomass burning in the tropics: impact on atmospheric chemistry and biogeochemical cycles. *Science*, *250*(4988), 1669–1678. https://www.ncbi.nlm.nih.gov/pubmed/17734705

Dai, T., Cheng, Y., Suzuki, K., Goto, D., Kikuchi, M., Schutgens, N. A. J., et al. (2019). Hourly aerosol assimilation of Himawari-8 AOT using the four-dimensional Local Ensemble Transform Kalman Filter. *Journal of Advances in Modeling Earth Systems*, *11*(3), 680–711.

De Groot, W. J., Field, R. D., Brady, M. A., Roswintiarti, O., & Mohamad, M. (2007). Development of the Indonesian and

Malaysian fire danger rating systems. *Mitigation and Adaptation Strategies for Global Change*, *12*(1), 165–180.

DEQ (2017). *Appendix E Plume Rise Diagnostic Evaluation, Crop Residue Burning Ozone State Implementation Plan Revision Amendment - Additional Photochemical Modeling Analysis*. https://www.regulations.gov/document/EPA-R10-OAR-2017-0566-0005

Di Giuseppe, F., Rémy, S., Pappenberger, F., & Wetterhall, F. (2018). Using the Fire Weather Index (FWI) to improve the estimation of fire emissions from fire radiative power (FRP) observations. *Atmospheric Chemistry and Physics*, *18*(8), 5359–5370.

Dickman, C., & McDonald, T. (2020). Some personal reflections on the present and future of Australia's fauna in an increasingly fire-prone continent. *Ecological Management & Restoration*, *21*(2), 86–96.

Doubleday, A., Schulte, J., Sheppard, L., Kadlec, M., Dhammapala, R., Fox, J., et al. (2020). Mortality associated with wildfire smoke exposure in Washington state, 2006–2017: A case-crossover study. *Environmental Health*, *19*(1), 4. https://www.ncbi.nlm.nih.gov/pubmed/31931820

Draxler, R. R. (1999). *HYSPLIT4 user's guide*. NOAA Tech. Memo. ERL ARL-230. Silver Spring, MD: NOAA Air Resources Laboratory.

Draxler, R. R., & Hess, G. (1997). *Description of the HYSPLIT4 modeling system*. NOAA Tech. Memo. ERL ARL-224.

Draxler, R. R., & Hess, G. (1998). An overview of the HYSPLIT_4 modeling system for trajectories. *Australian Meteorological Magazine*, *47*(4), 295–308.

Drury, S. A., Larkin, N. S., Strand, T. T., Huang, S., Strenfel, S. J., Banwell, E. M., et al. (2014). Intercomparison of fire size, fuel loading, fuel consumption, and smoke emissions estimates on the 2006 Tripod Fire, Washington, USA. *Fire Ecology*, *10*(1), 56–83.

Emmerson, K. M., Cope, M. E., Galbally, I. E., Lee, S., & Nelson, P. F. (2018). Isoprene and monoterpene emissions in south-east Australia: Comparison of a multi-layer canopy model with MEGAN and with atmospheric observations. *Atmospheric Chemistry and Physics*, *18*(10), 7539–7556.

Fan, K., Dhammapala, R., Harrington, K., Lamb, B., & Lee, Y. (2023). Machine learning-based ozone and PM2.5 forecasting: Application to multiple AQS sites in the Pacific Northwest. *Frontiers in Big Data*, *6*.

Fehsenfeld, F., Calvert, J., Fall, R., Goldan, P., Guenther, A. B., Hewitt, C. N., et al. (1992). Emissions of volatile organic compounds from vegetation and the implications for atmospheric chemistry. *Global Biogeochemical Cycles*, *6*(4), 389–430.

Field, R. D. (2020). Evaluation of Global Fire Weather Database reanalysis and short-term forecast products. *Natural Hazards and Earth System Sciences*, *20*(4), 1123–1147.

Field, R. D., Van Der Werf, G. R., Fanin, T., Fetzer, E. J., Fuller, R., Jethva, H., et al. (2016). Indonesian fire activity and smoke pollution in 2015 show persistent nonlinear sensitivity to El Niño-induced drought. *Proceedings of the National Academy of Sciences*, *113*(33), 9204–9209.

Flemming, J., Huijnen, V., Arteta, J., Bechtold, P., Beljaars, A., Blechschmidt, A. M., et al. (2015). Tropospheric chemistry in the Integrated Forecasting System of ECMWF. *Geoscientific Model Development*, *8*(4), 975–1003.

Freitas, S. R., Longo, K. M., Chatfield, R., Latham, D., Silva Dias, M., Andreae, M., et al. (2007). Including the sub-grid scale plume rise of vegetation fires in low resolution atmospheric transport models. *Atmospheric Chemistry and Physics*, *7*(13), 3385–3398. .

Freitas, S. R., Longo, K. M., Trentmann, J., & Latham, D.(2010) Technical note: Sensitivity of 1-D smoke plume rise models to the inclusion of environmental wind drag. *Atmospheric Chemistry and Physics*, *10*, 585–594. https://doi.org/10.5194/acp-10-585-2010

Fromm, M., Lindsey, D. T., Servranckx, R., Yue, G., Trickl, T., Sica, R., et al. (2010). The untold story of pyrocumulonimbus. *Bulletin of the American Meteorological Society*, *91*(9), 1193–1210.

Gan, R. W., Ford, B., Lassman, W., Pfister, G., Vaidyanathan, A., Fischer, E., et al. (2017). Comparison of wildfire smoke estimation methods and associations with cardiopulmonary-related hospital admissions. *Geohealth*, *1*(3), 122–136. https://www.ncbi.nlm.nih.gov/pubmed/28868515

Gettelman, A., Mills, M. J., Kinnison, D. E., Garcia, R. R., Smith, A. K., Marsh, D. R., et al. (2019). The Whole Atmosphere Community Climate Model Version 6 (WACCM6). *Journal of Geophysical Research: Atmospheres*, *124*(23), 12380–12403.

Girard, C., Plante, A., Desgagné, M., McTaggart-Cowan, R., Côté, J., Charron, M., et al. (2014). Staggered vertical discretization of the Canadian Environmental Multiscale (GEM) model using a coordinate of the log-hydrostatic-pressure type. *Monthly Weather Review*, *142*(3), 1183–1196.

Granier, C., Darras, S., van der Gon, H. D., Jana, D., Elguindi, N., Bo, G., et al. (2019). *The Copernicus atmosphere monitoring service global and regional emissions (April 2019 version)*. Copernicus Atmosphere Monitoring Service,

Grell, G., Freitas, S. R., Stuefer, M., & Fast, J. (2011). Inclusion of biomass burning in WRF-Chem: Impact of wildfires on weather forecasts. *Atmospheric Chemistry and Physics*, *11*(11), 5289–5303.

Griffin, D., Sioris, C., Chen, J., Dickson, N., Kovachik, A., de Graaf, M., et al. (2020). The 2018 fire season in North America as seen by TROPOMI: Aerosol layer height intercomparisons and evaluation of model-derived plume heights. *Atmospheric Measurement Techniques*, *13*(3), 1427–1445.

Guenther, A., Karl, T., Harley, P., Wiedinmyer, C., Palmer, P. I., & Geron, C. (2006). Estimates of global terrestrial isoprene emissions using MEGAN (Model of Emissions of Gases and Aerosols from Nature). *Atmospheric Chemistry and Physics*, *6*(11), 3181–3210.

Guérette, E.-A., Chang, L. T.-C., Cope, M. E., Duc, H. N., Emmerson, K. M., Monk, K., et al. (2020). Evaluation of regional air quality models over Sydney, Australia: Part 2, Comparison of PM2.5 and ozone. *Atmosphere*, *11*(3).

Guth, J., Josse, B., Marécal, V., Joly, M., & Hamer, P. (2016). First implementation of secondary inorganic aerosols in the MOCAGE version R2.15.0 chemistry transport model. *Geoscientific Model Development*, *9*(1), 137–160.

Hansen, A. B., Witham, C. S., Chong, W. M., Kendall, E., Chew, B. N., Gan, C., et al. (2019). Haze in Singapore: Source attribution of biomass burning PM 10 from Southeast Asia. *Atmospheric Chemistry and Physics*, *19*(8), 5363–5385.

Hatch, L. E., Yokelson, R. J., Stockwell, C. E., Veres, P. R., Simpson, I. J., Blake, D. R., et al. (2017). Multi-instrument comparison and compilation of non-methane organic gas emissions from biomass burning and implications for smoke-derived secondary organic aerosol precursors. *Atmospheric Chemistry and Physics*, *17*(2), 1471–1489.

Henderson, S. B., & Johnston, F. H. (2012). Measures of forest fire smoke exposure and their associations with respiratory health outcomes. *Current Opinion in Allergy and Clinical Immunology*, *12*(3), 221–227. https://www.ncbi.nlm.nih.gov/pubmed/22475995

Hertwig, D., Burgin, L., Gan, C., Hort, M., Jones, A., Shaw, F., et al. (2015). Development and demonstration of a Lagrangian dispersion modeling system for real-time prediction of smoke haze pollution from biomass burning in Southeast Asia. *Journal of Geophysical Research: Atmospheres*, *120*(24), 12605–12630.

Hodshire, A. L., Bian, Q., Ramnarine, E., Lonsdale, C. R., Alvarado, M. J., Kreidenweis, S. M., et al. (2019). More than emissions and chemistry: Fire size, dilution, and background aerosol also greatly influence near-field biomass burning aerosol aging. *Journal of Geophysical Research: Atmospheres*, *124*(10), 5589–5611.

Holben, B. N., Eck, T. F., Slutsker, I. A., Tanre, D., Buis, J., Setzer, A., et al. (1998). AERONET: A federated instrument network and data archive for aerosol characterization. *Remote Sensing of Environment*, *66*(1), 1–16.

Horsley, J. A., Broome, R. A., Johnston, F. H., Cope, M., & Morgan, G. G. (2018). Health burden associated with fire smoke in Sydney, 2001–2013. *Medical Journal of Australia*, *208*(7), 309–310. https://www.ncbi.nlm.nih.gov/pubmed/29642818

Hu, X., Yu, C., Tian, D., Ruminski, M., Robertson, K., Waller, L. A., et al. (2016). Comparison of the Hazard Mapping System (HMS) fire product to ground-based fire records in Georgia, USA. *Journal of Geophysical Research: Atmospheres*, *121*(6), 2901–2910. http://dx.doi.org/10.1002/2015JD024448

Hu, Y., Ai, H. H., Odman, M. T., Vaidyanathan, A., & Russell, A. G. (2019). Development of a WebGIS-based analysis tool for human health protection from the impacts of prescribed fire smoke in southeastern USA. *International Journal of Environmental Research and Public Health*, *16*(11), 1981. https://www.mdpi.com/1660-4601/16/11/1981

Huang, J., McQueen, J., Wilczak, J., Djalalova, I., Stajner, I., Shafran, P., et al. (2017). Improving NOAA NAQFC PM2.5 predictions with a bias correction approach. *Weather and Forecasting*, *32*(2), 407–421. https://journals.ametsoc.org/view/journals/wefo/32/2/waf-d-16-0118_1.xml

Huang, R., Hu, Y., Russell, A. G., Mulholland, J. A., & Odman, M. T. (2019). The impacts of prescribed fire on PM2.5 air quality and human health: application to asthma-related emergency room visits in Georgia, USA. *International Journal of Environmental Research and Public Health*, *16*(13), 2312. https://www.mdpi.com/1660-4601/16/13/2312

Huang, R., Zhang, X., Chan, D., Kondragunta, S., Russell Armistead, G., & Odman, M. T. (2018). Burned area comparisons between prescribed burning permits in southeastern United States and two satellite-derived products. *Journal of Geophysical Research: Atmospheres*, *123*(9), 4746–4757. https://doi.org/10.1029/2017JD028217

Hyer, E. J., Reid, J. S., Prins, E. M., Hoffman, J. P., Schmidt, C. C., Miettinen, J. I., et al. (2013). Patterns of fire activity over Indonesia and Malaysia from polar and geostationary satellite observations. *Atmospheric Research*, *122*, 504–519.

Jaffe, D. A., O'Neill, S. M., Larkin, N. K., Holder, A. L., Peterson, D. L., Halofsky, J. E., et al. (2020). Wildfire and prescribed burning impacts on air quality in the United States. *Journal of the Air & Waste Management Association* , *70*(6), 583–615. https://www.ncbi.nlm.nih.gov/pubmed/32240055

Janssens-Maenhout, G., Dentener, F., Van Aardenne, J., Monni, S., Pagliari, V., Orlandini, L., et al. (2012). *EDGAR-HTAP: A harmonized gridded air pollution emission dataset based on national inventories*. European Commission Publications Office, Ispra, Italy, EUR report No EUR, 25229, 40.

Jiang, N., Riley, M., Scorgie, Y., Betts, A., Kirkwood, J., Duc, H., et al. (2015). *Enhancing air quality forecast in New South Wales*. Paper presented at the The Clean Air Society of Australia and New Zealand (CASANZ) Conference, Melbourne, Australia.

Johnson, B. T., Brooks, M. E., Walters, D., Woodward, S., Christopher, S., & Schepanski, K. (2011). Assessment of the Met Office dust forecast model using observations from the GERBILS campaign. *Quarterly Journal of the Royal Meteorological Society*, *137*(658), 1131–1148.

Johnston, F. H., Borchers-Arriagada, N., Morgan, G. G., Jalaludin, B., Palmer, A. J., Williamson, G. J., et al. (2021). Unprecedented health costs of smoke-related PM 2.5 from the 2019–20 Australian megafires. *Nature Sustainability*, *4*(1), 42–47.

Johnston, F. H., Henderson, S. B., Chen, Y., Randerson, J. T., Marlier, M., Defries, R. S., et al. (2012). Estimated global mortality attributable to smoke from landscape fires. *Environmental Health Perspect*, *120*(5), 695–701. https://www.ncbi.nlm.nih.gov/pubmed/22456494

June, N., Vaughan, J., Lee, Y., & Lamb, B. K. (2021). Operational bias correction for PM2. 5 using the AIRPACT air quality forecast system in the Pacific Northwest. *Journal of the Air & Waste Management Association*, *71*(4), 515–527.

Kaiser, J. W., Heil, A., Andreae, M. O., Benedetti, A., Chubarova, N., Jones, L., et al. (2012). Biomass burning emissions estimated with a global fire assimilation system based on observed fire radiative power. *Biogeosciences*, *9*(1), 527–554.

Khade, V. M., Hansen, J. A., Reid, J. S., & Westphal, D. L. (2013). Ensemble filter based estimation of spatially distributed parameters in a mesoscale dust model: Experiments with simulated and real data. *Atmospheric Chemistry and Physics*, *13*(6), 3481–3500.

Klose, M., Jorba, O., Gonçalves Ageitos, M., Escribano, J., Dawson, M. L., Obiso, V., et al. (2021). Mineral dust cycle in the Multiscale Online Nonhydrostatic AtmospheRe CHemistry model (MONARCH) version 2.0. *Geoscientific Model Development*, *14*(10), 6403–6444.

Koster, R. D., Darmenov, A. S., & da Silva, A. M. (2015). *The Quick Fire Emissions Dataset (QFED): Documentation of versions 2.1, 2.2 and 2.4*. Volume *38*; Technical Report Series on Global Modeling and Data Assimilation.

Lahm, P. W., & Larkin, N. K. (2020). The U.S. Interagency wildland fire air quality response program. *Environmental Managment*, *7*.

Larkin, N. K., O'Neill, S. M., Solomon, R., Raffuse, S., Strand, T., Sullivan, D. C., et al. (2009). The BlueSky smoke modeling framework. *International Journal of Wildland Fire*, *18*(8).

Larkin, N. K., Raffuse, S. M., Huang, S., Pavlovic, N., Lahm, P., & Rao, V. (2020). The Comprehensive Fire Information Reconciled Emissions (CFIRE) inventory: Wildland fire emissions developed for the 2011 and 2014 U.S. National Emissions Inventory. *Journal of the Air & Waste Management Association*, *70*(11), 1165–1185. https://www.ncbi.nlm.nih.gov/pubmed/32915705

Larkin, N. K., Raffuse, S. M., & Strand, T. M. (2014). Wildland fire emissions, carbon, and climate: U.S. emissions inventories. *Forest Ecology and Management*, *317*, 61–69.

Larkin, N. K., Strand, T. M., Drury, S. A., Raffuse, S. M., Solomon, R. C., O'Neill, S. M., et al. (2012). *Phase 1 of the Smoke and Emissions Model Intercomparison Project (SEMIP): creation of SEMIP and evaluation of current models*. Final Report to the Joint Fire Science Program, Project #08-1-6-10. Seattle, WA, USA. https://www.firescience.gov

Lawson, S. J., Cope, M., Lee, S., Galbally, I. E., Ristovski, Z., & Keywood, M. D. (2017). Biomass burning at Cape Grim: Exploring photochemistry using multi-scale modelling. *Atmospheric Chemistry and Physics*, *17*(19), 11707–11726.

Lee, P., McQueen, J., Stajner, I., Huang, J., Pan, L., Tong, D., et al. (2017). NAQFC developmental forecast guidance for fine particulate matter (PM2.5). *Weather and Forecasting*, *32*(1), 343–360. https://journals.ametsoc.org/view/journals/wefo/32/1/waf-d-15-0163_1.xml

Lee, S. Y., Gan, C., & Chew, B. N. (2016). Visibility deterioration and hygroscopic growth of biomass burning aerosols over a tropical coastal city: A case study over Singapore's airport. *Atmospheric Science Letters*, *17*(12), 624–629.

Li, M., Zhang, Q., Kurokawa, J.-I., Woo, J.-H., He, K., Lu, Z., et al. (2017). MIX: A mosaic Asian anthropogenic emission inventory under the international collaboration framework of the MICS-Asia and HTAP. *Atmospheric Chemistry and Physics*, *17*(2), 935–963.

Li, Y., Tong, D. Q., Ngan, F., Cohen, M. D., Stein, A. F., Kondragunta, S., et al. (2020). Ensemble PM2.5 forecasting during the 2018 Camp Fire Event using the HYSPLIT transport and dispersion model. *Journal of Geophysical Research: Atmospheres*, *125*(15), e2020JD032768. https://agupubs.onlinelibrary.wiley.com/doi/abs/10.1029/2020JD032768

Linn, R., Reisner, J., Colman, J. J., & Winterkamp, J. (2002). Studying wildfire behavior using FIRETEC. *International Journal of Wildland Fire*, *11*(4).

Lipner, E. M., O'Dell, K., Brey, S. J., Ford, B., Pierce, J. R., Fischer, E. V., et al. (2019). The associations between clinical respiratory outcomes and ambient wildfire smoke exposure among pediatric asthma patients at National Jewish Health, 2012–2015. *Geohealth*, *3*(6), 146–159. https://www.ncbi.nlm.nih.gov/pubmed/32159037

Liu, Y. (2014). A regression model for smoke plume rise of prescribed fires using meteorological conditions. *Journal of Applied Meteorology and Climatology*, *53*(8), 1961–1975.

Lynch, P., Reid, J. S., Westphal, D. L., Zhang, J., Hogan, T. F., Hyer, E. J., et al. (2016). An 11-year global gridded aerosol optical thickness reanalysis (v1.0) for atmospheric and climate sciences. *Geoscientific Model Development*, *9*(4), 1489–1522.

Majumder, S., Guan, Y., Reich, B. J., O'Neill, S., & Rappold, A. G. (2021). Statistical downscaling with spatial misalignment: Application to wildland fire PM2.5 concentration forecasting. *Journal of Agricultural, Biological and Environmental Statistics*, *26*(1), 23–44. https://www.ncbi.nlm.nih.gov/pubmed/33867783

Makar, P. A., Akingunola, A., Chen, J., Pabla, B., Gong, W., Stroud, C., et al. (2020). Forest fire aerosol-weather feedbacks over western North America using a high-resolution, fully coupled, air-quality model. *Atmospheric Chemistry and Physics Discussions*, 1–55.

Mandel, J., Amram, S., Beezley, J., Kelman, G., Kochanski, A., Kondratenko, V., et al. (2014). Recent advances and applications of WRF-SFIRE. *Natural Hazards and Earth System Sciences*, *14*(10), 2829–2845.

Marécal, V., Peuch, V. H., Andersson, C., Andersson, S., Arteta, J., Beekmann, M., et al. (2015). A regional air quality forecasting system over Europe: The MACC-II daily ensemble production. *Geoscientific Model Development*, *8*(9), 2777–2813.

Marsha, A., & Larkin, N. K. (2022). Evaluating satellite fire detection products and an ensemble approach for estimating burned area in the United States. *Fire*, *5*(5), 147.

Mass, C. F., Albright, M., Ovens, D., Steed, R., Maciver, M., Grimit, E., et al. (2003). Regional environmental prediction over the Pacific Northwest. *Bulletin of the American Meteorological Society*, *84*(10), 1353–1366. https://journals.ametsoc.org/view/journals/bams/84/10/bams-84-10-1353.xml

Matz, C. J., Egyed, M., Xi, G., Racine, J., Pavlovic, R., Rittmaster, R., et al. (2020). Health impact analysis of PM2.5 from wildfire smoke in Canada (2013–2015, 2017–2018). *Science of the Total Environment*, *725*, 138506. https://www.ncbi.nlm.nih.gov/pubmed/32302851

McKenzie, D., Raymond, C., Kellogg, L.-K., Norheim, R., Andreu, A., Bayard, A., et al. (2007). Mapping fuels at multiple scales: Landscape application of the Fuel Characteristic Classification System. *Canadian Journal of Forest Research*, *37*(12), 2421–2437.

Mell, W., Jenkins, M. A., Gould, J., & Cheney, P. (2007). A physics-based approach to modelling grassland fires. *International Journal of Wildland Fire*, *16*(1).

Meyer, C. M., Luhar, A. K., & Mitchell, R. M. (2008). Biomass burning emissions over northern Australia constrained by aerosol measurements: I. Modelling the distribution of hourly emissions. *Atmospheric Environment*, *42*(7), 1629–1646.

Miettinen, J., Shi, C., Tan, W. J., & Liew, S. C. (2012). 2010 land cover map of insular Southeast Asia in 250-m spatial resolution. *Remote Sensing Letters*, *3*(1), 11–20.

Moisseeva, N., & Stull, R. (2021). Wildfire smoke-plume rise: A simple energy balance parameterization. *Atmospheric Chemistry and Physics*, *21*(3), 1407–1425.

Monk, K., Chang, L. T.-C., Barthelemy, X., Fuchs, D., Trieu, T., Duc, H., et al. (2019). *Development and evaluation of a smoke emission module in NSW Operational Air Quality Forecast Modelling system*. Paper presented at the the 6th International Fire Behavior and Fuels Conference, Sydney, Australia. https://www.researchgate.net/publication/332872310_Development_and_evaluation_of_asmoke_emission_module_in_NSW_operational_air_quality_forecast_modelling_system

Moore, C. T., Jr., Randall, D., Mavko, M., Morris, R., Koo, B., Fitch, M., et al. (2013). *Deterministic and empirical assessment of smoke's contribution to ozone (DEASCO3)*. Final report, Joint Fire Sciences Program Project # 11-1-6-6.

Moran, M. D., Lupu, A., Zhang, J., Savic-Jovcic, V., & Gravel, S. (2018). A comprehensive performance evaluation of the next generation of the Canadian operational regional air quality deterministic prediction system. In *Air pollution modeling and its application XXV 35* (pp. 75–81). Springer International Publishing.

Moran, M. D., Menard, S., & Anselmo, D. (2019). *Regional Air Quality Deterministic Prediction System (RAQDPS): Update from version 020.2 to version 021. Technical Note*. Montreal, Quebec, Canada. http://collaboration.cmc.ec.gc.ca/cmc/cmoi/product_guide/docs/tech_notes/technote_raqdps-021_20190703_e.pdf

Moran, M. D., Menard, S., Pavlovic, R., Anselmo, D., Antonopoulos, S., Makar, P., et al. (2014). Recent advances in Canada's national operational AQ forecasting system. *Air Pollution Modeling and Its Application XXII*, 215–220.

Munoz-Alpizar, R., Ménard, S., Menelaou, K., Keita, S., Pavlovic, R., Moran, M. D., et al. (2020). *Regional Air Quality Deterministic Prediction System with Near-Real-Time Wildfire Emissions (RAQDPSFW): Upgrade from version 020.2 to version 021. Technical note*. Montreal, Quebec, Canada.

Munoz-Alpizar, R., Pavlovic, R., Moran, M. D., Chen, J., Gravel, S., Henderson, S. B., et al. (2017). Multi-year (2013–2016) PM2. 5 wildfire pollution exposure over North America as determined from operational air quality forecasts. *Atmosphere*, *8*(9), 179.

Munson, J., Vaughan, J. K., Lamb, B. K., & Lee, Y. (2021). Decadal evaluation of the AIRPACT Regional Air Quality Forecast System in the Pacific Northwest from 2009–2018. https://doi.org/10.31223/X5J61T

Napelenok, S., Cohan, D., Odman, M. T., & Tonse, S. (2008). Extension and evaluation of sensitivity analysis capabilities in a photochemical model. *Environmental Modelling & Software*, *23*(8), 994–999.

NRCan (2020). *Canadian Wildland Fire Information System*. https://cwfis.cfs.nrcan.gc.ca/home

O'Neill, S. M., Diao, M., Raffuse, S., Al-Hamdan, M., Barik, M., Jia, Y., et al. (2021). A Multi-analysis approach for estimating regional health impacts from the 2017 northern California wildfires. *Journal of the Air & Waste Management Association*. https://www.ncbi.nlm.nih.gov/pubmed/33630725

O'Neill, S. M., & Lamb, B. K. (2005). Intercomparison of the community multiscale air quality model and CALGRID using process analysis. *Environmental Science & Technology*, *39*(15), 5742–5753. https://www.ncbi.nlm.nih.gov/pubmed/16124311

O'Neill, S. M., Lamb, B. K., Chen, J., Claiborn, C., Finn, D., Otterson, S., et al. (2006). Modeling ozone and aerosol formation and transport in the pacific northwest with the community Multi-Scale Air Quality (CMAQ) modeling system. *Environmental Science & Technology*, *40*(4), 1286–1299. https://www.ncbi.nlm.nih.gov/pubmed/16572788

Odman, M., Huang, R., Pophale, A., Sakhpara, R., Hu, Y., Russell, A., et al. (2018). Forecasting the impacts of prescribed fires for dynamic air quality management. *Atmosphere*, *9*(6), 220. http://www.mdpi.com/2073-4433/9/6/220

Ovens, D., & Mass, C. (2020). *On-line WRF change log*. University of Washington, Department of Atmospheric Science, Pacific Northwest Environmental Forecasts and Observations. https://a.atmos.washington.edu/mm5rt/log.html

Page, S. E., Rieley, J. O., & Banks, C. J. (2011). Global and regional importance of the tropical peatland carbon pool. *Global Change Biology*, *17*(2), 798–818.

Pagowski, M., & Grell, G. A. (2012). Experiments with the assimilation of fine aerosols using an ensemble Kalman filter. *Journal of Geophysical Research: Atmospheres*, *117*(D21).

Pan, L., Tong, D., Lee, P., Kim, H. C., & Chai, T. (2014). Assessment of NOx and O3 forecasting performances in the U.S. National Air Quality Forecasting Capability before and after the 2012 major emissions updates. *Atmospheric Environment*, *95*, 610–619.

Pan, X., Ichoku, C., Chin, M., Bian, H., Darmenov, A., Colarco, P., et al. (2020). Six global biomass burning emission datasets: Intercomparison and application in one global aerosol model. *Atmospheric Chemistry and Physics*, *20*(2), 969–994.

Paugam, R., Wooster, M., Freitas, S., & Val Martin, M. (2016). A review of approaches to estimate wildfire plume injection height within large-scale atmospheric chemical transport models. *Atmospheric Chemistry and Physics*, *16*(2), 907–925.

Pavlovic, R., Chen, J., Anderson, K., Moran, M. D., Beaulieu, P. A., Davignon, D., et al. (2016). The FireWork air quality forecast system with near-real-time biomass burning emissions: Recent developments and evaluation of performance for the 2015 North American wildfire season. *Journal of the Air & Waste Management Association*, *66*(9), 819–841. https://www.ncbi.nlm.nih.gov/pubmed/26934496

Pavlovic, R., Moran, M. D., Gilbert, S., Davignon, D., Bouchet, V., Stajner, I., et al. (2018). *Multi-model air quality performance analysis over North America for ECCC, NOAA/NWS and CAMS operational forecast systems*. https://atmosphere.copernicus.eu/sites/default/files/2018-11/2_3rd_ECCC_NOAA_ECMWF_v06.pdf

Peterson, D. A., Campbell, J. R., Hyer, E. J., Fromm, M. D., Kablick, G. P., III, Cossuth, J. H., et al. (2018). Wildfire-driven thunderstorms cause a volcano-like stratospheric injection of smoke. *npj Climate and Atmospheric Science*, *1*. https://www.ncbi.nlm.nih.gov/pubmed/31360778

Peterson, D. A., Hyer, E. J., Campbell, J. R., Solbrig, J. E., & Fromm, M. D. (2017). A conceptual model for development of intense pyrocumulonimbus in western North America. *Monthly Weather Review*, *145*(6), 2235–2255. https://journals.ametsoc.org/view/journals/mwre/145/6/mwr-d-16-0232.1.xml

Powers, J. G., Klemp, J. B., Skamarock, W. C., Davis, C. A., Dudhia, J., Gill, D. O., et al. (2017). The weather research and forecasting model: Overview, system efforts, and future directions. *Bulletin of the American Meteorological Society*, *98*(8), 1717–1737.

Prichard, S., Larkin, N. K., Ottmar, R., French, N. H., Baker, K., Brown, T., et al. (2019). The fire and smoke model evaluation experiment: A plan for integrated, large fire-atmosphere field campaigns. *Atmosphere*, *10*(2), 66.

Pyne, S. J. (2001). *Fire: A brief history*. University of Washington Press.

Raffuse, S. M., Craig, K. J., Larkin, N. K., Strand, T. T., Sullivan, D. C., Wheeler, N. J., et al. (2012). An evaluation of modeled plume injection height with satellite-derived observed plume height. *Atmosphere*, *3*(1), 103–123.

Raffuse, S. M., Pryden, D. A., Sullivan, D. C., Larkin, N. K., Strand, T., & Solomon, R. (2009). *SMARTFIRE algorithm description*. https://doi.org/10.1007/978-1-935704-23-2_3

Randles, C. A., Da Silva, A. M., Buchard, V., Colarco, P. R., Darmenov, A., Govindaraju, R., et al. (2017). The MERRA-2 aerosol reanalysis, 1980 onward; Part I: System description and data assimilation evaluation. *Journal of Climate*, *30*(17), 6823–6850. https://www.ncbi.nlm.nih.gov/pubmed/29576684

Rappold, A. G., Cascio, W. E., Kilaru, V. J., Stone, S. L., Neas, L. M., Devlin, R. B., et al. (2012). Cardio-respiratory outcomes associated with exposure to wildfire smoke are modified by measures of community health. *Environmental Health*, *11*, 71. https://www.ncbi.nlm.nih.gov/pubmed/23006928

Reid, C. E., Brauer, M., Johnston, F. H., Jerrett, M., Balmes, J. R., & Elliott, C. T. (2016). Critical review of health impacts of wildfire smoke exposure. *Environmental Health Perspect*, *124*(9), 1334–1343. https://www.ncbi.nlm.nih.gov/pubmed/27082891

Reid, J. S., Benedetti, A., Colarco, P. R., & Hansen, J. A. (2011). International Operational Aerosol Observability Workshop. *Bulletin of the American Meteorological Society*, *92*(6), ES21–ES24.

Reid, J. S., Hyer, E. J., Prins, E. M., Westphal, D. L., Zhang, J., Wang, J., et al. (2009). Global monitoring and forecasting of biomass-burning smoke: description of and lessons from the fire locating and modeling of burning emissions (FLAMBE) program. *IEEE Journal of Selected Topics in Applied Earth Observations and Remote Sensing*, *2*(3), 144–162.

Reisen, F., Meyer, C., Weston, C., & Volkova, L. (2018). Ground-based field measurements of PM2. 5 emission factors from flaming and smoldering combustion in eucalypt forests. *Journal of Geophysical Research: Atmospheres*, *123*(15), 8301–8314.

Remer, L. A., Kaufman, Y. J., Tanré, D., Mattoo, S., Chu, D. A., Martins, J. V., et al. (2005). The MODIS aerosol algorithm, products, and validation. *Journal of the Atmospheric Sciences*, *62*(4), 947–973. https://journals.ametsoc.org/view/journals/atsc/62/4/jas3385.1.xml

Rémy, S., Kipling, Z., Flemming, J., Boucher, O., Nabat, P., Michou, M., et al. (2019). Description and evaluation of the tropospheric aerosol scheme in the European Centre for Medium-Range Weather Forecasts (ECMWF) Integrated Forecasting System (IFS-AER, cycle 45R1). *Geoscientific Model Development*, *12*(11), 4627–4659.

Richter, D., Lamb, B., Westburg, H., Vaughan, J., & Gross, M. (2004). *Development of simple dispersion model for simulation of air toxics in urban areas*. Paper presented at the ASME 2004 Heat Transfer/Fluids Engineering Summer Conference.

Roberts, G., Wooster, M. J., & Lagoudakis, E. (2009). Annual and diurnal african biomass burning temporal dynamics. *Biogeosciences*, *6*(5), 849–866.

Rubin, J. I., Reid, J. S., Hansen, J. A., Anderson, J. L., Holben, B. N., Xian, P., et al. (2017). Assimilation of AERONET and MODIS AOT observations using variational and ensemble data assimilation methods and its impact on aerosol forecasting skill. *Journal of Geophysical Research: Atmospheres*, *122*(9), 4967–4992.

Salinas, S. V., Chew, B. N., & Liew, S. C. (2009). Retrievals of aerosol optical depth and Ångström exponent from ground-based Sun-photometer data of Singapore. *Applied optics*, *48*(8), 1473–1484.

Sayer, A. M., Hsu, N. C., Lee, J., Kim, W. V., Dubovik, O., Dutcher, et al. (2018). Validation of SOAR VIIRS over-water aerosol retrievals and context within the global satellite aerosol data record. *Journal of Geophysical Research: Atmospheres*, *123*(23).

Schutgens, N. A. J., Miyoshi, T., Takemura, T., & Nakajima, T. (2010). Sensitivity tests for an ensemble Kalman filter for aerosol assimilation. *Atmospheric Chemistry and Physics*, *10*(14), 6583–6600.

Schwartz, C. S., Liu, Z., Lin, H.-C., & Cetola, J. D. (2014). Assimilating aerosol observations with a "hybrid" variational-ensemble data assimilation system. *Journal of Geophysical Research: Atmospheres*, *119*(7), 4043–4069.

Sekiyama, T. T., Tanaka, T. Y., Shimizu, A., & Miyoshi, T. (2010). Data assimilation of CALIPSO aerosol observations. *Atmospheric Chemistry and Physics*, *10*(1), 39–49.

Sessions, W. R., Reid, J. S., Benedetti, A., Colarco, P. R., da Silva, A., Lu, S., et al. (2015). Development towards a global operational aerosol consensus: basic climatological characteristics of the International Cooperative for Aerosol Prediction Multi-Model Ensemble (ICAP-MME). *Atmospheric Chemistry and Physics*, *15*(1), 335–362.

Sindelarova, K., Granier, C., Bouarar, I., Guenther, A., Tilmes, S., Stavrakou, T., et al. (2014). Global data set of biogenic VOC emissions calculated by the MEGAN model over the last 30 years. *Atmospheric Chemistry and Physics*, *14*(17), 9317–9341.

Singapore (1997). Regional haze action plan. https://cil.nus.edu.sg/wp-content/uploads/formidable/18/1997-Regional-Haze-Action-Plan.pdf

Skamarock, C., Klemp, B., Dudhia, J., Gill, O., Liu, Z., Berner, J., et al. (2019). *A description of the advanced research WRF model version 4*. National Center for Atmospheric Research: Boulder, CO, USA, 145(145), 550. NCAR Technical Note NCAR/TN-556+STR

Skamarock, W. C. (2004). Evaluating mesoscale NWP models using kinetic energy spectra. *Monthly Weather Review*, *132*(12), 3019–3032.

Smirnov, A., Holben, B., Eck, T., Dubovik, O., & Slutsker, I. (2000). Cloud-screening and quality control algorithms for the AERONET database. *Remote Sensing of Environment*, *73*(3), 337–349.

Snow, J. A., Dennison, J. B., Jaffe, D. A., Price, H. U., Vaughan, J. K., & Lamb, B. (2003). Aircraft and surface observations of air quality in Puget Sound and a comparison to a regional model. *Atmospheric Environment*, *37*(28), 4019–4032.

So, R., Teakles, A., Baik, J., Vingarzan, R., & Jones, K. (2018). Development of visibility forecasting modeling framework for the Lower Fraser Valley of British Columbia using Canada's Regional Air Quality Deterministic Prediction System.

Journal of the Air & Waste Management Association, *68*(5), 446–462. https://www.ncbi.nlm.nih.gov/pubmed/29341857

Sofiev, M., Ermakova, T., & Vankevich, R. (2012). Evaluation of the smoke-injection height from wild-land fires using remote-sensing data. *Atmospheric Chemistry and Physics*, *12*(4), 1995–2006.

Sofiev, M., Vankevich, R., Ermakova, T., & Hakkarainen, J. (2013). Global mapping of maximum emission heights and resulting vertical profiles of wildfire emissions. *Atmospheric Chemistry and Physics*, *13*(14), 7039–7052.

Sofiev, M., Vankevich, R., Lotjonen, M., Prank, M., Petukhov, V., Ermakova, T., et al. (2009). An operational system for the assimilation of the satellite information on wild-land fires for the needs of air quality modelling and forecasting. *Atmospheric Chemistry and Physics*, *9*(18), 6833–6847.

Sofiev, M., Vira, J., Kouznetsov, R., Prank, M., Soares, J., & Genikhovich, E. (2015). Construction of the SILAM Eulerian atmospheric dispersion model based on the advection algorithm of Michael Galperin. *Geoscientific Model Development*, *8*(11), 3497–3522.

Stein, A., Draxler, R. R., Rolph, G. D., Stunder, B. J., Cohen, M., & Ngan, F. (2015). NOAA's HYSPLIT atmospheric transport and dispersion modeling system. *Bulletin of the American Meteorological Society*, *96*(12), 2059–2077.

Stieb, D. M., Burnett, R. T., Smith-Doiron, M., Brion, O., Shin, H. H., & Economou, V. (2008). A new multipollutant, no-threshold air quality health index based on short-term associations observed in daily time-series analyses. *Journal of the Air & Waste Management Association*, *58*(3), 435–450. https://www.ncbi.nlm.nih.gov/pubmed/18376646

Sullivan, A., Surawski, N., Crawford, D., Hurley, R., Volkova, L., Weston, C., et al. (2018). Effect of woody debris on the rate of spread of surface fires in forest fuels in a combustion wind tunnel. *Forest Ecology and Management*, *424*, 236–245.

Tanaka, T. Y., & Ogi, A. (2018). On the upgrade of the JMA's global aeolian dust forecasting model. *Sokko Jiho*, *84*, 20.

Tang, Y., Bian, H., Tao, Z., Oman, L. D., Tong, D., Lee, P., et al. (2021). Comparison of chemical lateral boundary conditions for air quality predictions over the contiguous United States during pollutant intrusion events. *Atmospheric Chemistry and Physics*, *21*(4), 2527–2550.

Tong, D. Q., Lamsal, L., Pan, L., Ding, C., Kim, H., Lee, P., et al. (2015). Long-term NOx trends over large cities in the United States during the great recession: Comparison of satellite retrievals, ground observations, and emission inventories. *Atmospheric Environment*, *107*, 70–84.

Tong, D. Q., & Tang, Y. (2018). Advancing air quality forecasting to protect human health. *Environmental Managers* (October 2018).

Tymstra, C., Stocks, B. J., Cai, X., & Flannigan, M. D. (2020). Wildfire management in Canada: Review, challenges and opportunities. *Progress in Disaster Science*, *5*.

Urbanski, S. P., Hao, W. M., & Baker, S. (2009). Chemical composition of wildland fire emissions. In A. Bytnerowicz, M. J. Arbaugh, A. R. Riebau, & C. Andersen (Eds.), *Wildland fires and air pollution (pp. 79–107)*. Developments in Environmental Science, Vol. 8. Elsevier. https://doi.org/10.1016/S1474-8177(08)00004-1

U.S. EPA, (2017). *2017 National Emissions Inventory (NEI) data*. https://www.epa.gov/air-emissions-inventories/2017-national-emissions-inventory-nei-data

Vaughan, J., Lamb, B., Frei, C., Wilson, R., Bowman, C., Figueroa-Kaminsky, C., et al. (2004). A numerical daily air quality forecast system for the Pacific Northwest. *Bulletin of the American Meteorological Society*, *85*(4), 549–562. https://journals.ametsoc.org/view/journals/bams/85/4/bams-85-4-549.xml

Volkova, L., Meyer, C. P. M., Haverd, V., & Weston, C. J. (2018). A data-model fusion methodology for mapping bushfire fuels for smoke emissions forecasting in forested landscapes of south-eastern Australia. *Journal of Environmental Management*, *222*, 21–29. https://www.ncbi.nlm.nih.gov/pubmed/29800860

Watt, S., Chang, L. T.-C., Jiang, N., Fuchs, D., Barthelemy, X., Scorgie, Y., et al. (2017). *Using Hysplit to forecast smoke plumes during hazard reduction burns in New South Wales*. Paper presented at the The Clean Air Society of Australia and New Zealand (CASANZ) Conference, Brisbane, Australia.

Wettstein, Z. S., Hoshiko, S., Fahimi, J., Harrison, R. J., Cascio, W. E., & Rappold, A. G. (2018). Cardiovascular and cerebrovascular emergency department visits associated with wildfire smoke exposure in California in 2015. *Journal of the American Heart Association*, *7*(8). https://www.ncbi.nlm.nih.gov/pubmed/29643111

Williamson, G. J., & Lucani, C. (2020). AQVx: An interactive visual display system for air pollution and public health. *Frontiers in Public Health*, *8*, 85.

WMO. (2017). *Revised Manual on the Global Data-Processing and Forecasting System*. http://www.wmo.int/pages/prog/www/DPS/documents/Manual-GDPFS-Jul2017.pdf

WMO (2018). *Vegetation Fire and Smoke Pollution Warning and Advisory System (VFSP-WAS): Concept note and expert recommendations*. https://library.wmo.int/opac/index.php?lvl=notice_display&id=20244

WMO (2020). *Training materials and best practices for chemical weather/air quality forecasting* (ETR- No. 26). https://library.wmo.int/index.php?lvl=notice_display&id=21801#.YKteKqgzZjE

Wotton, B. M., Alexander, M. E., & Taylor, S. W. (2009). *Updates and revisions to the 1992 Canadian forest fire behavior prediction system. Sault Ste*. Marie, ON, Canada: Great Lakes Forestry Centre. https://www.researchgate.net/profile/Sw-Taylor/publication/274383139_Updates_and_Revisions_to_the_1992_Canadian_Forest_Fire_Behavior_Prediction_System_Sault_Ste/links/551da6e20cf213ef063e84ea/Updates-and-Revisions-to-the-1992-Canadian-Forest-Fire-Behavior-Prediction-System-Sault-Ste.pdf

Wotton, B. M., Flannigan, M. D., & Marshall, G. A. (2017). Potential climate change impacts on fire intensity and key wildfire suppression thresholds in Canada. *Environmental Research Letters*, *12*(9).

Xi, Y., Kshirsagar, A. V., Wade, T. J., Richardson, D. B., Brookhart, M. A., Wyatt, L., et al. (2020). Mortality in U.S. hemodialysis patients following exposure to wildfire smoke. *Journal of the American Society of Nephrology*, *31*(8), 1824–1835. https://www.ncbi.nlm.nih.gov/pubmed/32675302

Xian, P., Reid, J. S., Hyer, E. J., Sampson, C. R., Rubin, J. I., Ades, M., et al. (2019). Current state of the global operational aerosol multi-model ensemble: An update from the International Cooperative for Aerosol Prediction (ICAP). *Quarterly Journal of the Royal Meteorological Society*, *145*(Suppl 1), 176–209. https://www.ncbi.nlm.nih.gov/pubmed/31787783

Ye, X., Arab, P., Ahmadov, R., James, E., Grell, G. A., Pierce, B., et al. (2021). Evaluation and intercomparison of wildfire smoke forecasts from multiple modeling systems for the 2019 Williams Flats fire. *Atmospheric Chemistry and Physics Discussions*, 1–69.

Zhang, J., Campbell, J. R., Reid, J. S., Westphal, D. L., Baker, N. L., Campbell, W. F., et al. (2011). Evaluating the impact of assimilating CALIOP-derived aerosol extinction profiles on a global mass transport model. *Geophysical Research Letters*, *38*(14), n/a-n/a.

Zhang, J., Reid, J. S., Westphal, D. L., Baker, N. L., & Hyer, E. J. (2008). A system for operational aerosol optical depth data assimilation over global oceans. *Journal of Geophysical Research: Atmospheres*, *113*(D10).

Zhang, J., Spurr, R. J., Reid, J. S., Xian, P., Colarco, P. R., Campbell, J. R., . . . & Baker, N. L. (2021). Development of an Ozone Monitoring Instrument (OMI) aerosol index (AI) data assimilation scheme for aerosol modeling over bright surfaces–a step toward direct radiance assimilation in the UV spectrum. *Geoscientific Model Development*, *14*(1), 27–42.

Zhang, L., et al. (2020). Development of GEFS-aerosols into NOAA's Unified Forecast System (UFS). *Geoscientific Model Development*. 2020 UFS Workshop.

Zhang, L., Montuoro, R., McKeen, S. A., Baker, B., Bhattacharjee, P. S., Grell, G. A., et al. (2022). Development and evaluation of the Aerosol Forecast Member in the National Center for Environment Prediction (NCEP)'s Global Ensemble Forecast System (GEFS-Aerosols v1). *Geoscientific Model Development*, *15*(13), 5337–5369.

Zhang, X., Kondragunta, S., & Roy, D. P. (2014). Interannual variation in biomass burning and fire seasonality derived from geostationary satellite data across the contiguous United States from 1995 to 2011. *Journal of Geophysical Research: Biogeosciences*, *119*(6), 1147–1162.

Zhang, X., Kondragunta, S., Ram, J., Schmidt, C., & Huang, H.-C. (2012). Near-real-time global biomass burning emissions product from geostationary satellite constellation. *Journal of Geophysical Research: Atmospheres*, *117*(D14).

Zou, Y., O'Neill, S. M., Larkin, N. K., Alvarado, E. C., Solomon, R., Mass, C., et al. (2019). Machine Learning-Based Integration of High-Resolution Wildfire Smoke Simulations and Observations for Regional Health Impact Assessment. *International Journal of Environmental Research and Public Health*, *16*(12). https://www.ncbi.nlm.nih.gov/pubmed/31212933

Part III

From Concentrations to Health Outcomes

10

Assessing Smoke Exposure in Space and Time

Patricia D. Koman[1] and Nancy H. F. French[2]

ABSTRACT

Numerous toxic air pollutants compose smoke from biomass burning, including fine particulate matter ($PM_{2.5}$), nitrogen oxide (NO_x) and volatile organic carbon (VOC) emissions that contribute to ozone formation, carbon monoxide (CO), polycyclic aromatic hydrocarbons (PAH), formaldehyde, and metals. This chapter provides a conceptual framework and vocabulary related to characterizing human exposures to smoke, focusing on inhalation exposure at individual and population levels. We briefly describe fundamental concepts of exposure assessment relevant to air pollution from biomass burning. Studying exposure requires knowledge of both the receptors (e.g., how the stressor acts on the human body, relevant periods of susceptibility) and the ways in which the stressor presents at this interface (e.g., fate and transport of chemical agents, social patterns that affect exposures and access to mitigation, time and activity patterns). We present considerations for quantifying exposure to air pollution, specifically for developing geospatial assessments of exposure, the types of exposure metrics, and associated reporting error and uncertainty. An area of active research is to what extent $PM_{2.5}$ exposure from smoke is associated with various types of health outcomes as compared with fossil fuel combustion emissions and all other sources and if similar populations are susceptible. Finally, we discuss research needs for quantifying exposure to fire pollution.

10.1. FUNDAMENTALS OF SMOKE EXPOSURE ESTIMATION

10.1.1. Introduction to Assessing Landscape Burning Smoke Risks to Population Health

Biomass burning smoke degrades air quality for broad portions of the Earth, exposing large populations to elevated levels of particulate matter (PM), ozone (O_3), and other pollutants known to cause harm to humans (Bernard et al., 2001; Holm et al., 2021; Pratt et al., 2019). According to the World Health Organization (WHO; World Health Organization, 2021), beyond fatalities, fires, and the resulting smoke and ash, can cause the following: Bodily injuries; eye, nose, throat, and lung irritation; decreased lung function, including coughing and wheezing; pulmonary inflammation, bronchitis, exacerbations of asthma, and other lung diseases like COPD; and exacerbation of cardiovascular diseases, such as heart failure.

Smoke from biomass burning is a dynamic mixture of gases, aerosols, and solids, changing temporally, spatially, and chemically as it disperses from its source. As discussed in Chapter 7, toxic air pollutants include PM, nitrogen oxides (NO_x) and volatile organic carbon (VOC) emissions that contribute to ozone (O_3) formation, carbon monoxide (CO), polycyclic aromatic hydrocarbons (PAH), formaldehyde, and metals (Cascio, 2018; Naeher et al., 2007). Smoke emission composition at the source is dependent on combustion conditions and fuel, while its variation across space as it transports is highly

[1] *Department of Environmental Health Sciences, University of Michigan, Ann Arbor, Michigan, USA*

[2] *Michigan Tech Research Institute, Michigan Technological University, Ann Arbor, Michigan, USA*

Landscape Fire, Smoke, and Health: Linking Biomass Burning Emissions to Human Well-Being, Geophysical Monograph 280, First Edition.
Edited by Tatiana V. Loboda, Nancy H. F. French, and Robin C. Puett.

DOI: 10.1002/9781119757030.ch10

influenced by atmospheric and meteorological factors (see Parts I and II of this volume). The chemical properties of air quality immediately proximate to a biomass burning event may differ substantially from areas far away due to chemical reactions that are sensitive to temperature and availability of sunlight. For example, PM can be directly emitted or formed secondarily in the atmosphere from precursors, while ozone, a powerful oxidant, is formed from combustion products as they break down in the atmosphere from sunlight. Therefore, biomass burning smoke content can be highly variable in space and time. Furthermore, the chemical signature of pollutants from landscape fires differs from those originating from fossil fuel combustion, and exposure scenarios differ as well, with substantially sharper peaks during wildfires versus typical urban ambient air pollution, which is relatively more constant at a regional scale (Liu et al., 2016; Thelen et al., 2013). Given this, approaches for quantifying and characterizing exposure to biomass burning smoke differ from fossil fuel derived pollution.

Smoke exposure assessment is an inherently a multidisciplinary and geospatial problem, which is fundamentally driven by the location and timing of fire events, the chemical composition, and where smoke is in relation to people. Approaching the topic in a geospatial framework provides access to the diverse set of powerful tools available from geospatial data science. To improve understanding of biomass burning smoke on air quality and health, characterizations of exposure in both space and time are needed for toxicological studies, from which a dose-response curve can be developed (e.g., for PAH toxicity) and for epidemiologic studies, which inform concentration-response functions (e.g., for PM with cardiovascular outcomes). The exposure assessment challenge is geospatial and temporal in nature, requiring attention to data and methods that can fully exploit the spatial and temporal variability of smoke originating from vegetation fires to characterize health risk.

10.1.2. Chapter Organization and Purpose

To understand smoke inhalation exposures, multiple disciplines are needed, including expertise in smoke production, air quality assessment, exposure mapping, and toxicological, epidemiological, and statistical analysis methods, as described in other chapters. Exposure data help scientists, forest managers, clinicians, public health practitioners, and decision makers to understand and quantify the nature and severity of health effects in various populations, including susceptible or highly exposed populations (U.S. Environmental Protection Agency, 2019a). Because of the multidisciplinary nature of the study of smoke exposures and risks as well as the broad set of facets of risk management, there is a need for a common vocabulary and understanding of exposure science frameworks presented in this chapter in the context of characterization of biomass burning smoke primarily in the United States. Accordingly, the purpose of this chapter is to provide a conceptual framework and vocabulary related to characterizing human exposures. The root causes (e.g., carbon and greenhouse gas emissions) of climate change, which contribute to changes in fire regimes and increases in human susceptibility and vulnerability, should be considered as part of the scope in fire exposure science; however, this chapter, as with the book overall, focuses on exposure by inhalation of air pollution. We focus on exposure to landscape fire smoke, rather than biomass smoke from household heating and cooking practiced in some regions. The scope of the chapter draws heavily on U.S. perspectives, although we acknowledge the importance of additional exposure contexts globally. Issues related to toxicology, health burden of disease globally, and specific epidemiologic evidence are discussed in Chapters 11, 12 and 13.

In addition, exposure to disaster events in general, such as landscape fires, has also been associated with mental health impacts including posttraumatic stress disorder (PTSD) from anticipation or experiences of the disaster, sheltering (short-term or prolonged periods), evacuation, illness, loss of property or life in a community, family, or among responders and fire fighters (National Academies of Sciences Engineering and Medicine, 2020). Fire damage has been linked to water pollution and associated health impacts. These exposure considerations, while important, require methods beyond the scope of this discussion.

This chapter has three main parts. First, in this section, we briefly describe fundamental concepts of exposure assessment relevant to understanding air pollution from unplanned wildland fire, prescribed fire, and agricultural burning. Next, we consider how to quantify exposure to biomass burning pollution, specifically for geospatial assessments of exposure and the types of exposure metrics, with associated reporting error. Finally, we discuss research needs for fire air pollution exposure assessment.

10.1.3. Exposure and Risk Assessment Frameworks and Concepts

This section summarizes some of the fundamental human exposure frameworks and concepts as applied to studying the human health impacts of wildland fire. There are two general structures to exposure assessment for human health: (1) chemical risk characterization, starting with hazard identification for a specific chemical pollutant); and (2) population health approaches, starting with population health surveillance and assessing and addressing fundamental causes, including environmental pollutant exposures and other social determinants of health, through policies, systems, and continuous improvement (Centers for Disease Control, 2020; PHNCI, 2020).

Exposure Assessment as Part of Chemical Risk Characterization

Exposure assessment is often part of a risk characterization (Fig. 10.1). Exposure is a proxy for the dose, which incorporates the concept of intake into the body (e.g., via inhalation) over a certain period of time (U.S. Environmental Protection Agency, 2019b, 2020; Zartarian et al., 2005). Modern exposure science is transforming from a traditional focus on discrete exposures to integrated approaches that consider exposures from source to dose; on multiple levels of integration, including temporal, spatial, and biological scales; and to multiple stressors (National Research Council, 2012). In this way, exposures can be scaled from molecular systems to individuals, populations, and ecosystems.

Human exposure is often framed as a border condition: the contact of a stressor with the body at the border between the human body (or receptor) and the outside environment. In the United States, the National Research Council (NRC) 2012 report entitled, "Exposure Science in the 21st Century: A Vision and a Strategy" defined exposure sciences as "the collection and analysis of quantitative and qualitative information needed to understand the nature of contact between receptors and physical, chemical, or biologic stressors" (National Research Council, 2012). Contact between stressor and receptor at an interface may encompass simple structures, such as air pollution in lungs, or complex or geographically large systems. In some cases, upstream root causes of the stressors should also be considered in assessing hazard and risk, such as the case of climate change health impacts where the emissions of CO_2 influence the climate system, which includes components that are fundamental to human health and well-being.

In the Guidelines for Human Exposure Assessment, U.S. EPA identified the primary purpose of an exposure assessment, which is to estimate exposure or dose. This estimate is then combined with chemical-specific exposure response (from human studies) or dose-response data (often from animal studies) to estimate risk (Fig. 10.1) (U.S. Environmental Protection Agency, 2019a). An exposure assessment describes contact (including upstream factors) with stressors as they occur related to intrinsic (e.g., at various life stages) and extrinsic factors related to differences in exposure or risk. An exposure assessment provides data to help decision makers and the public understand and quantify health outcomes as they occur or are projected to occur in various populations.

As discussed in the U.S. EPA's guidelines, exposure assessments typically answer three basic questions:

1. What are the characteristics of exposure (e.g., magnitude, frequency, duration, route of contact with the human body)? This question is also influenced by the characteristics of susceptible or exposed individuals.
2. How can exposure be reduced? Exposure assessment analyzes the individuals exposed and identifies the sources, routes, and pathways for exposure either prospectively or retrospectively to aid mitigation of harmful exposures.
3. To what extent has exposure changed over time? Exposure assessments inform public health by monitoring status and trends in exposure over time. Public health authorities can identify emerging and ongoing health risks, quantify health disparities, and evaluate risk mitigation actions.

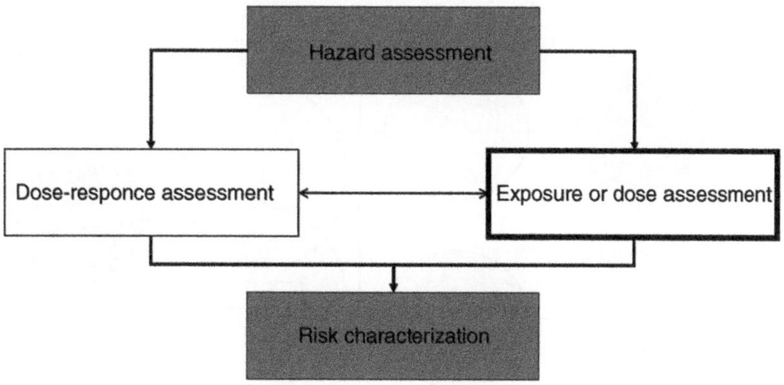

Figure 10.1 Exposure assessment is one component of environmental pollutant risk characterization as presented by the National Research Council (NRC) Science and Decisions Report (2009). After a hazard is identified, the exposure or dose assessment (right box) establishes the amount of a pollutant encountered by the receptor(s). The dose-response assessment (left box) defines the amount of a pollutant needed to elicit a specific response. These assessment data can be combined into a risk characterization.

Characterizing exposure requires knowledge of both the receptors (e.g., hypotheses about how the stressor acts on the human body, relevant periods of susceptibility) and the ways in which the stressor presents at this interface (e.g., fate and transport of chemical agents, climatic systems, social patterns that affect exposures and access to mitigation, human time and activity patterns). An increasing but still developing literature suggests that certain populations are disproportionately affected by biomass burning smoke due to intrinsic factors such as sex (Liu et al., 2017; Reid et al., 2016b), pregnancy status (Koman et al., 2018), and age (e.g., Ignotti et al., 2010; Nunes et al., 2013; Resnick et al., 2015; Sarnat et al., 2000; Shaposhnikov et al., 2014) although there are discrepancies as to which age groups are more vulnerable (Reid et al., 2016a). Extrinsic human factors should also be considered, and these can be socially patterned such as marginalized groups and residential settlement patterns, occupation, socioeconomic status, dietary habits, and access to mitigating factors and health care (Morello-Frosch et al., 2011).

Several influential science policy reports highlighted the need for more systematic consideration of combined risks from multiple chemical and social environmental stressors (Institute of Medicine, 1999; National Research Council, 2009; Thurston et al., 2020; Thurston et al., 2017). These recommendations bring forward concepts from the population-health approach, including an understanding of the intersection of social systems and biological/physical systems as it relates to risk. A joint European Respiratory Society/American Thoracic Society (ATS) policy statement notes that health equity and environmental justice considerations are relevant to population risk assessments (Thurston et al. 2017), and the U.S. EPA's exposure guidelines acknowledge these goals (U.S. Environmental Protection Agency, 2019a).

The concept of exposures within a dynamic system, as depicted in Figure 10.2, can serve as a useful framework for wildland fire smoke exposure assessment due to the complexity of biomass burning as a source of pollution and the multiple factors that influence outcomes of smoke exposure. The National Research Council (NRC) committee considered that the study of exposure extends beyond the exposure events themselves (i.e., the interface between stressor and humans) to observe and characterize the multiple processes that affect the transport and transformation of agents from their source to a dose at a target internal organ, tissue, or toxicity pathway associated with a disease process (National Research Council, 2009). This definition builds on recommendations in the NRC's 2009 Science and Decisions report, which recommends that exposures from all pathways (e.g., oral, inhalation, dermal) of stressors into the body be considered in aggregate and that cumulative exposures (e.g., multiple stressors over time that impact the same organ systems) be analyzed as a best practice in quantitative chemical risk assessment.

Population Health Approaches to Exposure Assessment

In contrast to the risk characterization approach, a public-health-based approach starts with health surveillance and equity and then incorporates the links between exposure and population health with the goal of promoting the health of all people in all communities. This approach has been outlined by the World Health

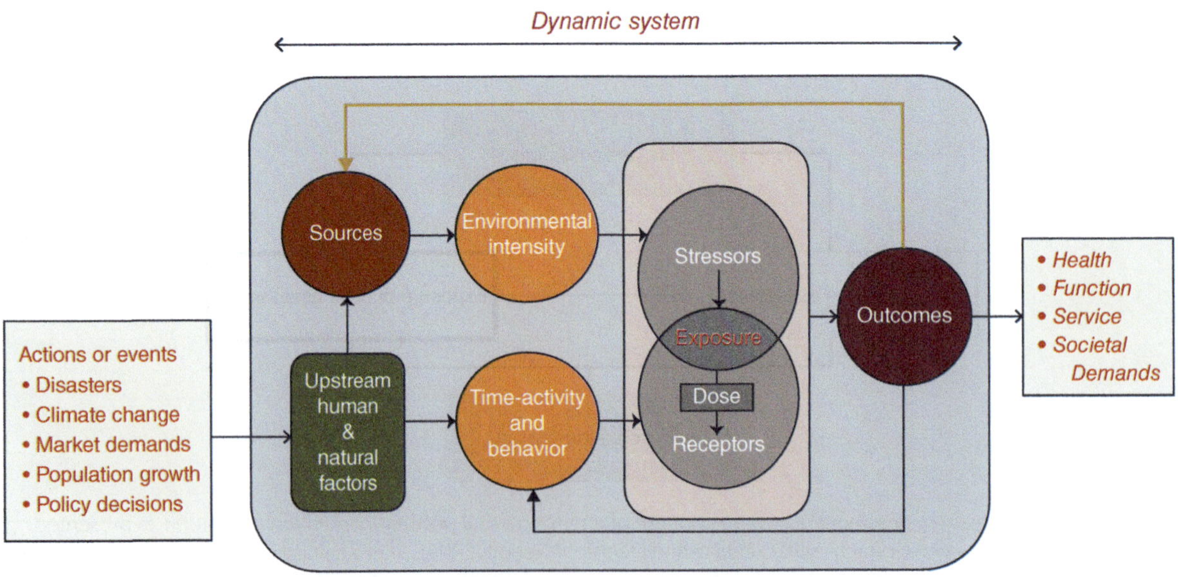

Figure 10.2 Conceptual Framework for Exposure Assessment, which serves as a useful framework for biomass burning smoke exposure assessment (National Research Council, 2012).

Organization and the U.S. Centers for Disease Control and Prevention (since the 1990s) and updated in 2020 by the Public Health National Center for Innovations. The World Health Organization defines essential public health functions as the indispensable set of actions, under the primary responsibility of a nation, that are fundamental for achieving the goal of public health, which is to improve, promote, protect, and restore the health of the population through collective action. The population-centered or public health approach starts with the health status of a population, utilizes aggregate and cumulative exposures, tests for health and exposure disparities that may exist for marginalized groups, and evaluates systems-level interventions to promote health (Centers for Disease Control, 2020).

Using either chemical risk assessment or population health approaches, exposure assessment can be a step in a regulatory risk assessment process, an element of a research study, part of a public warning system, or part of public health or occupational health surveillance. Additionally, the extent and scope of the assessment will depend on the goal of the exposure assessment and availability and quality of data. An exposure assessment should describe the sources, routes, pathways, and uncertainty in the assessment to be useful for individual or population health studies (Zartarian et al., 2005).

10.2. CONSIDERATIONS FOR CHARACTERIZING BIOMASS BURNING SMOKE EXPOSURE

10.2.1. Quantifying Exposure

In the two approaches discussed above, exposure to air pollution from biomass burning would need characterization, such as the magnitude, frequency, duration of contact with exposed individuals. Personal exposure, for example, can be assessed by integrating exposures over time and across the microenvironments in which a person spends time (e.g., indoors and outdoors). In some settings, household biomass burning would need to be accounted for as part of the personal exposure to smoke. Exposure can also be estimated geospatially through air quality monitoring or modeling, as discussed below.

In developed countries, most time is spent indoors; therefore, the infiltration of ambient sourced pollution such as biomass burning smoke is important. Infiltration can be estimated, and the infiltration factor depends on both the penetration of the pollutant indoors, indoor sources, and the ventilation and deposition characteristics that influence indoor pollutant concentration (Rosofsky et al., 2019; Rosofsky et al., 2018). Because of PM's infiltration into indoor environments and the regional nature of all-source PM, ambient concentrations have been demonstrated to be an adequate proxy or exposure surrogate for ambient-source PM in epidemiologic studies (Özkaynak et al., 2013; Sarnat et al., 2000; U.S. Environmental Protection Agency, 2019b). For ozone, indoor levels are generally lower than outdoor levels due to surface reactions, air conditioning, and the degree to which ozone penetrates the building envelope (Chen et al., 2012; Sarnat et al., 2013; U.S. Environmental Protection Agency, 2020). Assuming an absence of devices that generate ozone, such as certain ozone-generating household air purifiers and some consumer products, indoor/outdoor ratios in the United States generally ranged from 0.1 to 0.4 (Chen et al., 2012). With the COVID pandemic, commercial and residential buildings may be updating their HVAC filtration systems and outdoor air circulation, and people may be modifying their time spent indoors compared with outdoors (Allen & Ibrahim, 2021).

Personal exposure to biomass burning pollutants also varies as a function of a number of intrinsic and extrinsic factors. Extrinsic factors include location (e.g., proximity to fire-prone areas), occupation, activities within a location, socioeconomic status, consumer behavior, dietary habits, and other lifestyle choices (U.S. Environmental Protection Agency, 2019a). Intrinsic factors such as life stage, health status, sex, and genetic differences are important factors that help determine dose. For example, baseline lung function and tidal volume are related to age, sex, and body size (U.S. Environmental Protection Agency, 2011). Behaviors relative to intrinsic factors such as life stage can be particularly influential determinants for exposure, especially for infants and toddlers and for the embryo/fetus during pregnancy (U.S. Environmental Protection Agency, 2014).

Considerations in assessing population-level biomass burning smoke exposure include factors such as pollutant concentrations, the chemical composition, number, and surface area of particles; types of gases; frequency, intensity and duration (or temporal variation) of fires and their emissions; and spatial variation and dispersion. These factors combined with information on the number of individuals who are exposed, critical physiological time periods of exposure, the susceptibility of exposed populations, the extent to which populations can protect themselves from the exposure, and the ability of exposed individuals to recover or access care provide a more comprehensive view of human health risks from biomass burning smoke. Development of exposure metrics for human health research often requires assumptions and proxies when direct exposure factors are unavailable. Studies must also weigh appropriate control or comparison groups and likely sources and extent of exposure measurement error.

Quantifying smoke exposure requires addressing multiple components, such as an indicator (direct or indirect measure or proxy of exposure), type of estimation technique, length of relevant exposure such as acute or chronic, averaging time lag (time from exposure to health effect), and the extent to which smoke exposures may be related to other factors (e.g., meteorological, population mobility, occupation). These attributes of exposure can relate to prior evidence of health risks or available methods. For example, an indicator of smoke exposure, $PM_{2.5}$, as measured by monitors operated by U.S. state and local governments for routine air pollution surveillance may be used as the exposure metric in an epidemiologic study.

While relationships between exposure and health outcomes at the personal or population level can sometimes be established through quantitative assessment, the causal factor, such as presence of a specific chemical component, may be unknown or difficult to measure for large, mobile populations spread over large geographic areas. To address these uncertainties, controlled lab experiments may be performed using animal models to study some health hypotheses such as those involving mechanisms of action, immune response, rare exposure or outcome conditions, long lag times, or the need to isolate one factor, as discussed in Chapter 11.

10.2.2. Biomass Burning Smoke Composition

The *concentration of a pollutant* is defined as the mass of the pollutant in a given volume of air (e.g., PM in μg/m^3). Concentrations observed in outdoor locations accessible to the public are referred to as *ambient concentrations*. In air pollution epidemiologic studies, ambient concentrations are often used as a proxy for dose of a pollutant inhaled over a specified time period (Zartarian et al., 2005). However, dose and personal exposures of concentrations near the breathing zone are not routinely measured in population epidemiology (although dose may be more common in animal studies or occupational medicine). As a result of the lack of these measures, exposure surrogates are used to represent exposure, such as $PM_{2.5}$ concentration measured *in situ* at a fixed site ambient monitor or a geospatially modeled concentration of ozone compared with a residential location. These techniques are discussed in Chapters 3, 6, and 8.

The chemistry of air pollution is also relevant to risk. Biomass burning emissions include directly emitted PM, NO_x, and VOC, which contribute to ozone formation, carbon monoxide (CO), PAHs, formaldehyde, acrolein, benzene, and metals (Adetona et al., 2016; Akagi et al., 2011; Andreae, 2019; Brey & Fischer, 2016; Naeher et al., 2007; O'Dell et al., 2020; Voulgarakis & Field, 2015). An important component of wildfire smoke relevant to human health is PM, with a higher proportion of ultrafine particles and higher levels of PAHs and VOCs in wildland fire smoke compared with ambient pollution from all sources in the United States (Voulgarakis & Field, 2015) although the concentrations can vary as the emissions age (O'Dell et al., 2020). (Na & Cocker, 2008).

Some fires encroach on the human-built environment adding the burning of plastics and other building materials or vehicles and structures to the challenges of exposure assessment. Toxic chemicals may be released from the burning of household or industrial goods, but how far these can disperse is much less well understood and they are likely to be relatively local concerns (Voulgarakis & Field, 2015). In addition, biomass burning smoke constituents may include different types of biota, including fungal spores, allergens, and other pathogens (Kobziar & Thompson, 2020). Consideration of these complex aspects of smoke is beyond the scope of this chapter but should be kept in mind as emissions and exposure to biomass burning smoke science progresses.

Fine Particulate Matter

PM can be directly emitted or formed secondarily in the atmosphere from combustion of all types, and notably from transportation and stationary sources (U.S. Environmental Protection Agency, 2019b). Worldwide, higher exposures to fine particulate matter (aerodynamic diameter < 2.5 μm or $PM_{2.5}$) from all sources are known to increase respiratory and cardiovascular morbidity and mortality (U.S. Environmental Protection Agency, 2019b; World Health Organization, 2013). In addition, increased $PM_{2.5}$ exposure from all sources has been associated with adverse maternal outcomes (e.g., preeclampsia, gestational hypertension, gestational diabetes) and birth outcomes such as fetal growth restriction, low birth weight, and prematurity, as well as cancer and cognitive decline in older adults (U.S. Environmental Protection Agency, 2019a, 2019b).

More than 44 million people around the world are exposed to elevated annual average $PM_{2.5}$ due to landscape fire emissions, defined by concentrations of $PM_{2.5}$ > 55 μg/m^3 (Roberts & Wooster, 2021). Thus, PM has been the focus of much of the exposure science and epidemiologic research related to biomass burning emissions. Short-term exposure to PM from wildland fire has been associated with premature mortality and hospital admissions as well as other adverse impacts to health and well-being (Abdo et al., 2019; Amjad et al., 2021; Liu et al., 2015; Liu et al., 2017; Rappold et al., 2017). However, fire-specific PM exposure health research is not as well established as all-source PM exposure health research. As a result, some studies have investigated the comparability of health risks from fire-specific $PM_{2.5}$ exposures with those from $PM_{2.5}$ from fossil fuel combustion and other sources. A research study focusing on PM specifically from wildfire smoke reported additional human health effects for PM smoke exposures compared with exposures of fine particles generated from all other sources (DeFlorio-Barker et al., 2019). Occupational and household biomass burning studies of smoke exposure provide additional support (Adetona et al., 2016). Household biomass burning has been associated with adverse birth outcomes in newborns (Patel et al., 2015) and higher rates of lower respiratory infections and pneumonias among children (Balmes, 2019). The evidence for health risks related to wildland and prescribed fire smoke are discussed further in Chapters 12 and 13.

Mechanistically, inhalation of PM from smoke can be expected to increase respiratory and cardiovascular risks due to direct deposition in the lungs leading to local oxidative stress and inflammation that can potentially spill over into the systemic circulation (Black et al., 2017; Holm et al., 2021). For example, when wildfire PM was directly instilled into rodent lungs, the oxidative stress and inflammatory responses were more robust compared with instillation of PM from other sources (Wegesser et al., 2010).

However, because PM from landscape fire smoke is a complex mixture, there is less certainty that associated health risks are the same as with PM from fossil fuel burning. Biomass burning produces high levels of PM, but the chemical signature of PM from landscape fires differs from PM originating from fossil fuel combustion. Exposure scenarios differ as well, with substantially sharper peaks related to wildfires versus typical urban ambient air pollution, which is relatively more constant at a regional scale. For example, $PM_{2.5}$ concentrations during several California wildfires have been estimated to be between 3 to 31 times the level of the health-based 24 hr $PM_{2.5}$ standard established by the U.S. Environmental Protection Agency (EPA) with large segments of the population exposed to smoke (Koman et al., 2019; Thelen et al., 2013).

Researchers are also studying to what extent similar populations are susceptible to PM from fires as compared with all sources or fossil fuel combustion. Associated air pollution effects from all sources are generally small magnitude risks spread over large populations with widespread and constant exposures, with adverse impacts such as hospitalizations and mortality (Mills et al., 2018). To detect these generally low-frequency effects with standard time-series epidemiology techniques, for example, requires the observation of a relatively large exposed population, which has implications for the practicalities of estimating exposures in the study of PM from fire smoke.

Findings from the research-to-date related to the association of fire-specific $PM_{2.5}$ exposure with adverse health outcomes (Reid et al., 2016a) and the potential for higher future exposure levels have prompted some decision makers to take steps to reduce exposures even as the evidence base is incomplete. A health-protective perspective would be to assume that all-source $PM_{2.5}$-related risks apply until we are able to distinguish among the various source types of $PM_{2.5}$ with respect to various health risks.

Ozone

Ozone is a known respiratory irritant (U.S. Environmental Protection Agency, 2020). Fires do not directly emit ozone, but rather it is secondarily formed from precursors (NO_x and VOCs) (Jaffe & Wigder, 2012). In contrast to PM, which is a complex mixture, ozone is the same chemical regardless of the source of its precursor gases. Therefore, ozone-related health risks would be similar regardless of the precursor sources because the air pollutant ozone is chemically the same; thus, in contrast to PM, the attributable fraction of ozone-related risk from landscape fires depends largely upon the fire emissions (i.e., NO_x and VOCs) and resulting ozone concentrations in places where people are exposed.

For pollutants like O_3 and some ambient-sourced PM, indoor concentrations can be lower than outdoor concentrations because these pollutants are unstable or filtered out (U.S. Environmental Protection Agency, 2019b). For other pollutants, like SO_2, CO, and NO_x, indoor and outdoor levels may be similar since these pollutants are relatively stable gases that are not removed by filters.

10.2.3. Timing of Exposure

Temporal considerations for smoke exposure include length of exposure, frequency of exposure, and the lag in onset of symptoms (e.g., an individual is exposed to fire smoke on one day and after a period of time, the changes or symptoms of disease may become apparent). Exposures to smoke air pollution can be chronic or long-term (e.g., weeks, season, multiple years) or acute or short-term (e.g., hours, days). Both long- and short-term PM and ozone exposures from all sources have been associated with adverse health consequences. Because ozone is the same chemical, it is possible to rely on the ozone-related health literature for temporal aspects. In contrast, PM is a mixture, which is different for fire-sourced PM compared with all sources; thus, fire-specific PM composition over space and time may deviate from all-source PM (Doubleday et al., 2020; Magzamen et al., 2021). More research is necessary to develop appropriate understanding of the temporal aspects and concentration-response functions, as shown by the uncertainties in cardiovascular disease risk for fire-specific sources. Studies of long-term exposures to unplanned and prescribed fires are lacking at this time.

Generally, epidemiologic studies consider and may control for other temporally correlated influences on health, such as temperature and ambient air pollution from other sources. The timing of mitigation techniques can also influence personal exposure (e.g., air conditioning/filtering, evacuation, school closings, behavior change from warning systems, PPE use such as wearing N95 masks, or sheltering indoors or in climate-controlled areas).

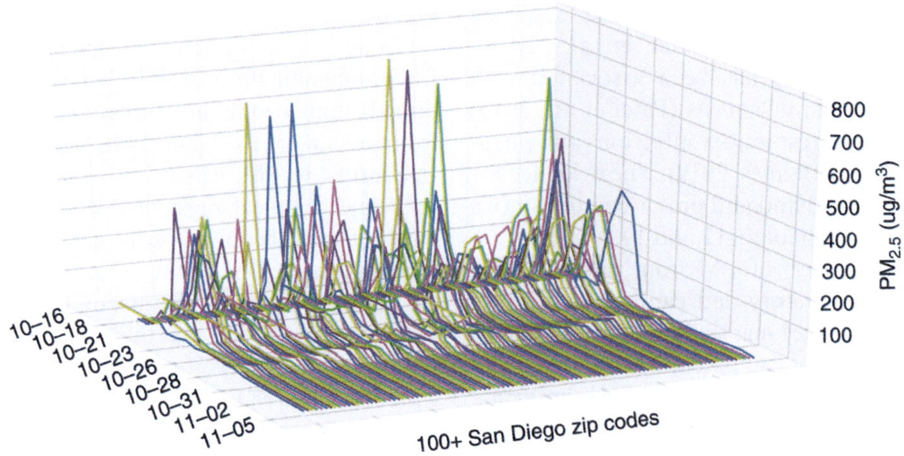

Figure 10.3 Modeled daily $PM_{2.5}$ concentration from Thelen et al. (2013) illustrating $PM_{2.5}$ smokewaves across postal zip codes for 20 days in October of 2007 in San Diego County, California, USA.

Knowledge of specific periods of human development is also critical for understanding the temporal aspects of the relationship of exposure and disease. Understanding the relationship of exposure and vulnerable periods of development (i.e., weeks of pregnancy or exposures during childhood) is important. From a public health perspective, it has been documented that environmental chemicals can have differing effects on lung development depending on the timing of exposure both in utero and through childhood, with some effects persisting (or only becoming apparent after) many years (Miller & Marty, 2010; Rosa et al., 2020). Global evidence regarding chronic smoke exposures among children is available from children who are exposed to biomass burning in their homes (Balmes, 2019; Patel et al., 2015) and from wildland fire smoke in the United States (Holm et al., 2021). However, effects from chronic exposure may take longer to become apparent and are not as well characterized.

Often, in contrast to pollution from other sources, the emissions from fires and contributing meteorology can be quite variable over time as well as geospatially. When wildfire smoke moves through an area, it does not affect all portions of the air shed area equally, which contributes to temporal and geospatial heterogeneity in exposure, and may not be adequately captured by air quality monitors relied on in general air pollution epidemiology. Mitigation steps taken during an active fire event that may affect smoke concentrations and penetration to indoor spaces also modify the exposure profile. In the United States, many EPA $PM_{2.5}$ *in situ* monitors measure only every third or sixth day, but wildfire smoke can change dramatically from day to day, and areas close to a monitor might get a different concentration than what the monitor detects (National Academies of Sciences Engineering and Medicine, 2020). As a result, other types of measurements (e.g., remote sensing or modeling) have emerged to better characterize the temporal and geospatial aspects of ambient smoke concentrations. Figure 10.3 shows daily modeled concentrations of $PM_{2.5}$ by zip code in San Diego, California, USA, from Thelen et al. (2013) showing the peak-driven and variable nature of smoke concentrations coming from unplanned fires.

While daily or hourly concentration metrics of fire-related exposure surrogates are typically used to assess exposure, observed fire-specific peak air pollution levels have lasted many days followed by drastic drop-offs with near zero smoke concentrations for many days (Fig. 10.3) (Hutchinson et al., 2018; Koman et al., 2019; Thelen et al., 2013). If exposure is averaged over a monthly or annual time period, as is often done to assess all source air pollution-associated health outcomes, the mean concentration is often low compared with peak concentrations or even daily or hourly averages and may not adequately represent the relevant exposure to wildfire smoke. To capture this exposure profile, the term *smokewave* has been used, analogous to a heat wave in which air temperature spikes for a short period, and defined as at least two consecutive days with wildfire-specific pollution concentrations greater than some threshold (Koman et al., 2019; Liu et al., 2016).

Little is known about whether PM exposure has a threshold or what the appropriate cut point level would be to select for a health-relevant concentration. For all-source PM and ozone, the pollutants are generally known to be associated with health effects among susceptible groups at levels below current U.S. regulatory standards. The impact of timing of smoke exposures is also poorly understood. For example, research has shown that heat waves earlier in the year exert more health effects than later-in-season heat waves (Liss et al., 2017), and similar phenomena need investigation for wildland and prescribed fire smoke episodes. Greater understanding is needed of the smoke exposure impacts on public health during a fire season, as well as across many years, to assess the impact of adaptive responses that could influence exposures.

10.2.4. Spatial Scale Challenges

Smoke can transport and contribute regionally and internationally to population exposures. In some parts of the world, area burned by fires is projected to decrease (Mortiz et al., 2012), while other areas may increase, such as in the western United States and Australia. The projected increases in wildfire area burned globally could lead to more frequent or longer episodes of wildfire smoke across broad regions, with implications for human exposures (Spracklen et al., 2007; Spracklen et al., 2009).

People can be exposed in residential areas or occupational settings close to fires or from distant areas where smoke plumes travel. A systematic review characterizing air quality impacts for wildland and prescribed fires included nine studies of wildfires in North America ranging in size from approximately 5,200 hectares for the French fire to over 113,400 hectares for the Cedar fires in California (Navarro et al., 2018; Viswanathan et al., 2006). In these studies, mean daily $PM_{2.5}$ concentrations from wildland fires ranged from 8.7 to 121 ug/m^3, with a maximum of 1,659 ug/m^3 (Navarro et al., 2018). Evidence from studies of prescribed burns supports that higher smoke levels are present closer to fire sites, although these measurements may not be available for unplanned fires.

Because emissions can travel significant distances and transform secondarily in the atmosphere to form PM and ozone (Jaffe & Wigder, 2012), the contributions of smoke to ambient pollution concentrations can be substantial. For example, in the western United States, variations from year to year in organic carbon component annual $PM_{2.5}$ concentrations are driven by wildfires in summer (Jaffe et al., 2008; Spracklen et al., 2007). Partitioning exposures and health effects specific to wildfire PM is challenging given that air pollutants originate from both wildfire and other sources. Meteorology and atmospheric conditions affect which compounds travel from a fire site, and therefore affect the exposures to the surrounding populace.

To characterize the geographical extent of smoke, in addition to the use of routine ambient air monitors, remote sensing and modeling have been used for more complete coverage of concentrations over large areas. Figure 10.4 shows county-level $PM_{2.5}$ modeled concentrations in the largest counties in California in a high five year (Koman et al., 2019). Figure 10.5 shows maps of modeled $PM_{2.5}$ from fires and all sources of pollution, showing the yearly variability in geographic extent (Koman et al., 2022).

Compared with characterizing air pollution from all sources, measuring smoke concentrations poses unique

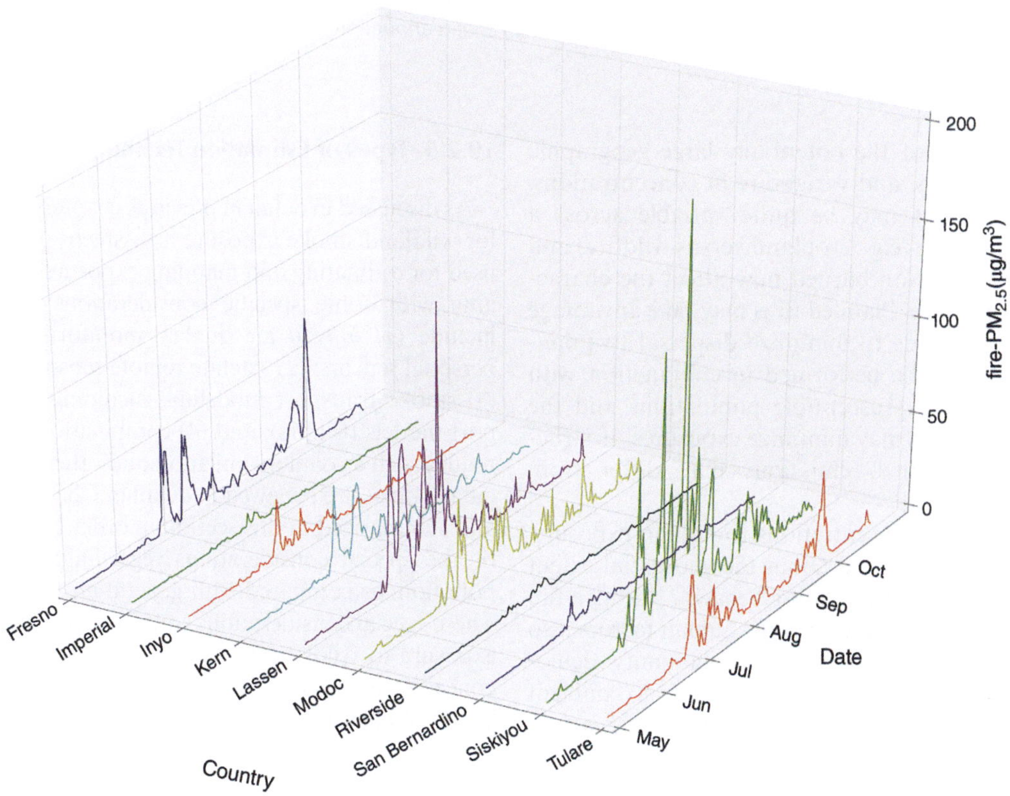

Figure 10.4 Mean daily Community Multiscale Air Quality (CMAQ) model-derived fire $PM_{2.5}$ for the 10 largest counties (by area) in California from May to November 2008, illustrating spatial and temporal fluctuation in fire $PM_{2.5}$ (Koman et al., 2019).

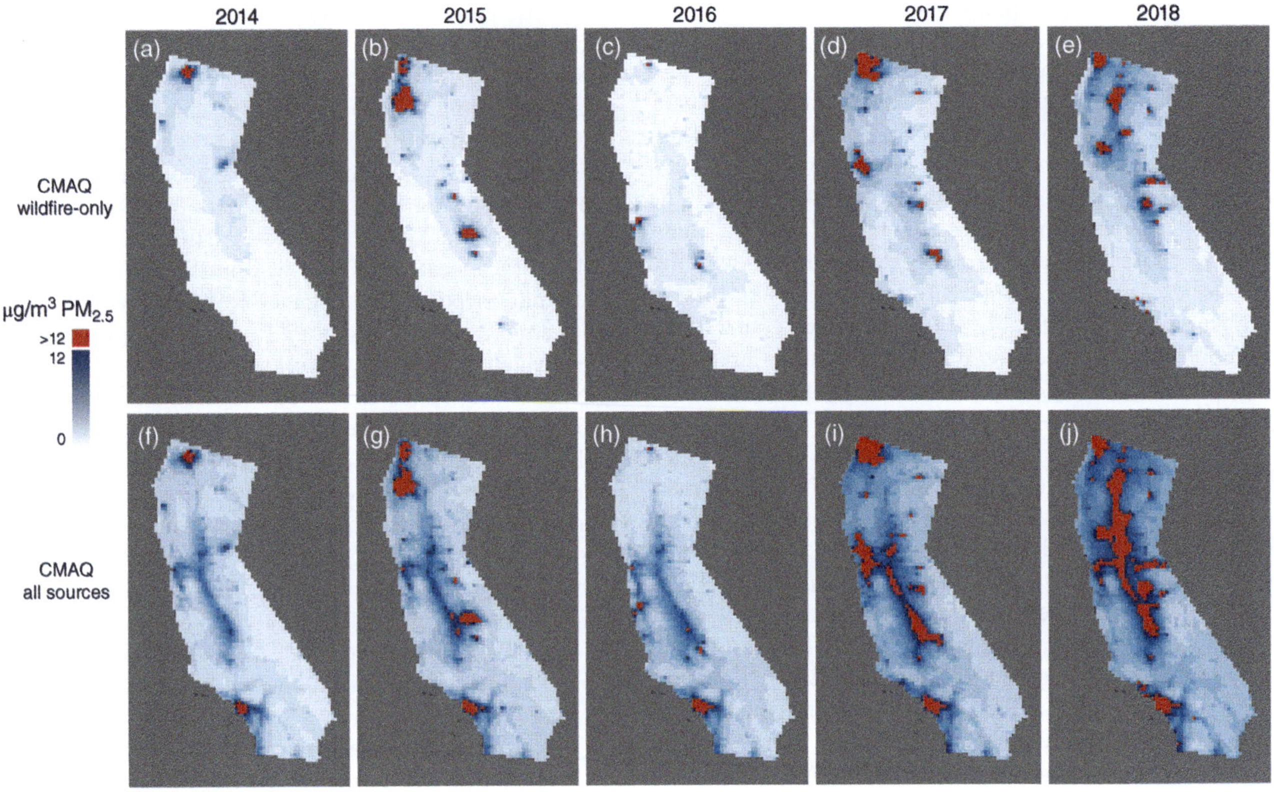

Figure 10.5 Community Multiscale Air Quality (CMAQ) model-derived mean annual $PM_{2.5}$ for California by 12 km grid: (a–e) 2014 to 2018, fire-only emissions sources; (f–j) 2014 to 2018, all sources. 12 μg/m³ represents the National Ambient Air Quality Standards (NAAQS) level for mean annual all-source $PM_{2.5}$ (Koman et al., 2022).

challenges because of the potentially large geographic ranges, remote areas, and variability in concentrations. First, fire emissions may be quite variable across a region. The fuel type (e.g., cropland versus wildfire) and condition of vegetation burned may affect the characteristics of the smoke. Planned fires may take advantage of weather conditions to minimize dispersal to populated areas or may be performed in conjunction with advance warnings to susceptible populations and the general public, which may minimize exposures. If structures are also burned, emissions may differ from vegetation-only fires.

In relating smoke concentrations to human exposure, more information is needed about the geospatial extent and spatial clustering of adverting behavior during a fire event. In addition, exposure-relevant human factors also have geospatial relationships (e.g., community demographics, housing characteristics that relate to ambient pollution infiltration, poverty and socioeconomic factors that can influence averting behavior) (Rappold et al., 2017). Drawing conclusions regarding the exposure-response relationship is complicated by issues related to exposure assessment.

10.2.5. Types of Estimation Techniques

As discussed in detail in previous chapters, the main tools for wildland smoke exposure assessment parallel methods used for estimating and mapping exposure to other pollutants with some specific considerations. Methods used include (1) *in situ* air quality monitoring stations and personal sensors; (2) satellite remote sensing of pollutants; (3) smoke transport modeling, including chemical transport models (CTMs); and (4) combinations of these techniques such as data assimilation and other methods. These data sources are reviewed in detail in Chapters 6 through 9 and summarized in this section in order to point out some of the special considerations regarding exposure to fire pollution, as compared with general air pollution sources. These special considerations come into play for quantifying exposure to wildland fire smoke and the relationships of concentration to dose for the purposes of understanding health impacts from wildland fire smoke.

Monitoring

Human exposure to pollution is often assessed with data collected at air quality monitoring stations. As described in Chapter 6, a *fixed-site or in situ* monitor (i.e., a monitor with a fixed position) is a measuring device used to estimate ambient air pollution average concentrations and their trends over neighborhood and urban scales. In the United States, federal agencies along with state, local, and Tribal governments maintain networks of air quality monitoring stations to provide information on locations and times when air is unhealthy, to assess compliance with regulatory requirements, and to use for research. These stations typically use Federal Reference Method (FRM) or equivalent devices to measure U.S. EPA-established criteria air pollutants including O_3, $PM_{2.5}$, PM_{10}, CO, NO_2, and SO_2, although all of these pollutants are rarely measured at the same location. The equipment at monitoring sites varies from location to location (e.g., sites can monitor from one to potentially dozens of pollutants), and the number of these monitors has been increasing over time. Most sites have basic meteorological instruments, such as wind direction and wind speed sensors.

Air pollution monitoring data are sometimes used as proxies for population exposure. These monitor data are often used for validation studies of other air quality mapping techniques such as modeling or remote sensing, as reviewed in Chapters 7 through 9. In the United States, federal reference or equivalent method monitoring data are sometimes also used in epidemiologic studies for estimating population exposures and in public warning systems. Health science researchers may use fixed-site monitor data as exposure surrogates for a variety of reasons. First, instrument error is typically small and the data quality assessment procedures are well documented for U.S. government-operated monitors. Additionally, an ambient monitor may provide a comprehensive set of measurements, which can be used as an exposure surrogate for a relatively large population. The ambient monitor approach is the least data-intensive approach among all exposure estimation methods because it requires data from only a single monitor to represent exposures for a large area (about 100 km^2), the values are often freely available on U.S. government websites, and they do not require sophisticated data processing compared with modeling results (U.S. Environmental Protection Agency, 2019b; Zanobetti & Schwartz, 2009). For some health studies, capturing variation over time for a restricted geographic area (i.e., one compact city) may be more important than the need to capture spatial variation. Ambient monitor data can provide useful information regarding trends in population exposure estimates over time and can be included in more sophisticated exposure estimation models used in epidemiologic research.

Monitors serve different purposes and thus provide different types of information in relation to exposure. Some sites are selected to represent population exposure (a.k.a., *population sites*). Some monitors quantify concentrations related to source-specific impacts from major industrial sources (*hotspot* or *source-oriented* sites) or from vehicle traffic (*near-road* sites), and more information about time and activity patterns may be necessary to fully characterize exposures. Some monitors are sited to provide upwind or foundational concentrations for pollutants that are transported into the area (*background* or *transport-oriented* sites). In the United States, there are also long-operated monitors that provide data for trend assessments or national-level assessments. Finally, special study monitors are sometimes deployed for weeks to months or years for research purposes or specific air quality concerns, including fire events. Chapter 6 describes other major networks in Australia, Canada, and Europe, but many areas globally lack ground-based air quality monitoring stations.

Gaps in the monitoring networks pose challenges to estimating exposure from fire smoke (Lassman et al., 2017). The data are limited to the location sampled, the schedule of sampling (which may not be every day), and lack of source attribution information. While some air pollutants, such as all-source PM and ozone, are ubiquitous and well mixed across large areas covered by air monitors in the United States, for example, wildland fire smoke can be locally concentrated and occur in areas with few monitoring stations. Studies don't always report the proximity of monitors to the fires. With the exception of the U.S. Interagency Monitoring of Protected Visual Environments (IMPROVE) network, most ambient air monitoring stations in the United States are focused in urban and other densely populated areas creating a lack of information for more remote regions where wildland or prescribed fire may occur. Depending on the smoke plume direction, even mega fires, such as King and Rim fires, can largely miss monitoring sites (Burley et al., 2016). During wildland fire smoke events, the nearest monitoring site may not accurately represent local pollution concentrations and potential exposures due to complex terrain and localized smoke that causes large gradients in smoke concentrations over space and time (Jaffe et al., 2020). Statistically based models might attempt to create a map surface of air pollution concentrations from monitors, but these techniques rely heavily on air quality monitored data, and they are limited to providing information where monitoring station data exist.

Another limitation for fixed site monitors is that smoke often originates in sparsely populated areas and can be transported very long distances, which also means the pollutants measured cannot be easily traced back to the source. Source attribution for these sites can be achieved with some inference and with atmospheric back trajectory methods, but attribution can be confounded by other sources, which can be well mixed by complex physical (winds, deposition) and chemical processes that are not fully captured in the dispersion model. Use of routine monitor network data for wildfire smoke exposure, therefore, may not fully capture the extent of impacts or concentrations in particular areas and times that were not monitored.

Increasingly, fire response teams are deploying additional short-term fixed site monitors (special study monitors) and sensors during fire outbreaks to better characterize possible exposures and to inform the public. Challenges remain with power for monitors, siting the monitors in short time frames, internet connectivity, interpreting the data relative to human population exposures, reliability of non-FRM or equivalent devices, and the highly dynamic and variable nature of the fires and their smoke.

Personal exposure monitors or sensors are becoming increasingly popular, especially for PM, but often the data quality may be less well characterized or more variable (Badura et al., 2018; Zou et al., 2020). PM sensors are widely used by the public in the United States, Canada, and Europe as a source of air quality data (ambient and indoor air) during wildfires, although the sensors' performance during smoke-impacted conditions has not been thoroughly evaluated (Holder et al., 2020; see for example https://maps.sensor.community/#2/-0.4/-1.3).

Wearable or mobile sensors are being rapidly improved, but currently they are pragmatic for limited exposure studies of short duration and with relatively small samples of people or areas compared with other techniques. These optically based sensors can also provide information about occupational exposures and can be paired with other types of GIS-based equipment or diaries for time-activity data. Power availability, sensor device memory capacity, internet connectivity or availability remain important technical issues for their widespread use. The sensors may be limited in the types of pollutants they measure, and ongoing research is comparing these sensors with more robust *in situ* monitors. However, these sensors show promise globally in areas without other methods (Awokola et al., 2020).

Many organizations, including U.S. EPA and U.S. Forest Service, are involved in research in this rapidly evolving area (https://www.epa.gov/air-sensor-toolbox/technical-approaches-sensor-data-airnow-fire-and-smoke-map).

Remote Sensing

Satellite-based sensing of the atmosphere provides a way to characterize the 3D composition of the atmosphere (Hammer et al., 2020). Several sensing systems that can quantify gas and aerosol concentrations are operated by U.S. and other space and weather agencies, and new technologies are under development (Chance et al., 2013; Lineberry, 2016; Zoogman et al., 2014). As described in Chapter 7, satellite data can be used to estimate total-column O_3 and O_3 precursors, including NO_2, and some VOC species, including formaldehyde and glyoxal. Total column information is also available for aerosols, through sensing of aerosol optical depth (AOD) and aerosol precursors SO_2 and NO_2. Surface level concentrations of these pollutants are often estimated with the total column data using observations or assumptions about vertical distribution, including empirically derived assumptions, from photochemical grid model simulations (Duncan et al., 2014) based on ground-based sampling or sensing systems. Satellite sensing systems used for characterizing surface air quality are reviewed by Martin (2008). Surface level estimates of pollutants relevant for wildfire impacts based on satellite data provide spatial information that is not available from surface-based routine ambient air quality monitoring networks. Satellite data can be integrated with transport model predictions through data assimilation methods, as reviewed in Chapter 7.

Satellite sensing of the atmosphere can provide a unique opportunity to understand hemispheric and regional scale presence of elevated pollution events, such as wildfires (Morton et al., 2013). However, remote sensing data used without supportive information or used in conjunction with model-based predictions (see Chapter 7) can have many limitations. Satellite data provide little specificity about the origins of surface level pollutants, are reliant on general assumptions or sparse data to apportion observed pollution to the surface level, and have limited spatial resolution and coverage that limit exposure assessment. Polar-orbiting satellites pass over a given location every few days or can miss opportunities due to atmospheric conditions (e.g., clouds; Duncan et al., 2014), thus limiting understanding the temporal aspects of population exposures. Given that wildfire plumes can move over short time intervals, the temporal sampling can be an issue in characterizing exposure.

Optical satellite imaging (images of plumes from space) provides an additional avenue for characterizing smoke and has been used to complement atmospheric sensing in two ways. First, data from Multi-angle Imaging SpectroRadiometer (MISR) on the Terra satellite platform images at various angles in visible and near-IR spectral bands. Its data have been used to derive information on smoke plume structure, including assessing plume height and density (Val Martin et al., 2010) and to validate plume rise information for smoke modeling (Val Martin et al., 2012). In addition, operational products derived from satellite imagery can provide data on smoke trajectories and densities. This information is compiled within the National Oceanographic and Atmospheric Administration (NOAA) Hazard Mapping System (HMS). NOAA-HMS data are derived by interpretation of images from multiple sensors, including the Suomi-NPP-VIIRS, GOES, and NASA-Terra/Aqua data satellite data (McNamara et al., 2004). These data provide an avenue to improve smoke models and for validating smoke dispersion information but, similar to column-integrated AOD observations, do not provide surface level pollutant concentrations.

Satellite-derived information, from both atmospheric sensors and imaging systems, can provide important information in assisting in source attribution for data collected at monitoring stations (Chapter 6), for example. Where fire events are large enough, it may be feasible to qualitatively infer the origin of secondarily formed pollutants at hemispheric and regional scales but a quantitative estimate would not be feasible. For instance, satellites at times may provide credible source-receptor relationships for visible pollution such as wildfires or large dust events but quantifying the amount of surface level O_3 or $PM_{2.5}$ from a specific fire or dust event at a downwind receptor location is not possible with remote sensing alone. Satellite visible smoke plumes have been used as part of weight-of-evidence demonstrations showing a relationship from a specific fire event and downwind O_3 impacts by California and Kansas (http://www2.epa.gov/air-quality-analysis/exceptional-events-requirements-reference-guide). Health studies have also used visible smoke information to assign regions and time periods as exposed to smoke (Johnston et al., 2012). Some studies employed satellite metrics of aerosol loading in the atmosphere to classify regions or days as fire exposed or not fire exposed using satellite-based fire radiative power estimates. These dichotomizations simplify exposure and could miss gradation in effects associated with concentrations of smoke exposure in a population during a wildfire event, so they have value primarily to support more precise quantitative methods.

Modeling

Air quality models have been used for decades to understand exposures to air pollutants (Russell, 1997), but the use of air quality models for wildland and prescribed fire pollution is relatively new (Reid et al., 2016a). As described in Chapter 7, model-based simulations of air pollutants add valuable information for understanding smoke exposures past and future, for addressing exposure based on pollutant origin, and potentially for deposition related to water quality exposure pathways. Models can provide more complete, wide-scale geospatial and temporal projections and coverage, overcoming some of the limitation of observational approaches.

A wide variety of air quality models have been developed to support scientific and regulatory exposure assessments of primary and secondary pollutants (e.g., O_3 and secondary components of PM), as described in Chapter 7. Eulerian grid models add the ability to compute chemical transformations based on complex interactions with ambient atmospheric chemistry and physical conditions. The full chemistry models are called *photochemical transport models* or alternatively *chemical transport models* (CTMs) and provide a unique ability to map the chemical concentrations of materials that originate from a source (including wildland and prescribed fire) as well as what is created in the atmosphere across space and time to give a fuller exposure estimate for a geographic area (e.g., Fig. 10.6).

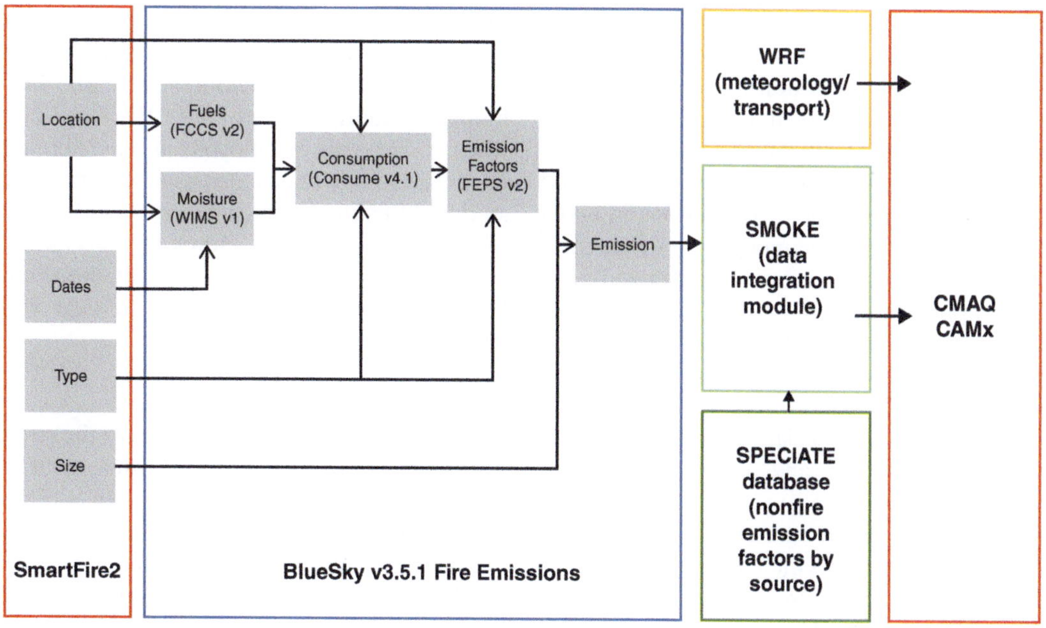

Figure 10.6 Modeling framework (published with permission Koman et al., 2019).

The fire-related emissions components (SmartFire2 and BlueSky v 3.5.1, orange and blue boxes) are combined with meteorology (yellow box), emissions from other sources (Sparse Matrix Operator Kernel Emissions modeling system (SMOKE)) and their chemical composition (SPECIATE) (green boxes) as inputs to the Community Multiscale Air Quality (CMAQ) chemical transport models (red box). Key modeling elements include the following: SmartFire2; FCCS: Fuel Characteristic Classification System; WIMS: Weather Information Management System; Consume; FEPS: Fire Emission Production Simulator; and WRF: Weather Research and Forecasting model.

Air quality models require specialized emissions inputs. As described in Chapter 5, a key requirement for modeling air quality is developing data on the type, intensity, location, height, and timing of pollutant emissions from all sources (anthropogenic and natural) including wildland fire. These data are required to quantify and map concentrations and to attribute the concentrations to various sources to understand specific exposure conditions. For wildland fires, more specialized models or model sequences may also be used (Rappold et al., 2014).

For wildfire smoke, state-of-the-art Eulerian CTMs add to the understanding of ambient exposures based on the type of fire, the location of fire, and other aspects of pollution not available with the simpler Lagrangian models or ambient monitored data alone. A CTM approach can be used to directly attribute emissions to a fire (or other sources) and provide both an estimate of total air pollution from all sources and that of wildfire-specific sources in a way not possible through sampling and sensing (Baker et al., 2016; Kansas Department of Health and Environment, 2012). CTMs are typically applied over grid sizes of 1 km or larger, depending upon the application (while grid resolutions of less than 10 km are used over urban areas, continental scale applications typically are done at about 10–40 km grid resolution, and global scale applications with larger grids still). Nested grids are used to achieve a range of resolutions in many applications. Outputs from these models are often compared with monitor data to correct or validate results but rarely use it as input (Baker et al., 2015). Some of these tools provide sufficient spatial and temporal information on O_3 and $PM_{2.5}$ to support health exposure assessments, as described in Chapter 7.

Atmospheric modeling is especially useful in assessing exposure to wildfire smoke and understanding the impacts of smoke on health because modeling data are available where monitors are not present, and the models provide information on the specific types of fire sources (e.g., wildfire, specific types of fire, or prescribed burns) for assessing health impacts. Using atmospheric modeling opens opportunities for assessing future wildfire-health scenarios by providing the data for predictive modeling, evaluating benefits and trade-offs of mitigation strategies and scenarios (Liu et al., 2016) and vulnerability mapping. Models can support forecasting and public health information warning systems alone or in combination of approaches such as data fusion, assimilation, or machine learning or artificial intelligence (Reid et al., 2015).

Modeling has limitations for exposure assessments. Modeling may provide information about ambient concentrations, but not fully account for indoor exposures where people in the United States spend most of their time. In addition, results from CTMs can be biased and subject to various errors due to inputs and model parameterizations, but factors leading to simulation errors continue to be identified and reduced (Baek et al., 2011; Baker & Woody, 2017; Barsanti et al., 2013; Foley et al., 2010; Yu et al., 2014). For instance, PM chemistry modules in CMAQ have been updated to address limitations in modeling secondary organic PM and nitrate chemistry. At a regional scale, CMAQ-modeled fire $PM_{2.5}$ has been shown to compare well to both aircraft plume transects and remotely sensed aerosol optical depth (Baker et al., 2018), though, in close proximity to wildland fires, a study of two large wildfires showed a tendency toward overestimation of $PM_{2.5}$ in comparison with a surface monitor network (Baker et al., 2016). Validation may be difficult with sparse monitoring data. Other limitations can include large grid sizes that lack specificity, uncertainties regarding emissions inputs, and uncertainties in modeling certain pollutants (e.g., ultrafine particles, toxics). Model horizontal and vertical grid definition can be a limitation, although studies have detected minimal differences in estimated exposures based on varying grid resolution between 4 and 36 km sized grid cells for all-source air pollution (Thompson et al., 2014). Specific modeling configurations must therefore be considered when CTMs are employed in epidemiologic studies. Dispersion modeling is also used in exposure assessment studies, often in conjunction with CTMs to provide fine-scale spatial resolution. However, dispersion models are still limited when applied in mountainous environments because they are not designed to deal with complex topography.

Integrated and Hybrid Approaches

A critical challenge is to maximize the strengths of air quality model data for public health assessment purposes and supplement that information with additional sources of data to best characterize and quantify source-receptor relationships and identify systematic biases. A growing number of studies have done this by combining information from ground-level *in situ* models and/or satellite data and air quality models. While anthropogenic air pollutants are often ubiquitous and well mixed across areas covered by *in situ* air quality monitors, wildland fire smoke can be locally concentrated and occur in areas with few monitoring stations. Therefore, the use of routine monitor network data for fire smoke exposure is not adequate because it does not fully capture the extent of impacts in relevant locations and time frames. Satellite-based remote sensing can provide some information on PM concentrations, and can be an effective way to supplement ground-based monitoring station data (Gupta et al., 2018). Satellite estimates of total column and surface layer concentrations of O_3, $PM_{2.5}$, and key precursors, including SO_2, NH_3, NO_2, and some VOC species, provide a robust opportunity to evaluate regional and continental-scale photochemical model estimates of these pollutants to provide additional confidence in their use for estimating population exposure to wildland fire emissions.

For example, van Donkelaar et al. (2010) estimated ground-level $PM_{2.5}$ from full-column AOD measurements derived from the MODIS and MISR satellites by multiplying the AOD by the ratio of estimated annual $PM_{2.5}$ to estimated AOD from the GEOS-Chem model (van Donkelaar et al., 2010). Johnston et al. (2012) estimated ground-level $PM_{2.5}$ concentrations just from fire as a product of twice the $PM_{2.5}$ from a run of GEOS-Chem that included only emissions from fires plus the scaled $PM_{2.5}$ estimates from MISR AOD and MODIS AOD, thereby equally weighting a contribution from satellite AOD and CTM output to estimate global $PM_{2.5}$ emissions from fire (Johnston et al., 2012). Reid et al. (2015) used machine learning methods to select from a wide variety of statistical models to optimize how WRF-Chem output, satellite AOD measures, combined with land-use and meteorological variables could predict $PM_{2.5}$ at monitoring stations. They then used that model to predict spatiotemporal exposures at other locations that were used to estimate health effects from a wildfire episode (Reid et al., 2015).

Research is ongoing on methods to integrate AQ monitor network data with other methods, including satellite-based remote sensing (Gupta et al., 2018) and modeled pollutant concentrations (Liu et al., 2015), providing a way to calibrate these indirect methods with observations (Bergen & Szpiro, 2015). The use of geospatial modeling that includes multisource data assimilation from ground-based monitors, satellite data, and model-based information is complex, and it requires expertise outside of traditional air quality and atmospheric modeling as well as epidemiological and health science research. PM epidemiological studies are taking advantage of satellite data (e.g., Liu et al., 2009; Madrigano et al., 2013; Reid et al., 2015) mobile monitoring data (e.g., Bergen et al., 2013; Levy et al., 2014), models (e.g., Fann et al., 2018; Jerrett et al., 2017; Thelen et al., 2013), and combined techniques (e.g., Gan et al., 2017).

10.2.6. Estimating and Reporting Error and Uncertainty in Exposure

This section introduces some of the limitations of using ambient concentrations as proxies for exposure of populations to ambient-source air pollutants in the context of fire smoke. When exposure surrogates such as ambient concentrations are used for estimating exposure to wildland fire pollution in epidemiologic studies, exposure error or exposure misclassification can result. As described by the U.S. EPA's Particulate Matter Integrated Science Assessment (ISA), *exposure error* refers to the bias and uncertainty associated with using concentration metrics to represent the actual exposure of an individual or population (U.S. Environmental Protection Agency, 2019b).

Additionally, *exposure misclassification* refers to exposure error that occurs when exposure conditions such as location, timing, or population grouping are incorrectly assigned. An example would be assuming an individual is present at a residence and assigning ambient values based on residential address when in fact the individual may have evacuated the area to a place with lower smoke concentrations. Exposure misclassification due to exposure assignment methods and spatial and temporal variability in pollutant concentrations may be either differential (i.e., systematic), or nondifferential (i.e., random). *Differential misclassification* refers to the situation in which exposure errors differ between groups. An example of differential misclassification is the use of geocoding to estimate air pollution exposure by proximity to a smoke source like a prescribed burn, because concentrations decrease with distance from the area and are different upwind and downwind of a prescribed burn location. If a single or limited number of monitors is used, the upwind group may be systematically misclassified. *Nondifferential misclassification* refers to the situation in which exposure characterization has a similar probability of being misclassified across groups.

These types of exposure errors have implications as they are carried forward into health research studies or risk assessment. As described in the U.S. EPA PM ISA, air pollution exposure error has two components: (1) exposure measurement error derived from uncertainty in the metric being used to represent exposure; and (2) error arising from the use of a surrogate parameter instead of the true exposure, which may be unobservable (U.S. Environmental Protection Agency, 2019b). These types of errors may have different effects in the health studies for PM exposure from wildland fire sources compared with other sources of PM, so additional analyses are needed. US EPA defines *classical error* as error scattered around the true personal exposure and independent of the measured exposure. Classical error results in bias of the epidemiological risk estimate toward underestimating or not detecting a true association. Because variation in the measurements tends to be greater than variation in the true exposures, classical error typically biases the health effect risk estimate toward the null (no effect of the exposure). This error causes the health effect risk estimate to be underestimated. Classical error can also cause inflation or reduction of the standard error of the health effect estimate. As documented by the US EPA PM ISA, in the PM literature, exposure error typically produces *underestimation* of health effects in epidemiological studies of short-term and long-term PM exposure. Bias away from the null can sometimes occur for long-term exposure studies if the exposure technique (e.g., monitor or model) underestimates population exposure.

In addition to time and activity diaries, one technique used to better understand individual exposures is using personal monitoring devices. These techniques are used in occupational settings and in community settings. In all-source PM studies, personal monitoring has been used to estimate exposure, but because these intensive studies are often expensive and difficult to conduct, the small size of the study often limits its explanatory power for larger populations (Özkaynak et al., 2013). New methodologies to enhance personal sampling incorporate sensors, videotape, global positioning systems (GPS; Westerdahl et al., 2005), direction-related mapping software, or data from smartphones into their sampling protocols to estimate personal exposure by using simultaneous measures of concentration and time-activity data.

10.3. RESEARCH GAPS AND FUTURE DIRECTIONS

Quantifying exposure to wildland fire-derived pollution presents unique challenges that are currently being addressed at the forefront of human health and exposure science. At the time of publication, serious knowledge gaps in our understanding of the full effects of wildland fire smoke exposure remain, including the associations between wildfire smoke and preclinical changes in health outcomes, birth and maternal outcomes, cognitive health, and the interaction with mental health. Data to better quantify concentration-response functions for known (e.g., asthma, COPD) or suspected wildfire health impacts, and to determine the independent impacts of fire-derived PM versus other pollutant exposures require robust and reliable methods of quantifying pollutant exposures. Additional studies of exposure to the multiple pollutants associated with fire emissions are needed. Geospatial data about fire pollutant concentrations is required to assess differential effects due to geospatial fire characteristics and human vulnerability characteristics.

Research is necessary regarding the effects of long-term and repeated wildland fire smoke pollution exposure across life stages, including developing fetuses and children, and among highly exposed occupational groups (e.g., firefighters, agricultural and forestry workers). Cumulative and aggregate exposure techniques must be developed, including a robust wildland fire data collection and modeling framework to characterize and predict all emissions, toxic potencies, and durations of smoke exposure, and all exposure pathways that impact health to inform decisions about public health and fire management. Consistent monitoring and modeling strategies for all wildland fires, whether prescribed or naturally occurring, would allow the most robust comparative analysis; however, these strategies must be responsive to global needs. More data are needed about changes during the COVID pandemic in infiltration and indoor air quality as well as the effect of changes in behaviors on relevant exposures and risks.

There are many objectives for estimating smoke exposures including epidemiologic research, risk assessments, policy assessment, forecasting and public warning systems, guidance for clinical health practice, and occupational health. These objectives help to shape future research for exposure science. For example, research is needed to improve the understanding of how drivers of climate change can influence the likelihood of exposure to smoke in the future. More knowledge of smoke plume characteristics, chemical transformation, and transport is needed to improve air quality modeling representation of smoke events. Data about the effectiveness of warning systems and mitigation strategies on exposures in a variety of global settings would also provide valuable information.

Ideally, estimates for all exposures that impact health positively or negatively would be considered. Fire-specific exposures could include aggregating pathways such as inhalation; dermal exposure to soot and ash; water quality impacts; and psychological impacts from anticipation, duration, and recovery from the crisis. Information about multiple pollutant and the cumulative exposures over time are exposure science research needs. In this chapter, we focused mainly on inhalation of smoke-related air pollution, but research is needed on other pathways such as mental health impacts. In addition, existing models simulating different components of the Earth system (e.g., air, land, water) can be connected to inform a more comprehensive understanding of the drivers and consequences of wildland fires.

Another research need for smoke exposure assessment relates to characterizing the full range of human individual and population variability and groups susceptible to adverse consequences of environmental smoke exposures. Fully describing exposure variability ensures the protection of everyone, including those most susceptible (e.g., increased biological sensitivity) or vulnerable (e.g., less access to health care or mitigation strategies). The current scientific understanding of contributions to heightened sensitivity includes intrinsic factors (e.g., life stage, genetics, disease status, nutrition) and extrinsic factors (e.g., additional chemical exposures and nonchemical stressors, occupation, socioeconomic status). Extrinsic factors include social stressors, such as lack of access to health care facilities, healthy food, and educational opportunities. These socially patterned factors can further compound the negative effects of environmental exposures. Furthermore, environmental exposures at multiple critical windows can modify each other, leading to greater adverse consequences. Better accounting for these factors is consistent with recommendations of the National Academy of Sciences (National Research Council, 2009) and scientific articles (Axelrad et al., 2019; Bhat et al., 2017; Chiu & Rusyn, 2018), which conclude that default approaches to treatment of human variability in risk assessments need to be updated to better incorporate current knowledge regarding human variability and vulnerability factors.

The increasing prevalence of wildland fires near communities globally has heightened the need for an improved understanding of exposure to smoke. Limitations in exposure science research could bias epidemiologic studies, resulting in less than optimal public health protections. Exposure science improvement is thus crucial to informing decisions to improve public health.

REFERENCES

Abdo, M., Ward, I., O'Dell, K., Ford, B., Pierce, J. R., Fischer, E. V., et al. (2019). Impact of wildfire smoke on adverse pregnancy outcomes in Colorado, 2007–2015. *International Journal of Environmental Research and Public Health*, *16*(19), 3720. https://doi.org/10.3390/ijerph16193720

Adetona, O., Reinhardt, T. E., Domitrovich, J., Broyles, G., Adetona, A. M., Kleinman, M. T., et al. (2016). Review of the health effects of wildland fire smoke on wildland firefighters and the public. *Inhalation Toxicology*, *28*(3), 95–139. https://doi.org/10.3109/08958378.2016.1145771

Akagi, S. K., Craven, J. S., Taylor, J. W., McMeeking, G. R., Yokelson, R. J., Burling, I. R., et al. (2011). Evolution of trace gases and particles emitted by a chaparral fire in California. *Atmospheric Chemistry and Physics Discussions*, *11*(8), 22483–22544. https://doi.org/10.5194/acpd-11-22483-2011

Allen, J. G., & Ibrahim, A. M. (2021). Indoor air changes and potential implications for SARS-CoV-2 transmission. *JAMA*, *325*(20), 2112–2113. https://doi.org/10.1001/jama.2021.5053

Amjad, S., Chojecki, D., Osornio-Vargas, A., & Ospina, M. B. (2021). Wildfire exposure during pregnancy and the risk of adverse birth outcomes: A systematic review. *Environment International*, *156*, 106644. https://doi.org/10.1016/j.envint.2021.106644

Andreae, M. O. (2019). Emission of trace gases and aerosols from biomass burning: An updated assessment. *Atmospheric Chemistry and Physics*, *19*(13), 8523–8546. https://doi.org/10.5194/acp-19-8523-2019

Awokola, B. I., Okello, G., Mortimer, K. J., Jewell, C. P., Erhart, A., & Semple, S. (2020). Measuring air quality for advocacy in Africa (MA3): Feasibility and practicality of longitudinal ambient PM2.5 measurement using low-cost sensors. *International Journal of Environmental Research and Public Health*, *17*(19), 7243. https://doi.org/10.3390/ijerph17197243

Axelrad, D. A., Setzer, R. W., Bateson, T. F., DeVito, M., Dzubow, R. C., Fitzpatrick, J. W., et al. (2019). Methods for evaluating variability in human health dose-response characterization. *Human and Ecological Risk Assessment*, *25*, 1–24. https://doi.org/10.1080/10807039.2019.1615828

Badura, M., Batog, P., Drzeniecka-Osiadacz, A., & Modzel, P. (2018). Evaluation of low-cost sensors for ambient PM2.5 monitoring. *Journal of Sensors*, *2018*, 5096540. https://doi.org/10.1155/2018/5096540

Baek, J., Hu, Y., Odman, M. T., & Russell, A. G. (2011). Modeling secondary organic aerosol in CMAQ using multigenerational oxidation of semi-volatile organic compounds. *Journal of Geophysical Research: Atmospheres*, *116*(D22). https://doi.org/10.1029/2011JD015911

Baker, K., Carlton, A., Kleindienst, T., Offenberg, J., Beaver, M., Gentner, D., et al. (2015). Gas and aerosol carbon in California: Comparison of measurements and model predictions in Pasadena and Bakersfield. *Atmospheric Chemistry and Physics*, *15*(9), 5243–5258. https://doi.org/10.5194/acp-15-5243-2015

Baker, K., Woody, M., Tonnesen, G., Hutzell, W., Pye, H., Beaver, M., et al. (2016). Contribution of regional-scale fire events to ozone and PM 2.5 air quality estimated by photochemical modeling approaches. *Atmospheric Environment*, *140*, 539–554. https://doi.org/10.1016/j.atmosenv.2016.06.032

Baker, K. R., & Woody, M. C. (2017). Assessing model characterization of single source secondary pollutant impacts using 2013 SENEX field study measurements. *Environmental Science & Technology*. https://doi.org/10.1021/acs.est.6b05069

Baker, K. R., Woody, M. C., Valin, L., Szykman, J., Yates, E. L., Iraci, L. T., et al. (2018). Photochemical model evaluation of 2013 California wild fire air quality impacts using surface, aircraft, and satellite data. *Science of the Total Environment*, *637–638*, 1137–1149. https://doi.org/10.1016/j.scitotenv.2018.05.048

Balmes, J. R. (2019). Household air pollution from domestic combustion of solid fuels and health. *Journal of Allergy and Clinical Immunology*, *143*(6), 1979–1987. https://doi.org/10.1016/j.jaci.2019.04.016

Barsanti, K. C., Carlton, A. G., & Chung, S. H. (2013). Analyzing experimental data and model parameters: Implications for predictions of SOA using chemical transport models. *Atmospheric Chemistry and Physics*, *13*(23), 12073–12088. https://doi.org/10.5194/acp-13-12073-2013

Bergen, S., & Szpiro, A. A. (2015). Mitigating the impact of measurement error when using penalized regression to model exposure in two-stage air pollution epidemiology studies. *Environmental and Ecological Statistics*, *22*(3), 601–631. https://doi.org/10.1007/s10651-015-0314-y

Bergen, S., Sheppard, L., Sampson, P. D., Kim, S.-Y., Richards, M., Vedal, S., et al. (2013). A national prediction model for PM2.5 component exposures and measurement error-corrected health effect inference. *Environmental Health Perspectives*, *121*(9), 1017–1025. https://doi.org/10.1289/ehp.1206010

Bernard, S. M., Samet, J. M., Grambsch, A., Ebi, K. L., & Romieu, I. (2001). The potential impacts of climate variability and change on air pollution-related health effects in the United States. *Environmental Health Perspectives*, *109* Suppl 2, 199–209. https://doi.org/10.1289/ehp.109-1240667

Bhat, V. S., Meek, M. E. B., Valcke, M., English, C., Boobis, A., & Brown, R. (2017). Evolution of chemical-specific adjustment factors (CSAF) based on recent international experience; Increasing utility and facilitating regulatory acceptance. *Critical Reviews in Toxicology*, *47*(9), 729–749. https://doi.org/10.1080/10408444.2017.1303818

Black, C., Tesfaigzi, Y., Bassein, J. A., & Miller, L. A. (2017). Wildfire smoke exposure and human health: Significant gaps in research for a growing public health issue. *Environmental Toxicology and Pharmacology*, *55*, 186–195. https://doi.org/10.1016/j.etap.2017.08.022

Brey, S. J., & Fischer, E. V. (2016). Smoke in the city: How often and where does smoke impact summertime ozone in the United States? *Environmental Science & Technology*, *50*(3), 1288–1294. https://doi.org/10.1021/acs.est.5b05218

Burley, J. D., Bytnerowicz, A., Buhler, M., Zielinska, B., Schweizer, D., Cisneros, R., et al. (2016). Air quality at Devils Postpile National Monument, Sierra Nevada Mountains, California, USA. *Aerosol and Air Quality Research*, *16*(10), 2315–2332. https://doi.org/10.4209/aaqr.2016.02.0069

Cascio, W. E. (2018). Wildland fire smoke and human health. *Science of the Total Environment*, *624*, 586–595. https://doi.org/10.1016/j.scitotenv.2017.12.086

Centers for Disease Control (2020). *Ten essential public health services*. https://www.cdc.gov/publichealthgateway/publichealthservices/essentialhealthservices.html.

Chance, K. V., Liu, X., Suleiman, R. M., Flittner, D. E., Al-Saadi, J., & Janz, S. J. (2013). *Tropospheric emissions: Monitoring of pollution (TEMPO)*. Paper presented at the SPIE 8866, Earth Observing Systems XVIII. https://spie.org/Publications/Proceedings/Volume/8866

Chen, C., Zhao, B., & Weschler, C. J. (2012). Assessing the influence of indoor exposure to "outdoor ozone" on the relationship between ozone and short-term mortality in U.S. communities. *Environmental Health Perspectives*, *120*(2), 235–240. https://doi.org/10.1289/ehp.1103970

Chiu, W. A., & Rusyn, I. (2018). Advancing chemical risk assessment decision-making with population variability data: Challenges and opportunities. *Mammalian Genome*, *29*(1–2), 182–189. https://doi.org/10.1007/s00335-017-9731-6

DeFlorio-Barker, S., Crooks, J., Reyes, J., & Rappold, A. G. (2019). Cardiopulmonary effects of fine particulate matter exposure among older adults, during wildfire and non-wildfire periods, in the United States, 2008–2010. *Environmental Health Perspectives*, *127*(3), 37006. https://doi.org/10.1289/EHP3860

Doubleday, A., Schulte, J., Sheppard, L., Kadlec, M., Dhammapala, R., Fox, J., et al. (2020). Mortality associated with wildfire smoke exposure in Washington state, 2006–2017: A case-crossover study. *Environmental Health*, *19*. https://doi.org/10.1186/s12940-020-0559-2

Duncan, B. N., Prados, A. I., Lamsal, L. N., Liu, Y., Streets, D. G., Gupta, P., et al. (2014). Satellite data of atmospheric pollution for US air quality applications: Examples of applications, summary of data end-user resources, answers to FAQs, and common mistakes to avoid. *Atmospheric Environment*, *94*, 647–662. https://doi.org/10.1016/j.atmosenv.2014.05.061

Fann, N., Alman, B., Broome, R. A., Morgan, G. G., Johnston, F. H., Pouliot, G., et al. (2018). The health impacts and economic value of wildland fire episodes in the U.S.: 2008–2012. *Science of the Total Environment*, *610–611*, 802–809. https://doi.org/10.1016/j.scitotenv.2017.08.024

Foley, K., Roselle, S., Appel, K., Bhave, P., Pleim, J., Otte, T., et al. (2010). Incremental testing of the Community Multiscale Air Quality (CMAQ) modeling system version 4.7. *Geoscientific Model Development*, *3*(1), 205. https://doi.org/10.5194/gmd-3-205-2010

Gan, R. W., Ford, B., Lassman, W., Pfister, G., Vaidyanathan, A., Fischer, E., et al. (2017). Comparison of wildfire smoke estimation methods and associations with cardiopulmonary-related hospital admissions. *GeoHealth*, *1*(3), 122–136. https://doi.org/10.1002/2017GH000073

Gupta, P., Doraiswamy, P., Levy, R., Pikelnaya, O., Maibach, J., Feenstra, B., et al. (2018). Impact of California fires on local and regional air quality: The role of a low-cost sensor network and satellite observations. *GeoHealth*. https://doi.org/10.1029/2018gh000136

Hammer, M. S., van Donkelaar, A., Li, C., Lyapustin, A., Sayer, A. M., Hsu, N. C., et al. (2020). Global estimates and long-term trends of fine particulate matter concentrations (1998–2018). *Environmental Science & Technology*, *54*(13), 7879–7890. https://doi.org/10.1021/acs.est.0c01764

Holder, A. L., Mebust, A. K., Maghran, L. A., McGown, M. R., Stewart, K. E., Vallano, D. M., et al. (2020). Field evaluation of low-cost particulate matter sensors for measuring wildfire smoke. *Sensors*, *20*(17), 4796. https://doi.org/10.3390/s20174796

Holm, S. M., Miller, M. D., & Balmes, J. R. (2021). Health effects of wildfire smoke in children and public health tools: A narrative review. *Journal of Exposure Science & Environmental Epidemiology*, *31*(1), 1–20. https://doi.org/10.1038/s41370-020-00267-4

Hutchinson, J. A., Vargo, J., Milet, M., French, N. H. F., Billmire, M., Johnson, J., et al. (2018). The San Diego 2007 wildfires and Medi-Cal emergency department presentations, inpatient hospitalizations, and outpatient visits: An observational study of smoke exposure periods and a bidirectional case-crossover analysis. *PLoS Medicine*, *15*(7), e1002601. https://doi.org/10.1371/journal.pmed.1002601

Ignotti, E., Valente, J. G., Longo, K. M., Freitas, S. R., Hacon, S. d. S., & Artaxo Netto, P. (2010). Impact on human health of particulate matter emitted from burnings in the Brazilian Amazon region. *Revista de Saude Publica*, *44*(1), 121–130. https://doi.org/10.1590/S0034-89102010000100013

Institute of Medicine (1999). *Toward environmental justice: Research, education, and health policy needs*. Washington, DC: The National Academies Press.

Jaffe, D., Hafner, W., Chand, D., Westerling, A., & Spracklen, D. (2008). Interannual variations in PM2.5 due to wildfires in the western United States. *Environmental Science & Technology*, *42*(8), 2812–2818. https://doi.org/10.1021/es702755v

Jaffe, D. A., & Wigder, N. L. (2012). Ozone production from wildfires: A critical review. *Atmospheric Environment*, *51*, 1–10. https://doi.org/10.1016/j.atmosenv.2011.11.063

Jaffe, D. A., O'Neill, S. M., Larkin, N. K., Holder, A. L., Peterson, D. L., Halofsky, J. E., et al. (2020). Wildfire and prescribed burning impacts on air quality in the United States. *Journal of the Air & Waste Management Association* *70*(6), 583–615. https://doi.org/10.1080/10962247.2020.1749731

Jerrett, M., Turner, M. C., Beckerman, B. S., Pope, C. A., Donkelaar, A. V., Martin, R. V., et al. (2017). Comparing the health effects of ambient particulate matter estimated using ground-based versus remote sensing exposure estimates. *Environmental Health Perspectives*, *125*(4), 552–559. https://doi.org/10.1289/EHP575

Johnston, F. H., Henderson, S. B., Chen, Y., Randerson, J. T., Marlier, M., Defries, R. S., et al. (2012). Estimated global mortality attributable to smoke from landscape fires. *Environmental Health Perspectives*, *120*(5), 695–701. https://doi.org/10.1289/ehp.1104422

Kansas Department of Health and Environment (2012). *State of Kansas exceptional event demonstration package April 6, 12, 13, and29, 2011*. Department of Health and Environment, Division of Environment, Bureau of Air. 27 November 2012. http://www.epa.gov/ttn/analysis/docs/KDHE_ExEvents_final_042011.pdf

Kobziar, L. N., & Thompson, G. R. (2020). Wildfire smoke, a potential infectious agent. *Science*, *370*(6523), 1408–1410. https://doi.org/10.1126/science.abe8116

Koman, P. D., Billmire, M., Baker, K. R., Carter, J. M., Thelen, B. J., French, N. H. F., et al. (2022). Using wildland fire smoke modeling data in gerontological health research (California, 2007–2018). *Science of the Total Environment*, *838*, 156403. https://doi.org/10.1016/j.scitotenv.2022.156403

Koman, P. D., Billmire, M., Baker, K. R., de Majo, R., Anderson, F. J., Hoshiko, S., et al. (2019). Mapping modeled exposure of wildland fire smoke for human health studies in California. *Atmosphere* *10*(6), 308. https://doi.org/10.3390/atmos10060308

Koman, P. D., Hogan, K. A., Sampson, N., Mandell, R., Coombe, C. M., Tetteh, M. M., et al. (2018). Examining joint effects of air pollution exposure and social determinants of health in defining "at-risk" populations under the Clean Air Act: Susceptibility of pregnant women to hypertensive disorders of pregnancy. *World Medical & Health Policy*, *10*(1), 7–54. https://doi.org/10.1002/wmh3.257

Lassman, W., Ford, B., Gan, R. W., Pfister, G., Magzamen, S., Fischer, E. V., et al. (2017). Spatial and temporal estimates of population exposure to wildfire smoke during the Washington state 2012 wildfire season using blended model, satellite, and in situ data. *GeoHealth*, *1*(3), 106–121. https://doi.org/10.1002/2017GH000049

Levy, I., Mihele, C., Lu, G., Narayan, J., Hilker, N., & Brook, J. R. (2014). Elucidating multipollutant exposure across a complex metropolitan area by systematic deployment of a mobile laboratory. *Atmospheric Chemistry and Physics*, *14*(14), 7173–7193. https://doi.org/10.5194/acp-14-7173-2014

Lineberry, D. (2016). The future of monitoring air quality from space. https://www.nasa.gov/feature/langley/the-future-of-monitoring-air-quality-from-space, 6/27/2017

Liss, A., Wu, R., Chui, K. K. H., & Naumova, E. N. (2017). Heat-related hospitalizations in older adults: An amplified effect of the first seasonal heatwave. *Scientific Reports*, *7*, 39581. https://doi.org/10.1038/srep39581

Liu, J., Schaaf, C., Strahler, A., Jiao, Z., Shuai, Y., Zhang, Q., et al. (2009). Validation of Moderate Resolution Imaging Spectroradiometer (MODIS) albedo retrieval algorithm: Dependence of albedo on solar zenith angle. *Journal of Geophysical Research*, *114*, D01106. https://doi.org/10.1029/2008JD009969

Liu, J. C., Mickley, L. J., Sulprizio, M. P., Dominici, F., Yue, X., Ebisu, K., et al. (2016). Particulate air pollution from wildfires in the western US under climate change. *Climatic Change*, *138*(3–4), 655–666. https://doi.org/10.1007/s10584-016-1762-6

Liu, J. C., Pereira, G., Uhl, S. A., Bravo, M. A., & Bell, M. L. (2015). A systematic review of the physical health impacts from non-occupational exposure to wildfire smoke. *Environmental Research*, *136*, 120–132. https://doi.org/10.1016/j.envres.2014.10.015

Liu, J. C., Wilson, A., Mickley, L. J., Dominici, F., Ebisu, K., Wang, Y., et al. (2017). Wildfire-specific fine particulate matter and risk of hospital admissions in urban and rural counties. *Epidemiology*, *28*(1), 77–85. https://doi.org/10.1097/EDE.0000000000000556

Madrigano, J., Kloog, I., Goldberg, R., Coull, B. A., Mittleman, M. A., & Schwartz, J. (2013). Long-term exposure to PM2.5 and incidence of acute myocardial infarction. *Environmental Health Perspectives*, *121*(2), 192–196. https://doi.org/10.1289/ehp.1205284

Magzamen, S., Gan, R. W., Liu, J., O'Dell, K., Ford, B., Berg, K., et al. (2021). Differential cardiopulmonary health impacts of local and long-range transport of wildfire smoke. *GeoHealth*, *5*(3), e2020GH000330. https://doi.org/10.1029/2020GH000330

Martin, R. V. (2008). Satellite remote sensing of surface air quality. *Atmospheric Environment*, *42*(34), 7823–7843. https://doi.org/10.1016/j.atmosenv.2008.07.018

McNamara, D., Stephens, G., Ruminski, M., & Kasheta, T. (2004). *The Hazard Mapping System (HMS), NOAA multi-sensor fire and smoke detection program using environmental satellites*. Paper presented at the 13th Conference on Satellite Meteorology and Oceanography. https://ams.confex.com/ams/13SATMET/webprogram/start.html

Miller, M. D., & Marty, M. A. (2010). Impact of environmental chemicals on lung development. *Environmental Health Perspectives*, *118*(8), 1155–1164. https://doi.org/10.1289/ehp.0901856

Mills, D., Jones, R., Wobus, C., Ekstrom, J., Jantarasami, L., St Juliana, A., & Crimmins, A. (2018). Projecting age-stratified risk of exposure to inland flooding and wildfire smoke in the

United States under two climate scenarios. *Environmental Health Perspectives*, *126*(4), 047007. https://doi.org/10.1289/EHP2594

Morello-Frosch, R., Zuk, M., Jerrett, M., Shamasunder, B., & Kyle, A. D. (2011). Understanding the cumulative impacts of inequalities in environmental health: Implications for policy. *Health Affairs (Millwood)*, *30*(5), 879–887. https://doi.org/10.1377/hlthaff.2011.0153

Moritz, M. A., Parisien, M.-A., Batllori, E., Krawchuk, M. A., Van Dorn, J., Ganz, D. J., et al. (2012). Climate change and disruptions to global fire activity. *Ecosphere*, *3*(6), 1–22.

Morton, D. C., Collatz, G. J., Wang, D., Randerson, J. T., Giglio, L., & Chen, Y. (2013). Satellite-based assessment of climate controls on US burned area. *Biogeosciences*, *10*(1), 247. https://doi.org/10.5194/bg-10-247-2013

Na, K., & Cocker, D. R. (2008). Fine organic particle, formaldehyde, acetaldehyde concentrations under and after the influence of fire activity in the atmosphere of Riverside, *California. Environmental Research*, *108*(1), 7–14. https://doi.org/10.1016/j.envres.2008.04.004

Naeher, L. P., Brauer, M., Lipsett, M., Zelikoff, J. T., Simpson, C. D., Koenig, J. Q., et al. (2007). Woodsmoke health effects: A review. *Inhalation Toxicology*, *19*(1), 67–106. https://doi.org/10.1080/08958370600985875

National Academies of Sciences Engineering and Medicine (2020). *Implications of the California Wildfires for Health, communities, and preparedness: Proceedings of a workshop*. Washington, DC. https://doi.org/10.17226/25622v

National Research Council. (2009). *Science and decisions: Advancing risk assessment*. Washington, DC: The National Academies Press.

National Research Council. (2012). *Exposure science in the 21st century: A vision and a strategy*. Washington, DC. https://doi.org/10.17226/13507

Navarro, K. M., Schweizer, D., Balmes, J. R., & Cisneros, R. (2018). A review of community smoke exposure from wildfire compared to prescribed fire in the United States. *Atmosphere*, *9*(5), 185. https://doi.org/10.3390/atmos9050185

Nunes, K. V. R., Ignotti, E., & Hacon, S. D. S. (2013). Circulatory disease mortality rates in the elderly and exposure to PM2.5 generated by biomass burning in the Brazilian Amazon in 2005. *Cadernos de Saude Publica*, *29*(3), 589–598.

O'Dell, K., Hornbrook, R. S., Permar, W., Levin, E. J. T., Garofalo, L. A., Apel, E. C., et al. (2020). Hazardous air pollutants in fresh and aged western US wildfire smoke and implications for long-term exposure. *Environmental Science & Technology*, *54*(19), 11838–11847. https://doi.org/10.1021/acs.est.0c04497

Özkaynak, H., Baxter, L. K., Dionisio, K. L., & Burke, J. (2013). Air pollution exposure prediction approaches used in air pollution epidemiology studies. *Journal of Exposure Science & Environmental Epidemiology*, *23*(6), 566–572. https://doi.org/10.1038/jes.2013.15

Patel, A. B., Meleth, S., Pasha, O., Goudar, S. S., Esamai, F., Garces, A. L., et al. (2015). Impact of exposure to cooking fuels on stillbirths, perinatal, very early and late neonatal mortality: A multicenter prospective cohort study in rural communities in India, Pakistan, Kenya, Zambia and Guatemala. *Maternal Health, Neonatology and Perinatology*, *1*, 18. https://doi.org/10.1186/s40748-015-0019-0

PHNCI (2020). *Futures initiative: The 10 essential public health services*. https://phnci.org/national-frameworks/10-ephs.

Pratt, J. R., Gan, R. W., Ford, B., Brey, S., Pierce, J. R., Fischer, E. V., et al. (2019). A national burden assessment of estimated pediatric asthma emergency department visits that may be attributed to elevated ozone levels associated with the presence of smoke. *Environmental Monitoring and Assessment*, *191*(Suppl 2), 269. https://doi.org/10.1007/s10661-019-7420-5

Rappold, A. G., Fann, N. L., Crooks, J., Huang, J., Cascio, W. E., Devlin, R. B., et al. (2014). Forecast-based interventions can reduce the health and economic burden of wildfires. *Environmental Science & Technology*, *48*(18), 10571–10579. https://doi.org/10.1021/es5012725

Rappold, A. G., Reyes, J. M., Pouliot, G., Cascio, W. E., & Diaz-Sanchez, D. (2017). Community vulnerability to health impacts of wildland fire smoke exposure. *Environmental Science & Technology*, *51*(12), 6674–6682. https://doi.org/10.1021/acs.est.6b06200

Reid, C. E., Brauer, M., Johnston, F. H., Jerrett, M., Balmes, J. R., & Elliott, C. T. (2016a). Critical review of health impacts of wildfire smoke exposure. *Environmental Health Perspectives*, *124*(9), 1334–1343. https://doi.org/10.1289/ehp.1409277

Reid, C. E., Jerrett, M., Petersen, M. L., Pfister, G. G., Morefield, P. E., Tager, I. B., et al. (2015). Spatiotemporal prediction of fine particulate matter during the 2008 northern California wildfires using machine learning. *Environmental Science & Technology*, *49*(6), 3887–3896. https://doi.org/10.1021/es505846r

Reid, C. E., Jerrett, M., Tager, I. B., Petersen, M. L., Mann, J. K., & Balmes, J. R. (2016b). Differential respiratory health effects from the 2008 northern California wildfires: A spatiotemporal approach. *Environmental Research*, *150*, 227–235. https://doi.org/10.1016/j.envres.2016.06.012

Resnick, A., Woods, B., Krapfl, H., & Toth, B. (2015). Health outcomes associated with smoke exposure in Albuquerque, New Mexico, during the 2011 Wallow fire. *Journal of Public Health Management and Practice*, *21*, S55–S61.

Roberts, G., & Wooster, M. (2021). Global impact of landscape fire emissions on surface level PM2.5 concentrations, air quality exposure and population mortality. *Atmospheric Environment*. https://doi.org/10.1016/j.atmosenv.2021.118210

Rosa, M. J., Hair, G. M., Just, A. C., Kloog, I., Svensson, K., Pizano-Zarate, M. L., et al. (2020). Identifying critical windows of prenatal particulate matter (PM2.5) exposure and early childhood blood pressure. *Environmental Research*, *182*, 109073. https://doi.org/10.1016/j.envres.2019.109073

Rosofsky, A., Levy, J. I., Breen, M. S., Zanobetti, A., & Fabian, M. P. (2019). The impact of air exchange rate on ambient air pollution exposure and inequalities across all residential parcels in Massachusetts. *Journal of Exposure Science & Environmental Epidemiology*, *29*(4), 520–530. https://doi.org/10.1038/s41370-018-0068-3

Rosofsky, A., Levy, J. I., Zanobetti, A., Janulewicz, P., & Fabian, M. P. (2018). Temporal trends in air pollution exposure inequality in Massachusetts. *Environmental Research*, *161*, 76–86. https://doi.org/10.1016/j.envres.2017.10.028

Russell, A. (1997). Regional photochemical air quality modeling: Model formulations, history, and state of the science. *Annual Review of Energy and the Environment*, *22*(1), 537–588. https://doi.org/10.1146/annurev.energy.22.1.537

Sarnat, J. A., Koutrakis, P., & Suh, H. H. (2000). Assessing the relationship between personal particulate and gaseous exposures of senior citizens living in Baltimore, MD. *Journal of the Air & Waste Management Association 50*(7), 1184–1198. https://doi.org/10.1080/10473289.2000.10464165

Sarnat, J. A., Sarnat, S. E., Flanders, W. D., Chang, H. H., Mulholland, J., Baxter, L., et al. (2013). Spatiotemporally resolved air exchange rate as a modifier of acute air pollution-related morbidity in Atlanta. *Journal of Exposure Science & Environmental Epidemiology*, *23*(6), 606–615. https://doi.org/10.1038/jes.2013.32

Shaposhnikov, D., Revich, B., Bellander, T., Bedada, G. B., Bottai, M., Kharkova, T., et al. (2014). Mortality related to air pollution with the Moscow heat wave and wildfire of 2010. *Epidemiology*, *25*(3), 359–364. https://doi.org/10.1097/EDE.0000000000000090

Spracklen, D. V., Logan, J. A., Mickley, L. J., Park, R. J., Yevich, R., Westerling, A. L., et al. (2007). Wildfires drive interannual variability of organic carbon aerosol in the western U.S. in summer. *Geophysical Research Letters*, *34*, L16816. https://doi.org/10.0129/GL030037

Spracklen, D. V., Mickley, L. J., Logan, J. A., Hudman, R. C., Yevich, R., Flannigan, M. D., et al. (2009). Impacts of climate change from 2000 to 2050 on wildfire activity and carbonaceous aerosol concentrations in the western United States. *Journal of Geophysical Research: Atmospheres*, *114*(D20). https://doi.org/10.1029/2008JD010966

Thelen, B., French, N. H. F., Koziol, B. W., Billmire, M., Owen, R. C., Johnson, J., et al. (2013). Modeling acute respiratory illness during the 2007 San Diego wildland fires using a coupled emissions-transport system and general additive modeling. *Environmental Health*, *12*(1), 94. https://doi.org/10.1186/1476-069X-12-94

Thompson, T. M., Saari, R. K., & Selin, N. E. (2014). Air quality resolution for health impact assessment: influence of regional characteristics. *Atmospheric Chemistry and Physics*, *14*(2), 969–978. https://doi.org/10.5194/acp-14-969-2014

Thurston, G. D., Balmes, J. R., Garcia, E., Gilliland, F. D., Rice, M. B., Schikowski, T., et al. (2020). Outdoor air pollution and new-onset airway disease. An official American Thoracic Society workshop report. *Annals of the. American Thoracic Society*, *17*(4), 387–398. https://doi.org/10.1513/AnnalsATS.202001-046ST

Thurston, G. D., Kipen, H., Annesi-Maesano, I., Balmes, J., Brook, R. D., Cromar, K., et al. (2017). A joint ERS/ATS policy statement: what constitutes an adverse health effect of air pollution? *An analytical framework. European Respiratory Journal*, *49*(1), 1600419. https://doi.org/10.1183/13993003.00419-2016

U.S. Environmental Protection Agency (2011). *Exposure Factors Handbook, Chapter 6 Inhalation rates*. Washington, DC: https://www.epa.gov/expobox/about-exposure-factors-handbook.

U.S. Environmental Protection Agency (2014). *Child-specific exposure scenarios examples EPA/600/R-14-217F* (EPA/600/R-14-217F). Washington, DC: https://cfpub.epa.gov/ncea/efp/recordisplay.cfm?deid=262211.

U.S. Environmental Protection Agency (2019a). *Guidelines for human exposure assessment*. Washington, DC, EPA/100/B-19/001.

U.S. Environmental Protection Agency (2019b). *Integrated Science Assessment (ISA) for particulate matter EPA/600/R-19/188*. Washington, DC, EPA/600/R-19/188.

U.S. Environmental Protection Agency (2020). *Integrated Science Assessment (ISA) for ozone and related photochemical oxidants EPA/600/R-20/012*. Washington, DC, EPA/600/R-20/012.

Val Martin, M., Kahn, R. A., Logan, J. A., Paugam, R., Wooster, M., & Ichoku, C. (2012). Space-based observational constraints for 1-D fire smoke plume-rise models. *Journal of Geophysical Research: Atmospheres*, *117*(D22). https://doi.org/10.1029/2012JD018370

Val Martin, M., Logan, J. A., Kahn, R. A., Leung, F. Y., Nelson, D. L., & Diner, D. J. (2010). Smoke injection heights from fires in North America: Analysis of 5 years of satellite observations. *Atmospheric Chemistry and Physics*, *10*(4), 1491–1510. https://doi.org/10.5194/acp-10-1491-2010

van Donkelaar, A., R., Martin, V., Brauer, M., Kahn, R., Levy, R., Verduzco, C., & Villeneuve, P. J. (2010). Global estimates of ambient fine particulate matter concentrations from satellite-based aerosol optical depth: Development and application. *Environmental Health Perspectives*. https://doi.org/10.1289/ehp.0901623

Viswanathan, S., Eria, L., Diunugala, N., Johnson, J., & McClean, C. (2006). An analysis of effects of San Diego wildfire on ambient air quality. *Journal of Air and Waste Management*, *56*(1), 56–67. https://doi.org/10.1080/10473289.2006.10464439

Voulgarakis, A., & Field, R. D. (2015). Fire influences on atmospheric composition, air quality and climate. *Current Pollution Reports*, *1*(2), 70–81. https://doi.org/10.1007/s40726-015-0007-z

Wegesser, T. C., Franzi, L. M., Mitloehner, F. M., Eiguren-Fernandez, A., & Last, J. A. (2010). Lung antioxidant and cytokine responses to coarse and fine particulate matter from the great California wildfires of 2008. *Inhalation Toxicology*, *22*(7), 561–570. https://doi.org/10.3109/08958370903571849

Westerdahl, D., Fruin, S., Sax, T., Fine, P. M., & Sioutas, C. (2005). Mobile platform measurements of ultrafine particles and associated pollutant concentrations on freeways and residential streets in Los Angeles. *Atmospheric Environment*, *39*(20), 3597–3610. https://doi.org/10.1016/j.atmosenv.2005.02.034

World Health Organization (2013). *Pocket book of hospital care for children: Guidelines for the management of common childhood illnesses*: World Health Organization.

World Health Organization. (2021). *Wildfires*. https://www.carbonbrief.org/explainer-how-climate-change-is-affecting-wildfires-around-the-world, 08/27/2021

Yu, S., Mathur, R., Pleim, J., Wong, D., Gilliam, R., Alapaty, K., et al. (2014). Aerosol indirect effect on the grid-scale clouds in the two-way coupled WRF-CMAQ: Model description, development, evaluation and regional analysis. *Atmospheric Chemistry and Physics*, *14*(20), 11247–11285. https://doi.org/10.5194/acp-14-11247-2014

Zanobetti, A., & Schwartz, J. (2009). The effect of fine and coarse particulate air pollution on mortality: A national analysis. *Environmental Health Perspectives*, *117*(6), 898–903. https://doi.org/10.1289/ehp.0800108

Zartarian, V., Bahadori, T., & McKone, T. (2005). Adoption of an official ISEA glossary. *Journal of Exposure Analysis and Environmental Epidemiology*, *15*(1), 1–5. https://doi.org/10.1038/sj.jea.7500411

Zoogman, P., Jacob, D. J., Chance, K., Liu, X., Lin, M., Fiore, A., et al. (2014). Monitoring high-ozone events in the US intermountain west using TEMPO geostationary satellite observations. *Atmospheric Chemistry and Physics*, *14*(12), 6261–6271. https://doi.org/10.5194/acp-14-6261-2014

Zou, Y., Young, M., Chen, J., Liu, J., May, A., & Clark, J. D. (2020). Examining the functional range of commercially available low-cost airborne particle sensors and consequences for monitoring of indoor air quality in residences. *Indoor Air*, *30*(2), 213–234. https://doi.org/10.1111/ina.12621

11

Wildfire Smoke Toxicology and Health

Luke Montrose[1], Adam Schuller[1], Savannah M. D'Evelyn[2], and Christopher Migliaccio[3]

ABSTRACT

With wildfires increasing in activity in the western United States and around the world, there is an immediate need to understand the toxic effects of the smoke. This chapter will provide a background of toxicology and apply principal concepts such as dose, duration, and frequency to help define the potential effects of smoke exposure. Characteristics that influence toxicity will be discussed, which include particle size, source, temperature, and the mixture of chemical constituents. An overview of the routes of exposure, mechanisms of action, toxicokinetics, and the role of the immune system will all be covered. The importance and mutual benefits of in vitro, ex vivo, and in vivo studies will be discussed. Finally, the chapter concludes by outlining knowledge gaps and research needs.

11.1. INTRODUCTION TO THE BASIC ELEMENTS OF TOXICOLOGY

To understand how complex chemical mixtures like wildfire smoke impact human health, or to compare and contrast the impacts of similar agents (e.g., wildfire smoke versus urban air pollution), some toxicology fundamentals should first be introduced. The cornerstone concept of toxicology is "the dose makes the poison," a phrase that is credited to Philippus Aureolus Theophrastus Bombastus von Hohenheim better known as Paracelsus (Grandjean, 2016). This concept posits that any chemical agent, whether snake venom, a pharmaceutical drug, or table salt, at some threshold amount can become deleterious for one's health. In addition to dose, this chapter will also introduce two additional fundamental concepts in toxicology, duration and frequency, which ultimately contribute to the dose. Note that the introduction and fundamentals of toxicology sections of this chapter are meant to establish a foundation of knowledge for those new to this field.

This chapter will explore wildfire smoke toxicity through occupational and public health lenses. Certain characteristics are particularly important to recognize, including particle size, source, and the mixture of chemical constituents. This chapter will also briefly highlight some human factors that can contribute to wildfire smoke toxicity such as age, genetics, and comorbidities. Route of exposure and mechanism of action are key concepts in toxicology that help explain how wildfire smoke enters the body and what it can potentially affect once inside. Accordingly, defense mechanisms, toxicokinetics, and the role of the immune system will be covered. Model systems are important to all toxicology research and thus we will outline the state of science in terms of in vitro, ex vivo and in vivo studies. Finally, the chapter will conclude with a discussion of future research needs, challenges and opportunities.

[1]*School of Public and Population Health, Boise State University, Boise, Idaho, USA*
[2]*Department of Environmental and Occupational Health Sciences, University of Washington, Seattle, Washington, USA*
[3]*Department of Biomedical and Pharmaceutical Health Sciences, University of Montana, Missoula, Montana, USA*

Landscape Fire, Smoke, and Health: Linking Biomass Burning Emissions to Human Well-Being, Geophysical Monograph 280, First Edition.
Edited by Tatiana V. Loboda, Nancy H. F. French, and Robin C. Puett.

DOI: 10.1002/9781119757030.ch11

11.1.1. Fundamental Concepts of Toxicology: Dose, Duration, and Frequency

Dose

In the medical literature, *dose* is defined as the amount of a therapeutic drug that is administered at a given time, and a *toxic dose* is the sum of a chemical that will "produce a harmful or untoward effect" (Tsatsakis et al., 2018). Thus, the ability to measure the amount of a chemical entering the body is critical in determining the dose of that chemical. An *exposure,* by contrast, is just one of many substances an individual comes into contact with as they move through the world. The concentration of such substances in the environment can be measured. In the occupational setting, this could take the form of an industrial hygienist placing an air monitor near a specific workstation. Because the dose is how much of this exposure enters into the body, we cannot simply assume that the external concentration is equal to the amount that is deposited in the lungs or other target tissues, and this is the clear distinction between dose and exposure. Instead, toxic principles such as absorption, distribution, metabolism, and excretion (ADME) must be applied and this adds to the difficulty of attributing dose from a measured exposure. For most occupational and public health studies, an exposure rather than a dose will be discussed.

Another complicating factor to understanding dosage of a substance, like wildfire smoke, is recognizing and appreciating the complexity of mixtures. Wildfire smoke is composed of a mixture of particulate matter (PM) and toxic gasses, which can vary by temperature, fuel type, and distance from source, among other factors (Black et al., 2017). Given that there can be thousands of individual chemical constituents in smoke from wildfires, and that this mixture is dynamic throughout the life of a fire and variable among fires, each of these factors requires reconsideration of the questions "What is the dose?" and "Dose of what?" Of all the wildfire smoke constituents, PM is often cited as the most abundant, considering weight by volume, and is also one the of the U.S. Environmental Protection Agency's (U.S. EPA's) six criteria air pollutants for ambient air quality (U.S. EPA, 2014). Thus, PM is often used as the primary proxy measurement of exposure. Having a standardized method of measurement can be helpful in both research and in policy making. However, appreciation of the many other chemical constituents and their potential health effects is important. The health effects of PM will be briefly discussed in the section on size fractions within this chapter as well as Chapters 10, 12, and 13. Other important constituents of wildfire smoke include carbon monoxide (CO), nitrogen oxides (NOx), and polycyclic aromatic hydrocarbons (PAHs), just to name a few. CO, at elevated levels, forms carboxyhemoglobin (COHb), which can increase incidence of adverse health outcomes, such as chest pain among those with preexisting comorbidities like heart disease (Anderson et al., 1973). From research using a canine model, we know that NOx binds to hemoglobin resulting in the production of methemoglobin, which can impact enzyme function and cause vascular membrane injury (Brizio-Molteni et al., 1984). Among those with asthma, NOx exposure can lead to bronchoconstriction (Zelikoff et al., 2002). PAHs have been found to adsorb onto PM particles resulting in immunosuppressive and cancerous effects in animals and potentially in humans (Bauer et al., 2022; Navarro et al., 2017).

In the scientific literature, exposure to wildfire-related air pollutants is commonly measured by local environmental air monitors. These could include a government monitoring station, a temporary monitor installed proximally to an active wildfire, or personal air monitors. The latter is somewhat less common due to a number of factors including cost, access, transport burden, and timely distribution. Discussion of passive versus active samplers and parsing out the merits of real-time or cumulative exposure monitoring are outside this chapter's scope. These methods all have advantages and disadvantages that need to be weighed based on specific exposure circumstances. Most notably, all of these methods are limited as a proxy for dose.

Compared with environmental monitoring, the appropriate use of biomarkers may be able to more accurately measure internal dose (Adetona et al., 2017). Distinguishing between biomarkers of exposure and biomarkers of effect is important. Biological monitoring methods for wildfire smoke exposure include chemicals exhaled in breath (e.g., CO), urinary markers (e.g., levoglucosan, methoxyphenols, or PAHs) as well as blood markers (e.g., carboxyhemoglobin) (Bergauff et al., 2010; Gill & Britz-McKibbin, 2020; Migliaccio et al., 2009; Schmidt et al., 2021). The use of such biomarkers is not always straightforward as variability may be introduced by diet, age, lifestyle, or other habits (e.g., smoking). Very few studies published to date adequately consider such factors, and thus the utility of the biomarker data is somewhat limited. Nevertheless, future studies can improve upon the collection of these factors and enhance the accuracy of biomarkers as a proxy for internal dose.

Duration

Duration is the amount of time an individual is exposed to a given substance. In the literature on chemical agent exposure (e.g., acid), you might see duration described as *contact time*. With regard to wildfire smoke, duration could be quantified differently depending on the situation. In the community setting, a town could be exposed to elevated wildfire smoke PM for multiple hours or days. Occupationally, wildland firefighters spend a certain number of hours on the fireline each year, and these hours add up over their career. Duration is not a mutually exclusive term from dose, but rather a contributing factor. In other words, the longer an individual is in contact with an exposure, the larger the likely dose will become. Another aspect of duration is to consider the time between repeated exposures. Among those exposed to wildfire smoke annually, little is currently known about the ability of lungs or other tissues to return to baseline after a "washout period"; however, this should be explored.

Duration of exposure is a particularly interesting concept to consider when we think about unique toxicologic profiles of ambient PM versus wildfire PM. Background or regional ambient PM in North American and European cities is often a relatively consistent, exposure that combines with more variable localized PM, both of which are typically chronic exposures for the residential populations of large, urban areas, that have been a main focus of air pollution epidemiology research. By contrast, epidemiological studies investigating wildfire smoke exposure are generally investigating high-dose acute exposures that may occur once annually or repeatedly and sporadically during a specific time of year to sparse, rural, and migratory populations, as well as larger, more dense populations in metropolitan areas. Notably, the fire season is growing in length (Liu et al., 2016). Given that the health effects of PM exposure are variable based on a number of factors including duration (Kampa & Castanas, 2008), the ultimate challenge will be understanding how these high-dose acute and/or episodic events differ in toxicity.

Frequency

Like duration, *frequency* is not mutually exclusive from dose, but rather an important contributing factor. Frequency, or the repetitive nature of exposure, is an inherent feature that differentiates occupational and community exposure. Wildland firefighters are, by design, at greater risk for exposure due to the proximity to a fire to conduct suppression activities. While communities are impacted by smoke only by chance, wildland firefighters work purposefully near active fires for up to 14 consecutive days on what is termed a *roll*, working up to 16 consecutive hours each shift (Navarro, 2020). While the time that each firefighter spends on the fireline is largely dependent on the duties (Reinhardt & Broyles, 2019), the frequency of exposure for this occupational group as a whole is starkly different than for individuals who live in nearby communities. This does not discount the potential health risks of living in a community, especially as the fire season lengthens and the number of episodes increases. Moreover, we should also focus our attention on the communities that find themselves in a cyclical pattern of exposure: fire season in the summer and fall, temperature inversions in the winter in combination with residential wood burning, and prescribed burning events in the spring. This exposure paradigm would be distinctly different from a wildland firefighter's experience, and also unique relative to communities that have only a few smoke events throughout a year. Remarkably, communities can adapt, sometimes called becoming *smoke ready*, to wildfires in ways that build resilience and that should be addressed in exposure assessment. For example, buildings can be designated as clean air spaces, stand-alone air filtration devices can be distributed, and community members can wear appropriate protective equipment such as N95 masks. Depending on the circumstances, firefighters and other outdoor workers may not have the same opportunities to mitigate smoke exposures.

11.1.2. Characteristics That Impact Toxicity

Here we will discuss some of the characteristics that can vary with any biomass combustion-generated air pollution event, including wildfire smoke events. While not exhaustive, this section will discuss factors that help to explain some common toxicological responses in both humans and animals.

Size Fractions

Wildfires can generate particles ranging from ultrafine up to ash or debris that may be several centimeters in size. Most large wildfire smoke particles including ash will precipitate within close proximity of the active fire and pose no inhalation risk to downwind communities. In contrast, smaller more transient particles can travel long distances, even reaching other continents.

Particulate pollution can be categorized many different ways, with these systems being mostly arbitrary but having some basis in characteristics of suspension or deposition in the human respiratory tract. The historical timeline of air quality standards in various countries demonstrates how, over time, research can inform regulation. In 1971, the U.S. Environmental Protection Agency (EPA) regulated PM based on total suspended particles, whereas today they regulate based on aerodynamic diameter of 10 and 2.5 micrometers (μm) (PM_{10} and $PM_{2.5}$; U.S. EPA, 2016). These two size fractions are notable because they are the approximate cutoffs for deposition in upper respiratory tract and lower respiratory tract or alveolar spaces (Fig. 11.1), respectively. Australia has developed similar cutoffs, developed in 1998 with the National Environment Protection Measure for Ambient Air Quality (NEPC, 1998), which were recently updated to include $PM_{2.5}$ in addition to PM_{10}. The European Union has two overarching air quality policies, the Thematic Strategy on Air Pollution

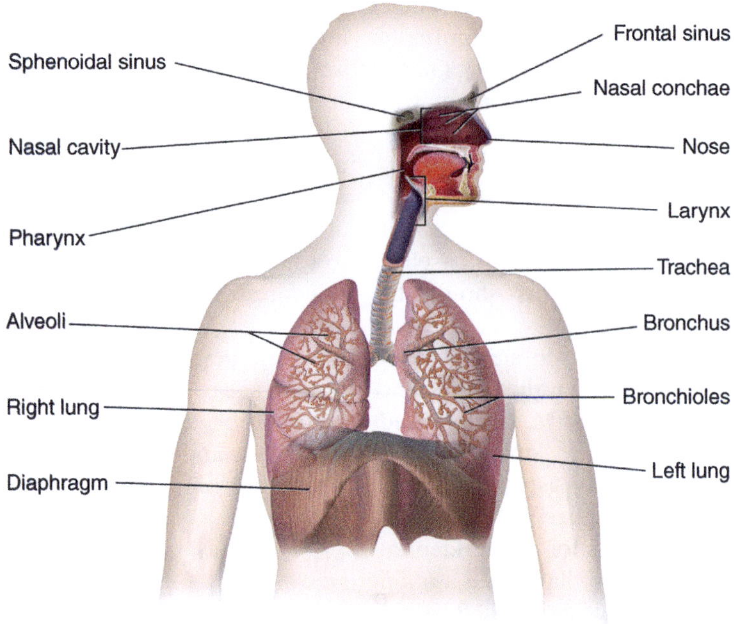

Figure 11.1 The respiratory system. https://commons.wikimedia.org/wiki/File:Blausen_0770_RespiratorySystem_02.png

and the National Emission Ceiling Directive, which were created in the early 2000s to set concentration standards for ambient air quality. These policies include standards for $PM_{2.5}$ and PM_{10} that are similar to U.S. EPA standards (Directive 2008/50/EC, 2008).

Sources and Fuels

With biomass combustion, "what is burning?" is an important question. Regarding wildfires, this could include some or all of the following: grass, brush, plant material on the forest floor (i.e., needles, duff, foliage), and trees, as well as manufactured materials (i.e., vehicles, structures). The ratio of these fuels can vary within a given ecosystem. Globally, fire suppression policies in certain regions have added to the availability of fuel (Curt et al., 2013; Westerling et al., 2014). As fires encroach on communities and communities encroach on the wilderness, building materials and other manmade products at this wildland-urban interface have also become a fuel. Wildfires in California, USA, during the 2018 and 2020 wildfire seasons, were particularly notable for their overlap into the wildland-urban interface according to the U.S. Forest Service (USDA Forest Service, 2019). Reducing smoke exposure risks from building materials and protecting homes that are bush fire prone is also a major goal in Australia (NSW Rural Fire Service, 2019).

Investigation of the toxicology profiles of different fuels is critical to help guide smoke mitigation and intervention activities. Researchers from the U.S. EPA conducted a trial of five different woodland fuels representing different U.S. geographical regions. The study found that toxicity, which was measured by quantifying the number of neutrophils infiltrating a mouse lung following exposure, differed by fuel source, and that mutagenicity, measured using the Ames Mutation assay, also varied by fuel source (Kim et al., 2018).

Temperature

Wildfire smoke is sometimes discussed as a static monolith of distinct chemicals. In reality, it is an extremely complicated and dynamic array of organic and inorganic compounds resulting from the unique combustion chemistry occurring under a set of specific conditions (Leonard et al., 2007). If the conditions or chemistry surrounding a fire event changes, then the resulting constituents of wildfire smoke will be different. Temperature is one of the most important conditions that contributes to combustion of wildland vegetation. Incomplete combustion of wildland vegetation occurs at temperatures between 300°C and 500°C and results in more organic carbon particles than is the case with more complete combustion (Kocbach Bølling et al., 2009). This slow, low temperature burning of solid material is known as *smoldering*, which is in contrast to the more homogeneous combustion of gasses at higher temperatures known as *flaming* (Santoso et al., 2019). Temperatures conducive to flaming are approximately 640°C, with the most favorable conditions for complete combustion being around 900°C (Seltenrich, 2018). Under flaming, oxygen-starved conditions, smoke products contain more elemental carbon aggregates; at higher temperatures with more oxygen, complete combustion produces mostly inorganic ash particles including alkali salts, chlorides, and carbonates (Kocbach Bølling et al., 2009).

Several studies have attempted to address how smoke toxicity is influenced by flaming versus smoldering conditions. In a study that burned eucalyptus, peat, and oak, researchers found that temperature overall had a dramatic effect on the smoke product for all sources (Hargrove et al., 2019). Smoldering conditions resulted in 10 times greater PM compared with flaming, whereas flaming conditions resulted in CO_2 and NO_2 levels that were 1,000 times higher compared with smoldering. In a separate study that evaluated oak, peat, pine needles, and pine wood, the researchers also demonstrated that smoldering conditions produce more PM, while flaming conditions produce more gases (Kim et al., 2018). In the studies cited above, the groups exposed mice under variable conditions and found that toxicity metrics differed with both burn temperature and fuel type (Hargrove et al., 2019; Kim et al., 2019; Kim et al., 2018).

Aging and Distance

Wildfires in the United States primarily affect forests on the West Coast, and data suggest the downwind areas most impacted by this smoke are the Rocky Mountain states (McClure & Jaffe, 2018). Intuitively, one could assume that as the smoke plume rises from a fire and is transported tens to thousands of miles away, that the PM concentration of the plume would decrease as dispersion and dilution occurs and, thus, the toxic potential of the wildfire smoke would presumably be diminished. This conclusion is accurate, but incomplete. Indeed, dilution of particles increases with distance (Radke et al., 1995). These changes to particles that occur with transport are referred to as *smoke aging*. However, photochemical processes with the aid of ultraviolet radiation from the Sun along with chemical mixing in the atmosphere can also alter the chemical makeup of the smoke (Wong et al., 2019). Interim findings from the PyroTRACH project in Greece suggest that aged smoke particles may actually be more toxic than fresh smoke particles (Gray et al., n.d.).

One product that is primarily generated by smoldering fires is brown carbon airborne particles, also called *tar balls* (Chakrabarty et al., 2010), which become particularly abundant in aged wildfire smoke (Girotto et al., 2018). Identification of brown carbon particles may be important for distinguishing it from black carbon, which is a signature of urban PM (Martenies et al., 2021). Brown carbon is a water-soluble organic component of wildfire smoke and Pardo et al. (2020) found that it increased the inflammatory and oxidative stress response in mice, was toxic to bronchial epithelial cells, and also induced an apoptotic pathway.

Mixtures

Wildfire smoke exposures do not occur in isolation from other sources of contamination. Accordingly, researchers must appreciate and attempt to account for potential interactions (whether they are additive, synergistic, agonistic, or antagonistic) between wildfire smoke and other possible air pollution exposures. As mentioned previously in this chapter, defining wildfire smoke alone can be problematic given the dynamic nature of wildfire smoke production. Though perhaps an immense task, the addition of other PM sources is the reality of ambient air exposures.

Studying coexposures is particularly important for those who work primarily outside, including wildland firefighters, agriculture, construction, landscape, utility, and facility maintenance workers (Navarro, 2020). Some of these coexposures could include silica, farm-related pesticides, traffic-related air pollution (TRAP), or diesel exhaust. There have been many studies that observed the effects of ambient air on health in both humans and animals, and some have likely overlapped with wildfire season. However, very few have been intentionally designed to consider the health effects associated with mixtures involving wildfire smoke among occupational workers. Reinhardt and Broyles (2019) found that a wildland firefighters' exposure to smoke and silica is more dependent on the task they perform than on the amount of time on the fireline.

Human Factors

A comprehensive discussion of human health effects and factors that may increase risk for wildfire smoke effects can be found in Chapters 10, 12, 13 of this volume, but it is important within the context of toxicology to point out these factors. In addition to the dose, duration, or frequency of exposure are the human factors that can influence toxicity. Factors that could make an individual more vulnerable, sensitive, or susceptible to smoke exposure include such preexisting variables as disease, lifestyle, ethnicity, and age. Genetic and epigenetic factors may also play a role but are outside the scope of this review (Black et al., 2017; Prunicki et al., 2020; Reid et al., 2016; Schuller & Montrose, 2020). Future studies could also focus on the vulnerability of communities, which might include an assessment of adaptive capacity and resilience.

11.2. ROUTES OF EXPOSURE

Humans are exposed to all the constituents of wildfire smoke through dermal absorption, inhalation, and ingestion. While the major route of exposure is inhalation through the pulmonary system, there are increasingly studied and notable extrapulmonary effects including neurological, reproductive, and cardiovascular health effects of wildfire smoke exposure (Brook et al., 2010; Fu et al., 2019; Klepac et al., 2018). The mechanisms of these effects, as well as the mechanisms of transport, are areas of on-going research. This section will briefly discuss all routes of exposure as well as tissue-specific mechanisms (barriers, cells, soluble factors) of toxicity.

11.2.1. Respiratory Exposure Route

The most common and significant route of exposure to wildfire smoke is via the respiratory system. This system includes the nasal passage, trachea, bronchi, and alveolar spaces, as well as a wide variety of cell types (Fig. 11.1).

The trachea divides into two main bronchi, which then subdivide into secondary bronchi, tertiary bronchi, and ultimately bronchopulmonary segments, lung sections separated by connective tissue. These tertiary bronchi further divide into fourth, fifth, and sixth order segmental bronchi eventually culminating in terminal bronchioles, which give rise to multiple alveolar ducts and five to six alveolar sacs with each duct (Bucher & Reid, 1961).

The science behind particle deposition in the respiratory system is complex and includes "inertial impaction, gravitational sedimentation, and Brownian diffusion, and to a lesser extent turbulent flows, interception, and electrostatic precipitation" (Darquenne, 2012, p.140). Particles larger than PM_{10} will be deposited and cleared by the anatomy and physiology of nostrils (Fig. 11.2). The coarse fraction, particles less than 10 but greater than 2.5 micrometers (μm), can evade impaction in the nasal conchae, but will be deposited in the upper respiratory where they are removed via mucociliary clearance (Bustamante-Marin & Ostrowski, 2017). The fine fraction, or $PM_{2.5}$, are at least 30 times smaller than the diameter of a human hair and bypass both the nasal conchae as well as the mucociliary escalator allowing for deposition in the lower airways and alveolar space. The last line of defense for fine ($PM_{2.5}$) and PM with an aerodynamic diameter less than 0.1 μm (ultrafine) is endocytosis by epithelial cells, or phagocytosis by two types of specialized innate immune cells, macrophages and polymorphonuclear neutrophils (Reynolds, 1985). Some particles avoid phagocytosis and remain in the alveolar space causing inflammation, while others may diffuse through the lung-blood barrier and gain access to other tissues via the body's circulatory system (Schraufnagel, 2020).

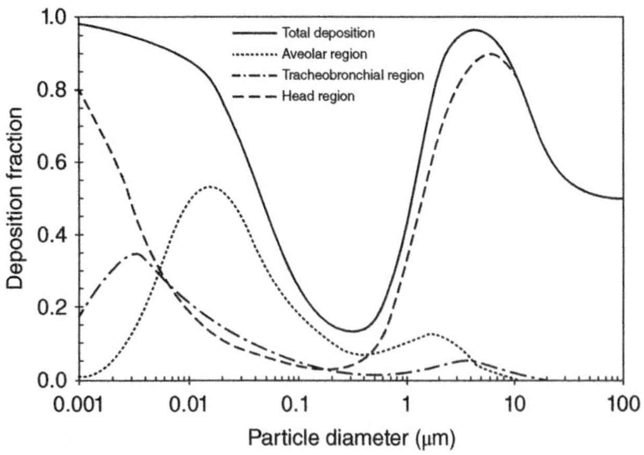

Figure 11.2 Probability of pulmonary deposition by size fraction (used with permission from Mack et al., 2020; https://www.ncbi.nlm.nih.gov/pmc/articles/PMC7553137/).

The U.S. EPA and its global counterparts categorize $PM_{2.5}$ as an atmospheric contaminant when determining pollution levels, and thus most respiratory studies focus on this size fraction. While many studies revolve around atmospheric $PM_{2.5}$ exposures and subsequent health effects (Brook et al., 2010; Brunekreef & Holgate, 2002; Pope, 2002; Shah et al., 2013), most of these involve chronic exposures (i.e., living in urban areas with elevated pollution) and the subsequent effects on at-risk groups such as those with cardiovascular disease, asthma, or COPD (Carey et al., 2013; Faustini et al., 2013; Karakatsani et al., 2012; Kelly & Fussell, 2011). These studies have illustrated the health complications for these groups including increased morbidity and mortality with exacerbations of pathologies. While many air quality indices (including the EPA's AQI) are mainly calculated using $PM_{2.5}$ levels (Air Quality Index, n.d.), there are two major reasons for independently assessing particle exposures from wildfire smoke and then comparing with the body of urban PM literature: duration of exposures and chemical content. Wildfire smoke exposures are usually higher in particle concentration and relatively shorter in duration compared with ambient PM exposures, suggesting the potential for distinct health effects.

11.2.2. Dermal Exposure Route

The skin is the largest organ of the body and is key to protection from environmental insults, including microbial pathogens, airborne particulates or chemicals, and helps with regulation of body temperature. Based on the extensive surface area, the skin is subject to significant environmental exposures and has potential for subsequent adverse effects. While the field of wood-smoke research is increasing in the number of both human and animal studies, none to date have specifically addressed the effects of smoke-derived $PM_{2.5}$ on the dermal system (Schwartz et al., 2020).

11.2.3. Ingestion Exposure Route

There are no direct studies to date assessing the contribution of ingested particles from wildfire smoke. However, gastrointestinal exposures via the mucociliary escalator can occur (Lee et al., 2020). Additional occupation-related exposures are possible through the ingestion route. For example, wildland firefighters could plausibly ingest smoke PM while eating given the lack of hygiene supplies in the field (Navarro et al., 2021).

11.3. TARGET ORGANS AND EFFECTS

This section describes (1) the areas of the body that are impacted by wildfire smoke and (2) the toxic effects that have been demonstrated or observed. Ambient PM effects will also be discussed where wildfire smoke data are lacking.

11.3.1. Respiratory Effects

The effects of ambient PM in the respiratory system have been examined. For example, research has shown increased asthma morbidity associated with increased exposures to certain air pollutants including ambient $PM_{2.5}$ (U.S. EPA, 2018). The lung is a major interface with the environment and requires a high level of immune regulation to limit reactivity. Exposure to a variety of environmental agents, both human-made and natural, can result in an increased susceptibility to infectious disease (Ciencewicki & Jaspers, 2007; de Perio et al., 2020; Huang et al., 2016). Inhaled diesel exhaust, cigarette smoke, and mining PM have been linked with an increased susceptibility to respiratory infections in both humans and animal models (Antonini et al., 2000; Arredouani et al., 2004; Castranova et al., 2001; Martin et al., 2006; McDonald et al., 2004; Ross & Murray, 2004; Zelikoff et al., 2002). In developing countries, chronic exposures to residential wood smoke (i.e., home heating, cooking) have correlated with increased incidence of respiratory infections (Akunne et al., 2006; Bautista et al., 2009; Bruce et al., 2000; Kilabuko et al., 2007; Mishra, 2003; Mishra & Retherford, 1997; Smith et al., 2000). In addition, population studies have found increased acute respiratory illnesses (ARI) in children in homes where biomass burning (i.e., wood) is the primary method of cooking (Akunne et al., 2006; Bautista et al., 2009).

Mechanistic evaluation of PM-induced susceptibility to respiratory infection has focused on prominent cell types including the alveolar macrophage and alveolar epithelial cells (Migliaccio et al., 2013; Sada-Ovalle et al., 2018; Sun et al., 2021). While wood-smoke-induced pulmonary inflammation has produced mixed results (Swiston et al., 2008), there are ambient PM studies that demonstrate the adverse effects of PM on the ability of macrophages to phagocytize and clear apoptotic cells and particles, a process referred to as efferocytosis (Boada-Romero et al., 2020). A significant portion of research on the respiratory effects of smoke has focused on the alveolar macrophage (Migliaccio et al., 2013; Sada-Ovalle et al., 2018). Macrophages are key players in both innate and adaptive immune responses and, with the recognition of multiple functional subsets (i.e., M1 and M2), macrophages are considered to be plastic in that they can alter their phenotype based on the current environment (Bain et al., 2016; Gordon, 2003). Trained immunity in macrophages is described as long-term functional changes resulting in exposures to particular stimulations (Netea et al., 2020). While unknown, there is a strong potential that wood-smoke exposures could induce trained immunity, thus affecting local macrophage populations and skewing subsequent responses. Most studies assessing potential roles of epithelial cells in air pollution effects focus on cellular damage and are in vitro with very few utilizing wood-smoke particles (Lee et al., 2008; Zeglinski et al., 2019).

11.3.2. Dermal Effects

Despite the lack of wildfire smoke-specific studies, there have been a number of studies assessing the dermal effects of exposure to ambient $PM_{2.5}$ (Dijkhoff et al., 2020; Magnani et al., 2016; Vierkötter et al., 2010). These include skin aging (Vierkötter et al., 2010), skin damage (Magnani et al., 2016), and effects on cholesterol metabolism (Liao et al., 2020) with the induction of oxidative stress considered a contributor to some or all effects (Jin et al., 2018; Piao et al., 2018). Due to the significant barrier and the size of particles (air quality studies historically have focused on $PM_{2.5}$), there are generally two main potential routes of penetration (hair follicles/pores, and across the stratum corneum) and studies are concentrated on the effects to the epidermal layer (Jin et al., 2018). In addition, the chemistry of ambient $PM_{2.5}$ is such that there are a number of constituents that have been shown to interact with the aryl hydrocarbon receptor (AhR), a transcription factor expressed throughout the body including skin cells, which can affect genes involving metabolism, inflammation, and oxidative stress (Napolitano & Patruno, 2018; Peng et al., 2019).

11.3.3. Extrapulmonary Effects

The effects of wildfire smoke on cardiovascular health outcomes have been examined, and, despite the amount of quality data, the results remain mixed (Black et al., 2017; Cascio, 2018; Chen et al., 2021; Hanigan et al., 2008; Henderson et al., 2011; Lee et al., 2009; Navarro et al., 2019; Rappold et al., 2011; Reid et al., 2016; Schranz et al., 2010; Stowell et al., 2019). Clinical and toxicological studies have highlighted plausible mechanisms, such as systemic inflammation, oxidative stress, extracellular vesicle release of microRNAs, and nervous system imbalance (Chen et al., 2021). Given the proximity of the heart to the lungs and transfer of blood as a vehicle, direct exposure to smoke PM via translocation is also a plausible exposure route (Cascio, 2018), which has been demonstrated in a mouse model (Furuyama et al., 2009). As noted in the dermal effects section (11.3.2), cholesterol metabolism can also be affected by exposures to air pollution, and this could indirectly impact cardiovascular health (Carson et al., 2020).

In addition to cardiovascular effects, wood-smoke PM that traverses the lungs into circulation and may interact with other organs will be essential to characterize moving forward. For example, Milton and White (2020) and Schuller and Montrose (2020) reviewed the literature on plausibility of wildfire smoke PM effects on the brain and nervous system, leveraging research on urban PM and brain effects given the paucity of wildfire smoke PM literature (Milton & White, 2020; Schuller & Montrose, 2020). This is a growing area of research that is supported by epidemiological data but is in need of more mechanistic laboratory studies. Epidemiologic literature on potential links between exposure to wildfire and other adverse health impacts are examined in Chapters 12 and 13, however we underscore the need for additional mechanistic and toxicologic research to elucidate further the pathophysiologic pathways. Research to date has precluded determining dose effects of wildfire smoke exposures due to such factors as limited information on duration, frequency, characteristics that impact toxicity, and variability of individuals.

11.4. MODEL SYSTEMS OF WILDFIRE SMOKE TOXICOLOGY

11.4.1. Wildfire Smoke Collection and Exposure

Understanding the toxicological effects of wildfire smoke exposure is dependent on the ability to collect ambient smoke or to create wood-smoke PM in a laboratory setting. Collection of nonsmoke ambient PM can be done using a versatile aerosol concentration enrichment system (VACES; Ferguson et al., 2013), a high volume cascade impactor (Demokritou et al., 2002), or similar collection devices. PM collection during wildfire season may capture some smoke particles; however, depending on the location, this sampling may be diluted or mixed with other sources. Ambient collection of fresh smoke alone is difficult due to the hazardous nature of wildfires. Thus, the majority of studies in the following section is conducted with experimental wood-smoke exposure models (Kim et al., 2018). While useful in understanding the toxicology of the specific type of fuel, these experimental approaches are limited in that they are missing the mosaic of fuels in a naturally occurring fire, as well as conditions, such as high temperature, low humidity, wind, and variable topography.

11.4.2. In Vitro Models

In vitro experimentation has many benefits in that it is relatively inexpensive, quick, and can give specific insight into disease mechanisms. These experiments are often used as a screening tool to look for initial toxicological effects before moving into an ex vivo or in vivo model. Furthermore, these systems can later facilitate pinpointing of a specific mechanism or pathway.

To measure the effects of air pollution in vitro, PM particles or extracts are resuspended in liquid and added to cell culture medium. Alternatively, the air liquid interface allows particles to be directly applied to the cells (Wang et al., 2019). There are limitations to liquid suspensions in that particles are more likely to aggregate, which has an effect on particle distribution to the cell. Aerosol exposures are much less likely to aggregate. Additionally, in the case of lab-created particulates, exposure can occur immediately after or simultaneously with combustion. This temporal aspect is important when considering the stability or volatility of certain chemical constituents. To date, there are still relatively few studies that have used in vitro methodologies to investigate the effect of wildfire smoke exposure on human health (Black et al., 2017; Dong et al., 2017). There are even fewer studies that measure outcomes from whole smoke exposure (including gaseous products), instead focusing on exposure outcomes from isolated smoke PM.

Two recent review papers, published on the health effects and toxicity of smoke exposure, addressed in vitro exposure studies through 2017 (Black et al., 2017; Dong et al., 2017). Black et al. focused on studies that came out of ambient smoke collected from the 2008 Northern California wildfires (Black et al., 2017). Authors highlighted distinct differences to nonsmoke pollutant exposures, but concluded that more information is needed on the chemical composition of smoke in order to further elucidate mechanisms (Black et al., 2017). A review by Dong et al. (2017) covered in vitro exposures of smoke from bushfires, prescribed burns, and experimental smoke. Taken together, these reviews suggest smaller particles are more toxic than larger particles, however, they also highlight the need for standardization of protocols for more effective comparisons.

A 2020 study by Ihantola et al. (2020) introduced both PM and the gaseous phase of smoke by using a combustion methodology for exposure to wood-smoke PM and gases combined (complete combustion emissions) as well as to a HEPA-filtered combustion aerosol (without PM). Authors exposed both macrophages and epithelial cells at the air-liquid interface to these three separate smoke constituents. With enhanced deposition of PM, cytotoxicity (measured by lactate dehydrogenase release) was increased in exposed macrophages. The only comparison that demonstrated a difference between exposure methods was an increase in interleukin (IL)-8 with HEPA-filtered combustion in comparison to complete combustion exposures. This work represents the rare study that looks at the effects of gaseous compounds alone, and thus more work needs to be done to understand which components of smoke are leading to observed effects.

Although studies on collected ambient smoke PM are lacking, studies have investigated the toxicological effects of experimental wood-smoke PM. In vitro experiments with smoke particles have demonstrated functional changes in cells such as a blocked autophagy pathway or decreased macrophage functions consistent with COPD (Migliaccio et al., 2013; Roscioli et al., 2018; Sada-Ovalle et al., 2018), decreased lymphocyte activation consistent with bacterial sensitivity (Migliaccio et al., 2013), DNA damage (Corsini et al., 2013; Danielsen et al., 2011; Leonard et al., 2000), and epigenetic changes linked to asthma, lung cancer, and other diseases that have been associated with PM exposure (Heßelbach et al., 2017). Deering-Rice et al. (2018) compared the combustion of different types of wood with diesel exhaust particles (DEP), and identified a transient receptor potential (TRP) channel (TRPV3), a marker for pneumotoxicity, as activated by wildfire smoke. The results from this study were dependent on the specific wood type burned, introducing the need for a comparison to different PM sources, as well as a need to elucidate fuel-specific effects. Filling this gap is particularly important given that wildfires are moving closer to the wildland-urban interface, adding building materials to the list of potential fuels burned.

A small number of studies have attempted to address this mosaic of smoke PM by collecting ambient PM during a wildfire event and exposing these particles in vitro. Jalava et al. (2006) collected long-range transport PM from wildfires outside of Helsinki and exposed a macrophage cell line to a range of PM-size fractions (see section 11.1.2.4 on aging and distance). Exposure to all size fractions demonstrated significant increases in NO production, and cytotoxicity as measured by the (3-(4,5-dimethylthiazol-2-yl)-2,5-diphenyltetrazolium bromide (MTT) colorimetric assay. Fractionate particles produced differential cytokine responses, with exposure to larger particles ($PM_{10-2.5}$) yielding a more robust response when compared with smaller particles. Environmental particle collection by Young et al.(2021) also demonstrated increased cytokine release with exposure to settled ash collected from a wildfire in Sonoma County, California, USA. Although significant, ash is importantly a nonrespirable size fraction. These bioassays demonstrated increases in COX-2, IL-8, and CYP1a1 as well as an increased activation of the aryl hydrocarbon receptor (AhR).

11.4.3. In Vivo Models

In order to better understand the dynamic health impacts attributed to wildfire smoke exposure, the application of in vivo models must be considered. While epidemiologic studies have motivated efforts to further explore the effects of wildfire smoke inhalation on body systems, these studies are often unable to account for important confounding variables (i.e., medical history, tobacco smoke exposure). For this reason, use of inhalation exposure paradigms is key when combining with existing biological model organisms and/or molecular manipulations. This is especially promising in the context of comparative systems biology where researchers hope to better characterize the phenotype of humans exposed to wildfire smoke by assessing the pathologic changes that animal models undergo following simulated subjection to wildfire smoke.

While literature detailing the study of wildfire-smoke-induced health impacts in vivo is not extensive, the correlation between increased inhalation of ambient PM and respiratory morbidity and mortality is well established (Walter et al., 2021; Zhang et al., 2021). Research is growing in the in vivo arena, concurrent with climate-change-induced increase in frequency and duration of wildfire events (Ford et al., 2018). These novel data are suggestive of extrapulmonary risks that require more thorough exploration (Oudin et al., 2018). To accomplish this, three major experimental exposure routes have been identified for laboratory animal models: intranasal/intratracheal instillation, nose-only inhalation, and whole-body exposure. Each of these methods has associated drawbacks, which should be considered when developing a research strategy. In consideration of direct instillation via micropipette, this type of exposure may not reflect ambient inhalation events, especially when considering that animals are often sedated to perform the exposure. This technique is also associated with additional cost of training (e.g., using a pilot cohort of animals and dye) as well as the use of the anesthetic agent. While widely used in many experimental settings, this technique may be best reserved for studies seeking to mimic high-level occupational exposure events or for laboratory environments that are not capable of conducting in-house biomass combustion for particle generation. By conducting a nose-only exposure paradigm, animals are subjected to increased handling and confinement, which can induce aberrant stress responses (Thomson et al., 2009). On the other hand, a whole-body exposure strategy allows for deposition of particulate matter in the fur of the animal and may result in undesired oral consumption in addition to nasal inhalation. These inhalation exposure techniques are better suited for simulation of ambient PM exposure, and can even be optimized to reduce inhalation mortality for long-term study duration using kerosene to minimize carbon monoxide concentration (Zhu et al., 2012).

Another study design parameter to consider is the species of the animal model used. Many labs employ commonly used laboratory rat or mouse strains due to the preexistence of housing facilities, low cost of implementation, and short life span for chronic studies. One major advantage of using smaller rodents for these studies is that these organisms are obligate nasal breathers. However, there are certainly anatomic differences between small rodents and associated human counterparts (Rydell-Törmänen & Johnson, 2019). Other research groups have more recently studied wildfire smoke exposure in higher order mammals such as rhesus macaques (Black et al., 2017) and canines (Calderón-Garcidueñas et al., 2003). While these animals share more similarity to human respiratory anatomy, challenges such as cost, husbandry, life span, and ethical considerations complicate their widespread usage.

Ethical considerations surround the sacrifice of a life to gain information that will ultimately influence future health directives. For this reason, mindfulness of the "Three Rs" (reduction, replacement, and refinement) is important, as is their implementation in the design of any animal experiment (Fenwick et al., 2009). Consideration of the benefits of sharing tissue with collaborators or banking tissue for future use is also worthwhile. Further, accounting for the drawbacks that each type of simulated laboratory exposure has in regards to mimicking human experience is critical. In summary, considerable evidence warrants future studies using in vivo models, so long as the study design and exposure paradigm pros and cons are well considered in planning each experiment.

11.5. FUTURE RESEARCH NEEDS

Several areas with respect to wildfire toxicology and health require increased research. First, a greater understanding of the various chemical constituents in wildfire smoke is needed. Several studies have characterized the distinct components of collected ambient wildfire smoke (O'Dell et al., 2020; Selimovic et al., 2019; Sillanpää et al., 2005). These studies have recognized that the gaseous phase of smoke is a significant component. However, to our knowledge, only one study has isolated this gas phase to examine gas-specific toxicology (Ihantola et al., 2020). Just as ambient gases (NOx, CO, etc.) have been associated with adverse health effects, this type of exposure through wildfire smoke could have detrimental effects and needs to be further investigated. Second, epidemiologic studies have typically used administrative data (i.e., emergency department visits and hospitalizations) to examine the relationship of acute wildfire exposures with respiratory and cardiovascular diagnoses (Alman et al., 2016; Cascio, 2018; Haikerwal et al., 2016; Hutchinson et al., 2018; Liu et al., 2015). However, research is limited for physiological systems beyond the respiratory and cardiovascular pathways (Chapters 12, 13), delayed health impacts from acute exposures (Orr et al., 2020), long- term morbidity (Schuller & Montrose, 2020; Sosedova et al., 2020), and assessment of effects from chronic, cumulative and mixtures of exposures. This research is expensive, complex, and time-consuming; but these studies are necessary to protect both public and occupational health. Last, there is a need for establishing a more refined PM exposure classification system. Evolving multidisciplinary research topics often lack a common language. More than just identifying consistent terminology, establishing a PM exposure classification system could help researchers design animal studies that are human-relevant and human studies that are more applicable to realistic wildfire exposures. Classifications will need to include exposure duration, frequency (i.e., episodic, annual) as well as higher resolved exposure levels to allow for continuity and comparison among studies. A shift from assigning the same exposure level to an entire population or a period of time to more temporally and spatially specific and variable community and individual-level exposures will also be necessary in future epidemiologic studies.

ACKNOWLEDGMENTS

Luke Montrose and Adam Schuller were supported by the Boise State University COBRE (P20GM109095). Luke Montrose was also supported by Idaho INBRE (P20GM103408). Savannah D'Evelyn was supported by T32 ES015459, T32 ES007059, and T32 HL007013.

REFERENCES

Adetona, O., Simpson, C. D., Li, Z., Sjodin, A., Calafat, A. M., & Naeher, L. P. (2017). Hydroxylated polycyclic aromatic hydrocarbons as biomarkers of exposure to wood smoke in wildland firefighters. *Journal of Exposure Science & Environmental Epidemiology*, *27*(1), 78–83.

Air Quality Index (n.d.). Air Quality Index (AQI) basics. https://www.airnow.gov/aqi/aqi-basics/

Akunne, A. F., Louis, V. R., Sanon, M., & Sauerborn, R. (2006). Biomass solid fuel and acute respiratory infections: The ventilation factor. *International Journal of Hygiene and Environmental Health*, *209*(5), 445–450.

Alman, B. L., Pfister, G., Hao, H., Stowell, J., Hu, X., Liu, Y., et al. (2016). The association of wildfire smoke with respiratory and cardiovascular emergency department visits in Colorado in 2012: A case crossover study. *Environmental Health*, *15*(1), 64.

Anderson, E. W., Andelman, R. J., Strauch, J. M., Fortuin, N. J., & Knelson, J. H. (1973). Effect of low-level carbon monoxide exposure on onset and duration of angina pectoris. A study in ten patients with ischemic heart disease. *Annals of Internal Medicine*, *79*(1), 46–50.

Antonini, J. M., Roberts, J. R., Yang, H. M., Barger, M. W., Ramsey, D., Castranova, V., et al. (2000). Effect of silica inhalation on the pulmonary clearance of a bacterial pathogen in Fischer 344 rats. *Lung*, *178*(6), 341–350.

Arredouani, M., Yang, Z., Ning, Y., Qin, G., Soininen, R., Tryggvason, K., et al. (2004). The scavenger receptor MARCO is required for lung defense against pneumococcal pneumonia and inhaled particles. *The Journal of Experimental Medicine*, *200*(2), 267–272.

Bain, C. C., Hawley, C. A., Garner, H., Scott, C. L., Schridde, A., Steers, N. J., et al. (2016). Long-lived self-renewing bone marrow derived macrophages displace embryo-derived cells to inhabit adult serous cavities. *Nature Communications*, *7*, ncomms11852.

Bauer, A. K., Siegrist, K. J., Wolff, M., Nield, L., Brüning, T., Upham, B. L., et al. (2022). The carcinogenic properties of overlooked yet prevalent polycyclic aromatic hydrocarbons in human lung epithelial cells. *Toxics*, *10*(1), 28. https://www.ncbi.nlm.nih.gov/pmc/articles/PMC8779510/

Bautista, L. E., Correa, A., Baumgartner, J., Breysse, P., & Matanoski, G. M. (2009). Indoor charcoal smoke and acute respiratory infections in young children in the Dominican Republic. *American Journal of Epidemiology*, *169*(5), 572–580.

Bergauff, M. A., Ward, T. J., Noonan, C. W., Migliaccio, C. T., Simpson, C. D., Evanoski, A. R., et al. (2010). Urinary levoglucosan as a biomarker of wood smoke: results of human exposure studies. *Journal of Exposure Science & Environmental Epidemiology*, *20*(4), 385–392.

Black, C., Tesfaigzi, Y., Bassein, J. A., & Miller, L. A. (2017). Wildfire smoke exposure and human health: Significant gaps in research for a growing public health issue. *Environmental Toxicology and Pharmacology*, *55*, 186–195.

Boada-Romero, E., Martinez, J., Heckmann, B. L., & Green, D. R. (2020). The clearance of dead cells by efferocytosis. *Nature Reviews Molecular Cell Biology*, *21*(7), 398–414.

Brizio-Molteni, L., Piano, G., Rice, P. L., Warpeha, R., Fresco, R., Solliday, N. H., et al. (1984). Effect of wood combustion smoke inhalation on angiotensin-1-converting enzyme in the dog. *Annals of Clinical and Laboratory Science*, *14*(5), 381–389.

Brook, R. D., Rajagopalan, S., Pope, C. A., Brook, J. R., Bhatnagar, A., Diez-Roux, A. V., et al. (2010). Particulate matter air pollution and cardiovascular disease. *Circulation*, *121*(21), 2331–2378.

Bruce, N., Perez-Padilla, R., & Albalak, R. (2000). Indoor air pollution in developing countries: A major environmental and public health challenge. *Bulletin of the World Health Organization*, *78*(9), 1078–1092.

Brunekreef, B., & Holgate, S. T. (2002). Air pollution and health. *The Lancet*, *360*(9341), 1233–1242.

Bucher, U., & Reid, L. (1961). Development of the intrasegmental bronchial tree: The pattern of branching and development of cartilage at various stages of intra-uterine life. *Thorax*, *16*(3), 207.

Bustamante-Marin, X. M., & Ostrowski, L. E. (2017). Cilia and mucociliary clearance. *Cold Spring Harbor Perspectives in Biology*, *9*(4), a028241.

Calderón-Garcidueñas, L., Maronpot, R. R., Torres-Jardon, R., Henríquez-Roldán, C., Schoonhoven, R., Acuña-Ayala, H., et al. (2003). DNA damage in nasal and brain tissues of canines exposed to air pollutants is associated with evidence of chronic brain inflammation and neurodegeneration. *Toxicologic Pathology*, *31*(5), 524–538.

Carey, I. M., Atkinson, R. W., Kent, A. J., Staa, T., Cook, D. G., & Anderson, H. R. (2013). Mortality associations with long-term exposure to outdoor air pollution in a national English cohort. *AmericanJournal of Respiratory and Critical Care Medicine*, *187*(11), 1226–1233.

Carson, J. A. S., Lichtenstein, A. H., Anderson, C. A. M., Appel, L. J., Kris-Etherton, P. M., Meyer, K. A., et al. (2020). Dietary cholesterol and cardiovascular risk: A science advisory from the American Heart Association. *Circulation*, *141*(3), e39–e53.

Cascio, W. E. (2018). Wildland fire smoke and human health. *The Science of the Total Environment*, *624*, 586–595.

Castranova, V., Ma, J. Y., Yang, H. M., Antonini, J. M., Butterworth, L., Barger, M. W., et al. (2001). Effect of exposure to diesel exhaust particles on the susceptibility of the lung to infection. *Environmental Health Perspectives*, *109* (Suppl 4), 609–612.

Chakrabarty, R. K., Moosmüller, H., Chen, L. W., Lewis, K., Arnott, W. P., Mazzoleni, C., et al. (2010). Brown carbon in tar balls from smoldering biomass combustion. *Atmospheric Chemistry and Physics*, *10*(13), 6363–6370.

Chen, H., Samet, J. M., Bromberg, P. A., & Tong, H. (2021). Cardiovascular health impacts of wildfire smoke exposure. *Particle and Fibre Toxicology*, *18*(1), 1–22.

Ciencewicki, J., & Jaspers, I. (2007). Air pollution and respiratory viral infection. *Inhalation Toxicology*, *19*(14), 1135–1146.

Corsini, E., Budello, S., Marabini, L., Galbiati, V., Piazzalunga, A., Barbieri, P., et al. (2013). Comparison of wood smoke PM2. 5 obtained from the combustion of FIR and beech pellets on inflammation and DNA damage in A549 and THP-1 human cell lines. *Archives of Toxicology*, *87*(12), 2187–2199.

Curt, T., Borgniet, L., & Bouillon, C. (2013). Wildfire frequency varies with the size and shape of fuel types in southeastern France: implications for environmental management. *Journal of Environmental Management*, *117*, 150–161.

Danielsen, P. H., Møller, P., Jensen, K. A., Sharma, A. K., Wallin, H., Bossi, R., et al. (2011). Oxidative stress, DNA damage, and inflammation induced by ambient air and wood smoke particulate matter in human A549 and THP-1 cell lines. *Chemical Research in Toxicology*, *24*(2), 168–184.

Darquenne, C. (2012). Aerosol deposition in health and disease. *Journal of Aerosol Medicine and Pulmonary Drug Delivery*, *25*(3), 140–147.

Deering-Rice, C. E., Nguyen, N., Lu, Z., Cox, J. E., Shapiro, D., Romero, E. G., et al. (2018). Activation of TRPV3 by wood smoke particles and roles in pneumotoxicity. *Chemical Research in Toxicology*, *31*(5), 291–301.

Demokritou, P., Kavouras, I. G., Ferguson, S. T., & Koutrakis, P. (2002). Development of a high volume cascade impactor for toxicological and chemical characterization studies. *Aerosol Science & Technology*, *36*(9), 925–933.

de Perio, M. A., Kobayashi, M., & Wortham, J. M. (2020). Occupational Respiratory Infections. *Clinics in Chest Medicine*, *41*(4), 739–751.

Dijkhoff, I. M., Drasler, B., Karakocak, B. B., Petri-Fink, A., Valacchi, G., Eeman, M., et al. (2020). Impact of airborne particulate matter on skin: a systematic review from epidemiology to in vitro studies. *Particle and Fibre Toxicology*, *17*(1), 35.

Directive 2008/50/EC (2008). Directive 2008/50/EC of the European Parliament and of the Council of 21 May 2008 on ambient air quality and cleaner air for Europe, 152 *Congressional Record*.

Dong, T. T. T., Hinwood, A. L., Callan, A. C., Zosky, G., & Stock, W. D. (2017). In vitro assessment of the toxicity of bushfire emissions: A review. *Science of the Total Environment*, *603–604*, 268–278.

Faustini, A., Stafoggia, M., Colais, P., Berti, G., Bisanti, L., Cadum, E., et al. (2013). Air pollution and multiple acute respiratory outcomes. *European Respiratory Journal*, *42*(2), 304–313.

Fenwick, N., Griffin, G., & Gauthier, C. (2009). The welfare of animals used in science: How the "Three Rs" ethic guides improvements. *The Canadian Veterinary Journal*, *50*(5), 523–530.

Ferguson, M. D., Migliaccio, C., & Ward, T. (2013). Comparison of how ambient PMc and PM2. 5 influence the inflammatory potential. *Inhalation Toxicology*, *25*(14), 766–773.

Ford, B., Val Martin, M., Zelasky, S. E., Fischer, E. V., Anenberg, S. C., Heald, C. L., et al. (2018). Future Fire Impacts on Smoke Concentrations, Visibility, and Health in the Contiguous United States. *GeoHealth*, *2*(8), 229–247. https://onlinelibrary.wiley.com/doi/abs/10.1029/2018GH000144

Fu, P., Guo, X., Cheung, F. M. H., & Yung, K. K. L. (2019). The association between PM2. 5 exposure and neurological disorders: a systematic review and meta-analysis. *Science of the Total Environment*, *655*, 1240–1248.

Furuyama, A., Kanno, S., Kobayashi, T., & Hirano, S. (2009). Extrapulmonary translocation of intratracheally instilled fine and ultrafine particles via direct and alveolar macrophage-associated routes. *Archives of Toxicology*, *83*(5), 429–437.

Gill, B., & Britz-McKibbin, P. (2020). Biomonitoring of smoke exposure in firefighters: A review. *Current Opinion in Environmental Science & Health*, *15*, 57–65.

Girotto, G., China, S., Bhandari, J., Gorkowski, K., Scarnato, B. V., Capek, T., et al. (2018). Fractal-like Tar Ball Aggregates from Wildfire Smoke. *Environmental Science & Technology Letters*, *5*(6), 360–365.

Gordon, S. (2003). Alternative activation of macrophages. *Nature Reviews Immunology*, *3*(1), 23–35.

Grandjean, P. (2016). Paracelsus revisited: The dose concept in a complex world. *Basic & Clinical Pharmacology & Toxicology*, *119*(2), 126–132.

Gray, R., Magazine, F. H., Research, H. T. E., & Magazine, I. (n.d.). Four times more toxic: How wildfire smoke ages over time. https://phys.org/news/2020-07-toxic-wildfire-ages.html

Haikerwal, A., Akram, M., Sim, M. R., Meyer, M., Abramson, M. J., & Dennekamp, M. (2016). Fine particulate matter (PM2.5) exposure during a prolonged wildfire period and emergency department visits for asthma. *Respirology (Carlton, Vic.)*, *21*(1), 88–94.

Hanigan, I. C., Johnston, F. H., & Morgan, G. G. (2008). Vegetation fire smoke, indigenous status and cardio-respiratory hospital admissions in Darwin, Australia, 1996–2005: A time-series study. *Environmental Health*, *7*, 42.

Hargrove, M. M., Kim, Y. H., King, C., Wood, C. E., Gilmour, M. I., Dye, J. A., et al. (2019). Smoldering and Flaming Biomass Wood Smoke Inhibit Respiratory Responses in Mice. *Inhalation Toxicology*, *31*(6), 236–247.

Henderson, S. B., Brauer, M., Macnab, Y. C., & Kennedy, S. M. (2011). Three measures of forest fire smoke exposure and their associations with respiratory and cardiovascular health outcomes in a population-based cohort. *Environmental Health Perspectives*, *119*(9), 1266–1271.

Heßelbach, K., Kim, G.-J., Flemming, S., Häupl, T., Bonin, M., Dornhof, R., et al. (2017). Disease relevant modifications of the methylome and transcriptome by particulate matter (PM2.5) from biomass combustion. *Epigenetics*, *12*(9), 779–792.

Huang, L., Zhou, L., Chen, J., Chen, K., Liu, Y., Chen, X., et al. (2016). Acute effects of air pollution on influenza-like illness in Nanjing, China: a population-based study. *Chemosphere*, *147*, 180–187.

Hutchinson, J. A., Vargo, J., Milet, M., French, N. H. F., Billmire, M., Johnson, J., et al. (2018). The San Diego 2007 wildfires and Medi-Cal emergency department presentations, inpatient hospitalizations, and outpatient visits: An observational study of smoke exposure periods and a bidirectional case-crossover analysis. *PLoS Medicine*, *15*(7), e1002601.

Ihantola, T., Di Bucchianico, S., Happo, M., Ihalainen, M., Uski, O., Bauer, S., et al. (2020). Influence of wood species on toxicity of log-wood stove combustion aerosols: a parallel animal and air-liquid interface cell exposure study on spruce and pine smoke. *Particle and Fibre Toxicology*, *17*(1), 27.

Jalava, P. I., Salonen, R. O., Hälinen, A. I., Penttinen, P., Pennanen, A. S., Sillanpää, M., et al. (2006). In vitro inflammatory and cytotoxic effects of size-segregated particulate samples collected during long-range transport of wildfire smoke to Helsinki. *Toxicology and Applied Pharmacology*, *215*(3), 341–353.

Jin, S.-P., Li, Z., Choi, E. K., Lee, S., Kim, Y. K., Seo, E. Y., et al. (2018). Urban particulate matter in air pollution penetrates into the barrier-disrupted skin and produces ROS-dependent cutaneous inflammatory response in vivo. *Journal of Dermatological Science*, *S0923-1811*(0918)30202–30200.

Kampa, M., & Castanas, E. (2008). Human health effects of air pollution. *Environmental Pollution*, *151*(2), 362–367.

Karakatsani, A., Analitis, A., Perifanou, D., Ayres, J. G., Harrison, R. M., Kotronarou, A., et al. (2012). Particulate matter air pollution and respiratory symptoms in individuals having either asthma or chronic obstructive pulmonary disease: a European multicentre panel study. *Environmental Health*, *11*, 75.

Kelly, F. J., & Fussell, J. C. (2011). Air pollution and airway disease. *Clinical & Experimental Allergy*, *41*(8), 1059–1071.

Kilabuko, J. H., Matsuki, H., & Nakai, S. (2007). Air quality and acute respiratory illness in biomass fuel using homes in Bagamoyo, Tanzania. *International Journal of Environmental Research and Public Health*, *4*(1), 39–44.

Kim, Y. H., King, C., Krantz, T., Hargrove, M. M., George, I. J., McGee, J.,. et al. (2019). The role of fuel type and combustion phase on the toxicity of biomass smoke following inhalation exposure in mice. *Archives of Toxicology*, *93*(6), 1501–1513.

Kim, Y. H., Warren, S. H., Krantz, Q. T., King, C., Jaskot, R., Preston, W. T., et al. (2018). Mutagenicity and lung toxicity of smoldering vs. flaming emissions from various biomass fuels: Implications for health effects from wildland fires. *Environmental Health Perspectives*, *126*(1). https://www.ncbi.nlm.nih.gov/pmc/articles/PMC6039157/

Klepac, P., Locatelli, I., Korošec, S., Künzli, N., & Kukec, A. (2018). Ambient air pollution and pregnancy outcomes: A comprehensive review and identification of environmental public health challenges. *Environmental Research*, *167*, 144–159.

Kocbach Bølling, A., Pagels, J., Yttri, K. E., Barregard, L., Sallsten, G., Schwarze, P. E., et al. (2009). Health effects of residential wood smoke particles: the importance of combustion conditions and physicochemical particle properties. *Particle and Fibre Toxicology*, *6*, 29.

Lee, T.-S., Falter, K., Meyer, P., Mott, J., & Gwynn, C. (2009). Risk factors associated with clinic visits during the 1999 forest fires near the Hoopa Valley Indian Reservation, California, USA. *International Journal of Environmental Health Research*, *19*(5), 315–327.

Lee, T.-S., Liu, Y.-J., Tang, G.-J., Yien, H.-W., Wu, Y.-L., & Kou, Y. R. (2008). Wood smoke extract promotes both apoptosis and proliferation in rat alveolar epithelial type II cells: The role of oxidative stress and heme oxygenase-1*. *Critical Care Medicine*, *36*(9), 2597–2606. https://journals.lww.com/ccmjournal/Abstract/2008/09000/Wood_smoke_extract_promotes_both_apoptosis_and.17.aspx

Lee, W., Kim, J., Lim, S.-S., Kim, Y., Ahn, Y.-S., & Yoon, J.-H. (2020). External airborne-agent exposure increase risk of digestive tract cancer. *Scientific Reports*, *10*(1), 8617. doi:10.1038/s41598-020-65312-6

Leonard, S. S., Castranova, V., Chen, B. T., Schwegler-Berry, D., Hoover, M., Piacitelli, C., et al. (2007). Particle size-dependent radical generation from wildland fire smoke. *Toxicology*, *236*(1), 103–113.

Leonard, S. S., Wang, S., Shi, X., Jordan, B. S., Castranova, V., & Dubick, M. A. (2000). Wood smoke particles generate free radicals and cause lipid peroxidation, DNA damage, NFkappaB activation and TNF-alpha release in macrophages. *Toxicology*, *150*(1–3), 147–157.

Liao, Z., Nie, J., & Sun, P. (2020). The impact of particulate matter (PM2.5) on skin barrier revealed by transcriptome analysis: Focusing on cholesterol metabolism. *Toxicology Reports*, *7*, 1–9.

Liu, J. C., Mickley, L. J., Sulprizio, M. P., Dominici, F., Yue, X., Ebisu, K., et al. (2016). Particulate air pollution from wildfires in the Western US under climate change. *Climatic Change*, *138*(3–4), 655–666. doi:10.1007/s10584-016-1762-6

Liu, J. C., Pereira, G., Uhl, S. A., Bravo, M. A., & Bell, M. L. (2015). A systematic review of the physical health impacts from non-occupational exposure to wildfire smoke. *Environmental Research*, *136*, 120–132. doi:http://dx.doi.org/10.1016/j.envres.2014.10.015

Magnani, N. D., Muresan, X. M., Belmonte, G., Cervellati, F., Sticozzi, C., Pecorelli, A., et al. (2016). Skin damage mechanisms related to airborne particulate matter exposure. *Toxicological Sciences*, *149*(1), 227–236.

Martenies, S. E., Hoskovec, L., Wilson, A., Allshouse, W. B., Adgate, J. L., Dabelea, D., et al. (2021). Assessing the impact of wildfires on the use of black carbon as an indicator of traffic exposures in environmental epidemiology studies. *GeoHealth*, *5*(6), e2020GH000347.

Martin, R. J., Wexler, R. B., Day, B. J., Harbeck, R. J., Pinkerton, K. E., & Chu, H. W. (2006). Interaction between cigarette smoke and mycoplasma infection: a murine model. *COPD*, *3*(1), 3–8.

McClure, C. D., & Jaffe, D. A. (2018). US particulate matter air quality improves except in wildfire-prone areas. http://www.pnas.org/site/aboutpnas/licenses.xhtml Published under the PNAS license.]. *Proceedings of the National Academy of Sciences*, *115*(31), 7901–7906. https://www.pnas.org/content/115/31/7901

McDonald, J. D., Harrod, K. S., Seagrave, J., Seilkop, S. K., & Mauderly, J. L. (2004). Effects of low sulfur fuel and a catalyzed particle trap on the composition and toxicity of diesel emissions. *Environmental Health Perspectives*, *112*(13), 1307–1312.

Migliaccio, C. T., Bergauff, M. A., Palmer, C. P., Jessop, F., Noonan, C. W., & Ward, T. J. (2009). Urinary levoglucosan as a biomarker of wood smoke exposure: observations in a mouse model and in children. *Environmental Health Perspectives*, *117*(1), 74–79.

Migliaccio, C. T., Kobos, E., King, Q. O., Porter, V., Jessop, F., & Ward, T. (2013). Adverse effects of wood smoke PM2. 5 exposure on macrophage functions. *Inhalation toxicology*, *25*(2), 67–76.

Milton, L. A., & White, A. R. (2020). The potential impact of bushfire smoke on brain health. *Neurochemistry International*, *139*, 104796.

Mishra, V. (2003). Indoor air pollution from biomass combustion and acute respiratory illness in preschool age children in Zimbabwe. *International Journal of Epidemiology*, *32*(5), 847–853.

Mishra, V. K., & Retherford, R. D. (1997). Cooking smoke increases the risk of acute respiratory infection in children. *National Family Health Survey Bulletin, Sep.*(*8*), 1–4. PMID: 12293014

Napolitano, M., & Patruno, C. (2018). Aryl hydrocarbon receptor (AhR) a possible target for the treatment of skin disease. *Medical Hypotheses*, *116*, 96–100.

Navarro, K. (2020). Working in smoke:: Wildfire impacts on the health of firefighters and outdoor workers and mitigation strategies. *Clinics in Chest Medicine*, *41*(4), 763–769.

Navarro, K. M., Cisneros, R., Noth, E. M., Balmes, J. R., & Hammond, S. K. (2017). Occupational exposure to polycyclic aromatic hydrocarbon of wildland firefighters at prescribed and wildland fires. *Environmental Science & Technology*, *51*(11), 6461–6469.

Navarro, K. M., Clark, K. A., Hardt, D. J., Reid, C. E., Lahm, P. W., Domitrovich, J. W., et al. (2021). Wildland firefighter exposure to smoke and COVID-19: A new risk on the fire line. *Science of the Total Environment*, *760*, 144296. https://www.sciencedirect.com/science/article/pii/S004896972037827X

Navarro, K. M., Kleinman, M. T., Mackay, C. E., Reinhardt, T. E., Balmes, J. R., Broyles, G. A., et al. (2019). Wildland firefighter smoke exposure and risk of lung cancer and cardiovascular disease mortality. *Environmental Research*, *173*, 462–468.

NEPC (1998). N. E. P. C. National environment protection (ambient air quality) measure. http://www.nepc.gov.au/nepms/ambient-air-quality

Netea, M. G., Domínguez-Andrés, J., Barreiro, L. B., Chavakis, T., Divangahi, M., Fuchs, E., et al. (2020). Defining trained immunity and its role in health and disease. *Nature Reviews Immunology*, *20*(6), 375–388.

NSW Rural Fire Service (2019). Planning for bush fire protection. https://www.rfs.nsw.gov.au/plan-and-prepare/building-in-a-bush-fire-area/planning-for-bush-fire-protection

O'Dell, K., Hornbrook, R. S., Permar, W., Levin, E. J. T., Garofalo, L. A., Apel, E. C., et al. (2020). Hazardous Air Pollutants in Fresh and Aged Western US Wildfire Smoke and Implications for Long-Term Exposure. *Environmental Science & Technology*, *54*(19), 11838–11847.

Orr, A., A. L. Migliaccio, C., Buford, M., Ballou, S., & Migliaccio, C. T. (2020). Sustained effects on lung function in community members following exposure to hazardous PM2.5 levels from wildfire smoke. *Toxics*, *8*(3), 53. https://www.ncbi.nlm.nih.gov/pmc/articles/PMC7560437/

Oudin, A., Segersson, D., Adolfsson, R., & Forsberg, B. (2018). Association between air pollution from residential wood burning and dementia incidence in a longitudinal study in Northern Sweden. *PLOS ONE*, *13*(6), e0198283.

Pardo, M., Li, C., He, Q., Levin-Zaidman, S., Tsoory, M., Yu, Q., et al. (2020). Mechanisms of lung toxicity induced by biomass burning aerosols. *Particle and Fibre Toxicology*, *17*(1), 4.

Peng, F., Tsuji, G., Zhang, J.-Z., Chen, Z., & Furue, M. (2019). Potential role of PM2.5 in melanogenesis. *Environment International*, *132*, 105063.

Piao, M. J., Ahn, M. J., Kang, K. A., Ryu, Y. S., Hyun, Y. J., Shilnikova, K., et al. (2018). Particulate matter 2.5 damages skin cells by inducing oxidative stress, subcellular organelle dysfunction, and apoptosis. *Archives of Toxicology*, *92*(6), 2077–2091.

Pope, C. A., III (2002). Lung cancer, cardiopulmonary mortality, and long-term exposure to fine particulate air pollution. *JAMA*, *287*(9), 1132.

Prunicki, M., Miller, S., Hopkins, A., Poulin, M., Movassagh, H., Yan, L., et al. (2020). Wildfire smoke exposure is associated with decreased methylation of the PDL2 gene. *The Journal of Immunology*, *204*(1 Supplement), 146.117–146.117.

Radke, L. F., Hegg, A. S., Hobbs, P. V., & Penner, J. E. (1995). Effects of aging on the smoke from a large forest fire. *Atmospheric Research*, *38*(1), 315–332.

Rappold, A. G., Stone, S. L., Cascio, W. E., Neas, L. M., Kilaru, V. J., Carraway, M. S., et al. (2011). Peat bog wildfire smoke exposure in rural North Carolina is associated with cardiopulmonary emergency department visits assessed through syndromic surveillance. *Environmental Health Perspectives*, *119*(10), 1415–1420.

Reid, C. E., Brauer, M., Johnston, F. H., Jerrett, M., Balmes, J. R., & Elliott, C. T. (2016). Critical review of health impacts of wildfire smoke exposure. *Environmental Health Perspectives*, *124*(9), 1334–1343.

Reinhardt, T. E., & Broyles, G. (2019). Factors affecting smoke and crystalline silica exposure among wildland firefighters. *Journal of Occupational and Environmental Hygiene*, *16*(2), 151–164.

Reynolds, H. Y. (1985). Phagocytic defense in the lung. *Antibiotics and Chemotherapy*, *36*, 74–87.

Roscioli, E., Hamon, R., Lester, S. E., Jersmann, H., Reynolds, P. N., & Hodge, S. (2018). Airway epithelial cells exposed to wildfire smoke extract exhibit dysregulated autophagy and barrier dysfunction consistent with COPD. *Respiratory Research*, *19*(1), 1–13.

Ross, M. H., & Murray, J. (2004). Occupational respiratory disease in mining. *Occupational Medicine (Oxford, England)*, *54*(5), 304–310.

Rydell-Törmänen, K., & Johnson, J. R. (2019). The applicability of mouse models to the study of human disease. *Methods in Molecular Biology*, *1940*, 3–22.

Sada-Ovalle, I., Chávez-Galán, L., Vasquez, L., Aldriguetti, S., Rosas-Perez, I., Ramiréz-Venegas, A., et al. (2018). Macrophage exposure to polycyclic aromatic hydrocarbons from wood smoke reduces the ability to control growth of *Mycobacterium tuberculosis*. *Frontiers in Medicine*, *5*. https://www.ncbi.nlm.nih.gov/pmc/articles/PMC6243050/

Santoso, M. A., Christensen, E. G., Yang, J., & Rein, G. (2019). Review of the transition from smouldering to flaming combustion in wildfires. *Frontiers in Mechanical Engineering*, *5*. https://www.frontiersin.org/articles/10.3389/fmech.2019.00049/full

Schmidt, A. J., Borras, E., Nguyen, A., Kenyon, N. J., & Davis, C. E. (2021). 23255 devices engineered to collect exhaled breath condensate (EBC) and their applications. *Journal of Clinical and Translational Science*, *5*(s1), 4.

Schranz, C. I., Castillo, E. M., & Vilke, G. M. (2010). The 2007 San Diego wildfire impact on the Emergency Department of the University of California, San Diego Hospital System. *Prehospital and Disaster Medicine*, *25*(5), 472–476.

Schraufnagel, D. E. (2020). The health effects of ultrafine particles. *Experimental & Molecular Medicine*, *52*(3), 311–317.

Schuller, A., & Montrose, L. (2020). Influence of woodsmoke exposure on molecular mechanisms underlying Alzheimer's disease: Existing literature and gaps in our understanding. *Epigenetics Insights*, *13*, 2516865720954873.

Schwartz, C., Bølling, A. K., & Carlsten, C. (2020). Controlled human exposures to wood smoke: A synthesis of the evidence. *Particle and Fibre Toxicology*, *17*(1), 1–17.

Selimovic, V., Yokelson, R. J., McMeeking, G. R., & Coefield, S. (2019). In situ measurements of trace gases, PM, and aerosol optical properties during the 2017 NW US wildfire smoke event. *Atmospheric Chemistry and Physics*, *19*(6), 3905–3926.

Seltenrich, N. (2018). Flavors of fire: Assessing the relative toxicity of smoke from different types of wildfires. *Environmental Health Perspectives*, *126*(4). https://www.ncbi.nlm.nih.gov/pmc/articles/PMC6071823/

Shah, A. S. V., Langrish, J. P., Nair, H., McAllister, D. A., Hunter, A. L., Donaldson, K., et al. (2013). Global association of air pollution and heart failure: A systematic review and meta-analysis. *Lancet*, *382*(9897), 1039–1048.

Sillanpää, M., Saarikoski, S., Hillamo, R., Pennanen, A., Makkonen, U., Spolnik, Z., et al. (2005). Chemical composition, mass size distribution and source analysis of long-range transported wildfire smokes in Helsinki. *Science of the Ttotal Environment*, *350*(1–3), 119–135.

Smith, K. R., Samet, J. M., Romieu, I., & Bruce, N. (2000). Indoor air pollution in developing countries and acute lower respiratory infections in children. *Thorax*, *55*(6), 518–532.

Sosedova, L. M., Vokina, V. A., Novikov, M. A., Rukavishnikov, V. S., Andreeva, E. S., Zhurba, O. M., et al. (2020). Paternal biomass smoke exposure in rats produces behavioral and cognitive alterations in the offspring. *Toxics*, *9*(1), 3. https://www.ncbi.nlm.nih.gov/pmc/articles/PMC7823662/

Stowell, J. D., Geng, G., Saikawa, E., Chang, H. H., Fu, J., Yang, C.-E., et al. (2019). Associations of wildfire smoke PM2. 5 exposure with cardiorespiratory events in Colorado 2011–2014. *Environment International*, *133*, 105151.

Sun, Z., Ji, N., Jiang, J., Tao, Y., Zhang, E., Yang, X., et al. (2021). Fine particulate matter (PM2.5) promotes CD146 expression in alveolar epithelial cells and *Cryptococcus neoformans* pulmonary Infection. *Frontiers in Microbiology*, *11*, 525976. https://www.ncbi.nlm.nih.gov/pmc/articles/PMC7848894/

Swiston, J. R., Davidson, W., Attridge, S., Li, G. T., Brauer, M., & Eeden, S. F. (2008). Wood smoke exposure induces a pulmonary and systemic inflammatory response in firefighters. *The European Respiratory Journal*, *32*(1), 129–138.

Thomson, E. M., Williams, A., Yauk, C. L., & Vincent, R. (2009). Impact of nose-only exposure system on pulmonary gene expression. *Inhalation Toxicology*, *21* (Suppl 1), 74–82.

Tsatsakis, A. M., Vassilopoulou, L., Kovatsi, L., Tsitsimpikou, C., Karamanou, M., Leon, G., et al. (2018). The dose response principle from philosophy to modern toxicology: The impact of ancient philosophy and medicine in modern toxicology science. *Toxicology Reports*, *5*, 1107–1113.

USDA Forest Service (2019). Most California wildfire is in wildland-urban interface area with less fuel, more people. *ScienceDaily,* (24 September). www.sciencedaily.com/releases/2019/09/190924131527.htm

U.S. EPA (2014). *NAAQS Table*. https://www.epa.gov/criteria-air-pollutants/naaqs-table

U.S. EPA (2016). Table of historical particulate matter (PM) National Ambient Air Quality Standards (NAAQS). https://www.epa.gov/pm-pollution/table-historical-particulate-matter-pm-national-ambient-air-quality-standards-naaqs

U.S. EPA (2018). The links between air pollution and childhood asthma. https://www.epa.gov/sciencematters/links-between-air-pollution-and-childhood-asthma

Vierkötter, A., Schikowski, T., Ranft, U., Sugiri, D., Matsui, M., Krämer, U., et al. (2010). Airborne particle exposure and extrinsic skin aging. *Journal of Investigative Dermatology*, *130*(12), 2719–2726.

Walter, C. M., Schneider-Futschik, E. K., Lansbury, N. L., Sly, P. D., Head, B. W., & Knibbs, L. D. (2021). The health impacts of ambient air pollution in Australia: A systematic literature review. *Internal Medicine Journal*, *51*(10), 1567–1579.

Wang, R., Chen, R., Wang, Y., Chen, L., Qiao, J., Bai, R., et al. (2019). Complex to simple: In vitro exposure of particulate matter simulated at the air-liquid interface discloses the health impacts of major air pollutants. *Chemosphere*, *223*, 263–274.

Westerling, A., Brown, T., Schoennagel, T., Swetnam, T., Turner, M., & Veblen, T. (2014). Briefing: Climate and wildfire in western U.S. forests. In V. Sample et al. (Eds.), *Forest conservation and management in the Anthropocene: Conference proceedings* (pp. 81–102). Proceedings, RMRS-P-71. Fort Collins, CO: U.S. Department of Agriculture, Forest Service. Rocky Mountain Research Station.

Wong, J. P. S., Tsagkaraki, M., Tsiodra, I., Mihalopoulos, N., Violaki, K., Kanakidou, M., et al. (2019). Atmospheric evolution of molecular-weight-separated brown carbon from biomass burning. *Atmospheric Chemistry and Physics*, *19*(11), 7319–7334.

Young, T. M., Black, G. P., Wong, L., Bloszies, C. S., Fiehn, O., He, G., et al. (2021). Identifying toxicologically significant compounds in urban wildfire ash using in vitro bioassays and high-resolution mass spectrometry. *Environmental Science & Technology*. https://pubs.acs.org/doi/pdf/10.1021/acs.est.0c06712

Zeglinski, M. R., Turner, C. T., Zeng, R., Schwartz, C., Santacruz, S., Pawluk, M. A., et al. (2019). Soluble wood smoke extract promotes barrier dysfunction in alveolar epithelial cells through a MAPK signaling pathway. *Scientific Reports*, *9*, 10027. https://www.ncbi.nlm.nih.gov/pmc/articles/PMC6624307/

Zelikoff, J. T., Chen, L. C., Cohen, M. D., & Schlesinger, R. B. (2002). The toxicology of inhaled woodsmoke. *Journal of Toxicology and Environmental Health. Part B, Critical Reviews*, *5*(3), 269–282.

Zhang, Z., Wang, J., Kwong, J. C., Burnett, R. T., van Donkelaar, A., Hystad, P., et al. (2021). Long-term exposure to air pollution and mortality in a prospective cohort: The Ontario health study. *Environment International*, *154*, 106570.

Zhu, F., Qiu, X., Wang, J., Jin, Y., Sun, Y., Lv, T., et al. (2012). A rat model of smoke inhalation injury. *Inhalation Toxicology*, *24*(6), 356–364.

12

Wildfire Smoke Exposures and Adult Health Outcomes

Miriam E. Marlier[1], Natalie Crnosija[2], and Tarik Benmarhnia[3]

ABSTRACT

Health outcomes attributable to landscape fire smoke pollution exposure are an important global health issue, especially as these events increase in frequency and intensity in the context of climate change and variability. Landscape fires include wildfires and land management fires and are associated with hundreds of thousands of deaths per year globally through the contribution to fine particulate matter (PM2.5) and other sources of air pollution. In this chapter, we describe different exposure assessment techniques that are used to estimate health outcomes from global to regional scales, including atmospheric models, remote sensing, and ground station observations. We then evaluate various methodological approaches used in the epidemiologic literature to assess mortality and respiratory and cardiovascular disease outcomes associated with smoke pollution exposure in adult populations, excluding occupational settings. We describe current gaps in the literature, which include understudied health outcomes, medically and socially vulnerable populations, geographic disparities in health studies, and co-occurring exposures. We conclude with a discussion of interventions to reduce the negative health outcomes associated with wildfire smoke pollution exposure, such as reducing landscape fire risk or limiting smoke exposure in indoor settings. In summation, we here present a contemporary overview of the literature regarding the complex, dynamic issue of landscape fire smoke exposure evaluation, epidemiology, and adult health outcomes.

12.1. GLOBAL BACKGROUND AND SIGNIFICANCE OF THE PROBLEM

12.1.1. Background

Landscape fires can have devastating impacts on human health through contributions to surface air pollution. Fires contribute to enhanced surface concentrations of fine particulate matter ($PM_{2.5}$; particles < 2.5 μm in diameter) and trace gases such as ozone (O_3), carbon monoxide (CO), and other pollutants. Severe fire events in Australia, the western United States, Indonesia, and the Amazon, which recently captured the world's attention have also exposed broad regional populations to dangerous levels of fire-contributed air pollution, hereafter referred to as smoke pollution (Fig. 12.1). In some regions of the world, increases in smoke pollution have negated other air quality improvements over past decades (McClure & Jaffe, 2018).

Understanding and documenting the health outcomes associated with smoke pollution exposure make up an important and growing public health issue. First, climate change is increasing the contribution of wildfires to smoke pollution in many regions. In the same vein, while wildfires are traditionally considered as acute events, their

[1] *Fielding School of Public Health, University of California Los Angeles, Los Angeles, California, USA*

[2] *School of Public Health, University of Maryland, College Park, Maryland, USA*

[3] *Scripps Institution of Oceanography, University of California San Diego, San Diego, California, USA*

Landscape Fire, Smoke, and Health: Linking Biomass Burning Emissions to Human Well-Being, Geophysical Monograph 280, First Edition.
Edited by Tatiana V. Loboda, Nancy H. F. French, and Robin C. Puett.

DOI: 10.1002/9781119757030.ch12

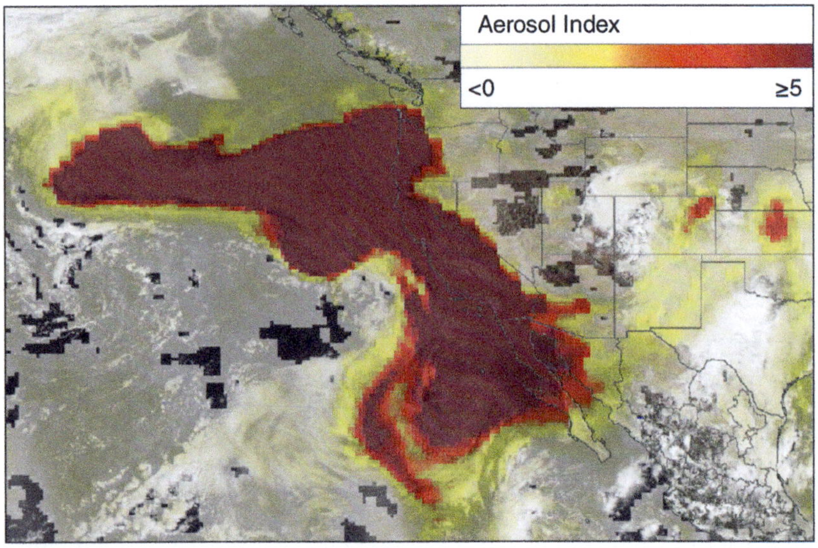

Figure 12.1 Aerosol Index from 10 September 2020 showing the presence of absorbing particles in the atmosphere across the western United States during a wildfire event. Observations from Ozone Mapping and Profiler Suite (OMPS)/Suomi National Polar-orbiting Partnership (S-NPP) (image courtesy of NASA Worldview).

staggering increase in prevalence and intensity is gradually constituting a subchronic environmental exposure, albeit with limited epidemiological evidence. Second, documenting which communities are particularly exposed and/or impacted and which underlying health conditions (e.g., diabetes, cardiovascular diseases) drive a higher susceptibility to smoke pollution is crucial to inform prevention efforts. Third, although evidence for smoke pollution and health outcomes has grown in the past two decades, this primarily comes from high income countries. Many other regions are exposed to smoke pollution, constituting a critical need for future research.

12.1.2. Smoke Pollution Health Burdens Around the World

Several studies have estimated global or regional health burdens to smoke $PM_{2.5}$ exposure. This provides useful information across several dimensions: (1) the relative contribution of smoke to the health burden of $PM_{2.5}$ compared with other sources of ambient pollution, (2) geographic and temporal variability in the health burden of smoke $PM_{2.5}$ exposure, and (3) sources of fire emissions that can inform intervention strategies.

Global exposure to smoke $PM_{2.5}$ from landscape fires is responsible for an estimated 340,000–680,000 deaths per year, amounting to between 8% and 21% of the total outdoor air pollution mortality burden (i.e., the total number of premature deaths directly attributable to this exposure) (Johnston et al., 2012; Roberts & Wooster, 2021). More than 44 million people around the world are exposed to unhealthy annual average $PM_{2.5}$ smoke pollution (> 55 μg/m^3) (Roberts & Wooster, 2021). However, there is significant spatial variability in smoke pollution sources that contribute to landscape fires (wildfires, deforestation, and forest degradation fires, savanna fires, agricultural fires, etc.) and the magnitude of public health burden.

In many tropical countries, fires associated with land use and drought conditions contribute to high levels of smoke pollution exposure. Johnston et al. (2012) and Roberts & Wooster (2021) highlighted sub-Saharan Africa and Southeast Asia as global hotspots in driving the attributable mortality burden of smoke $PM_{2.5}$ exposure. Recent estimates suggest nearly 10% of premature deaths due to $PM_{2.5}$ exposure in Brazil were linked to smoke pollution (Nawaz & Henze, 2020). Preventing vegetation fires in the Amazon Basin could avert approximately 17,000 premature deaths due to smoke $PM_{2.5}$ exposure (Butt et al., 2020). Another tropical fire hotspot is Indonesia, where severe fires in 2015 were linked to an estimated 44,000–100,000 premature deaths across equatorial Asia (Koplitz et al., 2016; Kiely et al., 2020) and the exposure of nearly 70 million people to unhealthy smoke pollution levels (Crippa et al., 2016).

Landscape fires also contribute to local and regional pollution in nonequatorial regions. Vegetation fires were linked to short-term $PM_{2.5}$ increases in southern and eastern Europe and low-to-moderate increases in daily $PM_{2.5}$ across the continent (Kollanus et al., 2017). Kollanus et al. (2017) estimated that 1,483 and 1,080 premature deaths across 27 countries in Europe in 2005 and 2008, respectively, were attributable to smoke $PM_{2.5}$. Across the United States, fires contribute to approximately 11% of $PM_{2.5}$ and 1% of O_3 on average but play a more important role in western states (Wilkins et al., 2018) and during extreme weather events (Wilkins et al., 2018; O'Dell et al., 2019). For instance, fires can contribute up to 50% of $PM_{2.5}$ in some parts of the western United States (Burke et al., 2021). In this region, nearly 50 million people over 2004–2009 were exposed to a smokewave event (more than 2 days with high smoke $PM_{2.5}$), with corresponding increases in respiratory hospital admissions (Liu et al., 2017b). A recent study estimated the number of asthma hospital admissions, emergency department visits, and premature deaths attributable to acute smoke $PM_{2.5}$ exposure across the United States using concentration-response functions (CRFs refer to the estimated dose-response between levels of $PM_{2.5}$ and the risk of observing a given health outcome of interest) specific to smoke $PM_{2.5}$ exposure as well as gas-phase hazardous air pollutants (HAPs) (O'Dell et al., 2021). They estimated that 216,000 deaths were attributable to wildfire smoke and that most of the burden took place outside the western United States as smoke typically traveled across the continent impacting a very large population.

Agricultural fires also contribute to consistent seasonal pollution enhancements in many parts of the world. For example, agricultural waste burning in central and west Africa is the dominant driver of smoke pollution across the continent, linked to 43,000 premature deaths per year (Bauer et al., 2019). Another important example is in India, where crop residue burning contributes to seasonal extreme pollution above World Health Organization guidelines in rural areas and urban centers (Cusworth et al., 2018; Liu et al., 2018). In many regions, additional research is required to separate agricultural fire contributions from other sources of pollution to quantify the health burden attributable to agricultural fires. However, it is worth mentioning that most studies described previously applied CRFs developed for all-source $PM_{2.5}$ or (more rarely) for smoke $PM_{2.5}$ specifically.

12.1.3. Scope of the Chapter

The aim of this chapter is to provide a current overview of the health consequences of smoke pollution exposure experienced by adults, excluding occupational settings (e.g., among firefighters) (McNamara et al., 2012; Navarro et al., 2017; Aisbett et al., 2012; Groot et al., 2019). The perinatal and child health burden is discussed in Chapter 13. We primarily document the health impacts associated with $PM_{2.5}$ smoke pollution exposure, but also briefly address other potentially synergistic consequences associated with the trauma of fire events, such as the emergence of mental health stressors. While $PM_{2.5}$ is the most investigated smoke constituent, it is important to note that other harmful compounds of smoke that impact human health exist such as ozone (O_3), volatile organic compounds (VOCs), carbon monoxide, lead, and other heavy metals and toxins that can be generated by burning biomass and flame contact with built structures. However, such pollutants may primarily impact populations directly impacted by the fire (not only through smoke) (O'Dell et al., 2021).

We first present a summary of the epidemiological literature on smoke pollution and adult health outcomes by synthesizing several recent reviews and additional studies. We consider the environmental justice implications of this phenomenon and the need to address differential susceptibility and exposure to biomass burning smoke pollution. We then discuss opportunities to improve our understanding of the adult health burden of smoke exposure more holistically. This includes how smoke pollution exposure estimates are matched to health data through different study designs, settings in which evidence is still lacking, and additional areas of future research, such as repeated smoke exposures and compounded impacts. We then present an overview of existing interventions to reduce the negative health outcomes associated with smoke pollution exposure. We conclude with a summary of research gaps and future directions.

12.2. OVERVIEW OF EPIDEMIOLOGIC EVIDENCE ON ADULT HEALTH OUTCOMES

12.2.1. Introduction

In the past few years, several literature reviews have been conducted with regards to the health impacts associated with landscape fire smoke pollution exposure (Reid et al., 2016; Liu et al., 2015; Henderson & Johnston, 2012; Arriagada et al., 2019; Youssouf et al., 2014; Black et al., 2017; Kochi et al., 2010; Naeher et al., 2007). $PM_{2.5}$ is one of the primary constituents of smoke pollution and is the focus of this section. Ambient $PM_{2.5}$ concentrations are monitored and regulated, as such particles are small enough to penetrate the respiratory system, interact with the circulation system, and further impact any organ in the body. Furthermore, $PM_{2.5}$ also impacts human health through systemic inflammation and activation of the autonomic nervous system (Brook et al., 2010). Evidence regarding the health effects of all-source $PM_{2.5}$ is vast, and several reviews have been published in the past decade (Feng et al., 2016; Atkinson et al., 2014; Fan et al., 2016; Xing et al., 2016), including both acute (e.g., asthma exacerbation, myocardial infarction, etc.) and chronic (atherosclerosis, dementia, lung cancer, etc.) effects. Smoke pollution concentrations are mostly considered as acute exposures in the epidemiological literature, but, as we discuss below, some long-term consequences may exist. The repeated nature of such events in the context of climate change makes such exposure more frequent and considering such exposures as sub-chronic in certain regions of the globe may be warranted in future studies.

The mechanisms through which $PM_{2.5}$ can impact human health (such as oxidative stress, alteration of the pulmonary immune system, and chronic inflammation) may differ according to particle composition (for more details, see Chapter 11). While it may seem reasonable to initially assume that smoke pollution may have similar toxicological mechanisms and impacts on human health compared with all-source $PM_{2.5}$, recent research justifies studying smoke as a separate exposure for several reasons. First, $PM_{2.5}$ concentrations during an extreme biomass burning event can be one order of magnitude larger or more when compared with typical exposure levels. This implies that epidemiological evidence from other sources of fine particles regarding the dose-response relationship, the types of symptoms, or which subgroups of the populations are susceptible may not extrapolate to such levels of exposure. Second, recent toxicological evidence suggests that smoke $PM_{2.5}$ may be more toxic than equal doses of $PM_{2.5}$ from other sources due to particle composition (Kim et al., 2018). Current air pollution guidelines or regulations do not distinguish by emissions source or chemical composition for $PM_{2.5}$. With these considerations, focusing on studies with specific smoke pollution exposures has led to dozens of epidemiological studies in the past two decades that we summarize here. We describe the types of health outcomes that have been investigated and the state of evidence regarding these outcomes. In Section 12.3, we will discuss health outcomes such as mental health for which evidence is still sparse.

12.2.2. Mortality

The evidence related to the impact of smoke pollution on acute premature mortality is relatively strong. Many studies have consistently found an increase in daily mortality during a wildfire event or in subsequent days (Doubleday et al., 2020). However, most of these studies focused on all-cause mortality or mortality for respiratory or CVD endpoints; studies about other cause-specific mortality outcomes are still lacking (Reid et al., 2016; Liu et al., 2015). In a recently published global analysis, Chen et al. (2021) found an annual average of 33,510 all-cause deaths to be attributable to smoke $PM_{2.5}$ pollution exposure using data from 749 cities in 43 countries.

12.2.3. Morbidity

Respiratory Diseases

Respiratory health outcomes have received the most attention in the published epidemiological literature. Various respiratory morbidity outcomes have been studied, including lung function, respiratory medication usage (Caamano-Isorna et al., 2011; Vora et al., 2007), physician visits, and emergency departments (ED) visits or hospital admissions for respiratory problems (Arriagada et al., 2019). When considering ED or physician visits for various respiratory outcomes, published studies strongly suggest a detrimental effect of smoke pollution. Among the specific respiratory outcomes, asthma has been extensively studied. A recent systematic review (Arriagada et al., 2019) focusing on asthma-related outcomes found consistent evidence for this outcome.

Fewer studies have examined changes in lung function (Arriagada et al., 2019). Amid mixed results, most studies were not able to identify the effect of smoke pollution. For medication usage, studies focused on various endpoints, such as medication use, initiation of oral steroid use, or medication for obstructive lung disease, also have inconsistent results (Reid et al., 2016). Finally, there is increasing evidence that wildfire smoke also exacerbates chronic obstructive pulmonary disease (COPD) (Gayle et al., 2021). More recent studies also investigated the role of smoke pollution on exacerbating respiratory infections or disease severity such as for seasonal influenza (Landguth et al., 2020); such connections between respiratory infectious diseases and wildfire are particularly relevant in the context of the COVID-19 pandemic (Henderson, 2020; Zhou et al., 2021).

Cardiovascular Diseases

Less epidemiological evidence exists for cardiovascular outcomes than for respiratory outcomes. Some studies have considered smoke pollution exposure and cardiovascular diseases (CVD) outcomes such as hypertension (Arbex et al., 2010). Most CVD studies assessed hospital admissions or ED visits for CVD causes (Reid et al., 2016). Fewer studies investigate specific CVD endpoints, like congestive heart failure (Rappold et al., 2012), ischemic heart disease (Crabbe, 2012), cardiac arrest (Dennekamp et al., 2015), or myocardial infarction (Haikerwal et al., 2015). However, the results of these studies focusing on CVD outcomes are mixed, with some studies identifying an increasing risk and other studies not detecting any effect.

12.2.4. Vulnerable Populations

While most studies conducted to date focused on the health impacts on the entire adult population, several have investigated whether certain population subgroups are more susceptible to the health impacts associated with a specific landscape fire event or smoke pollution more generally (Rappold et al., 2017; Liu et al., 2017a,b). Such work investigating effect modification by various sociodemographic characteristics is motivated by the large evidence on differential susceptibility for fine particles in general (i.e., from other sources of emission) (Forastiere et al., 2007). Indeed, the environmental justice literature has found that socioeconomic and racial and ethnic minorities suffer from a disproportionate burden of air pollution exposure in general, and $PM_{2.5}$ in particular (Hajat et al., 2015).

However, studies assessing the extent to which certain sociodemographic characteristics modify the smoke-pollution-health risk remain limited. Most studies investigating such differential susceptibility questions conducted stratified analyses or included an interaction term between smoke-pollution exposure and the sociodemographic variable of interest (Benmarhnia et al., 2021). Among these studies, most focused on age as a susceptibility factor (Haikerwal et al., 2015; Ignotti et al., 2010; Morgan et al., 2010). Some studies have shown that the risk for most health outcomes was higher among older populations (with various cutoffs across the studies such as > 65 or > 75 years old), but other studies found the opposite pattern or no evidence of such effect modification by age (Reid et al., 2016). Several studies assessed potential gender heterogeneity, but the results are mixed (Reid et al., 2016).

Differential susceptibility across socioeconomic and racial/ethnic groups, including individual race or ethnicity (Liu et al., 2017a,b), neighborhood socioeconomic status (SES) (Henderson et al., 2011), indigenous status (Hanigan et al., 2008), or proxies such as district-specific food consumption (Jayachandran, 2009) have also been considered. However, results from these studies were mixed, with some studies finding that low SES groups were more susceptible to wildfire and other studies found no differences among groups. Finally, other vulnerability factors included preexisting health conditions (using different proxies such as number of physician visits in the previous year; (Henderson et al., 2011)) but available evidence is inconclusive to date.

12.3. CONSIDERATIONS FOR FUTURE EPIDEMIOLOGICAL STUDIES

12.3.1. Geographic Disparities in Health Studies

More studies are needed in geographically underrepresented areas where wildfire smoke pollution is common and/or projected to increase in the future, especially where the public health infrastructure is more vulnerable (Chen et al., 2021). This emphasis should be placed on developing exposure and health operational capacity in parts of the world with high levels of smoke pollution, particularly in sub-Saharan Africa, given the majority of existing studies examined these impacts in North America or Australia (Voulgarakis & Field, 2015; Johnson et al., 2020). The geographical distribution of existing epidemiological studies highlighted an important gap, which is not unique to smoke pollution (Green et al., 2019). In Figure 12.2, we illustrate the discrepancies between where studies have been conducted thus far and where

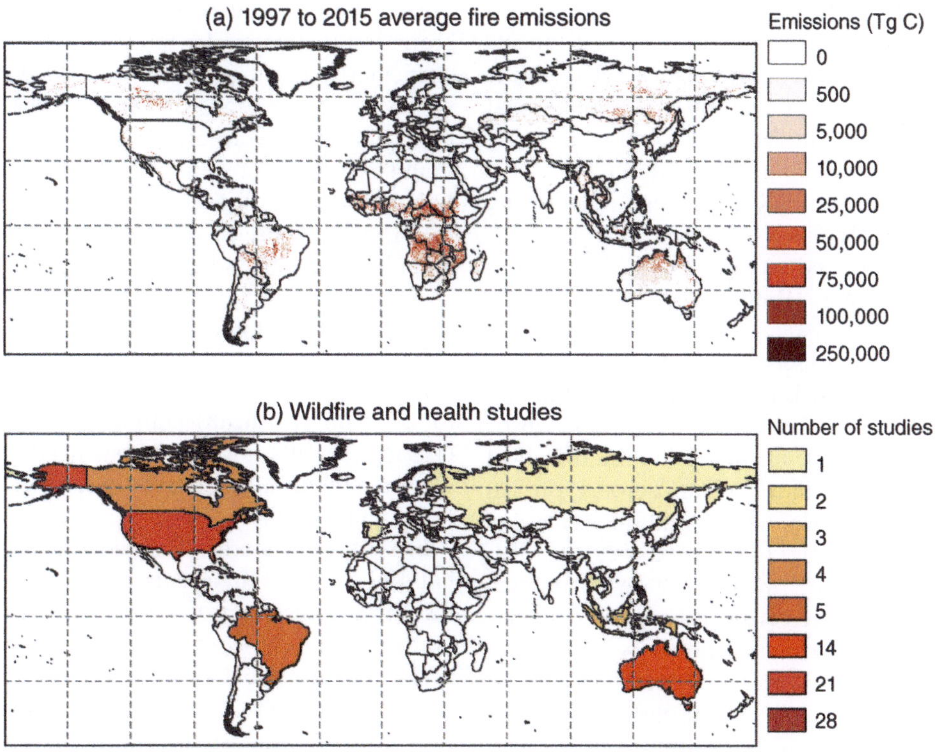

Figure 12.2 Maps comparing the number of all-source smoke-pollution-health studies conducted in each country with (a) the average fire emissions from 1997 to 2015 from the Global Fire Emissions Database (Giglio et al., 2018; Giglio et al., 2013). (b) The included studies are extracted from four recent systematic reviews (Reid et al., 2016; Liu et al., 2015; Henderson & Johnston, 2012; Arriagada et al., 2019).

most wildfire emissions take place. Epidemiological evidence is lacking in several regions where fires are an important source of emissions, especially in Low and Middle Income Countries (LMICs).

12.3.2. Exposure Estimates for Epidemiological Studies

Fire contributions to smoke pollution are estimated with various approaches, including atmospheric modeling, satellite-based techniques, ground station data, or blended methods that merge multiple information sources. Here, we briefly review the primary smoke exposure methods through the lens of providing recommendations for use in epidemiological studies. For an in-depth discussion of each of these methods, we refer the reader to Chapters 6–9.

The first category of exposure assessment is atmospheric models, which can be applied at global to local scales. Lelieveld et al. (2015) and Johnston et al. (2012) used global atmospheric models to quantify the smoke pollution health burden from multiple fires around the world compared with nonfire pollution sources. At smaller scales, atmospheric dispersion models can be used to track smoke pollution from individual fire events (Li et al., 2020).

Second, satellite observations, can monitor pollution during fire events. Satellite-based products include the National Environmental Satellite, Data, and Information Service (NESDIS), Hazard Mapping System (HMS) smoke plume data in North America, which connects observed smoke plumes to active fires (Burke et al., 2021; Brey et al. 2018; Schweizer et al., 2019). Aerosol Optical Depth (AOD) from the Moderate Resolution Imaging Spectroradiometer (MODIS) instrument on the Terra and Aqua satellites and the TROPOspheric Monitoring Instrument (TROPOMI) have been used to map atmospheric aerosol loading and infer surface $PM_{2.5}$ concentrations during fire events (Raffuse et al., 2013; Torres et al., 2020).

Third, ground station observations have been used as an input into blended models to replicate the spatial and temporal variability of smoke pollution. Low-cost sensor networks also show promise for informing the statistical relationship between satellite column aerosol optical depth and surface-level $PM_{2.5}$ during wildfire events due to their dense spatial coverage (Gupta et al., 2018).

Each of these exposure methods have their own strengths and weaknesses to consider for use in epidemiological studies. The first consideration is the spatial scale of exposure data sets. With coarser models or sparse ground station data, the ability to resolve peak smoke pollution concentrations may be reduced, which could result in an underestimate of health outcomes. An additional consideration is the assignment of a single exposure variable to an entire population, despite significant individual-level differences in exposure, such as across zip codes (Aguilera et al., 2020), and whether individuals spend the majority of their time indoors or outdoors. Second, when possible, we recommend that epidemiological studies use multiple exposure estimates to test the sensitivity of the studies to exposure methods. Blended models consider multiple sources of information to represent smoke $PM_{2.5}$ concentrations. For example, ground station monitors provide surface-level estimates of $PM_{2.5}$ at specific locations and/or time points. Satellite-based observations can be used to fill in some of the gaps in this spatial or temporal coverage. Cleland et al. (2021) recently compared the smoke pollution health burden using multiple exposure estimates (ground monitor, modeled, and blended). The authors found that the choice of exposure data set drove uncertainty in the resulting health burden estimate. In a review of 28 studies around the world that estimated PM associated with open burning, Johnson et al. (2020) found that blended approaches tend to have the best results by at least partially compensating for limitations associated with each individual approach. Lassman et al. (2017) also found more accurate wildfire $PM_{2.5}$ predictions from monitors relative to satellite AOD or atmospheric modeling simulations, but that blended techniques were more accurate if ground monitor density was low. Exploring multiple exposure data sets may not always be possible due to data or computational limitations. For example, in regions of the world that lack dense ground station networks, modeling or satellite studies are particularly useful (Bauer et al., 2019). Finally, recognizing the implications of exposure method for issuing public health guidance is critical. Fadadu et al. (2020), for example, found substantial variability with the magnitude and timing of peak smoke pollution derived from HMS satellite-derived smoke polygons of low, medium, and high intensity and ground station monitors.

12.3.3. Epidemiological Study Designs

It is first important to distinguish two approaches to evaluate the health impacts of smoke pollution: (1) single events and (2) repeated effects of long-term smoke pollution exposure over a long time span (e.g., multiple years).

First, several studies focused on an individual or a handful of major fire events and then evaluated whether health outcome changes were observed in affected areas (with or without control groups). Examples of such events include the October 2007 Southern California wildfires (Hutchinson et al., 2018), summer Russian wildfires in 2010 (Shaposhnikov et al., 2015), or Indonesia's forest fires of 1997 (Jayachandran, 2009). In such settings, authors relied on various study designs including case crossover designs (Hutchinson et al., 2018), interrupted time series designs (Leibel et al., 2020), or panel analyses (Kim et al., 2017). Such designs capitalize on the specific location and timing of the event of interest and formulate an identification strategy to compare observed outcome in the exposed group to a substitute for the counterfactual population (that was not exposed to the smoke pollution). Quasi-experimental designs, such as difference-in-differences, can also be employed for such research questions but remain underused for such type of events. Yet, they are a powerful alternative strategy to address various confounders that may or may not be measured while checking identification assumptions, such as parallel trends (Angrist & Pischke, 2008). In addition, simpler approaches have been implemented where excess events were estimated by comparing observed outcomes during the event of interest to outcome rates on a given calendar dates from previous years (Shaposhnikov et al., 2015). This technique has been frequently used in the context of extreme weather events, like extreme heat and hurricanes, to estimate excess mortality (Fouillet et al., 2006; Sandberg et al., 2019). However, such approaches do not typically control for any time-varying confounders, such as temperature or other sources of air pollution, which may lead to biased estimates.

Second, other studies focus instead on estimating the overall impact of smoke pollution by considering multiple years and focusing on various spatial scales (from single cities to an entire country). In this setting, studies rely on various techniques to estimate exposure to smoke pollution, such as atmospheric models or statistical techniques (for more details, see previous chapters). Accordingly, various study designs have employed, including ecological time series models or case crossover designs (Liu et al., 2016a, 2016b) and individual designs based on existing cohorts, nested case-control designs or ad hoc surveys (Kolbe & Gilchrist, 2009).

Finally, several studies (Aguilera et al., 2020; Reid et al., 2016) have investigated the spatial variability in the health impacts associated with wildfire events and found important heterogeneity of the geographical distribution of the impacts. Such studies remain rare, as compared with studies that aggregate the estimates spatially, but can provide estimates that can be particularly useful to identify vulnerable communities.

12.3.4. Understudied Health Outcomes

Certain health outcomes have been understudied. These include mental health outcomes in the adult population, such as psychological distress, solastalgia (i.e., the distress caused by environmental change) (Eisenman et al., 2015), changing psychological outcomes (Bryant et al., 2014) or mental and emotional well-being (Dodd et al., 2018). Investigating the short- and long-term impacts of landscape fires on mental health is particularly important to design interventions following such events and improve the resilience of affected communities. In a random digit dial survey of an area affected by multiple wildfires, Felix and Afifi (2015) found that those who were exposed to wildfire and were evacuated had poorer measured mental health and greater total fire stress than those who were not evacuated; relative to men, women had poorer mental health and greater total fire stress. More evidence regarding these links is currently needed and future studies capitalizing on self-reported mental health symptoms or medical claims are critical to the development of this field. Other issues such as diabetic (Xi et al., 2020) or ophthalmologic (Versura et al., 1999) outcomes, as well as injuries (Cleland et al., 2011) were investigated by few studies and more evidence is definitively warranted. Furthermore, given the emerging literature linking exposure to $PM_{2.5}$ and incident diabetes (Yang et al., 2020) and dementia (Weuve et al., 2021), there is a need to further investigate the impact of smoke pollution and these outcomes.

It is also particularly important to better understand which preexisting medical conditions constitute susceptibility factors for smoke pollution exposure. Apart from respiratory, CVD, or other chronic conditions, such as diabetes, it is necessary to investigate the extent to which individuals with dementia or Alzheimer's Disease Related Dementias (ADRD), or cancer survivors, for example, are more susceptible to poor health outcomes as compared with the general population. Documenting such preexisting susceptibility factors will inform existing preventive policies such as early warning systems by identifying which priority populations for interventions. Other plausible susceptibility factors have also received little attention to date and future epidemiological studies are critically needed. Such factors include metrics of health care access, background exposure to other sources of pollution, occupation (e.g., outdoor workers), and populations with physical disabilities.

12.3.5. Fires and the Built Environment

Landscape fires have the capacity to burn not only vegetation but also the built environments in which people live, work, and function (Gill et al., 2009). As a variety of substances are used in the construction and maintenance of these structures, these materials have varying capacity as fuel and have differing toxic potential when burned. Carratt et al.'s review (2017) noted that there was a spatial overlap of burned area and the prior application of pesticides in California; fire combatting chemical also can be found in these locations. Epidemiological data regarding the health effects of these ignited, potentially combined chemical exposures is lacking. Studies that address the acute and chronic health outcomes after landscape fire events generally do not address the chemical species of $PM_{2.5}$. More developed is the literature addressing the protection of the built environment from biomass burning events. Penman et al. (2014) used a Bayesian Network model to analyze the strategic use of fuel breaks, among other factors, in San Diego County, California. Found to be an effective strategy, the use of this model determined that high density communities, which tend to be at low elevations, were more susceptible to burning than were those at higher elevations, which tended to be less densely populated. Weather, too, contributed substantially to the wildfire's size and ability to travel and affect communities; the treatment of potential fuel had a minimal effect on the fire's ability to spread and endanger property. Housing density in wildfire vulnerable areas in the United States increased 1,350% between 1940 and 2010 (Strader, 2018), increasing the likelihood of the built environment's involvement in wildfire events. Wildfire adaptation has been investigated at the individual (McFarlane et al., 2011) and community (Schumann et al., 2020) level, both of which address the necessity of managing vulnerability at the wildland-urban interface.

12.3.6. Future Research Needs

Apart from knowledge gaps in relation to health outcomes and susceptibility factors, there are multiple avenues for future research. First, while most studies focused on $PM_{2.5}$ as the main component of biomass burning smoke exposure, other pollutants, such as O_3 or polycyclic aromatic hydrocarbons (PAHs) are generated in fire plumes. Recent studies have shown that wildfires generate increases in tropospheric O_3 levels through processes distinct from $PM_{2.5}$ (Reid et al., 2019; Watson et al., 2019). In future studies, it will be particularly important to understand how smoke pollutants other than $PM_{2.5}$ impact population health and study potential synergies among these pollutants.

Another important area of research relates to whether $PM_{2.5}$ smoke pollution affects health outcomes differently from $PM_{2.5}$ from other sources. While this pattern has been suggested by toxicological studies where differences in the composition led to higher effects of smoke $PM_{2.5}$ compared with ambient sources (Kim et al., 2018; Wegesser et al., 2009; Wegesser et al., 2010; Franzi et al., 2011), evidence at the population level is lacking. To the best of our knowledge, only two studies (DeFlorio-Barker et al., 2019; Kiser et al., 2020) focusing on asthma addressed this research question. Furthermore, in a recent study, it has been shown that smoke $PM_{2.5}$ can cause a greater impact on respiratory health than $PM_{2.5}$ from other sources (Aguilera et al., 2021). While further studies are needed to confirm these emerging findings, such patterns point to the need for air quality policies to consider the variability in $PM_{2.5}$ impacts on human health according to the sources of emissions.

Another area of research that deserves more attention is related to improvement in understanding the long-term impacts of smoke pollution on various outcomes (besides mental health as described above). Indeed, while wildfire has been considered traditionally as an acute environmental exposure, such instances are rapidly evolving in the context of climate change and variability (Burke et al., 2021; Ford et al., 2018). As the length of wildfire seasons increases (Jolly et al., 2015), the duration of exposure to extreme smoke pollution and corresponding health outcomes will likely increase. For example, in the western United States, future smoke $PM_{2.5}$ concentrations under multiple climate change scenarios suggests an increasing threat to public health, particularly for many vulnerable subpopulations (Ford et al., 2018; Spracklen et al., 2009; Liu et al., 2016). This motivates the need to better understand individual actions to reduce exposure as well as larger scale interventions to reduce wildfire emissions in order to reduce negative human health outcomes (see Section 12.4) (Xu et al., 2020).

It is also important to better understand the compounded impacts of smoke pollution with other contemporaneous risks. For example, smoke pollution and extreme heat events may co-occur as recently illustrated by Australian fires and the western United States in 2020. Smoke pollution and extreme heat share similar mechanisms through which they impact human health and several studies have demonstrated the synergistic effects of air pollution (not specific to smoke) and extreme heat (Anenberg et al., 2020). In addition, evidence about heat-related vulnerability identified similar population subgroups as for smoke pollution. Characterizing joint exposures to extreme heat and smoke pollution, as well as associated impacts, constitutes an important area for future research, especially while both events are expected to increase in intensity and frequency in our changing climate.

Finally, an important (and timely) area of investigation is related to the interactions between smoke pollution and infectious diseases. As discussed above, some recent studies have shown that biomass burning smoke may influence seasonal influenza incidence rates. For example, Landguth et al. (2020) concluded that increases in $PM_{2.5}$ concentrations during the wildfire season led to an increase in the influenza incidence in the following winter influenza season in Montana. In parallel, experimental studies showed that exposure to air pollutants, including

PM (Zelikoff et al., 2003) increased susceptibility to viral lung infections by affecting the immune system. A recent study suggested that other infectious agents, such as pathogenic fungus (e.g., causing coccidioidomycosis), may be transported by smoke pollution (Kobziar & Thompson, 2020). The interaction between air pollution and infectious diseases such as tuberculosis (Popovic et al., 2019) and coronavirus infection (e.g., SARS (Cui et al., 2003)) constitutes a novel area of research that is judicious in the context of the COVID-19 pandemic (Henderson, 2020; Benmarhnia, 2020; Villeneuve & Goldberg, 2020). Notably, a recent study by Zhou et al. (2021) found that the 2020 wildfires in Washington, Oregon, and California counties amplified the effect of short-term exposure of $PM_{2.5}$ on COVID-19 cases and deaths.

12.4. INTERVENTIONS TO REDUCE THE WILDFIRE'S IMPACT ON PUBLIC HEALTH

Our current understanding of the health outcomes associated with smoke pollution can help inform potential preventive strategies to protect public health. In this section, we provide a brief overview of several types of actions that exist to mitigate this health exposure.

12.4.1. Preemptive Power Outages

Certain strategies to reduce fire risk can also bring second-order health effects. For example, preemptive de-energization policies in California to reduce ignition sources associated with power lines during extreme fire weather conditions can disproportionately impact the health of communities with lower adaptive capacity (Abatzoglou et al., 2020). Power outages may lead to unintended health consequences, such as mental health outcomes, injuries, or heat-related illnesses (through air-conditioning interruption, for example). However, to the best of our knowledge, there is no empirical evidence regarding these issues, and we encourage future studies to address such connections.

12.4.2. Land Management

Another important area of future research is how land management interventions could reduce the public health burden of wildfire smoke pollution exposure through fuel and prescribed burning management (Schoennagel et al., 2017; Jaffe et al., 2020). These land management strategies will likely change health outcomes by altering the magnitude, frequency, timing, and duration of smoke exposure. For example, prescribed fires typically take place during lower atmospheric ventilation conditions to help control fire behavior, which can increase local exposure, whereas intense wildfires may be more likely to be injected higher into the atmosphere, with broader regional effects (Williamson et al., 2016). Across broader spatial and temporal scales, implementing low-level prescribed burning strategies could reduce the risk of extreme wildfire events and minimize large-scale smoke pollution impacts (Long & Tarnay, 2018). New research is needed to focus on the unique characteristics of prescribed fires as a coupled human-natural system (Hiers et al., 2020).

Schweizer et al. (2019) compared wildfire and prescribed burning smoke plumes in California using HMS observations and found that larger and more intense fires exposed more people per area burned because the smoke was transported over larger distances. Preliminary studies find worse health outcomes in children exposed to smoke from wildfires compared with prescribed burning (Prunicki et al., 2019). However, quantifying the health benefits of prescribed burning remains highly uncertain (Navarro et al., 2018). One primary source of this uncertainty is accurately estimating how low-level prescribed burns could offset the risk of future, higher emissions from extreme wildfire events (Schweizer & Cisneros, 2017). Another is due to measurement differences between wildfire and prescribed burning exposure estimates, with the proximity of sensors to fires often closer to prescribed burns than extreme wildfire events (Navarro et al., 2018).

12.4.3. Public Health Interventions

Limiting the number of people directly exposed to smoke pollution through evacuations is perhaps the most obvious intervention to protect public health (Abatzoglou et al., 2020; McCaffrey et al., 2018). In addition to populations directly exposed to wildfire threats, most of the smoke pollution health burden will be driven by regional exposures due to smoke transport. In this context, early warning systems (EWS) that aim to reduce a population's exposure to smoke pollution by collective or individual behavioral changes are crucial. Several models provide near-real-time or forecasted smoke $PM_{2.5}$ concentrations in the United States. For example, the CDC's National Environmental Public Health Tracking Network provides short-term smoke pollution forecasts to identify at-risk populations and strengthen public health preparedness (Vaidyanathan et al., 2018). Prior studies suggest that intervention advisories about low $PM_{2.5}$ concentration thresholds, coupled with strong public adherence, can effectively reduce risk (Rappold et al., 2014) in susceptible populations. In southern Australia, the Air Quality Visualization (AQVx) combines data to assess landscape fire-health effects from smoke exposure and to evaluate dispersion models, allowing targeted warning messages at a local scale (Williamson et al., 2020). For a discussion of other real-time and operational smoke forecasting systems, we refer the reader to Chapter 9.

Actions that take place during such EWS include modifications to work plans, school or business closures, and event cancelations. Individual behavioral changes are also urged when EWS are activated, such as usage of individual protections (N95 masks or respirators that filter particles), recommendations to stay indoors, limit physical activity, and reduce other activities that impact air quality, such as smoking, wood burning, or traffic emissions (Holm et al., 2021). Other long-term structural actions also exist to improve building resilience by improving mechanical ventilation systems to filter incoming air or providing air purifiers with a high efficiency (HEPA) filter. Studies that have evaluated the potential effectiveness of some of these interventions are rare (Yao et al., 2013) and more evidence is urgently needed. We strongly encourage future experimental studies that would compare the effectiveness of different actions on various populations as well as quasi-experimental studies as done in the context of other EWS (Benmarhnia et al., 2016; Chen et al., 2018).

12.5. CONCLUSIONS

In the context of climate change and variability, health outcomes associated with exposure to smoke pollution are an increasingly important global health issue. Understanding such outcomes in various locations and populations, and for different multiple health endpoints, is an urgent priority. In this chapter, we provide a contemporary overview of the epidemiological evidence for adult health outcomes related to smoke pollution exposure. While stronger evidence exists for associations between short-term exposures and all-cause mortality or respiratory morbidity, for example, additional studies are needed to address cardiovascular outcomes, the mental health burden, and vulnerable populations. Geographic disparities exist in existing adult epidemiological studies, which require additional information to better understand potential regional differences in health outcomes. We discuss how exposure to smoke pollution has been estimated, various methodological considerations for epidemiological study designs, and emerging evidence for several understudied health outcomes. Several opportunities exist to reduce smoke pollution exposure through land use interventions, early warning systems, and behavioral modifications. Taken together, while strong evidence exists for certain health outcomes and regions of the world, future studies will allow us to comprehensively understand the adult health burden of smoke pollution exposure by considering additional health outcomes, interactions among exposures, and additional opportunities to protect health.

ACKNOWLEDGMENTS

Miriam E. Marlier and Tarik Benmarhnia acknowledge support from the University of California Multicampus Research Program and Initiatives (MRPI, MRP-17-446315). Natalie Crnosija was supported by NRT-INFEWS: UMD Global STEWARDS (STEM Training at the Nexus of Energy, WAter Reuse and FooD Systems), which was awarded to the University of Maryland School of Public Health by the National Science Foundation, National Research Traineeship Program, Grant number 1828910.

REFERENCES

Abatzoglou, J. T., Smith, C. M., Swain, D. L., Ptak, T., & Kolden, C. A. (2020). Population exposure to pre-emptive de-energization aimed at averting wildfires in Northern California. *Environmental Research Letters*, *15*(9), 094046.

Aguilera, R., Gershunov, A., Corringham, T. W., & Benmarhnia, T. (2021). Wildfire smoke impacts respiratory health more than fine particles from other sources: Observational evidence from Southern California. *Nature Communications*, *12*(1), 1493.

Aguilera, R., Hansen, K., Gershunov, A., Ilango, S. D., Sheridan, P., & Benmarhnia, T. (2020). Respiratory hospitalizations and wildfire smoke: A spatiotemporal analysis of an extreme firestorm in San Diego County, *California. Environmental Epidemiology*, *4*(5), e114.

Aisbett, B., Wolkow, A., Sprajcer, M., & Ferguson, S. A. (2012). "Awake, smoky, and hot": Providing an evidence base for managing the risks associated with occupational stressors encountered by wildland firefighters. *Applied Ergonomics*, *43*(5), 916–925.

Anenberg, S. C., Haines, S., Wang, E., Nassikas, N., & Kinney, P. L. (2020). Synergistic health effects of air pollution, temperature, and pollen exposure: A systematic review of epidemiological evidence. *Environmental Health*, *19*(1), 1–19.

Angrist, J. D., & Pischke, J.-S. (2008). *Mostly harmless econometrics: An empiricist's companion*. Princeton University Press.

Arbex, M. A., Saldiva, P. H. N., Pereira, L. A. A., & Braga, A. L. F. (2010). Impact of outdoor biomass air pollution on hypertension hospital admissions. *Journal of Epidemiology & Community Health*, *64*(7), 573–579.

Arriagada, N. B., Horsley, J. A., Palmer, A. J., Morgan, G. G., Tham, R., & Johnston, F. H. (2019). Association between fire smoke fine particulate matter and asthma-related outcomes: Systematic review and meta-analysis. *Environmental Research*, *179*, 108777.

Atkinson, R., Kang, S., Anderson, H, Mills, I., & Walton, H. (2014). Epidemiological time series studies of PM2. 5 and daily mortality and hospital admissions: A systematic review and meta-analysis. *Thorax*, *69*(7), 660–665.

Bauer, S. E., Im, U., Mezuman, K., & Gao, C. Y. (2019). Desert dust, industrialization, and agricultural fires: Health impacts of outdoor air pollution in Africa. *Journal of Geophysical Research: Atmospheres*, *124*(7), 4104–4120.

Benmarhnia, T. (2020). Linkages between air pollution and the health burden from COVID-19: Methodological challenges and opportunities. *American Journal of Epidemiology*,*189*(11), 1238–1243.

Benmarhnia, T., Bailey, Z., Kaiser, D., Auger, N., King, N., & Kaufman, J. S. (2016). A difference-in-differences approach to assess the effect of a heat action plan on heat-related mortality, and differences in effectiveness according to sex, age, and socioeconomic status (Montreal, Quebec). *Environmental Health Perspectives*, *124*(11), 1694.

Benmarhnia, T., Hajat, A., & Kaufman, J. S. (2021). Inferential challenges when assessing racial/ethnic health disparities in environmental research. *Environmental Health*, *20*(1), 1–10.

Black, C., Tesfaigzi, Y., Bassein, J. A., & Miller, L. A. (2017). Wildfire smoke exposure and human health: Significant gaps in research for a growing public health issue. *Environmental Toxicology and Pharmacology*, *55*, 186–195.

Brey, S. J., Ruminski, M., Atwood, S. A., & Fischer, E. V. (2018). Connecting smoke plumes to sources using Hazard Mapping System (HMS) smoke and fire location data over North America. *Atmospheric Chemistry and Physics*, *18*(3), 1745–1761.

Brook, R. D., Rajagopalan, S., Pope, C. A,, et al. (2010). Particulate matter air pollution and cardiovascular disease an update to the scientific statement from the American Heart Association. *Circulation*, *121*(21), 2331–2378.

Bryant, R. A., Waters, E., Gibbs, L., et al. (2014). Psychological outcomes following the Victorian Black Saturday bushfires. *Australian & New Zealand Journal of Psychiatry*, *48*(7), 634–643.

Burke, M., Driscoll, A., Heft-Neal, S., Xue, J., Burney, J., & Wara, M. (2021). The changing risk and burden of wildfire in the United States. *Proceedings of the National Academy of Sciences*, *118*(2).

Butt, E. W., Conibear, L., Reddington, C. L., et al. (2020). Large air quality and human health impacts due to Amazon forest and vegetation fires. *Environmental Research Communications*, *2*(9), 095001.

Caamano-Isorna, F., Figueiras, A., Sastre,, I., Montes-Martínez, A., Taracido, M., & Piñeiro-Lamas, M. (2011). Respiratory and mental health effects of wildfires: An ecological study in Galician municipalities (north-west Spain). *Environmental Health*, *10*(1), 1–9.

Carratt, S. A., Flayer, C. H., Kossack, M. E., & Last, J. A. (2017). Pesticides, wildfire suppression chemicals, and California wildfires: A human health perspective. *Current Topics in Toxicology*, *13*, 1–12.

Chen, G., Guo, Y., Yue, X., et al. (2021). Mortality risk attributable to wildfire-related PM2.5 pollution: A global time series study in 749 locations. *The Lancet Planetary Health*,*5*(9), e579–e587.

Chen, H., Li, Q., Kaufman, J. S., et al. (2018). Effect of air quality alerts on human health: A regression discontinuity analysis in Toronto, Canada. *The Lancet Planetary Health*,*2*(1), e19–e26.

Chen, H., Samet, J. M., Bromberg, P. A., & Tong, H. (2021). Cardiovascular health impacts of wildfire smoke exposure. *Particle and Fibre Toxicology*,*18*(1),1–22.

Cleland, H. J., Proud, D., Spinks, A., & Wasiak, J. (2011). Multidisciplinary team response to a mass burn casualty event: Outcomes and implications. *Medical Journal of Australia*, *194*(11), 589–593.

Clelland, S. E., Serre, M. L., Rappold, A. G., & West, J. J. (2021). Estimating the acute health impacts of fire-originated PM2.5 exposure during the 2017 California Wildfires: Sensitivity to choices of inputs. *GeoHealth*, *5*(7), e2021GH000414.

Crabbe, H. (2012). Risk of respiratory and cardiovascular hospitalisation with exposure to bushfire particulates: New evidence from Darwin, Australia. *Environmental Geochemistry and Health*, *34*(6), 697–709.

Crippa, P., Castruccio, S., Archer-Nicholls, S., et al. (2016). Population exposure to hazardous air quality due to the 2015 fires in Equatorial Asia. *Scientific Reports*, *6*(1), 1–9.

Cui, Y., Zhang, Z.-F., Froines, J., et al. (2003). Air pollution and case fatality of SARS in the People's Republic of China: An ecologic study. *Environmental Health*, *2*(1), 1–5.

Cusworth, D. H., Mickley, L. J., Sulprizio, M. P., et al. (2018). Quantifying the influence of agricultural fires in northwest India on urban air pollution in Delhi, *India. Environmental Research Letters*, *13*(4), 044018.

DeFlorio-Barker, S., Crooks, J., Reyes, J., & Rappold, A. G. (2019). Cardiopulmonary effects of fine particulate matter exposure among older adults, during wildfire and non-wildfire periods, in the United States 2008–2010. *Environmental Health Perspectives*, *127*(3), 037006.

Dennekamp, M., Straney, L. D., Erbas, B., et al. (2015). Forest fire smoke exposures and out-of-hospital cardiac arrests in Melbourne, Australia: A case-crossover study. *Environmental Health Perspectives*, *123*(10), 959–964.

Dodd, W., Scott, P., Howard, C., et al. (2018). Lived experience of a record wildfire season in the Northwest Territories, *Canada. Canadian Journal of Public Health*, *109*(3), 327–337.

Doubleday, A., Schulte, J., Sheppard, L., et al. (2020). Mortality associated with wildfire smoke exposure in Washington state, 2006–2017: A case-crossover study. *Environmental Health*, *19*(1), 1–10.

Eisenman, D., McCaffrey, S., Donatello, I., & Marshal, G. (2015). An ecosystems and vulnerable populations perspective on solastalgia and psychological distress after a wildfire. *EcoHealth*, *12*(4), 602–610.

Fadadu, R. P., Balmes, J. R., & Holm, S. M. (2020). Differences in the estimation of wildfire-associated air pollution by satellite mapping of smoke plumes and ground-level monitoring. *International Journal of Environmental Research and Public Health*, *17*(21), 8164.

Fan, J., Li, S., Fan, C., Bai, Z., & Yang, K. (2016). The impact of PM2.5 on asthma emergency department visits: A systematic review and meta-analysis. *Environmental Science and Pollution Research*, *23*(1), 843–850.

Felix, E. D., & Afifi, W. (2015). The role of social support on mental health after multiple wildfire disasters. *Journal of Community Psychology*, *43*(2), 156–170.

Feng, S., Gao, D., Liao, F., Zhou, F., & Wang, X. (2016). The health effects of ambient PM2.5 and potential mechanisms. *Ecotoxicology and Environmental Safety*, *128*, 67–74.

Forastiere, F., Stafoggia, M., Tasco, C., et al. (2007). Socioeconomic status, particulate air pollution, and daily mortality: Differential exposure or differential susceptibility. *American Journal of Industrial Medicine*, *50*(3), 208–216.

Ford, B., Val Martin, M., Zelasky, S., et al. (2018). Future fire impacts on smoke concentrations, visibility, and health in the contiguous United States. *GeoHealth*, *2*(8), 229–247.

Fouillet, A., Rey, G., Laurent, F., et al. (2006). Excess mortality related to the August 2003 heat wave in France. *International Archives of Occupational and Environmental Health*, *80*(1), 16–24.

Franzi, L. M., Bratt, J. M., Williams, K. M., & Last, J. A. (2011). Why is particulate matter produced by wildfires toxic to lung macrophages? *Toxicology and Applied Pharmacology*, *257*(2), 182–188.

Gayle, A, V., Quint, J. K., & Fuertes, E. I. (2021). Understanding the relationships between environmental factors and exacerbations of COPD. *Expert Review of Respiratory Medicine*, *15*(1), 39–50.

Giglio, L., Randerson, J. T., & van der Werf, G. R. (2013). Analysis of daily, monthly, and annual burned area using the fourth-generation global fire emissions database (GFED4). *Journal of Geophysical Research*, *118*(1), 317–328.

Giglio, L., Randerson, J. T., & van der Werf, G. R. (2018). *Global fire emissions indicators, grids: 1997–2015*. Palisades, NY: NASA Socioeconomic Data and Applications Center (SEDAC).

Gill, A. M., & Stephens, S. L. (2009). Scientific and social challenges for the management of fire-prone wildland–urban interfaces. *Environmental Research Letters*, *4*(3), 034014.

Green, H., Bailey, J., Schwarz, L., Vanos, J., Ebi, K., & Benmarhnia, T. (2019). Impact of heat on mortality and morbidity in low and middle income countries: A review of the epidemiological evidence and considerations for future research. *Environmental Research*, *171*, 80–91.

Groot, E., Caturay, A., Khan, Y., & Copes, R. (2019). A systematic review of the health impacts of occupational exposure to wildland fires. *International Journal of Occupational Medicine and Anvironmental Health*, *32*(2), 121–140.

Gupta, P., Doraiswamy, P., & Levy, R., et al. (2018). Impact of California fires on local and regional air quality: The role of a low-cost sensor network and satellite observations. *GeoHealth*, *2*(6), 172–181.

Haikerwal, A., Akram, M., Del Monaco, A., et al. (2015), Impact of fine particulate matter (PM 2.5) exposure during wildfires on cardiovascular health outcomes. *Journal of the American Heart Association*, *4*(7), e001653.

Hajat, A., Hsia, C., & O'Neill, M. S. (2015). Socioeconomic disparities and air pollution exposure: A global review. *Current Environmental Health Reports*, *2*(4), 440–450.

Hanigan, I. C., Johnston, F. H., & Morgan, G. G. (2008). Vegetation fire smoke, indigenous status and cardio-respiratory hospital admissions in Darwin, Australia, 1996–2005: A time-series study. *Environmental Health*, *7*(1), 1–12.

Henderson, S. B. (2020). The COVID-19 pandemic and wildfire smoke: Potentially concomitant disasters. *American Journal of Public Health*, *110*(8), 1140–1142.

Henderson, S. B., & Johnston, F. H. (2012). Measures of forest fire smoke exposure and their associations with respiratory health outcomes. *Current Opinion in Allergy and Clinical Immunology*, *12*(3), 221–227.

Henderson, S. B., Brauer, M., MacNab, Y. C., & Kennedy, S. M. (2011). Three measures of forest fire smoke exposure and their associations with respiratory and cardiovascular health outcomes in a population-based cohort. *Environmental Health Perspectives*, *119*(9), 1266–1271.

Hiers, J. K., O'Brien J. J., Varner J. M., et al. (2020). Prescribed fire science: The case for a refined research agenda. *Fire Ecology*, *16*(1), 1–15.

Holm, S. M., Miller, M. D., & Balmes, J. R. (2021). Health effects of wildfire smoke in children and public health tools: A narrative review. *Journal of Exposure Science & Environmental Epidemiology*, *31*(1), 1–20.

Hutchinson, J. A., Vargo, J., Milet, M., et al. (2018). The San Diego 2007 wildfires and Medi-Cal emergency department presentations, inpatient hospitalizations, and outpatient visits: An observational study of smoke exposure periods and a bidirectional case-crossover analysis. *PLoS Medicine*, *15*(7), e1002601.

Ignotti, E., Valente, J. G., Longo, K. M., Freitas, S. R., Hacon, S., & Artaxo Netto, P. (2010). Impact on human health of particulate matter emitted from burnings in the Brazilian Amazon region. *Revista de Saude Publica*, *44*, 121–130.

Jaffe, D. A., O'Neill, S. M., Larkin, N. K., et al. (2020). Wildfire and prescribed burning impacts on air quality in the United States. *Journal of the Air & Waste Management Association*. *70*(6), 583–615.

Jayachandran, S. (2009). Air quality and early-life mortality evidence from Indonesia's wildfires. *Journal of Human Resources*, *44*(4), 916–954.

Johnson, A. L., Abramson, M. J., Dennekamp, M., Williamson, G. J, & Guo, Y. (2020). Particulate matter modelling techniques for epidemiological studies of open biomass fire smoke exposure: A review. *Air Quality, Atmosphere & Health*, *13*(1), 35–75.

Johnston, F. H., Henderson, S. B., Chen, Y., et al. (2012). Estimated global mortality attributable to smoke from landscape fires. *Environmental Health Perspectives*, *120*(5), 695–701.

Jolly, W. M., Cochrane, M. A., Freeborn, P. H., et al. (2015). Climate-induced variations in global wildfire danger from 1979 to 2013. *Nature Communications*, *6*(1), 1–11.

Kiely, L., Spracklen, D. V., Wiedinmyer, C., et al.(2020). Air quality and health impacts of vegetation and peat fires in Equatorial Asia during 2004–2015. *Environmental Research Letters*, *15*(9), 094054.

Kim, Y., Knowles, S., Manley, J., & Radoias, V. (2017). Long-run health consequences of air pollution: Evidence from Indonesia's forest fires of 1997. *Economics & Human Biology*, *26*, 186–198.

Kim, Y. H., Warren, S. H., Krantz, Q. T., et al.(2018). Mutagenicity and lung toxicity of smoldering vs. flaming emissions from various biomass fuels: Implications for health effects from wildland fires. *Environmental Health Perspectives*, *126*(1), 017011.

Kiser, D., Metcalf, W. J, Elhanan, G., et al. (2020). Particulate matter and emergency visits for asthma: A time-series study of their association in the presence and absence of wildfire smoke in Reno, Nevada, 2013–2018. *Environmental Health, 19*(1), 1–12.

Kobziar, L. N, & Thompson, G. R. (2020). Wildfire smoke, a potential infectious agent. *Science. 370*(6523), 1408–1410.

Kochi, I., Donovan, G. H., Champ, P. A., & Loomis, J. B. (2010). The economic cost of adverse health effects from wildfire-smoke exposure: A review. *International Journal of Wildland Fire, 19*(7), 803–817.

Kolbe, A., & Gilchrist, K. L. (2009). An extreme bushfire smoke pollution event: Health impacts and public health challenges. *New South Wales Public Health Bulletin, 20*(2), 19–23.

Kollanus, V., Prank, M., Gens, A., et al. (2017). Mortality due to vegetation fire-originated PM2.5 exposure in Europe: Assessment for the years 2005 and 2008. *Environmental Health Perspectives, 125*(1), 30–37.

Koplitz, S. N., Mickley, L. J., Marlier, M. E., et al. (2016). Public health impacts of the severe haze in Equatorial Asia in September–October 2015: Demonstration of a new framework for informing fire management strategies to reduce downwind smoke exposure. *Environmental Research Letters, 11*(9), 094023.

Landguth, E. L., Holden, Z. A., Graham, J., et al. (2020). The delayed effect of wildfire season particulate matter on subsequent influenza season in a mountain west region of the USA. *Environment international, 139*, 105668.

Lassman, W., Ford, B., Gan, R. W., et al. (2017). Spatial and temporal estimates of population exposure to wildfire smoke during the Washington state 2012 wildfire season using blended model, satellite, and in situ data. *GeoHealth, 1*(3), 106–121.

Leibel, S., Nguyen, M., Brick, W., et al.(2020). Increase in pediatric respiratory visits associated with Santa Ana wind–driven wildfire smoke and PM2.5 levels in San Diego County. *Annals of the American Thoracic Society, 17*(3), 313–320.

Lelieveld, J., Evans, J. S., Fnais, M., Giannadaki, D., & Pozzer, A. (2015). The contribution of outdoor air pollution sources to premature mortality on a global scale. *Nature. 525*(7569), 367–371.

Li, Y., Tong, D., Ngan, F., et al. (2020). Ensemble PM2.5 forecasting during the 2018 Camp Fire event using the HYSPLIT transport and dispersion model. *Journal of Geophysical Research: Atmospheres, 125*(15), e2020JD032768.

Liu, J. C., Mickley, L. J., Sulprizio, M. P., et al. (2016a). Future respiratory hospital admissions from wildfire smoke under climate change in the Western US. *Environmental Research Letters, 11*(12), 124018.

Liu, J. C., Mickley, L. J., Sulprizio, M. P., et al. (2016b). Particulate air pollution from wildfires in the Western US under climate change. *Climatic Change, 138*(3–4), 655–666.

Liu, J. C., Pereira, G., Uhl, S. A., Bravo, M. A., & Bell, M. L. (2015). A systematic review of the physical health impacts from non-occupational exposure to wildfire smoke. *Environmental Research, 136*, 120–132.

Liu, J. C., Wilson, A., Mickley, L. J., et al. (2017a). Who among the elderly is most vulnerable to exposure to and health risks of fine particulate matter from wildfire smoke? *American Journal of Epidemiology, 186*(6), 730–735.

Liu, J. C., Wilson, A., Mickley, L. J., et al. (2017b). Wildfire-specific fine particulate matter and risk of hospital admissions in urban and rural counties. *Epidemiology (Cambridge, Mass). 28*(1), 77.

Liu, T., Marlier, M. E., DeFries, R. S., et al. (2018). Seasonal impact of regional outdoor biomass burning on air pollution in three Indian cities: Delhi, Bengaluru, and Pune. *Atmospheric Environment. 172*, 83–92.

Long, J. W., Tarnay, L. W., & North, M. P. (2018). Aligning smoke management with ecological and public health goals. *Journal of Forestry, 116*(1), 76–86.

McCaffrey, S., Wilson, R., & Konar, A. (2018). Should I stay or should I go now? Or should I wait and see? Influences on wildfire evacuation decisions. *Risk Analysis, 38*(7), 1390–1404.

McClure, C. D., & Jaffe, D. A. (2018). U.S. particulate matter air quality improves except in wildfire-prone areas. *Proceedings of the National Academy of Sciences, 115*(31), 7901–7906.

McFarlane, B. L., McGee, T. K., & Faulkner, H. (2011). Complexity of homeowner wildfire risk mitigation: An integration of hazard theories. *International Journal of Wildland Fire, 20*(8), 921–931.

McNamara, M. L., Semmens, E. O., Gaskill, S., Palmer, C., Noonan, C. W., & Ward, T. J. (2012). Base camp personnel exposure to particulate matter during wildland fire suppression activities. *Journal of Occupational and Environmental Hygiene, 9*(3), 149–156.

Morgan, G., Sheppeard, V., Khalaj, B., et al. (2010). Effects of bushfire smoke on daily mortality and hospital admissions in Sydney, Australia. *Epidemiology*, 47–55.

Naeher, L. P., Brauer, M., Lipsett, M., et al. (2007). Woodsmoke health effects: A review. *Inhalation Toxicology, 19*(1), 67–106.

Navarro, K. M., Cisneros, R., Noth, E. M., Balmes, J. R., & Hammond, S. K. (2017). Occupational exposure to polycyclic aromatic hydrocarbon of wildland firefighters at prescribed and wildland fires. *Environmental Science & Technology, 51*(11), 6461–6469.

Navarro, K. M., Schweizer, D., Balmes, J. R., & Cisneros, R. (2018). A review of community smoke exposure from wildfire compared to prescribed fire in the United States. *Atmosphere, 9*(5), 185.

Nawaz, M., & Henze, D. (2020). Premature deaths in Brazil associated with long-term exposure to PM2. 5 from Amazon fires between 2016 and 2019. *GeoHealth, 4*(8), e2020GH000268.

O'Dell, K., Bilsback, K., Ford, B., et al. (2021). Estimated mortality and morbidity attributable to smoke plumes in the US: Not just a western US problem. *GeoHealth*, e2021GH000457.

O'Dell, K., Ford, B., Fischer, E. V., & Pierce, J. R. (2019). Contribution of wildland-fire smoke to US PM2.5 and its influence on recent trends. *Environmental Science & Technology, 53*(4), 1797–1804.

Penman, T. D., Collins, L., Syphard, A. D., Keeley, J. E., & Bradstock, R. A. (2014). Influence of fuels, weather and the built environment on the exposure of property to wildfire. *PLoS One, 9*(10), e111414.

Popovic, I., Magalhaes, R. J. S., Ge, E., et al. (2019). A systematic literature review and critical appraisal of epidemiological studies on outdoor air pollution and tuberculosis outcomes. *Environmental Research, 170*, 33–45.

Prunicki, M., Rodd Kelsey, J. L., Zhou, X., et al. (2019). The impact of prescribed fire versus wildfire on the immune and cardiovascular systems of children. *Allergy, 74*(10), 1989.

Raffuse, S. M., McCarthy, M. C., Craig, K. J., et al. (2013). High resolusion MODIS aerosol retrieval during wildfire events in California for use in exposure assessment. *Journal of Geophysical Research: Atmospheres, 118*(19), 11,242–211,255.

Rappold, A. G., Cascio, W. E., Kilaru, V. J., et al. (2012). Cardio-respiratory outcomes associated with exposure to wildfire smoke are modified by measures of community health. *Environmental Health, 11*(1), 1–9.

Rappold, A. G., Fann, N. L., Crooks, J., et al. (2014). Forecast-based interventions can reduce the health and economic burden of wildfires. *Environmental Science & Technology, 48*(18), 10571–10579.

Rappold, A. G., Reyes, J., Pouliot, G., Cascio, W. E., & Diaz-Sanchez, D. (2017). Community vulnerability to health impacts of wildland fire smoke exposure. *Environmental Science & Technology, 51*(12), 6674–6682.

Reid, C. E., Brauer, M., Johnston, F. H., Jerrett, M., Balmes, J. R., & Elliott, C. T. (2016). Critical review of health impacts of wildfire smoke exposure. *Environmental Health Perspectives, 124*(9), 1334–1343.

Reid, C. E., Considine, E. M., Watson, G. L., Telesca, D., Pfister, G. G., & Jerrett, M. (2019). Associations between respiratory health and ozone and fine particulate matter during a wildfire event. *Environment International, 129*, 291–298.

Reid, C. E., Jerrett, M., Tager, I. B., Petersen, M. L., Mann, J. K., & Balmes, J. R. (2016). Differential respiratory health effects from the 2008 northern California wildfires: A spatiotemporal approach. *Environmental Research, 150*, 227–235.

Roberts, G., & Wooster, M. (2021). Global impact of landscape fire emissions on surface level PM2.5 concentrations, air quality exposure and population mortality. *Atmospheric Environment, 118210.*

Sandberg, J., Santos-Burgoa, C., Roess, A., et al. (2019). All over the place?: Differences in and consistency of excess mortality estimates in Puerto Rico after Hurricane Maria. *Epidemiology, 30*(4), 549–552.

Schoennagel, T., Balch, J. K., Brenkert-Smith, H., et al. (2017). Adapt to more wildfire in western North American forests as climate changes. *Proceedings of the National Academy of Sciences, 114*(18), 4582–4590.

Schumann, R. L, III, Mockrin, M., Syphard, A. D., et al. (2020). Wildfire recovery as a "hot moment" for creating fire-adapted communities. *International Journal of Disaster Risk Reduction, 42*, 101354.

Schweizer, D., & Cisneros, R. (2017). Forest fire policy: Change conventional thinking of smoke management to prioritize long-term air quality and public health. *Air Quality, Atmosphere & Health, 10*(1), 33–36.

Schweizer, D., Preisler, H. K., & Cisneros, R. (2019). Assessing relative differences in smoke exposure from prescribed, managed, and full suppression wildland fire. *Air Quality, Atmosphere & Health, 12*(1), 87

Shaposhnikov, D., Revich, B., Bellander, T., et al. (2015). Long-term impact of moscow heat wave and wildfires on mortality. *Epidemiology, 26*(2), e21–e22.

Spracklen, D. V., Mickley, L. J., Logan, J. A., et al. (2009). Impacts of climate change from 2000 to 2050 on wildfire activity and carbonaceous aerosol concentrations in the western United States. *Journal of Geophysical Research: Atmospheres, 114*(D20).

Strader, S. M. (2018). Spatiotemporal changes in conterminous US wildfire exposure from 1940 to 2010. *Natural hazards, 92*(1), 543–565.

Torres, O., Jethva, H., Ahn, C., Jaross, G., & Loyola, D. G. (2020). TROPOMI aerosol products: Evaluation and observations of synoptic-scale carbonaceous aerosol plumes during 2018–2020. *Atmospheric Measurement Techniques, 13*(12), 6789–6806.

Vaidyanathan, A., Yip, F., & Garbe, P. (2018). Developing an online tool for identifying at-risk populations to wildfire smoke hazards. *Science of the Total Environment, 619*, 376–383.

Versura, P., Profazio, V., Cellini, M., Torreggiani, A., Caramazza, R. (1999). Eye discomfort and air pollution. *Ophthalmologica, 213*(2), 103–109.

Villeneuve, P. J., & Goldberg, M. S. (2020). Methodological considerations for epidemiological studies of air pollution and the SARS and COVID-19 coronavirus outbreaks. *Environmental Health Perspectives, 128*(9), 095001.

Vora, C., Renvall, M. J., Chao, P., Ferguson, P., & Ramsdell, J. W. (2011). 2007 San Diego wildfires and asthmatics. *Journal of Asthma, 48*(1), 75–78.

Voulgarakis, A., & Field, R. D. (2015). Fire influences on atmospheric composition, air quality and climate. *Current Pollution Reports, 1*(2), 70–81.

Watson, G. L., Telesca, D., Reid, C. E., Pfister, G. G., Jerrett, M. (2019). Machine learning models accurately predict ozone exposure during wildfire events. *Environmental Pollution, 254*, 112792.

Wegesser, T. C., Franzi, L. M., Mitloehner, F. M., Eiguren-Fernandez, A., & Last, J. A. (2010). Lung antioxidant and cytokine responses to coarse and fine particulate matter from the great California wildfires of 2008. *Inhalation Toxicology, 22*(7), 561–570.

Wegesser, T. C., Pinkerton, K. E., & Last, J. A. (2009). California wildfires of 2008: Coarse and fine particulate matter toxicity. *Environmental Health Perspectives, 117*(6), 893–897.

Weuve, J., Bennett, E. E., Ranker, L., et al. (2021). Exposure to air pollution in relation to risk of dementia and related outcomes: An updated systematic review of the epidemiological literature. *Environmental Health Perspectives, 129*(9), 096001.

Wilkins, J. L., Pouliot, G., Foley, K., Appel, W., & Pierce, T. (2018). The impact of US wildland fires on ozone and particulate matter: A comparison of measurements and CMAQ model predictions from 2008 to 2012. *International Journal of Wildland Fire, 27*(10), 684–698.

Williamson, G. J., & Lucani, C. (2020). AQVx: An interactive visual display system for air pollution and public health. *Frontiers in Public Health*, *8*, 85.

Williamson, G. J., Bowman, D. M. S., Price, O. F., Henderson, S., & Johnston F. (2016). A transdisciplinary approach to understanding the health effects of wildfire and prescribed fire smoke regimes. *Environmental Research Letters*, *11*(12), 125009.

Xi, Y., Kshirsagar, A. V., Wade, T. J., et al. (2020). Mortality in US hemodialysis patients following exposure to Wildfire smoke. *Journal of the American Society of Nephrology*, *31*(8), 1824–1835.

Xing, Y-F., Xu, Y-H., Shi, M-H., & Lian, Y-X. (2016). The impact of PM2.5 on the human respiratory system. *Journal of Thoracic Disease*, *8*(1), E69.

Xu, R., Yu, P., Abramson, M. J., et al. (2020). Wildfires, global climate change, and human health. *New England Journal of Medicine*, *383*(22), 2173–2181.

Yang, B.-Y., Fan, S., Thiering, E., et al. (2020). Ambient air pollution and diabetes: A systematic review and meta-analysis. *Environmental Research*, *180*, 108817.

Yao, J., Brauer, M., & Henderson, S. B.(2013). Evaluation of a wildfire smoke forecasting system as a tool for public health protection. *Environmental Health Perspectives*, *121*(10), 1142.

Youssouf, H., Liousse, C., Roblou, L., et al. (2014). Non-accidental health impacts of wildfire smoke. *International Journal of Environmental Research and Public Health*, *11*(11), 11772–11804.

Zelikoff, J. T., Chen, L. C., Cohen, M. D., et al. (2003). Effects of inhaled ambient particulate matter on pulmonary antimicrobial immune defense. *Inhalation Toxicology*, *15*(2), 131–150.

Zhou, X., Josey, K., Kamareddine, L., et al. (2021). Excess of COVID-19 cases and deaths due to fine particulate matter exposure during the 2020 wildfires in the United States. *Science Advances*, *7*(33), eabi8789.

13

Health Effects of Wildfire Smoke During Pregnancy and Childhood

Amy M. Padula[1] and Camille Raynes-Greenow[2]

ABSTRACT

Pregnant people and children are more susceptible to wildfire smoke owing to physiological changes and the rapid growth and development that occur during the critical periods of pregnancy and childhood. The aim of this chapter is to provide an overview of the epidemiological evidence that assesses the associations between wildfire smoke and health outcomes in the perinatal period and childhood. The primary pollutant assessed in this body of research is particulate matter less than 2.5 μm in aerodynamic diameter ($PM_{2.5}$) from wildfire sources. First, we focus on pregnancy and discuss the unique characteristics of the perinatal period and present a summary of studies of wildfire exposure in pregnancy and effects on maternal and newborn outcomes such as low birth weight, preterm birth, and mental health outcomes. We discuss the various methods used to access exposure and ascertain the outcomes and covariate adjustment in epidemiologic analyses. Second, we focus on exposure to wildfire smoke during childhood and review what is known on the effects of wildfire smoke on child health outcomes including health-service use, respiratory symptoms, and asthma. Our chapter concludes by highlighting the gaps in the literature and directions for future research and policy. This area of study on wildfire exposure and health effects during pregnancy and childhood is in a nascent stage and our knowledge in this is expected to change drastically over the next several years.

13.1. INTRODUCTION

As described in Chapter 12, wildfires pose enormous threats to human health (Reid et al., 2016; Black et al., 2017b) and some populations are more sensitive to health impacts from wildfires. Pregnancy and childhood are periods of increased vulnerability to wildfire smoke owing to physiological changes in the mother during pregnancy and the rapid growth and development that occurs in the fetus during gestation, during infancy, and throughout childhood. The scant literature on wildfire smoke exposure during pregnancy and childhood on health outcomes is bolstered by more abundant research on ambient air pollution.

The aim of this chapter is to provide an overview of the impact of wildfire smoke on perinatal and childhood health outcomes. Exposure assessment of wildfire smoke exposure in epidemiological studies was primarily based on exposure to a single fire event or fine particulate matter. Particulate matter, especially particulate matter less than 2.5 μm in aerodynamic diameter ($PM_{2.5}$) is a component of wildfire smoke, and is frequently measured as part of routine ambient air pollution monitoring and hence is a useful and objective measure in epidemiological studies to assess the effect of exposure. Therefore,

[1] *Department of Obstetrics, Gynecology and Reproductive Sciences, Program for Reproductive Health and the Environment, University of California San Francisco, San Francisco, California, USA*

[2] *Sydney School of Public Health, The University of Sydney, Sydney, Australia*

Landscape Fire, Smoke, and Health: Linking Biomass Burning Emissions to Human Well-Being, Geophysical Monograph 280, First Edition.
Edited by Tatiana V. Loboda, Nancy H. F. French, and Robin C. Puett.

DOI: 10.1002/9781119757030.ch13

much of the evidence that we discuss is based on exposure to $PM_{2.5}$ during periods of wildfire events, and the findings that we present are borrowed from the more extensive literature of ambient air pollution (including nonwildfire sources) and tobacco smoke to help inform our understanding of the potential perinatal and childhood health effects of wildfire smoke. More recently, studies have included more wildfire-specific metrics including smoke days from the Fire and Smoke product of the National Oceanic and Atmospheric Administration's Hazard Mapping System or wildfire-specific $PM_{2.5}$ from various modeling techniques but so far evidence is limited.

In this chapter, we first focus on pregnancy and the unique characteristics of the prenatal period. We present a summary of studies of wildfire exposure in pregnancy and effects on maternal and newborn outcomes such as low birth weight, preterm birth, birth defects, infant mortality, and mental health outcomes of the mother. We discuss various methods used to assess exposure, ascertain outcomes, and adjust for covariates in epidemiological analyses. We discuss methodological issues such as critical periods during pregnancy, seasonality, and coexposures of stress during wildfire events.

Second, we focus on childhood as another critical period of exposure to wildfires. We review what is known on the effects of wildfire smoke on child health outcomes including health-service use, respiratory symptoms, and asthma. Our chapter concludes by highlighting the gaps in the literature and future directions for research and policy. This area of study on wildfire exposure and health effects during pregnancy and childhood is in a nascent stage and our knowledge in this is expected to change drastically over the next several years.

13.2. PRENATAL EXPOSURE TO WILDFIRE SMOKE AND PERINATAL OUTCOMES

13.2.1. Pregnancy as a Period Vulnerable to Air Pollution

Fetal Growth and Development

Gestation is a period of rapid fetal growth and development, and the placenta serves multiple critical functions and acts as the primary metabolic barrier to environmental chemicals. Air pollutants, including $PM_{2.5}$, a well-studied and primary pollutant in wildfire smoke, have the potential to cross the placental barrier (Malley et al., 2017). Altered or impaired placental function is significantly linked with common pregnancy complications, lower birth weight, developmental delay, and diseases that manifest later in life such as hypertension, coronary heart disease, and type 2 diabetes (Maltepe & Fisher, 2015). Recent evidence suggests black carbon, a component of $PM_{2.5}$ formed from incomplete combustion, accumulates on the fetal side of the placenta (Bove et al., 2019).

Two of the most studied perinatal outcomes, fetal growth and preterm birth, are key indicators of public health. Fetal growth, as measured by birth weight for gestational age, has important influences on human health. Low birth weight (birth weight below 2,500 g) is associated with increased rates of cardiovascular and cerebrovascular diseases in adulthood (Barker, 2004) and more immediate adverse outcomes including neonatal mortality and childhood morbidity (Saigal & Doyle, 2008; Zhang et al., 2014). Preterm birth is an important contributor of perinatal morbidity and mortality, with potential lifelong health implications for being preterm (Diamanti-Kandarakis et al., 2009; President's Cancer Panel, 2010; Welshons et al., 2006; American College of Obstetricians and Gynecologists Committee on Health Care for Underserved Women et al., 2013) including pulmonary and neurodevelopmental outcomes (Behrman & Butler, 2007). In 2010, 2.7–3.4 million preterm births globally were estimated to be associated with $PM_{2.5}$ exposure (Malley et al., 2017). In California between 2007–2012, an estimated 6,974 preterm births were attributable to wildfire smoke (Heft-Neal et al., 2022). Annual percentages of preterm births attributable to wildfire smoke ranged from 1.8% in 2011 to 6.3% in 2008, a high wildfire year (Heft-Neal et al., 2022).

While research specific to wildfire smoke exposures is limited, numerous studies across different years and countries have found consistent associations between ambient $PM_{2.5}$ and measures of fetal growth (i.e., birth weight, low birth weight, and small for gestational age) (Bekkar et al., 2020; Klepac et al., 2018; Shah et al., 2011; Yuan et al., 2020; Stieb et al., 2012). Similarly, higher ambient air pollution levels have been consistently associated with a ~10% increase in risk of preterm birth (Li et al., 2017; Klepac et al., 2018). Furthermore, epidemiological studies demonstrate that exposures to high levels of $PM_{2.5}$ in the second trimester and the end of pregnancy are most critical with regard to risk of preterm birth (Padula et al., 2014; Chang et al., 2015) and associations are stronger for earlier (i.e., shorter gestation) preterm births (Padula et al., 2014).

Ambient air pollution exposures during pregnancy are associated with perinatal mortality (Li et al., 2021a) and sudden infant death syndrome (Chen et al., 2021). Higher levels of air pollution are associated with increased risk of several structural birth defects including neural tube defects (Padula et al., 2013a), heart defects (Padula et al., 2013b; Yang et al., 2021), and other defects (Padula et al., 2013c). Additionally, $PM_{2.5}$ exposure for mothers is associated with shorter telomere length in their offspring (Martens et al., 2017).

Cigarette smoke contains many of the same toxic compounds as those found in wildfire smoke including polycyclic aromatic hydrocarbons, carbon monoxide, and free radicals (Zeglinski et al., 2019). Cigarette smoking during pregnancy is associated with a range of offspring outcomes including IQ (Corrêa et al., 2021) and academic achievement in school (Kristjansson et al., 2018). Prenatal exposure to smoke from traditional cookstoves using solid fuel is linked to increased adverse birth outcomes (Amegah et al., 2014), decreased childhood pulmonary function, and increased childhood risk of pneumonia and hospitalization (Lee et al., 2019).

Socioeconomically disadvantaged populations often face heightened exposures to structural and individual-level stressors and bear a disproportionate burden of environmental exposures including air pollution. Research demonstrates stronger associations between air pollution and preterm birth in neighborhoods with lower socioeconomic status (Padula et al., 2014; Gray et al., 2014; Mekonnen et al., 2021). Those with low socioeconomic status may have greater exposure to air pollution in general and wildfire smoke in particular due to higher rates of outdoor work, occupancy in poor-quality housing, linguistic barriers to access information, limited ability to evacuate, and structural barriers to healthcare access (Kizer, 2020). This double jeopardy of exposure to pollutants and poverty may contribute to disparities in adverse pregnancy outcomes.

Maternal Health

Pregnancy is also a period of physiological changes in the mother (Varshavsky et al., 2020) that result in vulnerability to air pollution including wildfire smoke. For example, during pregnancy respiratory rate and cardiac output are increased on average by 40% and 50%, respectively (Hegewald & Crapo, 2011; Bobrowski, 2010; Tan & Tan, 2013; Koman et al., 2018). Additionally, changes in the immune system are subject to disruption by air pollution (Buxton et al., 2019). Smaller particles such as $PM_{2.5}$ can travel, penetrate deeper into the lung, and enter the bloodstream more easily than larger sized particulate matter, and higher concentrations of these small particles due to wildfire smoke result in more inhaled particles (Makkonen et al., 2010; Nance, 1993; U.S. EPA, 2009). This problem may be more dangerous as fires are occurring more often at the wildland-urban interface leading to structure burning, and, in turn, more toxic exposures. The toxicity of these wildland-urban interface fires are of increasing concern with respect to the chemical composition and subsequent health effects.

Prenatal exposures to ambient air pollution including $PM_{2.5}$ are associated with maternal outcomes including hypertensive disorders in pregnancy (i.e., gestational hypertension, preeclampsia, and eclampsia) and gestational diabetes (Pedersen et al., 2014; Weber et al., 2019; Padula et al., 2018; Padula et al., 2020). These conditions may be on the causal pathway between air pollution and adverse perinatal outcomes. Maternal conditions are not only critical to the fetus's health, but also heighten maternal susceptibility to future metabolic and cardiovascular disease (Kampmann et al., 2015; Holdsworth-Carson et al., 2010). Exposure to wildfire $PM_{2.5}$ may increase risk of pregnancy complications and adverse perinatal outcomes, particularly for vulnerable populations, which in turn could amplify existing reproductive and perinatal health disparities.

Potential biological pathways mediating the link between air pollution and adverse perinatal outcomes include oxidative stress (Moller et al., 2014), systemic inflammation (Panasevich et al., 2009), endothelial dysfunction (Wauters et al., 2013), and DNA damage (Risom et al., 2005; Gaskins et al., 2019), but less is known about the biological effects specifically of wildfire smoke. Maternal inflammatory responses are modified during pregnancy to establish and maintain a viable fetus and elevated inflammatory signals are linked with parturition (Palm et al., 2013). While the mechanisms remain unresolved, imbalances in antiinflammatory and proinflammatory mediators are associated with common pregnancy complications, such as preterm birth and fetal growth restriction. Additionally, PM_{10} has been associated with the proinflammatory cytokines in pregnancy (Buxton et al., 2019).

13.2.2. Methodological Considerations of Wildfire Smoke Exposure in Pregnancy

Quantification of adverse perinatal health effects attributable to wildfire smoke presents additional methodologic challenges including identification of critical windows of exposure. Studies must quantify drastic increases in air pollution from a wildfire event, which can range in duration from days to weeks at varying levels during pregnancy. Critical periods of exposure depend on the outcome of interest and may not be consistent for maternal morbidity, fetal loss, birth defects, fetal growth, gestational length, and infant mortality.

Wildfires occur seasonally, though the length of the season has grown in some regions of the world (e.g., western United States and southeastern Australia). Seasonal patterns of birth outcomes also occur, which could be a result of seasonal fluctuations in air pollution, or other factors such as demographics in the population and seasonal birth patterns (Currie & Schwandt, 2013). Wildfire events compound existing air pollution exposures from the usual sources, such as traffic. Owing to sources, meteorology and topography, ambient air pollution varies by season as well. Accounting for these existing sources of air pollution while evaluating additional wildfire impact is important to qualify the effects of wildfire smoke specifically.

Wildfires are natural experiments with combined environmental and psychosocial stressors. Families living near wildfires may face threatened or actual evacuation, which may increase psychosocial stress, previously linked to gestational hypertension and adverse birth outcomes (Muglia & Katz, 2010). The combined and heightened exposure to psychosocial stressors and environmental contaminants faced by pregnant people can affect their fetuses, with potentially irreversible adverse impacts on health and development. One limitation of wildfire research is a difficulty differentiating between the air pollution exposure and the psychosocial stressors to tease apart their individual contributions to adverse outcomes (Padula et al., 2020).

13.2.3. Epidemiological Evidence of Perinatal Health Effects of Wildfire Smoke

Numerous studies investigate the impact of ambient air pollution during pregnancy on adverse perinatal outcomes; however, fewer have examined the contribution of wildfire-specific air pollution exposure, especially beyond single wildfire events. Much remains unknown about potential risk of extremely high (and potentially more toxic) levels of exposure, critical windows of vulnerability, and longer-term effects of acute and prolonged exposure to wildfire smoke.

We summarize the studies that examine the effects of wildfire smoke exposure during pregnancy in relation to adverse birth outcomes including low birth weight, preterm birth, birth defects, and infant mortality as well as mental health outcomes (O'Donnell & Behie, 2015; Prass et al., 2012; O'Donnell & Behie, 2013; Jayachandran, 2009; Heft-Neal et al., 2022; Park et al., 2021; Requia et al., 2021a; Requia et al., 2021b). These studies have been conducted all over the world from wildfires in the western United States (Holstius et al., 2012; Abdo et al., 2019; Heft-Neal et al., 2022), Canada (Verstraeten et al., 2020), and Indonesia (Jayachandran, 2009); forest fires in South America (Prass et al., 2012; Requia et al., 2021a; Requia et al., 2021b); and bushfires in Australia (O'Donnell & Behie, 2015; O'Donnell & Behie, 2013). Additional evidence of the perinatal effects from landscape fires comes from low-income and middle-income countries (Li et al., 2021b). These studies employ a variety of epidemiological study designs including retrospective cohorts (Abdo et al., 2019; O'Donnell & Behie, 2015; O'Donnell & Behie, 2013; Heft-Neal et al., 2022), case control (Li et al., 2021b), case crossover (Requia et al., 2021b), time series (Holstius et al., 2012), and ecological study (Prass et al., 2012; Jayachandran, 2009; Requia et al., 2021a).

Low Birth Weight and Preterm Birth

One of the first studies to examine wildfire smoke in relation to birth weight during the 2003 Southern California fires, which included 886,034 mothers, compared those pregnant during the fires with those pregnant either before or after the fires (Holstius et al., 2012). It found that births where mothers were exposed to the wildfire had, on average, 9.7 g lower birth weights when exposed during the second trimester. Models adjusted for characteristics from the birth certificate including maternal age, race/ethnicity, education, parity, infant sex, and gestational age. Sensitivity analyses accounted for season, additional PM_{10} measurements, and length of peak period. Small decreases were observed in the first and third trimesters (3.3 g and 7.0 g, respectively) (Holstius et al., 2012).

An additional study examined prenatal exposure to wildfires in Colorado between 2007 and 2015 and risk of preterm birth and effects on birth weight in 535,895 pregnancies (Abdo et al., 2019). They examined $PM_{2.5}$ concentrations from the wildfire smoke at the ZIP code level during each trimester and found exposure in the second trimester was associated with 13% increased risk of preterm birth and exposure in the first trimester was associated with a decrease in birth weight of 5.7 g (Abdo et al., 2019). They adjusted for maternal factors (age, income, education, race/ethnicity, smoking, drinking, asthma, prenatal care) as well as temperature deviation, month, year, and other air pollution exposures including nonwildfire $PM_{2.5}$, PM_{10}, and ozone.

The largest study to date examined ~3 million births in California between 2006 and 2012 to estimate risk of preterm birth in relation to ZIP code level estimates of exposure based on satellite-based wildfire smoke plume boundaries and gridded estimates of surface $PM_{2.5}$ concentrations (Heft-Neal et al., 2022). This study found that each day of exposure to any wildfire smoke was associated with a 0.49% increase in risk of preterm birth. Estimates by trimester suggest stronger associations with exposure later in pregnancy and estimates by smoke intensity indicate that observed associations were driven by higher intensity smoke days.

Several studies conducted in the Brazilian Amazon examined prenatal exposure to deliberate but uncontrolled forest fires and other biomass burning. One study included a retrospective cohort study of 6,147 full-term, singleton births between 2004 and 2005 during periods of burning assessed with Brazilian Biomass Burning Emission Model based on hotspots identified by remote sensing (Candido da Silva et al., 2014). They found risk of term low birth weight ($< 2,500$ g with gestational age ≥ 37 weeks) was 50% higher when women were exposed to the highest quartile of either $PM_{2.5}$ in the second or third trimester, or CO in the second trimester. They adjusted for newborn's sex, mother's age and education, and prenatal care (Candido da Silva et al., 2014). An ecologic study used heat spots from satellite images as they varied by month and year between 2001 and 2005 in relation to birth weight among boys and girls (Prass et al., 2012). This study did not find conclusive evidence and was limited by the inability to adjust for confounding and imprecise exposure measurement. More recently, a study examined associations between wildfire-related $PM_{2.5}$ during trimesters of pregnancy and preterm birth using a case-crossover design (Requia et al., 2021b). They found increased odds of preterm birth in the southeast regions when wildfire occurred during the first trimester (odds ratio, OR = 1.41) and increased odds of preterm birth in the north region when wildfire exposure occurred during the second trimesters (OR = 1.05) (Requia et al., 2021b).

A few studies come from Australia, an area of regular wildfire occurrence. During the 2003 Canberra bushfires, one study found that male infants born in the most severely fire-affected area had significantly higher average birth weights than their less exposed peers and were also heavier than males born in the same areas in nonfire years (O'Donnell & Behie, 2015). Higher average weights were attributable to an increase in the number of macrosomic infants (i.e., > 4,000 g) (and potentially attributable to the stress of the wildfires in addition to the air pollution). There was no significant effect on the weight of female infants or on gestational age for either sex (O'Donnell & Behie, 2015). An additional study examined the bushfires in Victoria, Australia, in 2009 and found small but significant increases in preterm birth and decreases in birth weight for those exposed to the fires in their second or third trimester. They assigned exposure based on maternal residence in a fire-affected area, as defined by small administrative areas (Local Government Area), compared with women who resided in unaffected areas (O'Donnell & Behie, 2013).

One study examined fire-sourced $PM_{2.5}$ in relation to birth weight in 54 low-income and middle-income countries using demographic and health survey data. It used a sibling-matched case control study of 227,948 newborns between 2000 to 2014 (Li et al., 2021b), and mothers with two births during the study period were included. A 1 $\mu g/m^3$ increase in the concentration of fire sourced $PM_{2.5}$ was associated with a 2.17 g reduction in birth weight, a 2.80% increase in low birth weight risk, and an 11.68% increase in very low birth weight risk (Li et al., 2021b).

Congenital Anomalies

A study in Brazil examined wildfire smoke exposure in relation to congenital anomalies between 2001–2018 (Requia et al., 2021a). The number of wildfires within the boundaries of a mother's home municipality during each trimester was compared in relation to 12 categories of birth defects in a population of 16,825,497 births (7,595 cases). They found associations between exposure and cleft lip/cleft palate (OR: 1.007) during the second trimester of exposure, congenital anomalies of the respiratory system (OR: 1.013) in the second trimester of exposure, and congenital anomalies of the nervous system (OR: 1.002) during the first trimester of exposure for the regions south, north, and midwest, respectively (Requia et al., 2021a).

In California, wildfire smoke during the first trimester was associated with gastroschisis, an abdominal wall defect, when compared with no wildfire smoke (7.8 compared with 5.7 per 10,000 births (relative risk, RR: 1.28; Park et al., 2021). Although the exposure assessment and case ascertainment in these studies are likely to be subject to misclassification, these results prompt a need for further studies to confirm (Padula & Benmarhnia, 2022).

Infant Mortality and Pregnancy Loss

One study examined pollution from land-clearing fires that got out of control in 1997 in Indonesia and population-level infant mortality (Jayachandran, 2009). It estimated smoke exposure during prenatal and postnatal periods by month and district and gauged impact on fetal, infant and childhood mortality based on the 2000 census count. The study found that fire-related increases in air pollution were associated with a 1.2% decrease in cohort size. The under-age-three mortality during this period was 6%, approximately 20% higher than during other times. They found in utero exposure to have the largest impact on survival, particularly exposure in the third trimester, which was associated with the smaller cohort size (Jayachandran, 2009).

A case-control study of rhesus macaques during the Northern Californian Camp Fire in November 2018, which affected the peak breeding season at the California National Primate Research Center, provides evidence of pregnancy loss (Willson et al., 2021). All females in the study were maintained in outdoor fields and exposed to high levels of $PM_{2.5}$ during the wildfires. Archival records of approximately 5,000 conceptions from the previous 9 yr served as a comparison group. They observed increased pregnancy loss during the breeding season, especially for those whose exposure occurred during the first 60 days of gestation (Willson et al., 2021). Although early pregnancy loss is difficult to capture in epidemiologic studies, this study in macaques provides evidence of early gestational exposure and the potential for increased risk of pregnancy loss.

Mental Health Outcomes

The consequences of wildfires are beyond the smoke exposure, and include the psychological impact of the fire threat and trauma. One Canadian study of 200 women who experienced the Fort McMurray Wood Buffalo wildfire during or shortly before pregnancy examined distress. Greater peritraumatic distress and dissociative experiences correlated with more-severe posttraumatic-stress-syndrome-like symptoms. Greater social support satisfaction was associated with less-severe posttraumatic stress symptoms but only when peritraumatic distress was below average; at more severe levels of peritraumatic distress inventory, this psychosocial variable was not protective (Verstraeten et al., 2020).

13.2.4. Summary of Pregnancy Studies

The most conclusive findings from the studies summarized here so far were between wildfire smoke during pregnancy and both lower birth weight and increased risk of preterm birth. Additional infant outcomes, such as birth defects and infant mortality, were explored but are less definitive and subject to more extensive methodologic issues including exposure misclassification, exposure measurement error, critical periods of exposure, unmeasured confounding, and outcome ascertainment. The effect of stress due to the wildfire event has been associated with maternal mental health and may be an additional pathway by which wildfires may affect adverse outcomes, particularly preterm birth, as stress can result in the premature onset of labor. As more research is under way to help inform policy to protect those pregnant during wildfire events, precautions should be taken to reduce exposure during this vulnerable period for both the mother and child.

13.3. EXPOSURE TO WILDFIRE SMOKE AND CHILD HEALTH OUTCOMES

The epidemiological evidence from studies investigating wildfire smoke exposure on child health is heterogeneous and limited. Most research has been conducted in high-income countries (Australia, Europe, and North America), assesses, most commonly, acute outcomes close to the time of exposure (hospital presentations or admissions, or respiratory symptoms), and uses a mix of methods to assess outcome and exposure assessment that make estimation of summary effects difficult. There have been some studies conducted to examine longer-term outcomes; however, these mostly focus on wildfire as a traumatic event and the impact on mental health outcomes and are not necessarily wildfire smoke exposure investigations. As a body of evidence, there are many gaps in our understanding of the impact of wildfire smoke exposure and child health. Nonetheless most literature acknowledges that childhood is a recognized window of susceptibility, and that children have particular vulnerabilities to wildfire smoke due to their immature immune systems, developmental stages, and behavior (such as outside play), and that wildfire events are increasing and therefore more children will be impacted (Liu et al., 2016; Holm et al., 2021).

13.3.1. Vulnerability During Childhood

Early childhood is a vulnerable period of growth and development, and children are particularly sensitive to environmental exposures (Sheffield & Landrigan, 2011). Young children's smaller airways, smaller lung size, higher respiratory rates, increased metabolic rates, and underdeveloped immune systems compared with adults increases their exposure and vulnerability to respirable pollution (Leibel et al., 2020). Young children are also less likely to adopt risk mitigation behaviors, have less access to appropriately sized masks, and may spend a large proportion of the time during an exposure period where the air is insufficiently ventilated (in school), or outdoors (Sheffield & Landrigan, 2011; Holm et al., 2021), and for these reasons are at greater risk of exposure and more vulnerable to the risks.

Particulate matter, one of the most frequently measured solid particles found in air, are common in both ambient air and wildfire smoke. Levels of particulate matter and adverse health outcomes are linearly related, and the risk for adverse health increases as concentrations of particulate matter increase regardless of the source (Gan et al., 2017). Although the evidence from wildfire smoke is increasing (Reid et al., 2016), much of the earlier epidemiological evidence comes from studies of ambient air pollution (Stieb et al., 2012). Although these studies are useful for hypothesis development of the effect of wildfire smoke exposure, there are some important differences between these exposures. For example, the chemical composition of wildfire smoke is similar but not the same as other causes of air pollution, including traffic, agricultural burn-off, and household air pollution (Künzli et al., 2006). Within these, the composition may vary based on type of biomass burned, whether structures were also burned, and duration of the burning event. Wildfire smoke is more likely to have a shorter duration, whereas exposure to traffic pollution is more constant. Examining short-duration, unpredictable exposure on acute health outcomes is also challenging, as is examining the longer-term impacts into adolescence and adulthood. The impact on adult health after short-term exposure in childhood is also unknown (Shao et al., 2019). Animal models of early life exposure and longer-term morbidity are suggestive of immune dysregulation and changes to lung function that extend at least into adolescence (Black et al., 2017a). Currently, most of the epidemiological research investigating longer-term effects has investigated mental health, or psychological outcomes due to exposure to trauma (of wildfire threat), rather than other morbidity, although all health outcomes are important areas for research.

13.3.2. Epidemiological Evidence of Childhood Health Effects of Wildfire Smoke

Health Service Use

Health service use is a frequently assessed outcome for epidemiological studies of wildfire exposure on children. Health service use is defined mostly through hospital, or emergency department, presentations or admissions, and/or doctor or clinic visits in children during or related to a wildfire event or smoke exposure. These episodes of care are mostly related to respiratory causes, and health service use in general and specifically related to respiratory symptoms, which have been reported to increase during periods of wildfire smoke exposure and this evidence comes from a range of studies (Henry et al., 2021). A prospective cohort study of > 6,000 school-aged children, mostly aged 6–7 yr, collected annual respiratory health data and included wildfire exposure following the 2003 Californian wildfires (Künzli et al., 2006). All respiratory symptoms and doctor visits were increased in children affected by smoke by up to 33% (Künzli et al., 2006). There were differences by asthma diagnosis and children's age (e.g., high school children reported more eye symptoms than primary children, 44.0% versus 35.5%). Children with asthma reported more respiratory symptoms, but this was modified by adoption of preventative behaviors (mask wearing, staying indoors, and using air conditioners), and the number of reported days of fire smell indoors. In children who were the most exposed (> 6 days of smoke, and without masks), sneezing more than doubled (OR = 2.8; 95% CI: 2.3, 3.5). The results from this study are generalizable to a range of similar settings as it was a population-based study. A recent ecological study also from California, reported increases in pediatric (< 19 yr old) emergency hospital presentations by 30% when the $PM_{2.5}$ attributable to wildfires increased by 10 μg/m^3 (Aguilera et al., 2021). During the exposure period, younger children, those between 0 and 5 yr were the most likely to present to hospital (34.5% increase) (Aguilera et al., 2021). Respiratory presentations of cough increased 21.8% with a 10 μg/m^3 increase in $PM_{2.5}$ attributable to wildfire smoke (Aguilera et al., 2021). A study using medical insurance claims for episodes of care related to respiratory or cardiac outcomes for people aged 0 to 64 yr during a 6 month window that included the wildfire season in San Diego reported that children's attendance at outpatient clinics for respiratory conditions was greater during peak fire-exposure periods. Overall, children aged 0 to 4 yr accounted for 14.4% of all inpatient hospitalizations, 15.1% of emergency presentations, and 28.8% of outpatient visits. Children younger than 2 yr had higher risk of presenting, although the numbers of children in this age group were very small and varied by the specific respiratory condition for presentation (Hutchinson et al., 2018). A larger interrupted time series study, also from San Diego, used hospital data for all children (< 19 yr old) to quantify excess daily hospital visits for respiratory symptoms during a 10 day fire exposure period. Daily emergency department visits increased by 16.03 daily visits per 10,000 children, during the exposure window. Younger children aged 0 to 5 yr had the highest absolute daily visits of 7.3 daily visits. Children aged 6 to 12 yr had the most change in visits, which changed by 3.4 (95% CI: 2.3, 4.6) daily visits (Leibel et al., 2020). In a study to assess different simulation models using hospital respiratory admissions in Washington, the odds of respiratory hospitalizations in children (< 15 yr) increased by ~10% on days when the $PM_{2.5}$ increased by 10 μg/m^3, which was found consistently in all three simulation models (Gan et al., 2017). In contrast, an older population-based study used hospital admission and mortality data over 8 yr, from Sydney, Australia, that included 32 bushfire days and found bushfire-derived PM_{10} was associated with moderate decreases in asthma hospital presentations for children 3 days after the peak event (Morgan et al., 2010).

In epidemiological studies of the association between wildfire smoke and health, $PM_{2.5}$ is the most common measure of wildfire smoke exposure; however, increases in ozone also occur during wildfires (Pratt et al., 2019). Ozone is a secondary pollutant and fluctuates with temperature, time of day, and weather conditions, and wildfires are an additional source of ozone (Pratt et al., 2019). A 2019 study used analytical models to estimate the burden of pediatric emergency department visits attributable to elevated ozone associated with wildfire smoke in the United States, between 2005 and 2014. It found associations between ozone that were attributable to wildfire smoke and increases in emergency department visits of children <18 yr presenting for asthma (Pratt et al., 2019). The overall median number of visits for the 10 yr of the study period was 29 emergency department visits per child (interquartile range (IQR): 3–63). The number was higher for males, 18 visits (IQR: 2–40) than females, 10 visits (IQR: 1–20) (Pratt et al., 2019).

Respiratory Symptoms and Asthma Medication Use

Morbidity based on respiratory symptoms, exacerbation of existing disease, new disease diagnoses, and medicine use associated with wildfire exposure in children has mostly been examined through self-report of these outcomes.

A 2016 Spanish birth cohort study of 496 children, whose parents completed a telephone interview 2 weeks after a wildfire event, reported an increase of respiratory symptoms for all children. Children with existing asthma were 9 times more likely to report itchy or watery eyes (OR = 9.26; 95% CI: 2.14, 40.12), or 11 times more likely to report sneezing compared with children without asthma associated with smoke from wildfires (OR = 11.40; 95% CI: 2.01–64.52) (Vicedo-Cabrera et al., 2016). A 2009 longitudinal study in California of 465 non-asthmatic children aged 16 to 19 yr, who self-reported respiratory and eye symptoms following a wildfire event, demonstrated a link between duration of smoke exposure and respiratory symptoms (Mirabelli et al., 2009). For children who reported smelling smoke for 6 or more days (the exposure variable), the prevalence of respiratory symptoms was 60%, the prevalence for eye symptoms was 80% (Mirabelli et al., 2009). The Californian study by Künzli et al. (2006) reported that respiratory morbidity, including dry cough, wheeze, and eye symptoms, was more common in children exposed to wildfire smoke than to those not exposed and was higher in children with existing asthma. In a panel study of 235 people in Darwin, Australia, of whom ~50% were < 18 yr old, asthma-medication use (both reliever medication and initiation of oral steroids) was associated with low levels of PM_{10} and $PM_{2.5}$ derived from wildfire smoke (Johnston et al., 2006). A 2015 retrospective study of two cohorts of > 6,000 Californian children, where exposure to air pollution was determined by ZIP codes and distance to smoke plumes 1 yr before and 1 yr after wildfires, analyzed several asthma outcome measures. These included short-acting β-agonists-dispensing behavior. Dispensing behavior following wildfire exposure increased in all asthmatic children, but the result was inconsistent between the two cohorts (Tse et al., 2015). In an older and very small study (n = 32 children), asthmatics exposed to excessive PM_{10} from bushfire smoke for 7 days did not find any association with peak expiratory flow rates; however, among a subgroup without bronchial hyperreactivity, there was an inverse association between PM_{10} levels and evening peak expiratory flow rates (Jalaludin et al., 2004). In more recent work, 1,404 children aged 4 to 21 yr completed a Forced Expiratory Volume in 1 second test (FEV1) in a 3 day window after wildfire smoke exposure. Children aged 12 to 21 yr had poorer lung function on day one, which returned to normal function by day three. Children 4 to 11 yr did not demonstrate any change in lung function (Lipner et al., 2019).

There is limited data on the effect of wildfire smoke exposure on child mortality. A recent study estimated the under 5 child mortality attributable to wildfire and managed landscape burn smoke exposure. It reported an estimate of 104,279 child deaths a year, which accounts for 39% of the global annual mortality burden attributable to wildfire and landscape burns smoke (677,745 annual premature deaths) (Roberts & Wooster, 2021). This finding comes from a 2021 modeling study that used statistical methods to isolate the contribution of managed landscape burning to overall $PM_{2.5}$ and then calculated the number of people exposed, to estimate mortality by country. Most of these deaths occurred in low-income and middle-income countries, and the data suggest that landscape biomass burning is responsible for between 8% and 21% of the population weighted $PM_{2.5}$ concentration in many countries, making it the biggest contributor to all ambient air pollution. More accurate data and further testing of the modeling will increase the accuracy and precision of these estimates.

Other Vulnerabilities and Indirect Outcomes

Some studies have investigated potentially vulnerable population subgroups, including in obese children (Tse, 2015). Most studies investigating body mass index (BMI) as a risk factor in children have examined other air pollutant exposures, such as tobacco smoke or $PM_{2.5}$, and not specifically wildfire smoke exposure (Huang et al., 2019). Obesity is a risk factor for asthmatic disease, both incident disease and severity (Sin & Sutherland, 2008), and the prevalence of childhood obesity is also increasing (WHO, 2018), which suggests that there is an urgent need to conduct these investigations to better understand risk factors.

In two retrospective cohorts of children aged 5 to 17 with asthma in Southern California, 1,050 obese children were included, as described above (Tse et al., 2015). Overall, dispensing of short-acting ß-agonists was increased in the post-fire periods, and the greatest increase in dispensing was in children who were obese (16% increase); however, when this was restricted to children only in the fire ZIP codes, there was no difference in dispensing by BMI. Overall, there was no increase in hospital presentations or admissions, oral corticosteroid dispensing frequency, or new asthma diagnoses post wildfire (Tse et al., 2015).

Impacts from wildfire or prescribed burning on the immune system and cardiovascular health were investigated in a small convenience sample of 220 children (mean age 7 yr) in California, reported as a letter to the editor (Prunicki et al., 2019). Children (n = 32) exposed to wildfire pollution or prescribed burning (based on residence) demonstrated a difference in Th1 cells but not in Th2 or T regulatory cells, compared with children (n = 18) with no wildfire smoke exposure (Prunicki et al., 2019). This was a small study suggesting potential immune changes, though results need to be further investigated.

The effect on mental health from exposure to wildfire evidence has been assessed mostly through the lens of experiencing a traumatic event. Trauma related to wildfire exposure has been defined using the full range of experience, from living in the fire- or smoke-affected community, being evacuated, perception of threat, through to the death of a parent. Estimates of posttraumatic stress disorder, the most studied mental health outcome in this field, is estimated to be between 12% and 30%, although this varies considerably with the measures used, the age of the children, and definition of the trauma experience. In one cross-sectional study conducted in Greece, 1,339 adolescents aged 12 to 17 yr completed a self-report questionnaire of wildfire exposure and symptoms of threat or perceived threat, including anxiety and depression, 6 months after a wildfire event. Of those, 432 adolescents (29.4%) reported levels above the cut-off of posttraumatic stress disorder symptoms, and perception of life threat was a greater predictor of these symptoms than actual threat (Papadatou et al., 2012).

Academic achievement was assessed 4 yr after exposure to a bushfire disaster in a recent Victoria, Australia, population-based study of 24,642 children who were highly or moderately exposed to a major bushfire. Exposure was assessed through school proximity to the bushfire. Academic achievement was measured using the standardized annual literacy and numeracy test scores at two time points 2 yr apart, 2 and 4 yr after the bushfire for each child. There were reductions in the children's expected numeracy and literacy test scores based on the expected trajectory between the two test scores at approximately ages 9 and 13 yr. The hypothesized impact pathway was through cognitive deficits related to impacts on working memory as a result of trauma including loss of parent(s), which may have resulted in disruption to schooling and lack of parental support for reading at home (Gibbs et al., 2019). Another study, whose primary study factor was health effects in children from wildfires, used an existing cohort of > 6,000 children, 86% aged 6 to 7 yr old in California (Künzli et al., 2006). Exposure was measured both subjectively and objectively in two analytical models. Although academic achievement was not assessed, the authors found that missed school attendance was associated with increased respiratory symptoms especially in children with an asthma diagnosis (21.7% compared with 11.55%) (Künzli et al., 2006). School closure is a common outcome of wildfire, and in 2018 in California, more than a million children were impacted by school closures due to wildfires (Chalupka & Anderko, 2019).

During the bushfire event in New South Wales, Australia, of 2019, a natural experiment was conducted in the context of an ongoing cluster randomized controlled trial of a school-based physical activity intervention in children aged 8–10 yr old (Del Pozo Cruz et al., 2021). During the bushfire event, the schools received targeted health advice from the government health department to avoid all outdoor physical activity. Exposure was measured using air quality data measurements located near the schools participating in the trial during the bushfire, and physical activity was measured using accelerometers that the children were wearing for the trial. The authors reported a minimal impact on physical activity up to a tipping point in air quality when it was rated as "very poor" and the physical activity dropped sharply. This suggests that the public health advice did not change children's activity and was largely ignored until the air quality reached very high pollution levels. Further, if air quality is poor for an extended period due to wildfires, it may result in reduced physical activity and children may have detrimental impacts on health outcomes.

The impact on infant feeding has also been assessed through the pathway of exposure to emergency or disaster. A study from Canada included 115 women with infants or young children up to 36 months who were recruited through social media after being evacuated during the 2016 Fort McMurray wildfire. They reported a 30% reduction in breast-feeding before versus after the evacuation (x^2 = 29.62, P = < 0.001) (Deyoung et al., 2018). Responses from participants suggested that many factors resulted in reduced breast-feeding, including stress, perception of reduced milk supply, and lack of privacy.

13.3.3. Summary of Childhood Studies

The impact of wildfire exposure on child health is an emerging area for research investigation. Wildfire smoke exposure is a unique composition of toxic chemicals that presents uniquely with shorter durations subject to climate and geography resulting in unpredictable exposure patterns (Black et al., 2017b). The effect on children is a research gap that needs investigation. The current body of evidence is a mix of methodology and quality and has many of the same methodological issues as the pregnancy literature, that make drawing meaningful and generalizable conclusions difficult. Population-level data, with accurate exposure assessment, inclusion of important health outcomes, especially those with children as the focus, are needed. Evidence examined here suggest that respiratory disease, either symptoms or exacerbation of disease is a confirmed consequence from wildfire exposure in children, as is increased health-service use including hospitalizations. Evidence from ambient sources of $PM_{2.5}$ not reviewed in this chapter has more extensive investigation and more definitive results, and the adult research as discussed in earlier chapters is also more advanced and extensive, and suggests that there is much more to learn in relation to the impact on children.

13.4. RESEARCH GAPS AND FUTURE DIRECTIONS

There is a lack of well-designed epidemiologic studies that quantify the association between wildfire smoke exposure and risk of adverse maternal, fetal, and child outcomes. The few studies that have been conducted vary with respect to study design, size, exposure assessment, outcome ascertainment, and statistical methods including inclusion of important confounders. This heterogeneity results in difficulty to synthesize and evaluate results. Many studies used satellite-based imaging and $PM_{2.5}$ measurements, while others used proximity to fire events (using residential or school address), or self-report of the indoor smoke smell, which do not accurately measure individual exposure. Choice of exposure measurement can impact the accuracy of the results (Gan et al., 2017). More uniform approaches for exposure assessment methods are needed to quantify wildfire-specific air pollution and separate wildfire smoke from existing ambient air pollution to better elucidate the health effects of wildfires and the impact on the life-course; and these need to be relevant beyond high-income settings. Quantification of smokewaves that incorporate the intensity of wildfire-specific $PM_{2.5}$, the number of consecutive days in which it is experienced, and the cumulative smokewaves across a pregnancy or throughout childhood is needed to examine perinatal and childhood outcomes. Consistency across future studies is encouraged to allow for pooling and better comparison of estimates (Weber et al., 2019).

Future research could benefit from both large population studies across time and space, including prospective cohorts in wildfire-prone areas, and investigations of biological mechanisms. Accurate individual-level data regarding behaviors, time activity, housing characteristics, adoption of risk-mitigation strategies including use of indoor air-filter technologies are also needed to target effective interventions. Improved exposure assessment that accounts for the short-term increases attributable to wildfire in the context of the background air pollution will also provide more accurate estimates of the relationship between wildfire smoke and perinatal and child outcomes. Assessment are urgently needed of interventions to reduce exposure, such as use of indoor air-filters, personal masks, or behaviors that limit or reduce exposure specifically in communities impacted by wildfires that include pregnant populations and children. Pragmatic designs that assess real-life situations and are generalizable beyond the immediate study population ready for adoption will have immediate impact on policy. Further quantification is needed of the specific toxic exposures that may differ depending on the material being burned in the fires (i.e., biomass versus structure burning) to better understand the relationships between wildfires and adverse health outcomes.

Studies on wildfire smoke and maternal morbidity are particularly lacking. Exposure to traffic-related pollution is presumed to be a hazard to pregnant people for developing hypertensive disorders of pregnancy, according to a recent monograph by the U.S. National Toxicology Program (2019). This conclusion was based on results primarily of studies of $PM_{2.5}$ and NO_2. Although no studies have been conducted on wildfire smoke and hypertensive disorders in pregnancy or gestational diabetes, it is plausible that an association exists based on ambient air pollution studies (Kim et al., 2021; Zhang et al., 2020; Tang et al., 2020). Future studies are needed to quantify these relationships.

More studies of fetal loss and perinatal mortality are needed to examine wildfire smoke in relation to these outcomes and identify potential underestimates of the association between wildfire smoke and other perinatal outcomes such as preterm birth. Additional studies on birth defects with precise exposure assessment and case ascertainment are needed to conclude their risk with relation to wildfire smoke.

Tracking wildfire smoke exposure and effects in these vulnerable periods (periconception, perinatal, neonatal, and childhood) into adulthood will help us quantify the cumulative effects of wildfire smoke across the life course. For example, a recent study found that those born extremely preterm (< 28 weeks gestation) or extremely low birth weight (< 1,000 g) were at increased risk of adverse respiratory health outcomes following wildfire smoke exposure in early adulthood (Haikerwal et al., 2021).

The distribution of wildfire smoke by socioeconomic status is not as well studied; however, infiltration of smoke indoors may be related to age and quality of housing, posing a critical inequality to reduce personal exposure indoors. We need to consider additional factors that may increase susceptibility to wildfire smoke including issues of environmental justice, housing conditions, and structural racism that impacts socioeconomic inequalities and discrimination that results in higher toxic exposures (Gutschow et al., 2021). The intersection of pregnancy and childhood with environmental and structural racism should be highlighted when prioritizing research and policy to protect communities from wildfire smoke. Furthermore, there is a paucity of studies from low-income and middle-income countries (Jayachandran, 2009; Li et al., 2021b). The usual endpoints, such as health service use or disease exacerbation, are mostly irrelevant or immeasurable in those settings, and the other measures that are context specific will be required to develop a strong evidence base of the effect on childhood and pregnancy.

The psychological stress caused by wildfire events may affect maternal and subsequent birth and child outcomes. This presents a methodological issue to separate these correlated exposures. Further qualitative and quantitative data assessing the effect of wildfire threat stress may help disentangle the effects from stress via acute wildfire smoke exposure.

We recommend future studies address policy-relevant questions related to wildfire smoke exposures and perinatal and childhood health outcomes and that vulnerable populations including pregnant people and children be considered a priority when policies are developed.

In conclusion, numerous studies have found adverse health outcomes associated with wildfire smoke during pregnancy and childhood. This evidence is extrapolated from extensive research on ambient air pollution and adverse pregnancy and child health outcomes. Future research is needed to estimate wildfire smoke exposure more specifically coupled with more health endpoints for maternal, fetal, infant, and childhood morbidity.

ACKNOWLEDGMENTS

Research reported in this chapter was supported by the National Institute of Environmental Health Science of the National Institutes of Health under award number R01ES031261 (Padula).

Camille Raynes-Greenow was supported by a Robinson Fellowship from the University of Sydney, Australia.

REFERENCES

Abdo, M., Ward, I., O'Dell, K., Ford, B., Pierce, J. R., Fischer, E. V., et al. (2019). Impact of Wildfire Smoke on Adverse Pregnancy Outcomes in Colorado, 2007–2015. *International Journal of Environmental Research and Public Health*, *16*.

Aguilera, R., Corringham, T., Gershunov, A., Leibel, S. & Benmarhnia, T. (2021). Fine particles in wildfire smoke and pediatric respiratory health in California. *Pediatrics*, *147*.

Amegah, A. K., Quansah, R. & Jaakkola, J. J. (2014). Household air pollution from solid fuel use and risk of adverse pregnancy outcomes: A systematic review and meta-analysis of the empirical evidence. *PLoS One*, *9*, e113920.

American College of Obstetricians and Gynecologists Committee on Health Care for Underserved Women, American Society for Reproductive Medicine Practice Committee & The University of California San Francisco Program on Reproductive Health and the Environment (2013). Exposure to toxic environmental agents. Washington, DC: The American College of Obstetricians and Gynecologists.

Barker, D. J. (2004). The developmental origins of chronic adult disease. *Acta Paediatrica Supplement*, *93*, 26–33.

Behrman, R., & Butler, A. (2007). *Preterm birth: Causes, consequences, and prevention*. Institute of Medicine (US). Committee on understanding premature birth and assuring healthy outcomes. Washington, DC: National Academies Press.

Bekkar, B., Pacheco, S., Basu, R., & Denicola, N. (2020). Association of air pollution and heat exposure with preterm birth, low birth weight, and stillbirth in the US: A systematic review. *JAMA Network Open*, *3*, e208243.

Black, C., Gerriets, J. E., Fontaine, J. H., Harper, R. W., Kenyon, N. J., Tablin, F., et al. (2017a). Early life wildfire smoke exposure is associated with immune dysregulation and lung function decrements in adolescence. *American Journal of Respiratory Cell & Molecular Biology*, *56*, 657–666.

Black, C., Tesfaigzi, Y., Bassein, J. A., & Miller, L. A. (2017b). Wildfire smoke exposure and human health: Significant gaps in research for a growing public health issue. *Environmental Toxicology and Pharmacology*, *55*, 186–195.

Bobrowski, R. A. (2010). Pulmonary physiology in pregnancy. *Clinical Obstetrics and Gynecology*, *53*, 285–300.

Bove, H., Bongaerts, E., Slenders, E., Bijnens, E. M., Saenen, N. D., Gyselaers, W., et al. (2019). Ambient black carbon particles reach the fetal side of human placenta. *Nature Communications*, *10*, 3866.

Buxton, M. A., Meraz-Cruz, N., Sanchez, B. N., Gronlund, C. J., Foxman, B., Vadillo-Ortega, F., et al. (2019). Air pollution and inflammation: Findings from concurrent repeated measures of systemic and reproductive tract cytokines during term pregnancy in Mexico City. *Science of the Total Environment*, *681*, 235–241.

Candido Da Silva, A. M., Moi, G. P., Mattos, I. E., & Hacon, S. (2014). Low birth weight at term and the presence of fine particulate matter and carbon monoxide in the Brazilian Amazon: A population-based retrospective cohort study. *BMC Pregnancy Childbirth*, *14*, 309.

Chalupka, S., & Anderko, L. (2019). Climate change and schools: implications for children's health and safety. *Creative Nursing*, *25*, 249.

Chang, H. H., Warren, J. L., Darrow, L. A., Reich, B. J., & Waller, L. A. (2015). Assessment of critical exposure and outcome windows in time-to-event analysis with application to air pollution and preterm birth study. *Biostatistics (Oxford, England)*, *16*, 509–521.

Chen, Y.-T., Liu, C.-L., Chen, C.-J., Chen, M.-H., Chen, C.-Y., Tsao, P.-N., et al. (2021). Association between short-term exposure to air pollution and sudden infant death syndrome. *Chemosphere*, *271*, 129515.

Corrêa, M. L., Soares, P. S. M., Da Silva, B. G. C., Wehrmeister, F., Horta, B. L., & Menezes, A. M. B. (2021). Maternal smoking during pregnancy and intelligence quotient in offspring: A systematic review and meta-analysis. *NeuroToxicology*, *85*, 99–114.

Currie, J., & Schwandt, H. (2013). Within-mother analysis of seasonal patterns in health at birth. *Proceedings of the National Academy of Science, USA*, *110*, 12265–12270.

Del Pozo Cruz, B., Hartwig, T. B., Sanders, T., Noetel, M., Parker, P., Antczak, D., et al. (2021). The effects of the Australian bushfires on physical activity in children. *Environment International*, *146*, 106214.

Deyoung, S. E., Chase, J., Branco, M. P., & Park, B. (2018). The effect of mass evacuation on infant feeding: the case of the 2016 Fort McMurray wildfire. *Maternal and Child Health Journal*, *22*, 1826–1833.

Diamanti-Kandarakis, E., Bourguignon, J.-P., Giudice, L. C., Hauser, R., Prins, G. S., Soto, A. M., et al. (2009). Endocrine-disrupting chemicals: An Endocrine Society scientific statement. *Endocrine Reviews*, *30*, 293–342.

Gan, R. W., Ford, B., Lassman, W., Pfister, G., Vaidyanathan, A., Fischer, E., Volckens, J., et al. (2017). Comparison of wildfire smoke estimation methods and associations with cardiopulmonary-related hospital admissions. *Geohealth*, *1*, 122–136.

Gaskins, A. J., Fong, K. C., Abu Awad, Y., Di, Q., Minguez-Alarcon, L., Chavarro, J. E., et al. (2019). Time-varying exposure to air pollution and outcomes of in vitro fertilization among couples from a fertility clinic. *Environmental Health Perspectives*, *127*, 77002.

Gibbs, L., Nursey, J., Cook, J., Ireton, G., Alkemade, N., Roberts, M., et al. (2019). Delayed disaster impacts on academic performance of primary school children. *Child Development*, *90*, 1402–1412.

Gray, S. C., Edwards, S. E., Schultz, B. D., & Miranda, M. L. (2014). Assessing the impact of race, social factors and air pollution on birth outcomes: A population-based study. *Environmental Health*, *13*, 4.

Gutschow, B., Gray, B., Ragavan, M. I., Sheffield, P. E., Philipsborn, R. P., & Jee, S. H. (2021). The intersection of pediatrics, climate change, and structural racism: ensuring health equity through climate justice. *Current Problems in Pediatric and Adolescent Health Care*, *51*, 101028.

Haikerwal, A., Doyle, L. W., Wark, J. D., Irving, L., & Cheong, J. L. (2021). Wildfire smoke exposure and respiratory health outcomes in young adults born extremely preterm or extremely low birthweight. *Environmental Research*, *197*, 111159.

Heft-Neal, S., Driscoll, A., Yang, W., Shaw, G., & Burke, M. (2022). Associations Between wildfire smoke exposure during pregnancy and risk of preterm birth in California. *Environmental Research*, *203*, 111872.

Hegewald, M. J., & Crapo, R. O. (2011). Respiratory physiology in pregnancy. *Clinics in Chest Medicine*, *32*, 1–13.

Henry, S., Ospina, M. B., Dennett, L., & Hicks, A. (2021). Assessing The risk of respiratory-related healthcare visits associated with wildfire smoke exposure in children 0–18 years old: A systematic review. *International Journal of Environmental Research and Public Health*, *18*.

Holdsworth-Carson, S. J., Lim, R., Mitton, A., Whitehead, C., Rice, G. E., Permezel, M., et al. (2010). Peroxisome proliferator-activated receptors are altered in pathologies of the human placenta: gestational diabetes mellitus, intrauterine growth restriction and preeclampsia. *Placenta*, *31*, 222–229.

Holm, S. M., Miller, M. D., & Balmes, J. R. (2021). Health effects of wildfire smoke in children and public health tools: a narrative review. *Journal of Exposure Science & Environmental Epidemiology*, *31*, 1–20.

Holstius, D. M., Reid, C. E., Jesdale, B. M., & Morello-Frosch, R. (2012). Birth weight following pregnancy during the 2003 southern California wildfires. *Environmental Health Perspectives*, *120*, 1340–1345.

Huang, J. V., Leung, G. M., & Schooling, C. M. (2019). The association of air pollution with body mass index: Evidence from Hong Kong's "Children of 1997" birth cohort. *International Journal of Obesity*, *43*, 62–72.

Hutchinson, J. A., Vargo, J., Milet, M., French, N. H. F., Billmire, M., Johnson, J., et al. (2018). The San Diego 2007 wildfires and medi-cal emergency department presentations, inpatient hospitalizations, and outpatient visits: an observational study of smoke exposure periods and a bidirectional case-crossover analysis. *PLOS Medicine*, *15*, E1002601.

Jalaludin, B., Smith, M., O'Toole, B., & Leeder, S. (2004). Acute effects of bushfires on peak expiratory flow rates in children with wheeze: a time series analysis. *Australian and New Zealand Journal of Public Health*, *24*, 147.

Jayachandran, S. (2009). Air quality and early-life mortality evidence from Indonesia's wildfires. *Journal of Human Resources*, *44*, 916–954.

Johnston F., Webby, R., Pilotto, L., Bailie, R., Parry, D., & Halpin, S. (2006). Vegetation fires, particulate air pollution and asthma: A panel study in the Australian monsoon tropics. *International Journal of Environmental Research and Public Health*, *16*, 391.

Kampmann, U., Madsen, L. R., Skajaa, G. O., Iversen, D. S., Moeller, N., & Ovesen, P. (2015). Gestational diabetes: A clinical update. *World Journal of Diabetes*, *6*, 1065–1072.

Kim, J. H., Choi, Y. Y., Yoo, S. I., & Kang, D. R. (2021). Association between ambient air pollution and high-risk pregnancy: A 2015–2018 national population-based cohort study in Korea. *Environmental Research*, *197*, 110965.

Kizer, K. W. (2020). Extreme wildfires: A growing population health and planetary problem. *JAMA*, *324*(16), 1605–1606. https://doi.org/10.1001/jama.2020.19334

Klepac, P., Locatelli, I., Korosec, S., Kunzli, N., & Kukec, A. (2018). Ambient air pollution and pregnancy outcomes: A comprehensive review and identification of environmental public health challenges. *Environmental Research*, *167*, 144–159.

Koman, P. D., Hogan, K. A., Sampson, N., Mandell, R., Coombe, C. M., Tetteh, M. M., et al. (2018). Examining Joint effects of air pollution exposure and social determinants of health in defining "at-risk" populations under the Clean Air Act: Susceptibility of pregnant women to hypertensive disorders of pregnancy. *World Medical Health Policy*, *10*, 7–54.

Kristjansson, A. L., Thomas, S., Lilly, C. L., Thorisdottir, I. E., Allegrante, J. P., & Sigfusdottir, I. D. (2018). Maternal smoking during pregnancy and academic achievement of offspring over time: a registry data-based cohort study. *Preventive Medicine*, *113*, 74–79.

Künzli, N., Avol, E., Wu, J., Gauderman, W. J., Rappaport, E., Millstein, J., et al. (2006). Health effects of the 2003 southern California wildfires on children. *American Journal of Respiratory and Critical Care Medicine*, *174*, 1221–1228.

Lee, A. G., Kaali, S., Quinn, A., Delimini, R., Burkart, K., Opoku-Mensah, J., et al. (2019). Prenatal household air pollution is associated with impaired infant lung function with sex-specific effects: Evidence from graphs, a cluster randomized cookstove intervention trial. *American Journal of Respiratory and Critical Care Medicine*, *199*, 738–746.

Leibel, S., Nguyen, M., Brick, W., Parker, J., Ilango, S., Aguilera, R., et al. (2020). Increase in pediatric respiratory visits associated with Santa Ana wind-driven wildfire smoke and PM(2.5) levels in San Diego County. *Annals of the. American Thoracic Society*, *17*, 313–320.

Li, G., Li, L., Liu, D., Qin, J., & Zhu, H. (2021a). Effect of PM(2.5) pollution on perinatal mortality in China. *Scientific Reports*, *11*, 7596.

Li, J., Guan, T., Guo, Q., Geng, G., Wang, H., Guo, F., Li, J., et al. (2021b). Exposure to landscape fire smoke reduced birthweight in low- and middle-income countries: Findings from a siblings-matched case-control study. *Elife*, *10*.

Li, X., Huang, S., Jiao, A., Yang, X., Yun, J., Wang, Y., et al. (2017). Association between ambient fine particulate matter and preterm birth or term low birth weight: An updated systematic review and meta-analysis. *Environmental Pollution*, *227*, 596–605.

Lipner, E. M., O'Dell, K., Brey, S. J., Ford, B., Pierce, J. R., Fischer, E. V., et al. (2019). The associations between clinical respiratory outcomes and ambient wildfire smoke exposure among pediatric asthma patients at National Jewish Health, 2012–2015. *Geohealth*, *3*, 146–159.

Liu, J. C., Mickley, L. J., Sulprizio, M. P., Dominici, F., Yue, X., Ebisu, K., et al. (2016). Particulate air pollution from wildfires in the western US under climate change. *Climatic Change*, *138*, 655–666.

Makkonen, U., Hellen, H., Anttila, P., & Ferm, M. (2010). Size distribution and chemical composition of airborne particles in south-eastern Finland during different seasons and wildfire episodes in 2006. *Science of the Total Environment*, *408*, 644–651.

Malley, C. S., Kuylenstierna, J. C., Vallack, H. W., Henze, D. K., Blencowe, H., & Ashmore, M. R. (2017). Preterm birth associated with maternal fine particulate matter exposure: A global, regional and national assessment. *Environment International*, *101*, 173–182.

Maltepe, E., & Fisher, S. J. (2015). Placenta: The forgotten Oorgan. *Annual Review of Cell and Developmental Biology*, *31*, 523–552.

Martens, D. S., Cox, B., Janssen, B. G., Clemente, D. B. P., Gasparrini, A., Vanpoucke, C., et al. (2017). Prenatal air pollution and newborns' predisposition to accelerated biological aging. *JAMA Pediatrics*, *171*, 1160–1167.

Mekonnen, Z. K., Oehlert, J. W., Eskenazi, B., Shaw, G. M., Balmes, J. R., & Padula, A. M. (2021). The relationship between air pollutants and maternal socioeconomic factors on preterm birth in California urban counties. *Journal of Exposure Science and Environmental Epidemiology*, *31*, 503–513.

Mirabelli, M. C., Künzli, N., Avol, E., Gilliland, F. D., Gauderman, W. J., Mcconnell, R., et al. (2009). Respiratory symptoms following wildfire smoke exposure: Airway size as a susceptibility factor. *Epidemiology*, *20*, 451–459.

Moller, P., Danielsen, P. H., Karottki, D. G., Jantzen, K., Roursgaard, M., Klingberg, H., et al. (2014). Oxidative stress and inflammation generated dna damage by exposure to air pollution particles. *Mutation Research: Reviews in Mutation Research*, *762*, 133–166.

Morgan, G., Sheppeard, V., Khalaj, B., Ayyar, A., Lincoln, D., Jalaludin, B., et al. (2010). Effects of bushfire smoke on daily mortality and hospital admissions in Sydney, Australia. *Epidemiology*, *21*, 47–55.

Muglia, L. J., & Katz, M. (2010). The enigma of spontaneous preterm birth. *New England Journal of Medecine*, *362*, 529–535.

Nance, J. H. P., Radke. L., & Ward, D. (1993). Airborne measurements of gases and particles from an Alaskan wildfire. *Journal of Geophysical Research*, *98*, 14873–14882.

National Toxicology Program (2019). *Monograph on the systematic review of traffic-related air pollution and hypertensive disorders of pregnancy*. Research Triangle Park, NC: National Toxicology Program.

O'Donnell, M. H., & Behie, A. M. (2015). Effects of wildfire disaster exposure on male birth weight in an Australian population. *Evolution, Medicine, and Public Health*, 344–354.

O'Donnell, M. H., & Behie, A. M. (2013). Effects of bushfire stress on birth outcomes: A cohort study of the 2009 Victorian Black Saturday bushfires. *International Journal of Disaster Risk Reduction*, *5*, 98–106.

Padula, A. M., & Benmarhnia, T. (2022). Wildfires in pregnancy: potential threats to the newborn. *Paediatric and Perinatal Epidemiology*, *36*, 54–56.

Padula, A. M., Mortimer, K. M., Tager, I. B., Hammond, S. K., Lurmann, F. W., Yang, W., et al. (2014). Traffic-related air pollution and risk of preterm birth in the San Joaquin Valley of California. *Annals of Epidemiology*, *24*, 888–895e4.

Padula, A. M., Rivera-Nunez, Z., & Barrett, E. S. (2020). Combined impacts of prenatal environmental exposures and psychosocial stress on offspring health: Air pollution and metals. *Current Environmental Health Reports*, *7*, 89–100.

Padula, A. M., Tager, I. B., Carmichael, S. L., Hammond, S. K., Lurmann, F., & Shaw, G. M. (2013a). The association of ambient air pollution and traffic exposures with selected congenital anomalies in the San Joaquin Valley of California. *American Journal of Epidemiology*, *177*, 1074–1085. https://doi.org/10.1093/aje/kws367

Padula, A. M., Tager, I. B., Carmichael, S. L., Hammond, S. K., Yang, W., Lurmann, F., et al. (2013b). Ambient air pollution and traffic exposures and congenital heart defects in the San Joaquin Valley of California. *Paediatric and Perinatal Epidemiology*, *27*(4), 329–339. https://doi.org/10.1111/ppe.12055

Padula, A. M., Tager, I. B., Carmichael, S. L., Hammond, S. K., Yang, W., Lurmann, F. W., et al. (2013c). Traffic-related air pollution and selected birth defects in the San Joaquin Valley of California. *Birth Defects Research Part A: Clinical and Molecular Teratology*, *97*, 730–735.

Padula, A. M., Yang, W., Lurmann, F. W., Balmes, J., Hammond, S. K., & Shaw, G. M. (2018). Prenatal exposure to air pollution, maternal diabetes and preterm birth. *Environmental Research*, *170*, 160–167.

Palm, M., Axelsson, O., Wernroth, L., Larsson, A., & Basu, S. (2013). Involvement of inflammation in normal pregnancy. *Acta Obstetricia et Gynecologica Scandinavica*, *92*, 601–605.

Panasevich, S., Leander, K., Rosenlund, M., Ljungman, P., Bellander, T., De Faire, U., et al. (2009). Associations of long- and short-term air pollution exposure with markers of inflammation and coagulation in a population sample. *Occupational & Environmental Medicine*, *66*, 747–753.

Papadatou, D., Giannopoulou, I., Bitsakou, P., Bellali, T., Talias, M. A., & Tselepi, K. (2012). Adolescents' reactions after a wildfire disaster in Greece. *Journal of Traumatic Stress*, *25*, 57–63.

Park, B. Y., Boles, I., Monavvari, S., Patel, S., Alvarez, A., Phan, M., et al. (2021). The association between wildfire exposure in pregnancy and foetal gastroschisis: A population-based cohort study. *Paediatric and Perinatal Epidemiology*, *36*(1), 45–453. https://doi.org/10.1111/ppe.12823

Pedersen, M., Stayner, L., Slama, R., Sorensen, M., Figueras, F., Nieuwenhuijsen, M. J., et al. (2014). Ambient air pollution and pregnancy-induced hypertensive disorders: A systematic review and meta-analysis. *Hypertension*, *64*, 494–500.

Prass, T. S., Lopes, S. R., Dorea, J. G., Marques, R. C., & Brandao, K. G. (2012). Amazon forest fires between 2001 and 2006 and birth weight in Porto Velho. *Bulletin of Environmental Contamination and Toxicology*, *89*, 1–7.

Pratt, J. R., Gan, R. W., Ford, B., Brey, S., Pierce, J. R., Fischer, E. V., Magzamen, S. (2019). A national burden assessment of estimated pediatric asthma emergency department visits that may be attributed to elevated ozone levels associated with the presence of smoke. *Environmental Monitoring and Assessment*, *191*(Suppl 2), 269. https://doi.org/10.1007/s10661-019-7420-5

President's Cancer Panel (2010). *Reducing environmental cancer risk: What we can do now*. Bethesda: U.S. Department of Health and Human Services.

Prunicki, M., Kelsey, R., Lee, J., Zhou, X., Smith, E., Haddad, F., et al. (2019). The impact of prescribed fire versus wildfire on the immune and cardiovascular systems of children. *Allergy*, *74*, 1989–1991.

Reid, C. E., Brauer, M., Johnston, F. H., Jerrett, M., Balmes, J. R., & Elliott, C. T. (2016). Critical review of health impacts of wildfire smoke exposure. *Environmental Health Perspectives*, *124*(9), 1334–1343. https://doi.org/10.1289/ehp.1409277

Requia, W. J., Kill, E., Papatheodorou, S., Koutrakis, P., & Schwartz, J. D. (2021a). Prenatal exposure to wildfire-related air pollution and birth defects in Brazil. *Journal of Exposure Science and Environmental Epidemiology*, *32*(4), 596–603. https://doi.org/10.1038/s41370-021-00380-y

Requia, W. J., Papatheodorou, S., Koutrakis, P., Mukherjee, R., & Roig, H. L. (2021b). Increased preterm birth following maternal wildfire smoke exposure in Brazil. *International Journal of Hygiene and Environmental Health*, *240*, 113901.

Risom, L., Moller, P., & Loft, S. (2005). Oxidative stress-induced DNA damage by particulate air pollution. *Mutation Research*, *592*, 119–137.

Roberts, G., & Wooster, M. J. (2021). Global impact of landscape fire emissions on surface level PM2.5 concentrations, air quality exposure and population mortality. *Atmospheric Environment*, *252*.

Saigal, S., & Doyle, L. W. (2008). An overview of mortality and sequelae of preterm birth from infancy to adulthood. *Lancet*, *371*, 261–9.

Shah, P. S., Balkhair, T., & Knowledge Synthesis Group on Determinants of Preterm/LBW Births (2011). Air pollution and birth outcomes: A systematic review. *Environment International*, *37*, 498–516.

Shao, J. A., Wheeler, A., Zosky, G., & Johnston, F. (2019). Long-term impacts of prenatal and infant exposure to fine particulate matter on wheezing and asthma: A systematic review and meta-analysis. *Environmental Epidemiology*, *3*, E042.

Sheffield, P. E., & Landrigan, P. J. (2011). Global climate change and children health: Threats and strategies for prevention. *Environmental Health Perspectives*, *119*, 291–298.

Sin, D. D., & Sutherland, E. R. (2008). Obesity And The Lung: 4 · Obesity And Asthma. *Thorax*, *63*, 1018.

Stieb, D. M., Chen, L., Eshoul, M., & Judek, S. (2012). Ambient air pollution, birth weight and preterm birth: A systematic review and meta-analysis. *Environmental Research*, *117*, 100–111.

Tan, E. K., & Tan, E. L. (2013). Alterations in physiology and anatomy during pregnancy. *Best Practice & Research: Clinical Obstetrics & Gynaecology*, *27*, 791–802.

Tang, X., Zhou, J. B., Luo, F., Han, Y., Heianza, Y., Cardoso, M. A., et al. (2020). Air pollution and gestational diabetes mellitus: Evidence from cohort studies. *BMJ Open Diabetes Research and Care*, *8*.

Tse, K., Chen, L., Tse, M., Zuraw, B., & Christiansen, S. (2015). Effect of catastrophic wildfires on asthmatic outcomes in obese children: Breathing fire. *Annals of Allergy, Asthma & Immunology*, *114*, 308–311.E4.

U.S. EPA (2009). *Integrated science assessment for particulate matter (final report)*. Washington, D.C.: U.S. Environmental Protection Agency.

Varshavsky, J., Smith, A., Wang, A., Hom, E., Izano, M., Huang, H., et al. (2020). Heightened susceptibility: A review of how pregnancy and chemical exposures influence maternal health. *Reproductive Toxicology*, *92*, 14–56.

Verstraeten, B. S. E., Elgbeili, G., Hyde, A., King, S., & Olson, D. M. (2020). Maternal mental health after a wildfire: effects of social support in the Fort McMurray Wood Buffalo study. *Canadian Journal of Psychiatry*, *66*(8), 710–718.

Vicedo-Cabrera, A. M., Esplugues, A., Iñíguez, C., Estarlich, M., & Ballester, F. (2016). Health effects of the 2012 Valencia (Spain) wildfires on children in a cohort study. *Environmental Geochemistry and Health*, *38*, 703–712.

Wauters, A., Dreyfuss, C., Pochet, S., Hendrick, P., Berkenboom, G., Van De Borne, P., & Argacha, J. F. (2013). Acute exposure to diesel exhaust impairs nitric oxide-mediated endothelial vasomotor function by increasing endothelial oxidative stress. *Hypertension*, *62*, 352–358.

Weber, K. A., Yang, W., Lurmann, F., Hammond, S. K., Shaw, G. M., & Padula, A. M. (2019). Air pollution, maternal hypertensive disorders, and preterm birth. *Environmental Epidemiology*, *3*.

Welshons, W. V., Nagel, S. C., & Vom Saal, F. S. (2006). Large effects from small exposures. III. Endocrine mechanisms mediating effects of bisphenol A at levels of human exposure. *Endocrinology*, *147*, S56–S69.

WHO (2018). Monitoring health for the SDGs, sustainable development goals. Https://Apps.Who.Int/Iris/Bitstream/Handle/10665/272596/9789241565585-Eng.Pdf?Sequence=1&Isallowed=Y.

Willson, B. E., Gee, N. A., Willits, N. H., Li, L., Zhang, Q., Pinkerton, K. E., & Lasley, B. L. (2021). Effects of the 2018 Camp Fire on birth outcomes in non-human primates: Case-control study. *Reproductive Toxicology*, *105*, 128–135.

Yang, B. Y., Qu, Y., Guo, Y., Markevych, I., Heinrich, J., Bloom, M. S., Bai, Z., et al. (2021). Maternal exposure to ambient air pollution and congenital heart defects in China. *Environment International*, *153*, 106548.

Yuan, L., Zhang, Y., Wang, W., Chen, R., Liu, Y., Liu, C., Kan, H., et al. (2020). Critical windows for maternal fine particulate matter exposure and adverse birth outcomes: The Shanghai birth cohort study. *Chemosphere*, *240*, 124904.

Zeglinski, M. R., Turner, C. T., Zeng, R., Schwartz, C., Santacruz, S., Pawluk, M. A., Zhao, et al. (2019). Soluble wood smoke extract promotes barrier dysfunction in alveolar epithelial cells through a MAPK signaling pathway. *Scientific Reports*, *9*, 10027.

Zhang, H., Wang, Q., He, S., Wu, K., Ren, M., Dong, H., et al. (2020). Ambient air pollution and gestational diabetes mellitus: A review of evidence from biological mechanisms to population epidemiology. *Science of the Total Environment*, *719*, 137349.

Zhang, Z., Kris-Etherton, P. M., & Hartman, T. J. (2014). Birth weight and risk factors for cardiovascular disease and type 2 diabetes in US children and adolescents: 10 year results from Nhanes. *Maternal and Child Health Journal*, *18*, 1423–1432.

14

State of the Science and Future Directions: From Biomass Fire to Health Outcomes

Robin C. Puett[1], Nancy H. F. French[2], and Tatiana V. Loboda[3]

ABSTRACT

Global increases in the frequency and area consumed from biomass burning have escalated the need for more accurate and comprehensive fire detection, monitoring of emissions, mapping and prediction of air quality, as well as understanding and development of preventive/responsive measures for associated health outcomes. In this concluding chapter, we summarize current knowledge and knowledge gaps, as contributed by authors in multiple scientific disciplines. The book and this conclusion provide a wide-ranging foundational overview from fire science to epidemiology. This conclusion reflects the intention of each chapter to summarize the state of the science for those outside each discipline, while identifying future needs for those with specific disciplinary expertise. This volume is intended to enhance further collaborations among disciplines, practitioners, and policy makers for a more extensive response to this critical environmental and human health challenge.

14.1. OVERVIEW

This book serves as an introduction to the scientific methods and current research addressing the cascade of events from wildfire to human health and well-being impacts. A comprehensive overview was not achievable in one volume, however, we sought to provide valuable foundational knowledge and input from the multiple disciplines necessary to build a more thorough response to the increasing challenge of wildfire during climate change. Each chapter has aimed to provide novices an introduction and working knowledge, while also summarizing current knowledge and significant gaps for those with more topically relevant experience. By addressing the continuum of science from fire to health in this way, this book is designed to create bridges among disciplines to forward the entire field. The authors and editors aspire to stimulate multidisciplinary collaborations among researchers, field personnel, scientists, advocates, and policy makers to fill significant gaps in the literature, develop innovative predictive models, and position resources wisely for prevention, intervention, and response.

In this closing chapter, we summarize overarching main points, as well as those specific to each section and outline recommended next steps for advancing the study of wildfire impacts and developing solutions for individual and community well-being.

14.2. FROM FIRE TO EMISSIONS

Fire science and fire detection and monitoring are arguably the most long-established elements of the modeling chain. Fire records in the United States, for example, go back to the early decades of the twentieth century,

[1] *Institute for Applied Environmental Health, School of Public Health, University of Maryland, College Park, Maryland, USA*

[2] *Michigan Tech Research Institute, Michigan Technological University, Ann Arbor, Michigan, USA*

[3] *Department of Geographical Sciences, University of Maryland, College Park, Maryland, USA*

Landscape Fire, Smoke, and Health: Linking Biomass Burning Emissions to Human Well-Being, Geophysical Monograph 280, First Edition.
Edited by Tatiana V. Loboda, Nancy H. F. French, and Robin C. Puett.

DOI: 10.1002/9781119757030.ch14

although establishing connection to health outcomes have never been the primary reason for their existence. With the advance of space-based Earth observations, our knowledge of global dynamics of biomass burning has been growing exponentially. The first section of this book introduces the different types of biomass burning events, fire regime, and describes most common approaches for mapping and monitoring biomass burning events, distribution of fuels, and emissions modeling. We have substantially advanced our capability to detect ongoing burning in near real time, our understanding of post-fire impacts, including fire extent, fire intensity, fuel composition, and consumption, and, subsequently, composition and amount of emitted pollutants has reached historically unprecedented levels. The observational and modeling capabilities are built upon and are strongly enhanced by over half a century long theoretical and applied knowledge-building within fire ecology, a field that is concerned with both natural and management biomass burning. The growing body of knowledge that links health outcomes to biomass burning pollution has prompted development of a series of new instruments within the constellation of international Earth observing satellites, which further enhance already established capabilities and, even more important, offer new approaches to directly observe and measure pollutants in the atmosphere.

As a mature branch of science, fire ecology has an extensive and specific terminology, which, when used incorrectly or inconsistently along the modeling chain, leads to opaque science. Theory defines ecological and human factors that need to be considered while looking at pollutants. But the combination of those factors change continuously, creating new situations; and thus our methods and approaches should be constantly evolving.

Despite the tremendous advances allowed by the fire records developed from satellite observations, these data sets contain a substantial amount of error and uncertainty. The magnitude of these errors vary widely based on the geographic characteristics of biomass burning events, location of their occurrence, instrument design, and satellite orbit among other factors. Management fires, such as crop residue burning and prescribed burns, are usually strongly undercounted because they frequently don't burn long enough and have a small spatial footprint, intentionally set during unfavorable conditions for observations (e.g., cloudy days), and are kept relatively cool to maintain control. However, even near-identical biomass burning events could have a drastically different detectability if one occurs near the equator and the other in midlatitudes. The quality of data record varies strongly through time, rendering observations from different systems not entirely compatible. Advances in instrument design have allowed for large improvements in detectability of events. Geostationary satellites, with enhanced fire monitoring capabilities, are likely to offer major improvements to fire monitoring in low latitudes and midlatitudes. Although the spatial resolution of those observations remains suboptimal for detecting small management fires, the massively increased frequency of observations offers some hope.

The composition and amount of biomass, together with characteristics of the burning process itself, determines the amount and constituents of smoke produced in fire events. As with fire, satellite observations have allowed for large advancements in developing maps of fuels and their characteristics across space and time. Again similarly to fire monitoring, the uncertainty and error vary across space and time driven by differences in ecological settings, landscape complexity, ecological complexity, and the type of satellite observations and instruments. Airborne and space-borne lidar observations have offered a tremendous opportunity in recent years for 3D characterization of fuel beds. With the Global Ecosystem Dynamics Investigation (GEDI) mission on the International Space Stations (ISS) providing current coverage of low latitude and midlatitude, we are likely to see production of global data sets or enhanced country-specific data sets in the near future. Lidar observations from IceSat-2 will also bring some, although not equivalent to GEDI, capabilities for the high northern latitudes of the globe. Lidar observations of post-fire conditions also offer substantial improvements for estimating fuel consumption. Other approaches to fuel consumption modeling rely on models linking weather conditions at the time of the biomass burning event or modeled Fire Radiative Power, each of which has its own limitations and uncertainties. Unfortunately, without a free-flying lidar mission, which is not yet under consideration, these improvements in the characterization of pre- and post-fire fuels will be short-lived as GEDI is currently offline and estimated to return in 2024. Finally, more work is need to refine combustion emission factors in field experiments to account for a variety of fuel beds and burning conditions to improve models' predictive capabilities.

14.3. FROM EMISSIONS TO CONCENTRATIONS

The influence of biomass burning on the atmosphere and its impact on air quality is understood through concepts based in atmospheric science, the study of physical and chemical processes that describe the atmosphere and its spatial and temporal dynamics. Observations, models, and analytical methods to map smoke from biomass burning have developed from this broad, complex science discipline. In Part II of this book, we have provided a general description of the data, models, and techniques that atmospheric scientists use for studying, mapping, and predicting concentrations and spatial distribution of key constituents that define air quality. Biomass burning smoke is a unique air pollutant, in that it is created episodically and often in remote locations, but it can be transported, sometimes long distances, to where people live. Atmospheric physics and chemistry determine how fire emissions are distributed, both horizontally across landscapes and vertically above the ground.

Building the capacity to map and predict where smoke will intersect with people requires sophisticated geospatial and dispersion modeling methods. Smoke monitoring systems can provide directly measured data on air pollution concentrations as well as advanced analytical tools that are used to combine *in situ* measurements with remote-sensing-based measures of air quality. Coupled fire-weather models can integrate data on fire behavior with atmospheric processes. Currently operational forecasting systems around the world are just beginning to employ some of these research-level tools.

Mapping and forecasting smoke is now an important operational capability being conducted in many countries around the world to better protect people from the adverse effects of biomass burning pollution. However, the adaptation of modeling capabilities developed for other aspects of atmospheric modeling to landscape fire smoke is not fully mature. Most smoke models in use require broad assumptions regarding several key variables. These include using metrics of smoke injection based on power plant smokestacks, a poor surrogate for landscape fire events. General assumptions are also used regarding emission distribution from one day to the next and through the day-night cycle; few observations of fire activity and smoke production are made at subdaily time steps. The assumption that fire activity and smoke production will follow predictable temporal patterns introduces large errors at times of extreme fire behavior and areas of extensive dry flammable fuels. Improvements in models in the future will include improved understanding of the fine-scale characteristics of processes that have been described, so far, as generalized behavior. With fire smoke now seen as one of the primary sources of pollution, taking on the task of model improvement will become a clear priority for quantifying and responding to pollution exposure for people around the world.

14.4. FROM CONCENTRATIONS TO HEALTH OUTCOMES

The scientific literature regarding health impacts specific to wildfire smoke exposures is scant but rapidly growing, particularly in high-income countries. The particulate matter component of wildfire smoke and respiratory health have received most of the attention from toxicological and exposure studies as well as epidemiologic research. Epidemiologic studies have often relied on mortality and treatment visit data, examining mortality and morbidity associated with acute exposures. Toxicologic, etiologic, mechanistic, and epidemiologic research specific to wildfire smoke exposures is limited, forcing reliance on the larger, well-established body of research on general ambient particulate matter exposures. However, some evidence has accumulated for increased risks of preterm birth, low birth weight, and all cause-mortality and adverse respiratory health outcomes among children and adults. Whether the evidence is strongest for these health outcomes merely due to more research activity in these areas remains to be seen. Differences in study designs, populations, health outcomes, exposure assessment methods, and consideration of confounding factors across studies have made comparisons and conclusions challenging.

As presented in Chapters 2–9, numerous fire and smoke monitoring and modeling advances have forwarded the fields of air quality mapping, prediction, and estimation resulting from biomass burning. However, many of these advances have not been incorporated into health research to date. Unfortunately, health researchers without collaborators in the field of fire science modeling often use exposure estimation techniques that rely on sparse monitors with little variability throughout space or time. Another limitation is the overwhelming focus on the particulate matter component in wildfire smoke. Characterization of additional smoke constituents is a significant need, as is the development of toxicological profiles for various fuels (anthropogenic and natural), fire temperatures, aging and distance of constituents, and mixtures. Application of these results to epidemiologic studies is critical for a more comprehensive picture of human exposures, as well as consideration of different exposure pathways; impacts of cumulative, multiple and repeated chronic exposures; and critical time windows of exposure and periods for physiologic recovery.

Significant gaps in the investigation of potential related health outcomes include maternal, fetal, infant, and childhood morbidity; mental health outcomes; and myriad conditions that have been associated with general ambient air pollution (i.e., cognitive development and decline, infectious diseases, diabetes). In addition to considering a broader range of study designs, health data and outcomes, intermediate physiologic changes, physiologic adaptive capacity and resilience, delayed physiologic impacts, and utility of biomarkers should be addressed. The contribution of preexisting susceptibility due to comorbid conditions, psychological stress, or other related factors such as socioeconomic status and situations of environmental injustice have also been understudied.

Gaps are evident not only with respect to exposure assessment, health outcomes, and pathophysiology but also with the development and application of innovative, standardized, rigorous research methods that are reported clearly and completely in the literature. Consistency among studies, large population-based participant samples that allow for control of potential confounders, and longitudinal studies that allow for variation of exposure across space and time in wildfire-prone areas are needed for generalizable findings that can effect policy. Example specific research needs include: the identification of critical time periods of exposure specific to morbidity and subpopulation (i.e., pregnant women versus older adults), separation of the impacts of wildfire smoke exposures from background air pollutants and other physiological and psychological stressors, as well as the elucidation of the joint effects of pollutant mixtures and stressors.

14.5. SUMMARY

Chapters in this volume touch on several topics that are significant to forwarding the field, but comprehensive examinations were outside the scope of the current volume, including from local to global policy and economic impacts, occupational exposures, and prevention and intervention measures (i.e., ventilation, building infiltration, evacuation). Further research is needed for low and middle income countries and to identify factors necessary for increasing community-level and individual-level adaptive capacity. Thoughtful consideration and attention should also be given to improved characterization of uncertainty and error in all fire research. Additionally, the formation of multidisciplinary fire research teams including disciplines less represented in this volume (i.e., social-behavioral and policy) will enable the application of innovative and more comprehensive methods to address more complex, holistic, and far-reaching questions in order to build fire-resilient communities globally.

INDEX

Landscape Fire, Smoke, and Health: Linking Biomass Burning Emissions to Human Well-Being, Geophysical Monograph 280, First Edition.
Edited by Tatiana V. Loboda, Nancy H. F. French, and Robin C. Puett.

DOI: 10.1002/9781119757030.index